Recent Developments in Quantum Mechanics

MATHEMATICAL PHYSICS STUDIES

A SUPPLEMENTARY SERIES TO
LETTERS IN MATHEMATICAL PHYSICS

VOLUME 12

Recent Developments in Quantum Mechanics

Proceedings of the Brasov Conference, Poiana Brasov 1989, Romania

edited by

Anne Boutet de Monvel
University Paris VII, Paris, France

Petre Dita
Bucharest University, Bucharest, Romania

Gheorghe Nenciu
Institute of Atomic Physics,
Theoretical Physics Department, Bucharest, Romania

and

Radu Purice
Bucharest University, Bucharest, Romania

SPRINGER SCIENCE+BUSINESS MEDIA, B.V.

Library of Congress Cataloging-in-Publication Data

Braşov Conference (1989 : Poiana Braşov, Romania)
Recent developments in quantum mechanics : proceedings of the Braşov Conference, Poiana Braşov 1989, Romania / edited by Anne Boutet de Monvel ... [et al.].
p. cm. -- (Mathematical physics studies ; v. 12)
ISBN 978-94-010-5449-2 ISBN 978-94-011-3282-4 (eBook)
DOI 10.1007/978-94-011-3282-4
1. Quantum theory--Congresses. 2. Quantum Hall effect--Congresses. 3. Pseudodifferential operators--Congresses. 4. Mathematical physics--Congresses. I. Boutet de Monvel, Anne, 1948- . II. Title. III. Series.
QC173.96.B73 1991
530.1'2--dc20 91-6654
CIP

ISBN 978-94-010-5449-2

Printed on acid-free paper

CONTENTS

PREFACE

This book contains the invited lectures given at the 1989 Poiana Braşov Summer School on: "Recent Developments in Quantum Mechanics". The main aim of the School has been to present some expository talks reviewing the state of the art in some mathematical problems of Quantum Mechanics. A special emphasize was on the behaviour of Bloch electrons in electric and magnetic fields. In this context various techniques from semiclassical analysis and pseudodifferential operators are discussed. The volume also contains the "mini-course" on $\mathcal{D}$-modules and pseudodifferential operators given by prof. Louis Boutet de Monvel. Many other subjects of current interest in the Mathematical problems of Q.M. are also contained, as: stochastic methods, scattering theory, Supersymmetric Quantum Mechanics, etc.

We owe special thanks to all the invited lecturers for their contributions to the success of the School.

The School has been organized in the frame of the collaboration programme between the University of Paris VII and the Romanian Academy. We also thank UNESCO, CNRS and IAP for their financial help.

Many of our colleagues from the Institute of Atomic Physics helped us in various ways, and we thank them for that. We are especially grateful to Drs. V.Ceauşescu and V.Georgescu for their inestimable help.

Ms. Geta Uglai, the secretary of the School, handled the tedious work related to the School with high competence and efficiency, for which we owe her our warm gratitude.

The Editors

Bucharest

Ils
pendent

à
la verticale

Ils
écoutent

le sable
tomber

Ils
entendent

la nuit
s'écouler

Ils
attendent

le jour
pour chanter

LA VIE

et tout se termine
au fond du sablier

Brasov, Septembre 1989

LIST OF PARTICIPANTS

ADAM, G.	-	Institute of Atomic Physics, Bucharest
ADAM, Sanda	-	Institute of Atomic Physics, Bucharest
ALDEA, A.	-	Institute of Atomic Physics, Bucharest
ANGELESCU, N.	-	Institute of Atomic Physics, Bucharest
ANGHEL, M.	-	Institute of Atomic Physics, Bucharest
ANGHEL, V.	-	Institute of Nuclear Power Reactors, Pitesti
APOSTOL, M.	-	Institute of Atomic Physics, Bucharest
ARSU, G.	-	Institute of Mathematics, Bucharest
BADEA, M.	-	Institute of Atomic Physics, Bucharest
BANCIU, M.G.	-	Institute of Nuclear Power Reactors, Pitesti
BAUMGARTNER, B.	-	University of Vienna
BADESCU, L.	-	Institute of Mathematics, Bucharest
BANICA, C.	-	Institute of Mathematics, Bucharest
BARAN, A.	-	Institute of Mathematics, Bucharest
BALDEA, I.	-	Institute of Atomic Physics, Bucharest
BECIU, M.	-	Civil Engineering Institute, Bucharest
BENTOSELA, F.	-	Centre de Physique Théorique, Luminy
BERCEANU, S.	-	Institute of Atomic Physics, Bucharest
BOGATU, N.	-	Institute of Atomic Physics, Bucharest
BOLDEA, Venera	-	Institute of Atomic Physics, Bucharest
BORAC, Cleo-Marieta	-	University of Bucharest
BORAC, S.	-	University of Bucharest
BOUTET DE MONVEL, Anne	-	Université Paris VII, Paris
BOUTET DE MONVEL, L.	-	Université Paris VI, Paris
BRANZANESCU, V.	-	Polytechnical Institute, Bucharest
BUDAU, P.	-	University of Bucharest
BULBOACA, I.	-	Institute of Atomic Physics, Bucharest
BUNDARU, M.	-	Institute of Atomic Physics, Bucharest
BUSLAEV, V.S.	-	University of Leningrad
BUZATU, F.	-	Institute of Atomic Physics, Bucharest
CAPRINI, Irinel	-	Institute of Atomic Physics, Bucharest
CALIAN, Violeta	-	Institute of Nuclear Power Reactors, Pitesti
CEAUSESCU, M.	-	Institute of Atomic Physics, Bucharest
CEAUSESCU, V.	-	Institute of Atomic Physics, Bucharest
CHIRANOV, M.	-	ISEH, Focsani
CHIRANOV, Roxana	-	ISEH, Focsani
CIONGA, Aurelia	-	Institute of Atomic Physics, Bucharest
CIONGA, V.	-	Institute of Atomic Physics, Bucharest
COMBES, J.M.	-	Université de Toulon et du Var, La Garde
CORCOTOI, I.	-	Institute of Atomic Physics, Bucharest
CORNS, R.	-	Institute of Theoretical Physics, Leuven
COSTIN, O.	-	Institute of Atomic Physics, Bucharest
COSTIN, Rodica	-	INCREST, Bucharest
COSEREANU, M.	-	University of Bucharest
COTFAS, N.	-	High School, Bucharest
CRISTEA, O.	-	University of Cluj-Napoca

CULETU, H.	-	ICECHIM, Bucharest
CZYZ, J.	-	Institute of Mathematics, Warsaw
DAFINEI, I.	-	Institute of Atomic Physics, Bucharest
DAFINEI, Mirela	-	IPA, Bucharest
DAMIAN, G.	-	ITIM, Cluj-Napoca
DAUMER, F.	-	Université de Nantes
DANESCU, A.	-	Institute of Atomic Physics, Bucharest
DELION, D.	-	Institute of Atomic Physics, Bucharest
DEMUTH, M.	-	Institute of Mathematics, Berlin
DEREZINSKY, J.	-	University of Warsaw
DITA, P.	-	Institute of Atomic Physics, Bucharest
DITA, Sanda	-	Institute of Atomic Physics, Bucharest
DMITRIEVA, Ludmila	-	Institute of Mathematics, Leningrad
DOMITIAN, V.	-	High School, Ploiesti
DRAGOI, D.	-	High School, Bucharest
DUDAS, E.	-	Institute of Atomic Physics, Bucharest
DULEA, M.	-	Institute of Atomic Physics, Bucharest
DUMITRESCU, O.	-	Institute of Atomic Physics, Bucharest
DUPONT, P.	-	Institute of Theoretical Physics, Leuven
EXNER, P.	-	JINR, Dubna
FAZAKAS, A.	-	Institute of Atomic Physics, Bucharest
FAZAKAS, Elena	-	Institute of Atomic Physics, Bucharest
FROESE, R.	-	University of British Columbia, Vancouver
GARTNER, P.	-	Institute of Atomic Physics, Bucharest
GARBAN, C.	-	IJA, Dragasani
GEORGESCU, Edita	-	Institute of Atomic Physics, Bucharest
GEORGESCU, V.	-	Institute of Mathematics, Bucharest
GHINCULOV, A.	-	University of Bucharest
GHEORGHE, C.	-	Institute of Atomic Physics, Bucharest
GHEORGHIU, H.	-	Institute of Nuclear Power Reactors, Pitesti
GRAMA, N.	-	Institute of Atomic Physics, Bucharest
GRECU, D.	-	Institute of Atomic Physics, Bucharest
GRIGORE, R.D.	-	Institute of Atomic Physics, Bucharest
GRIGORESCU, M.	-	Institute of Atomic Physics, Bucharest
GROSSE, H.	-	University of Vienna
GROSU, Corina	-	CIMIC, Bucharest
GROSU, Marta	-	DEN, Bucharest
GRÜNFELD, C.P.	-	Institute of Atomic Physics, Bucharest
GUSSI, G.	-	Institute of Mathematics, Bucharest
HARABOR, Ana	-	University of Craiova
HARABOR, V.	-	University of Craiova
HESS-NIELSEN, N.	-	University of Aarhus, Aarhus
HOROI, M.	-	Institute of Atomic Physics, Bucharest
HRISTEA, M.R.	-	ICECHIM, Bucharest
IACOMI, M.	-	University of Bucharest
IANCU, E.	-	University of Bucharest

IFTIMOVICI, A. - Institute of Atomic Physics, Bucharest
ION, B.D. - Institute of Atomic Physics, Bucharest
IONESCU, L. - ICSIT, Bucharest
IORDANESCU, A. - Institute of Atomic Physics, Bucharest
IOSIFESCU, M. - Institute of Atomic Physics, Bucharest
ISAR, A. - Institute of Atomic Physics, Bucharest
IXARU, L. - Institute of Atomic Physics, Bucharest

JOYE, A. - EPFL, Lausanne

KARPIO, A. - University of Warsaw

LANDAU, L. - King's College, London
LASCAR, R. - Université Paris VI, Paris
LEMNETE, Luminita - ICTC, Bucharest
LOEFFELHOLZ, Y. - Karl Marx Universität, Leipzig
LUDU, A. - University of Bucharest

MALGRANGE, B. - Université de Grenoble
MANDA, H. - Institute of Mathematics, Bucharest
MANDACHE, Ana Maria - COS Tîrgoviste
MANDACHE, N. - COS Tîrgoviste
MANTOIU, M. - Institute of Mathematics, Bucharest
MAZILU, D. - Institute of Atomic Physics, Bucharest
MELNIKOVA, Natalia - Academy of Sciences, Moscow
MICU, L. - Institute of Atomic Physics, Bucharest
MIHALACHE, D. - Institute of Atomic Physics, Bucharest
MINEA, G. - Institute of Mathematics, Bucharest
MOCANU, Raluca - Institute of Atomic Physics, Bucharest
MOLNAR, I. - Institute of Mathematics, Bucharest
MUNTEANU, Cristina - ITIM, Cluj
MOT, P. - Institute of Medicine, Bucharest
NAKAMURA, S. - ETH, Zürich
NAJAITU, V. - Institute of Mathematics, Bucharest
NEMES, G. - Institute of Atomic Physics, Bucharest
NENCIU, Alexandrina - Institute of Atomic Physics, Bucharest
NENCIU, G. - Institute of Atomic Physics, Bucharest
NISTOR, P. - High School, Bacau

ONCICA, A. - Institute of Atomic Physics, Bucharest
ONCIUL, Daniela - Institute of Nuclear Power Reactors, Pitesti
OPREA, A. - Institute of Atomic Physics, Bucharest

PANEJAH, B.P. - Moscow
PANTEA, D. - Institute of Atomic Physics, Bucharest
PASCU, E. - Institute of Mathematics, Bucharest
PASCU, M. - Institute of Mathematics, Bucharest
PALIVAN, Cornelia - ICCF, Bucharest
PALIVAN, H. - Institute of Atomic Physics, Bucharest
PETRASCU, Anca - University of Bucharest
PETRASCU, M. - Institute of Atomic Physics, Bucharest
PISO, M. - Institute of Atomic Physics, Bucharest

PIRJOL, D.	-	University of Bucharest
POENARU, D.	-	Institute of Atomic Physics, Bucharest
POPP, O.T.	-	Institute of Atomic Physics, Bucharest
PRIBU, Mihaela	-	ICSIT, Bucharest
PURICE, R.	-	Institute of Mathematics, Bucharest
PUTA, M.	-	University of Timisoara
PYTEL, Z.	-	Institute of Physics, Warsaw
RADU, Gabriela	-	Institute of Nuclear Power Reactors, Pitesti
RADULESCU, O.	-	University of Bucharest
SALIU, L.	-	University of Craiova
SANDULESCU, A.	-	Institute of Atomic Physics, Bucharest
SANDULESCU, N.	-	Institute of Atomic Physics, Bucharest
SARU, D.	-	ICPE, Bucharest
SEILER, R.	-	Technische Universität, Berlin
SILISTEANU, I.	-	Institute of Atomic Physics, Bucharest
SIMINA, Anca	-	ICSIT, Bucharest
SJÖSTRAND, J.	-	Université Paris Sud, Orsay
SMONDIREV, M.	-	JINR, Dubna
SOBOLEV, S.	-	Institute of Mathematics, Leningrad
SOFONEA, M.	-	Institute of Mathematics, Bucharest
SPINEANU, F.	-	Institute of Atomic Physics, Bucharest
STANCIU, Sonia	-	University of Bucharest
STEINBRECHER, G.	-	University of Craiova
STOVICEK, P.	-	JINR, Dubna
STRATAN, G.	-	Institute of Atomic Physics, Bucharest
STRULECKAYA, Veronica	-	IEMB, Moscow
SUHANOV, V.	-	University of Leningrad
SVIRSCHEVSKI, Speranta	-	Institute of Atomic Physics, Bucharest
TATARU, L.	-	University of Craiova
TETA, A.	-	University of Napoli
THALLER, B.	-	University of Graz
TIBAR, M.	-	Institute of Mathematics, Bucharest
TOPOR-POP, Rodica	-	High School, Bucharest
TOPOR-POP, V.	-	Institute of Atomic Physics, Bucharest
TOPOR, Nadia Marina	-	ICPE, Bucharest
TRACHE, Maria	-	University of Bucharest
TRIFONOV, D.A.	-	Institute of Nuclear Research, Sofia
TRUMANN, A.	-	University College of Swansea
VLAD, Madalina	-	Institute of Atomic Physics, Bucharest
VULCANOV, D.	-	Polytechnical Institute, Timisoara
YAFAEV, D.	-	Institute for Mathematics, Leningrad
YAJIMA, K.	-	University of Tokyo
ZAVODSKY, P.	-	ITIM, Cluj
ZNOJIL, M.	-	Institute of Nuclear Physics, Prague

REVUE SUR LA THÉORIE DES $\mathcal{D}$-MODULES ET MODÈLES D'OPÉRATEURS PSEUDODIFFÉRENTIELS

A SURVEY ON THE THEORY OF $\mathcal{D}$-MODULES. MODELS FOR PSEUDODIFFERENTIAL OPERATORS

L. Boutet de Monvel

Université Paris VI, Dept. de Mathématiques
4, place Jussieu, 75252 Paris Cedex 05

Cet article résume une série de quatre conférences données à Brasov en septembre 1989 qui ont pour but de donner un aperçu de la théorie des $\mathcal{D}$-modules introduite par M.Sato, T.Kawai et M.Kashiwara pour "l'analyse algébrique" des équations aux dérivées partielles et de questions qui s'y rattachent. Accompagnée de la définition des microfonctions et des méthodes de localisation que rendent possibles les opérateurs microdifférentiels analytiques, elle fournit l'outil le plus puissant pour l'étude du formalisme algébrique des équations différentielles ou aux dérivées partielles. Cette théorie peut paraître inutile ou lourde si on ne pense qu'à une équation aux dérivées partielles particulière; mais elle devient très efficace et même indispensable pour la plupart des questions qui concernent les systèmes de plusieurs équations aux dérivées partielles, ainsi que dans toutes les questions concernant les changements de variables ou les restrictions pour de tels systèmes.

La théorie algébrique des $\mathcal{D}$-modules, et plus particulièrement des $\mathcal{D}$-modules holonômes, est aussi un outil de plus en plus utilisé dans la théorie des représentations des groupes.

Un des aspects de cette théorie est de donner des définitions et des notations commodes pour manipuler des systèmes d'équations différentielles, ou les transformer en systèmes équivalents mais d'expression plus simple; ainsi Sato a pu donner des formes microlocales canoniques simples et remarquables pour les systèmes "à caractéristiques simples".

Cette théorie se limite aux équations à coefficients analytiques, qui sont sans doute les plus importantes et dont seules les solutions ont des raisons de donner lieu à un calcul algébriquement intelligible; pour les systèmes "généraux" à coefficients C^∞ certaines

A. Boutet de Monvel et al. (eds.), Recent Developments in Quantum Mechanics, 1–31.

définitions sont plus faciles ou élémentaires, en particulier parce qu'on dispose de partitions de l'unité, mais on n'a pas de théorème de finitude ou de cohérence, encore moins le théorème d'involutivité (§4) de sorte que les théorèmes centraux de la théorie sont faux, et ces "méthodes algébriques" ne sont plus guère applicables, même si les mêmes idées servent dans les cas les plus importants.

Il n'était guère possible en quatre heures de donner un compte rendu complet avec démonstrations détaillées et je me suis limité à essayer de décrire aussi clairement que possible les définitions et à énoncer, le plus souvent sans démonstration, les résultats les plus frappants concernant les $\mathcal{D}$-modules. Le lecteur intéressé devra se reporter aux livres ou articles cités dans la bibliographie pour une étude plus complète. La théorie de Sato, Kawai, Kashiwara fait un usage constant de l'analyse complexe: faisceaux analytiques cohérents, et de l'algèbre homologique; en particulier un certain nombre de résultas ne s'énoncent bien que dans le cadre des catégories dérivées introduites par J.-L.Verdier. J'ai essayé de ne pas trop abuser du vocabulaire de ces théories et d'en rappeler à mesure la signification des termes les plus importants, mais l'étude des $\mathcal{D}$-modules suppose aussi un minimum de familiarité avec celles-ci.

J'ai terminé ces conférences par une description de "modèles" pour les opérateurs pseudodifférentiels qui peuvent donner une autre image géométrique ou analytique utile pour toutes ces questions. Je n'aborde les questions de microlocalisation que de façon détournée dans cette deuxième partie, en supposant que le lecteur a déjà un minimum de familiarité avec la théorie des opérateurs pseudodifférentiels. Celui-ci devra aussi se reporter aux articles cités dans la bibliographie pour des descriptions et démonstrations plus détaillées.

Chapitre 1 $\mathcal{D}$-MODULES

§1. FAISCEAUX $\mathcal{O}$, $\mathcal{D}$ SUR UNE VARIÉTÉ

a. Dans tout ce qui suit X désigne une variété analytique réelle ou complexe de dimension n. On note $\mathcal{O}$ le faisceau des fonctions analytiques: pour U ouvert de X, $\mathcal{O}(U)$ est l'algèbre des fonctions holomorphes sur U à valeurs dans C. Pour tout $x \in X$ l'algèbre $\mathcal{O}_x$ des germes en x de fonctions holomorphes est une algèbre locale noethérienne, régulière de dimension n; on peut en outre lui appliquer le théorème des fonctions implicites, et y résoudre des équations différentielles (ce qui la distingue des algèbres de type fini de la géométrie algébrique).

Le faisceau $\mathcal{O}$ est localement de cohomologie triviale: le théorème de Grauert montre que si U est de Stein ou "pseudoconvexe", en particulier si U est isomorphe à un ouvert convexe de C^n (en tant que variété complexe) ou si U est réel, les groupes de cohomologie $H^q(U,\mathcal{O})$ sont nuls pour q>0. Le théorème d'Oka affirme en outre que $\mathcal{O}$ est cohérent, autrement dit si u: $\mathcal{O}^p \to \mathcal{O}^q$ est un homomorphisme de faisceaux de $\mathcal{O}$-modules, le $\mathcal{O}$-module ker u est localement de type fini (i.e. si des sections locales $s_1,\ldots,s_N$ engendrent le module de germes $(\ker u)_x$ les s_i engendrent encore $(\ker u)_y$ aux points y assez voisins de x). On a alors une notion de $\mathcal{O}$-module cohérent: un faisceau $\mathcal{M}$ de $\mathcal{O}$-modules est cohérent s'il est localement engendré par un nombre fini de sections elles mêmes soumises localement à un nombre fini de relations; les faisceaux noyau et co-image (ker u et coker u) d'un homomorphisme de $\mathcal{O}$-modules cohérents sont cohérents.

b. On note $\mathcal{D}$ le faisceau des opérateurs différentiels à coefficients analytiques: pour U ouvert de X, $\mathcal{D}(U)$ est l'algèbre des homomorphismes de faisceaux $\mathcal{O}_U \to \mathcal{O}_U$ qui s'écrivent localement, après un choix de coordonnées locales, sous la forme d'une somme finie

$$f \to \sum a_\alpha(x)\,(\partial/\partial x)^\alpha f$$

où on a utilisé la notation usuelle $(\partial/\partial x)^\alpha f = \partial^{\alpha_1+\ldots+\alpha_n} f/\partial x_1^{\alpha_1}\ldots\partial x_n^{\alpha_n}$ pour les indices de dérivation, $\alpha=(\alpha_1,\ldots,\alpha_n)\in N^n$. De façon équivalente $\mathcal{D}$ est la C-algèbre d'endomorphismes de $\mathcal{O}$ engendrée par $\mathcal{O}$ et le faisceau $\mathcal{X}$ des champs de vecteurs analytiques (= le faisceau des dérivations de $\mathcal{O}$).

$\mathcal{D}$ est filtré; $\mathcal{D}^k$ est le sous-faisceau des opérateurs différentiels de degré $\leq k$. C'est un $\mathcal{O}$-module cohérent (localement libre de type fini). On sait que le commutateur [P,Q] de deux opérateurs de degrés p,q est au plus de degré p+q−1 (parce que c'est vrai si $p,q \leq 1$), et que la "partie principale" du produit PQ est le produit de celles de P et de Q. Il en résulte que l'algèbre graduée $\mathrm{gr}\,\mathcal{D} = \oplus \mathcal{D}^k/\mathcal{D}^{k-1}$ associée à $\mathcal{D}$ s'identifie à l'algèbre des fonctions polynomiales d'un covecteur $\xi \in T^*X$, à coefficients dans $\mathcal{O}$ — localement de la forme $\sum a_\alpha(x)\xi^\alpha$ (algèbre symétrique graduée $\mathcal{O}[TX]$).

Si X est ouvert dans un espace numérique $\mathbf{R}^n$ ou $\mathbf{C}^n$ (ou si on a choisi un système de coordonnées locales au voisinage d'un point de X) il est commode d'associer à un polynôme $a(x,\xi)=\sum a_\alpha(x)\xi^\alpha$ l'opérateur différentiel $a(x,\partial)=\sum a_\alpha(x)\partial^\alpha$ (on a noté ∂ la collection $\partial/\partial x_1,\ldots,\partial/\partial x_n$; les analystes, qui utilisent la transformation de Fourier et les opérateurs pseudodifférentiels utilisent plutôt $D = -i\partial$ (ce qui revient à se limiter aux ξ imaginaires purs). Avec cette convention on a pour les composés la formule $a(x,\partial)b(x,\partial) = c(x,\partial)$ avec

(1.1) $$c(x,\xi) = \sum 1/\alpha!\, (\partial/\partial\xi)^\alpha a\, (\partial/\partial x)^\alpha b$$

qui résulte de la formule de Leibniz. En particulier pour la partie principale (i.e. termes de plus haut degré) d'un commutateur:

(1.2) $$\sigma_{m+m'-1}(a\circ b) = \sum \partial a/\partial\xi_j\, \partial b/\partial x_j - \partial a/\partial x_j\, \partial b/\partial \xi_j = \{a,b\}$$

où intervient la forme symplectique canonique $\sum d\xi_j\, dx_j$ du fibré cotangent et le crochet de Poisson { , } associé.

§2. SYSTÈMES DIFFÉRENTIELS ET $\mathcal{D}$-MODULES

a. Le point de vue adopté ici est le suivant: pour un système d'équations différentielles on s'intéresse, avant toute tentative de résolution, aux équations elles mêmes et à ce qu'on peut en tirer après manipulation formelle. Classiquement un système d'équations différentielles se présente de la façon suivante: on se donne une suite de fibrés vectoriels analytiques sur X (dans la suite on utilisera la même lettre pour désigner un fibré E et le faisceau de ses sections holomorphes, qui est un $\mathcal{O}$-module localement libre), et une suite d'opérateurs différentiels $A=(a_j)$:

(1.3) $$A:\quad \ldots \to E^j \to E^{j+1} \to \ldots$$

telle que $A_{j+1} \circ A_j = 0$ (les A_j représentent les équations, et les relations entre celles ci). Ainsi le complexe de De Rham est le complexe

$$d: \quad \mathcal{O} \to \ldots \to \Omega^j \to \Omega^{j+1} \to \ldots$$

où Ω^j est le faisceau des formes extérieures de degré j et d la différentielle extérieure. Le laplacien $\Delta = \Sigma \partial^2 / \partial x_j^2$ correspond au complexe de longueur 2:

$$0 \to \mathcal{O} \xrightarrow{\Delta} \mathcal{O} \to 0.$$

A un tel complexe A on associe un complexe de $\mathcal{D}$-modules à droite (resp. à gauche) localement libres $\mathcal{A} = \mathrm{Diff}(\mathcal{O}, A)$ (resp. $\mathrm{Diff}(A, \mathcal{O})$): le j-ième $\mathcal{D}$-module est le faisceau des opérateurs différentiels de type $\mathcal{O} \to E^j$ (resp. $E^j \to \mathcal{O}$) et la différentielle est $P \to A \circ P$ (resp. $P \circ A$).

b. Exemples: L'équation de Laplace sur une variété X munie d'une métrique Riemannienne analytique correspond, comme on a dit, au complexe de longueur 2: $\mathcal{O} \xrightarrow{\Delta} \mathcal{O}$. Le complexe de $\mathcal{D}$-modules à droite correspondant est

$$\mathcal{D} \xrightarrow{\Delta} \mathcal{D}$$

dont la différentielle est la multiplication à gauche par Δ; ce complexe est équivalent au module $\mathcal{D}/\Delta\mathcal{D}$ (pour être plus correct, au complexe de $\mathcal{D}$-modules comportant comme seul module non nul $\mathcal{D}/\Delta\mathcal{D}$ en degré 1), i.e. c'en est une résolution.

Si Y est une hypersurface de X, d'équation y=0, le problème de Dirichlet est associé au système

(1.4) $$(\Delta, 1): \mathcal{O}_X \to \mathcal{O}_X \oplus \mathcal{O}_Y$$

où on a noté abusivement 1 l'opérateur de restriction; $\mathcal{O}_Y$ s'identifie à $\mathcal{O}_X / y\mathcal{O}_X$ comme $\mathcal{O}_X$-module. Pour rester dans le cadre "localement libre" ci-dessus, on peut le remplacer par le système équivalent

(1.5) $$\begin{bmatrix} \Delta & 0 \\ 1 & y \end{bmatrix}: \mathcal{O}_X \oplus \mathcal{O}_X \to \mathcal{O}_X \oplus \mathcal{O}_X.$$

Le complexe correspondant de $\mathcal{D}$-modules à droite est donné par la même matrice, en remplaçant $\mathcal{O}_X$ par $\mathcal{D}_X$, et il est aussi équivalent au seul $\mathcal{D}$-module $\mathcal{D}_X / \Delta_Y \mathcal{D}_X$.

c. Equivalence et quasi-isomorphismes

Dans ce qui précède nous avons utilisé le terme "équivalent" dans le sens le plus naïf: il existe des formules différentielles qui font passer des solutions d'un système à celles de l'autre (dans l'exemple du problème de Dirichlet, il revient au même de résoudre le problème $\Delta f=g$, $f=h$ sur Y, et l'équation $\Delta yu=v$, compte tenu que toute fonction h sur Y est restriction d'une fonction sur X, ce qui ramène au cas $f|Y=0$, et qu'alors f est divisible par y donc de la forme yu). La généralisation la plus immédiate serait de convenir que deux systèmes A et A' comme ci-dessus sont équivalents s'ils sont homotopes i.e. s'il existe une famille d'opérateurs différentiels $S = (S_j)$ (S_j: $E^j \to E^{j-1}$) telle que $AS+SA=Id$. En fait cette notion est insuffisante et on l'élargit en disant que deux complexes de $\mathscr{D}$-modules à droite A et A' sont équivalents s'il existe un morphisme de complexe $A \to A'$ qui induit un isomorphisme sur les faisceaux de cohomologie (ceux-ci sont des $\mathscr{D}$-modules à droite); la relation ainsi définie n'est pas une relation d'équivalence, et la relation d'équivalence engendrée est la *quasi-isomorphie.* La catégorie dérivée est la catégorie déduite de celle des complexes de $\mathscr{D}$-modules en inversant les quasi-isomorphismes (il y a des variantes où on se limite aux complexes bornés, ou bornés à droite ou à gauche); ce sont ses objets (à isomorphisme près) qui représentent bien notre notion de "classe d'équivalence" de systèmes différentiels.

d. Exemples (suite): Complexe de De Rham. Soit X une variété; le complexe de De Rham de X est le complexe de la différentielle extérieure opérant sur les formes différentielles:

(1.6) $$d: \quad 0 \to \mathcal{O} \to W^1 \to \ldots \to W^n \to 0.$$

Il lui correspond un complexe de $\mathscr{D}$-modules à droite DR^d et un complexe de $\mathscr{D}$-modules à gauche DR^g. Le complexe DR^g est une résolution localement libre du $\mathscr{D}$-module à gauche $\mathcal{O}$, autrement dit DR^g est (localement) exact en degré <0 et l'homomorphisme d'augmentation $P \to P(1)$ induit un isomorphisme de la cohomologie en degré 0 sur $\mathcal{O}$. De même DR^d est une résolution localement libre du $\mathscr{D}$-module à droite Ω^n des formes différentielles de degré maximum $n=\dim X$ (l'action de $\mathscr{D}$ est celle définie par la "transposition" des opérateurs différentiels; on a en particulier $\omega X = -L_X\omega$ si X est un champ de vecteurs, L_X la dérivée de Lie associée).

Un peu plus généralement soient X et Y deux variétés et $f: Y \to X$ une submersion (i.e. la différentielle de f est surjective en tout point de Y). On introduit alors le complexe de De Rham relatif $d_{Y/X}$ qui opère sur les formes différentielles relatives (ou verticales), et qu'on peut aussi interpréter comme la version avec paramètres (dans X) du précédent. Il lui correspond de même un complexe de $\mathcal{D}$-modules à droite $DR^{d}_{Y/X}$ et un complexe de $\mathcal{D}$-modules à gauche $DR^{g}_{Y/X}$; ce dernier est une résolution d'un $\mathcal{D}_Y$-module à gauche $\mathcal{D}_{Y\to X}$ que nous retrouverons plus loin sous le nom de "module de transfert". Localement on peut supposer que Y est un produit X×Z et $\mathcal{D}_{Y\to X}$ se décrit au moyen des opérateurs différentiels $P(x,z,\partial/\partial x)$ sans dérivation verticale.

e. Solutions. Inversement si $\mathcal{A}$ est le complexe de $\mathcal{D}$-modules à droite déduit de A comme ci-dessus, A s'identifie au produit tensoriel $A = \mathcal{A}\otimes_{\mathcal{D}}\mathcal{O}$ par l'application qui à un opérateur différentiel P de type $\mathcal{O}\to E^j$ associe P(1) (dans le cas des modules à gauche A s'identifie à $\mathrm{Hom}_{\mathcal{D}}(\mathcal{A},\mathcal{O})$).

Dans ce qui précède (sauf pour les exemples $\mathcal{D}/\Delta\mathcal{D}$, $\mathcal{D}/\Delta y\mathcal{D}$) seuls ont apparu des $\mathcal{D}$-modules libres ou localement libres; mais plus généralement on peut considérer des complexes de $\mathcal{D}$-modules quelconques. Ce sont les complexes cohérents, ou à cohomologie cohérente, qui représentent raisonnablement l'idée qu'on a d'un système différentiel, mais rien n'empêche de regarder aussi les autres comme étape intermédiaire. Pour en tirer l'analogue d'un système différentiel on remplace $\mathcal{A}\otimes_{\mathcal{D}}\mathcal{O}$ par $\mathcal{A}\overset{L}{\otimes}_{\mathcal{D}}\mathcal{O}$ (produit tensoriel total — ou dérivé, ou complété homologiquement) (resp. on remplace $\mathrm{Hom}_{\mathcal{D}}(\mathcal{A},\mathcal{O})$ par $\mathrm{RHom}_{\mathcal{D}}(\mathcal{A},\mathcal{O})$ dans le cas des modules à gauche): cela veut dire qu'on remplace un des facteurs $\mathcal{A}$ ou $\mathcal{O}$ par un complexe de $\mathcal{D}$-modules plats sur $\mathcal{D}$ (resp. $\mathcal{O}$ par un complexe de $\mathcal{D}$-modules injectifs); on peut en particulier pour ces opérations toujours remplacer $\mathcal{O}$ par le complexe de De Rham DR^{g} qui en est une résolution localement libre. Le complexe $\mathcal{A}\otimes^{L}\mathcal{O}$ est le complexe des solutions du système différentiel représenté par $\mathcal{A}$; il s'agit en fait d'un objet d'une catégorie dérivée dont le caractère le plus significatif est la suite des faisceaux de cohomologie, qui correspond bien en gros à notre idée des faisceaux de solutions ou d'obstructions à l'existence de solutions.

§3. COHÉRENCE ET BONNES FILTRATIONS

On a vu que $\mathcal{D}$ est muni d'une filtration naturelle $\mathcal{D}=\bigcup\mathcal{D}_k$, où $\mathcal{D}_k$ est l'ensemble des opérateurs d'ordre $\leq k$, et que pour cette filtration $\mathrm{gr}\mathcal{D}$ est l'algèbre de polynômes $\mathcal{O}[TX]$. L'algèbre $\mathrm{gr}\mathcal{D}$ est cohérente, et il en résulte facilement que $\mathcal{D}$ lui-même est un faisceau cohérent de C-algèbres.

Définition.- Soit M un $\mathcal{D}$-module. Une *bonne filtration* sur M est une filtration $M=\bigcup M_k$ telle que

(i) pour tout k, M_k est un $\mathcal{O}$-module cohérent, et $M_k = 0$ si $k \ll 0$

(ii) $M_k\mathcal{D}_p \subset M_{k+p}$, avec égalité si $k \gg 0$.

Si M est muni d'une bonne filtration le gradué grM est un gr$\mathscr{D}$-module cohérent (en particulier de type fini) et il en résulte, comme dans le cas de $\mathscr{D}$, que M est cohérent. Inversement soit M un $\mathscr{D}$-module cohérent. Alors localement M est isomorphe au quotient d'un $\mathscr{D}$-module libre $\mathscr{D}^p$ par un sous module de type fini N. Il résulte du théorème de Krull que les filtrations sur N et M induite par, ou quotient de celle de $\mathscr{D}^p$, sont de bonnes filtrations, donc M possède localement de bonnes filtrations; plus généralement si P est un $\mathscr{D}$-module bien filtré, et $N \subset P$ un sous module cohérent, il résulte du théorème de Krull que la filtration induite sur N et la filtration quotient sur P/N sont bonnes. [On ne sait pas si un $\mathscr{D}$-module cohérent en général possède une bonne filtration globale; c'est néanmoins vrai pour un $\mathscr{D}$-module "algébrique", ou pour un $\mathscr{D}$-module "holonôme régulier" comme il résulte d'un théorème difficile de Kawai et Kashiwara (voir la définition plus bas). De même on n'a pas démontré les théorèmes A et B de Cartan pour le faisceau $\mathscr{D}$.]

Soit $M: \ldots \to M^k \to M^{k+1} \to \ldots$ un complexe de $\mathscr{D}$-modules. Supposons les M^k munis de bonnes filtrations de sorte que la différentielle soit de degré 0. On a alors un complexe gradué grM. Il est élémentaire que M est exact si grM l'est, mais la réciproque est en général fausse (au mieux il existe une suite spectrale, convergente, $H^*(grM) \Rightarrow grH^*(M)$; un peu de la même façon, un opérateur peut être elliptique si on le considère comme opérateur d'un ordre donné n; il ne l'est alors certainement plus si on le considère comme opérateur d'ordre n+1). Nous dirons que le complexe M est bien filtré si $H^*(grM) = grH^*(M)$. Si M est un complexe de longueur finie, et si chaque M^k est bien filtrable, il existe aussi un bonne filtration (de complexe) sur M qu'on construit par exemple par récurrence descendante comme suit: supposons la filtration construite sur M^p pour $p>k$ (ce qui est le cas si k est assez grand pour qu'on ait $M^p = 0$ pour $p>k$) et construisons-la sur M^k. Notons u la différentielle et M^k_p une bonne filtration sur M^k; alors $M^k_{p+i} \cap u^{-1}(M^{k+1}_p)$ en est une autre, qui répond à la question pour i assez grand (cette construction marche si le complexe M est seulement borné à droite; elle est locale car elle suppose M de type fini; il y a une construction analogue par récurrence ascendante si M est borné à gauche).

Du point de vue des équations aux dérivées partielles, l'idée de préciser une filtration est étroitement apparentée à celle d'étudier ces équations dans des espaces de Sobolev dont on a précisé les exposants, ce qui est naturel en vue du fait que ce sont les méthodes d'estimées a priori qui ont donné en premier des résultats généraux d'existence ou d'unicité pour ces équations. De ce point de vue le fait de changer de filtration change un peu le problème (ce qui n'entre pas dans la notion d'équivalence suggérée plus haut); ainsi dans l'exemple du §1.b plus haut les deux descriptions du problème de Dirichlet ne suggèrent pas de façon évidente les mêmes inégalités a priori, particulièrement la seconde présentation $\mathscr{D}/\Delta y\mathscr{D}$. Mais pour une étude des propriétés algébriques des systèmes différentiels on peut, dans un premier temps, faire abstraction de ces inégalités.

§4. CARACTÉRISTIQUES, THÉORÈME DES SUPPORTS INVOLUTIFS

Soit M est un complexe cohérent de $\mathcal{D}$-modules. Par définition l'ensemble caractéristique de M est le support (au sens de la théorie des faisceaux) de gr $H^*(M)$, i.e. l'ensemble des points où le symbole du complexe d'opérateurs différentiels correspondant à M n'est pas exact, pour toute bonne filtration de M; on le note CarM; on démontre élémentairement que ce support ne dépend pas du choix d'une bonne filtration — d'abord pour deux filtrations M_k, M'_k telles que $M_k \subset M'_k \subset M_{k+1}$, cas auquel on se ramène localement par récurrence (1). Il résulte de la définition que CarM est un ensemble analytique, conique i.e. stable par les homothéties $(x,\xi)\to(x,\lambda\xi)$. M.Kashiwara a démontré le résultat fondamental suivant (voir [SKK] ou [K5] pour une démonstration; voir aussi [M1]):

Théorème.– Si M est cohérent, Car M est un ensemble analytique involutif.

L'analyticité est facile: Car M est le support d'un $\mathcal{O}$-module cohérent sur T^*X (il faut quand même vérifier, en suivant l'idée indiquée plus haut, que ce support ne dépend pas du choix d'une bonne filtration, bien que grM en dépende). Involutif signifie qu'en tout point lisse de Car M l'espace tangent TCar M contient son orthogonal pour la structure symplectique; ou encore que pour toutes fonctions analytiques f et g définies au voisinage d'un point de Car M, et nulles sur Car M, le crochet de Poisson $\{f,g\}$ est aussi nul sur Car M. L'exemple suivant rend le théorème plausible et intuitif: si $M=\mathcal{D}/\mathcal{J}$ où $\mathcal{J}$ est un idéal cohérent de $\mathcal{D}$, Car M a pour équations $\sigma(P)=0$, $P\in\mathcal{J}$; or pour $P,Q\in\mathcal{J}$ on a $[P,Q]\in\mathcal{J}$ et $\sigma([P,Q])=\{\sigma(P),\sigma(Q)\}\in\sigma(\mathcal{J})$: $\sigma(\mathcal{J})$ est stable par $\{\ ,\ \}$; ceci démontre le théorème dans le cas où $\sigma(\mathcal{J})$ contient (au moins sur un ouvert dense de Car M) un système d'équations transverses de Car M; la difficulté du théorème en général provient du fait que les éléments de $\sigma(\mathcal{J})$ peuvent s'annuler tous à un ordre plus élevé que 1 sur Car M.

Modules holonômes. En particulier Car M est toujours de dimension ≥ dimX. Un cas particulièrement intéressant est alors celui où Car M est de dimension minimale n = dimX, i.e. Car M est une variété Lagrangienne, c'est à dire involutive (ou isotrope) pour la forme symplectique canonique de T^*X, et de dimension dimX. On dit alors que M est holonôme, ou qu'on a affaire à un système holonôme d'équations différentielles (maximalement

---

(1) La théorie du faisceau $\mathcal{E}$ des opérateurs pseudodifférentiels analytiques, ou microdifférentiels, permet de localiser sur la fibré cotangent et de définir l'ensemble caractéristique (ou spectre singulier) de n'importe quel $\mathcal{D}$-module, comme support du $\mathcal{E}$-module sur T^*X qui s'en déduit. Sans hypothèse de cohérence ce support n'a pas de raison d'être involutif, ni même analytique. Dans le cas d'un $\mathcal{D}$-module cohérent cette définition se réduit à celle plus simple que nous avons indiqué.

surdéterminé dans la première terminologie japonaise). Un tel module est, localement, de longueur finie en tant que $\mathcal{D}$-module.

Parmi les modules holonômes une classe importante est constituée par les modules holonômes réguliers (qui généralisent pour les problèmes à plusieurs variables la notion d'équation différentielle à point singulier régulier): M est un tel module s'il possède (localement) une bonne filtration $M=\bigcup M_k$ telle que $M_k P \subset M_{k+m-1}$ pour tout opérateur différentiel P d'ordre m dont le symbole s'annule sur Car M. On appelle aussi fonction ou distribution holonôme une fonction (ou distribution) qui satisfait à un système holonôme d'équations aux dérivées partielles; ainsi les fonctions d'une variable satisfaisant à une équation différentielle — c'est à dire la plupart des fonctions usuelles, sont holonômes; la masse de Dirac ou plus généralement les distributions ponctuelles, les distributions définies par les parties finies d'intégrales de Hadamard sont holonômes (et régulières). Beaucoup de fonctions usuelles intervenant en analyse linéaire sont ainsi holonômes. Il convient néanmoins d'ajouter que cette notion, extrêmement fructueuse et bien adaptée à l'analyse linéaire, s'avère vite insuffisante pour l'analyse non linéaire; pour celle-ci la théorie d'Ecalle des fonctions résurgentes, dont on trouvera une présentation dans les exposés de B.Malgrange, est plus riche et prometteuse, même si elle reste pour l'instant confinée aux problèmes à une variable, dépendant éventuellement de paramètres.

Systèmes elliptiques. Si X est une variété réelle, on dit qu'un $\mathcal{D}_X$-module bien filtré M est elliptique si la variété caractéristique carM ne contient pas de covecteur réel non nul. Dans ce cas on montre que les groupes de cohomologie de SolM sont de dimension finie; la formule de Atiyah et Singer (que nous ne décrirons pas ici) permet de calculer la somme alternée de leurs dimensions en termes de propriétés géométriques du symbole grM. Il y a un résultat analogue lorsque X est une variété à bord complexe; dans ce cas on dit que M est elliptique si carM ne contient aucun covecteur non nul orthogonal au bord ∂X (i.e. orthogonal aux vecteurs holomorphes tangents au bord ∂X); exemple: X est la boule unité de $\mathbf{C}^n$ et M correspond à l'opérateur différentiel $\sum z_j \, \partial/\partial z_j$.

Exemples: 1) Le système de De Rham est holonôme; les caractéristiques sont $\xi=0$. En fait le $\mathcal{D}$-module à gauche correspondant au système de De Rham est, comme on a vu, le faisceau $\mathcal{O}$; on montre facilement que réciproquement tout $\mathcal{D}$-module cohérent dont l'ensemble caractéristique est la section nulle $\xi=0$ du fibré cotangent est localement somme directe d'un nombre fini de facteurs isomorphes à $\mathcal{O}$ (dans le cas de complexes, il faudrait décaler certains facteurs).

2) Si on se donne de plus sur la variété une métrique quadratique, on construit à partir du système de De Rham l'opérateur d+d* qui va des formes paires dans les formes impaires, lié à l'opérateur de Dirac dans le cas d'une métrique positive, ou aux équations de Maxwell dans le cas d'une métrique hyperbolique; le carré $dd^*+d^*d = -\Delta$ est l'opérateur de Laplace-Beltrami de la métrique. L'ensemble caractéristique de ces systèmes est le cône des covecteurs isotropes (imaginaire dans le cas d'une métrique positive); c'est une hypersurface et le $\mathcal{D}$-module correspondant n'est pas holonôme, sauf en dimension 1.

3) Les dérivées de la masse de Dirac en un point, et plus généralement les couches multiples de densité analytique portées par une sous variété analytique, engendrent un $\mathcal{D}$-module holonôme. Il en est de même des distributions définies par les parties finies d'intégrales de Hadamard.

Plus généralement si f est une fonction analytique réelle ≠ 0 (sur une variété réelle) de partie positive f_+, on sait (Lojaciewicz) que la fonction $s \to f_+^s$ se prolonge en un fonction méromorphe sur **C**, admettant seulement des pôles simples en les points rationnels d'une progression arithmétique négative. Dans le cas où f est un polynôme, Bernstein a donné une démonstration élémentaire de ceci, exploitant le fait que f^s engendre un module holonôme sur l'algèbre des opérateurs différentiels de n variables sur le corps C(s); cette démonstration a été adaptée et prolongée dans le cas où f est analytique par Kashiwara (cf [K2]). Alors en fait pour chaque s la "distribution" f_+^s ainsi définie par prolongement, ou le terme constant du développement en série de Laurent au voisinage de s lorsqu'il y a un pôle, engendre un $\mathcal{D}$-module holonôme régulier (cf. [K2], [K5]).

Ainsi par exemple la solution élémentaire "usuelle" de l'équation de Laplace sur $\mathbf{R}^n$: $E(x) = \mathrm{cste} \times \|x\|^{2-n}$ est de ce type et donc satisfait un système holonôme régulier d'équations aux dérivées partielles; de même pour la solution élémentaire "usuelle" (i.e. portée par le cône d'onde d'avenir) de l'équation des ondes; il s'agit d'ailleurs du même système holonôme, vu dans un autre système de coordonnées complexes.

4) La solution élémentaire de l'équation de la chaleur

$$G(x) = c^{ste} \times t^{-n/2} \exp(-x^2/4t)$$

satisfait aussi à un système holonôme d'équations aux dérivées partielles, cette fois non régulier; de même que la fonction exp(−1/t) est un exemple typique de fonction vérifiant une équation différentielle à point singulier irrégulier.

5) Si G est un groupe de Lie, on peut former le système d'équations aux dérivées partielles que vérifient les distributions caractères des représentations irréductibles de G: une telle distribution f est centrale (i.e. $gfg^{-1} = f$ pour tout $g \in G$), ce qui de façon infinitésimale se traduit par les équations différentielles

$$(L_\alpha - R_\alpha)f = 0$$

pour tout $\alpha \in \mathscr{G}$ (algèbre de Lie de G), où L_α resp. R_α est le champ de vecteurs invariant à gauche, resp. à droite, défini par α; d'autre part l'irréductibilité implique qu'on a, avec une constante λ_p convenable,

$$Pf = \lambda_p f$$

pour tout opérateur différentiel biinvariant P associé à un élément du centre de l'algèbre enveloppante de $\mathscr{G}$.

Lorsque G est le groupe des matrices GL_n on identifie le dual $\mathscr{G}^*$ à l'espace des matrices $\mathscr{G}=M_n$; les équations différentielles ainsi formées ont pour symbole (partie principale): $[\xi,g]=0$, $p(\xi)=0$ où ξ est un vecteur cotangent et p le symbole d'un élément du centre de $U(\mathscr{G})$, i.e. un polynôme invariant de ξ. Les équations $p(\xi)=0$ signifient que les coefficients du polynôme caractéristique de ξ sont nuls i.e. que ξ est une matrice nilpotente; la variété les matrices nilpotentes est de codimension n, ses point génériques (points des parties lisses des composantes irréductibles) sont les matrices dont les blocs de Jordan sont maximaux (un seul bloc par valeur propre); $\mathrm{ad}\xi$ est alors de rang maximal et $\dim \ker \mathrm{ad}\xi = n$, de sorte qu'au voisinage d'un tel point ξ les points (g,ξ) de la variété caractéristique (i.e. tels que $[\xi,g]=0$) forment une variété de dimension $\dim G=n^2$. La variété caractéristique définie par ces équations est donc Lagrangienne (c'est aussi la réunion des adhérences des variétés $T^*_Y G$, où Y parcourt l'ensemble des parties lisses des composantes irréductibles de la variété des points irréguliers de G, pour lesquels adg n'est pas de rang maximal). Ce système est donc holonôme. On montre facilement qu'il en est de même si G est réductif, en particulier s'il est semisimple. La théorie de Harish-Chandra montre qu'il est en outre régulier et que ses solutions sont localement intégrables. (Si G est résoluble, le système ci-dessus ne contient en général pas assez d'équations pour être holonôme.)

§5. IMAGES DIRECTES ET INVERSES

a. Solutions locales et globales. Au § 2 nous avons introduit le complexe $\mathrm{Sol}(\mathscr{A}) = \mathscr{A} \overset{L}{\otimes}_{\mathscr{D}} \mathscr{O}$ des solutions d'un système différentiel $\mathscr{A}$ ($\mathscr{D}$-module à droite, ou complexe de tels $\mathscr{D}$-modules, ou objet de la catégorie dérivée); c'est un élément de la catégorie dérivée de la catégorie des faisceaux de C-espaces vectoriels sur X, qui représente essentiellement le faisceau des solutions du système différentiel considéré et les faisceaux qui en dérivent cohomologiquement et qui sont indispensables si on veut pouvoir écrire de jolies formules. Par exemple si on considére une équation (vectorielle) P: $E_0 \rightarrow E_1$, identifié à un complexe de longueur 2 en degrés 0 et 1, le faisceau de cohomologie $H^0 = \ker P$ est le faisceau des solutions locales analytiques de l'équation Pf = 0, et le faisceau de cohomologie $H^1 = \operatorname{coker} P$ est le faisceau des obstructions à l'existence d'une solution analytique à l'équation Pf=g. (Si on formule la théorie à l'aide de $\mathcal{D}$-modules à gauche il faut remplacer $\mathcal{A} \otimes^{L}_{\mathcal{D}} \mathcal{O}$ par $R\mathrm{Hom}_{\mathcal{D}}(\mathcal{A}, \mathcal{O})$; les deux théories droite et gauche sont ainsi équivalentes, du moins tant qu'on se limite aux $\mathcal{D}$-modules cohérents.) Il est parfois utile de considérer les solutions du système dans un $\mathcal{D}$-module à gauche $\mathcal{N}$ autre que $\mathcal{O}$, par exemple le faisceau des fonctions C^∞ ou des distributions, ou celui des hyperfonctions qui a l'intérêt d'être flasque; ou encore un $\mathcal{D}$-modules holonome.

Les solutions globales d'un système différentiel doivent correspondre aux sections globales du faisceau Sol. En fait, pas plus que le produit tensoriel ⊗, le foncteur "sections globales" n'est exact, et on considère plutôt "l'hypercohomologie" du complexe Sol($\mathcal{A}$) (ou si on préfère le complexe $R\Gamma$Sol($\mathcal{A}$), qui est un élément de la catégorie dérivée de la catégorie des espaces vectoriels sur C, i.e. est déterminé par la suite des groupes de cohomologie [si F est un faisceau ou un complexe de faisceaux sur X, K une partie de X, $\Gamma_K(X,F)$ désigne le complexe des sections globales de F dans X à support dans K: $R\Gamma$ signifie qu'on a remplacé F par un complexe quasiisomorphe injectif (ou flasque), et est défini comme objet d'une catégorie dérivée].

b. Image directe. Module de transfert. Il est utile d'avoir une version relative, ou avec paramètres, du "complexe" des solutions; celle-ci est décrite par l'image directe des $\mathcal{D}$-modules, qui a été introduite par Kashiwara [K1], [K5], cf aussi [Bo 2]. Soient X, Y deux variétés analytiques, f: Y→X une application analytique. Si M est un $\mathcal{O}_Y$-module on définit le faisceau image directe f_*M qui est un faisceau de $\mathcal{O}_X$-modules (si U est un ouvert de X, $f_*M(U) = M(f^{-1}U)$); comme le foncteur "sections globales" ce foncteur n'est pas exact et on considère plus souvent le foncteur dérivé Rf_*. Si M est cohérent et f propre, Rf_* est cohérent d'après H.Grauert. Si M est un $\mathcal{D}_Y$-module (ou complexe de tels modules, représentant un système différentiel) l'image directe f_*M n'est pas un $\mathcal{D}_X$-module et il convient d'abord de définir un procédé pour changer de coefficients.

Le module de transfert $\mathcal{D}_{Y \rightarrow X}$ est le faisceau sur Y des opérateurs différentiels de type $\mathcal{O}_X \rightarrow \mathcal{O}_Y$, i.e. de la forme $u \rightarrow P_Y\big((Q_X\, u) \circ f\big)$ où P_Y est un opérateur différentiel sur Y et Q_X un tel opérateur sur X; c'est un $\mathcal{D}_Y$-module à gauche et un $f^{-1}(\mathcal{D}_X)$-module à droite, ceci traduisant le fait que si $A\in \mathcal{D}_{Y\to X}$, $P\in \mathcal{D}_Y$ et $Q\in \mathcal{D}_X$ le composé PAQ est bien défini. Lorsque f est une submersion on retrouve le module du §2d.

L'image directe f_+M est alors définie par

(1.7) $$f_+M = Rf_*(M\otimes^{L}_{\mathcal{D}_Y}\mathcal{D}_{Y\to X})$$

Lorsque X est réduit à un point on a $\mathcal{D}_{Y\to X} = \mathcal{O}_Y$ et f_* est le foncteur "sections globales", de sorte que la formule (2.1) représente l'hypercohomologie des solutions globales de M. Dans le cas général cette formule représente, de façon quantifiée, un système différentiel qui en gros décrit les relation différentielles sur la base X entre les solutions globales le long des fibres des solutions du système donné.

On définit de façon analogue l'image inverse des $\mathcal{D}$-modules. Ainsi si M est un $\mathcal{D}_X$-module à gauche, l'image inverse f^+M est le $\mathcal{D}_Y$- module

(1.8) $$f^+M = f^{-1}(\mathcal{D}_{Y\to X}\otimes_{f^{-1}\mathcal{D}_X} M).$$

Par exemple si Y est une hypersurface de X et f est l'injection canonique de Y dans X, l'image inverse représente le système différentiel satisfait par les données de Cauchy des solutions de M.

[Pour des $\mathcal{D}$-modules à droite il y a une définition semblable, mais qui n'a pas tout à fait la même interprétation: on remplace le module de transfert $\mathcal{D}_{Y\to X}$ par le symétrique $\mathcal{D}_{X\to Y}$, faisceau sur Y des "opérateurs différentiels de type $\Omega_X\to\Omega_Y$, où Ω désigne le faisceau des formes différentielles de degré maximum].

Dans de bons cas l'image directe ou inverse d'un $\mathcal{D}$-module cohérent (ou bien filtré) est cohérent; les théorèmes permettant une telle assertion sont apparentés aux théorèmes de cohérence de Grauert: il faut faire sur f une hypothèse de propreté, et/ou en outre sur M une hypothèse de non caractéristicité ou d'ellipticité relative. Ainsi dans [BM-M] nous montrons que sous une hypothèse d'ellipticité relative qui assure, comme l'ont montré C.Houzel et P.Schapira, qu'une image f^+M est cohérente et bien filtrée, on peut calculer l'élément de K-théorie défini par le gradué (symbole) $\mathrm{gr}f^+M$ par une formule qui généralise celle de Atiyah et Singer. Un résultat non trivial de M.Kashiwara montre que si f est l'injection canonique d'une sous variété Y dans X, l'image inverse f^+M est holonôme si M est holonôme.

Chapitre 2 MODÈLES D'ALGÈBRES D'OPD

Dans la section précédente nous avons présenté les résultats premiers de la théorie des $\mathcal{D}$-modules. Sans développer cette question nous avons par force fait allusion à la microlocalisation, qui apparaît en particulier dans la définition des caractéristiques d'un système différentiel: il s'agit du fait que toute cette analyse peut s'obtenir en recollant à partir d'une analyse locale sur l'espace des phases T^*X (fibré cotangent); ceci peut être réalisé de façon un peu formelle et algébrique au niveau des opérateurs eux mêmes, mais les notions correspondantes pour les distributions et leurs singularités ont été définies aussi. Nous ne pousserons pas ici plus loin cette théorie qui, dans le cadre analytique ci-dessus se décrit le plus commodément mais de façon un peu compliquée et élaborée en termes de cohomologie à supports du faisceau des fonctions holomorphes.

Des notions similaires ont été introduites plus simplement dans le cadre des opérateurs à coefficients C^∞: la théorie des opérateurs pseudodifférentiels sert précisément à cela. Elle a été développée sous sa forme actuelle à partir de 1965 par de nombreux auteurs, tout particulièrement Kohn et Nirenberg [Ko-N] et Hörmander [Hö]. Les opérateurs pseudodifférentiels modulo les opérateurs de degré $-\infty$ forment un faisceau sur les ouverts conique du fibré cotangent, dont les sections globales sont les opérateurs pseudodifférentiels; autrement dit ils répondent de façon satisfaisante à cette question de la localisation sur le fibré cotangent des opérateurs différentiel. Ils donnent lieu au calcul symbolique ébauché au chap I §1, essentiellement équivalent au "passage au gradué" à partir d'une bonne filtration" décrit au chapitre I; pour ce calcul nous renvoyons à la bibliographie (on en retrouvera la plus grande partie au §1). C'est dans ce cadre que s'inscrivent les modèles qui suivent, qui sont tous équivalents microlocalement au modèle pseudodifférentiel, mais en donnent des images nouvelles, permettant selon le problème de faire appel à d'autres théories que celle des distributions et equations aux dérivées partielles — particulièrement celle des fonctions holomorphes. (Les deux derniers modèles rendent une fois de plus vraisemblable, sans que cela donne lieu pour l'instant à des énoncés précis, l'idée qu'il y ait une relation entre les inégalités de l'analyse complexe comme par exemple celles de Phragmen-Lindeloff, et la géométrie symplectique, idée exploitée par ailleurs par M.Gromov.)

§1. MODÈLES D'OPÉRATEURS MICROLOCAUX

1.1 Modèle des OPD "polynomiaux"

On se place sur $\mathbf{R}^n$. Le cotangent $T^*\mathbf{R}^n = \mathbf{R}^{2n}$ est muni de la forme symplectique canonique $\sum d\xi_j dx_j$, qui est homogène de degré 1 si on compte que x et ξ sont homogènes de degré 1/2.

On note $\mathcal{H}^\infty$, resp. $\mathcal{H}^{-\infty}$ l'espace de Schwartz des fonctions C^∞ à décroissance rapide, resp. des distributions tempérées sur $\mathbf{R}^n$. Le noyau de Schwartz d'un opérateur continu $\mathcal{H}^\infty \to \mathcal{H}^{-\infty}$ est une distribution tempérée sur $\mathbf{R}^{2n}$. Si $a \in \mathcal{H}^{-\infty}(\mathbf{R}^{2n})$ on note a(x,D), ou Op(a), le composé de l'opérateur de noyau $(2\pi)^{-n} e^{ix.\xi}$ a et de la transformation de Fourier. Ainsi si a est une fonction à croissance polynomiale en (x,ξ) on a la formule usuelle:

$$(2.1) \qquad a(x,D)f(x) = (2\pi)^{-n} \int e^{ix.\xi}\, a(x,\xi) \hat{f}(\xi)\, d\xi.$$

Ici nous nous intéressons aux mêmes opérateurs que dans la théorie usuelle, mais en repérant différemment les degrés: nous dirons que a(x,D) est un opérateur pseudodifférentiel polynomial d'ordre m et de type δ si le symbole a(x,ξ) appartient à l'analogue S^m_δ d'un espace de symboles de Hörmander:

(2.2) pour m, δ réels on note S^m_δ l'ensemble des fonctions $a(x,\xi) \in C^\infty(\mathbf{R}^{2n})$ telles que pour tout indice de dérivation α (en x et ξ) on ait, avec une constante c_α convenable:

$$|\partial^\alpha a| \leq c_\alpha <x\xi>^{m+(\delta-1/2)|\alpha|}$$

où on a posé $<x\xi> = 1+|x|^2+|\xi|^2$, poids de degré 1 sur $\mathbf{R}^{2n}$.

Avec cette convention x et ξ sont de degré 1/2.

Pour abréger on notera souvent $S^m = S^m_0$. Comme dans [Hö] on peut généraliser en prenant des poids plus variés pour définir le degré m, ou des "métriques" plus générales pour définir les longueurs des dérivations ∂_x, ∂_ξ; nous ne décrirons pas ceci ici. L'espace correspondant d'opérateurs est noté OPS^m_δ. Nous ne le considérerons que dans le cas $0 \leq \delta < 1$ (pour δ<0 on vérifie élémentairement qu'un élément de S^m_δ est somme d'un polynôme et d'un symbole de degré $-\infty$; pour d≥1 l'espace S^m_δ n'a plus grand-chose à voir avec l'analyse microlocale). Dans le cas δ=0 les éléments de OPS^m_0 sont l'analogue des opérateurs pseudodifférentiels "classiques" d'ordre m. Ces opérateurs satisfont au calcul symbolique usuel: si $a \in S^m_\delta$ et $b \in S^{m'}_{\delta'}$ avec $\delta+\delta'<1$, on a

$a(x,D)b(x,D)=c(x,D)$, avec $c\in S^{m+m'}_{\delta''}$ ($\delta''=\sup(\delta,\delta')$); c a pour développement asymptotique le D.A. usuel de la théorie des opérateurs pseudodifférentiels

(2.3) $\qquad c(\text{symbole total de } a\circ b) \approx \sum i^{-\alpha}/\alpha!\ \partial_\xi^\alpha a\ \partial_x^\alpha b.$

En particulier pour le symbole principal ($\sigma_m(a(x,D)$ = a modulo les symboles de degré $\le m-1$) on a

(2.4) $\qquad \sigma_{m+m'}(AB) = \sigma_m(A)\ \sigma_{m'}(b)$

(2.5) $\qquad \sigma_{m+m'-1}([A,B]) = -i\{\sigma_m(A),\sigma_{m'}(B)\}$

où { , } est le crochet de Poisson associé à la forme symplectique canonique $\sum dx_j\, d\xi_j$.

Le symbole principal de A détermine A modulo les opérateurs de degré inférieur. S'il est inversible (i.e. si A est elliptique), A est inversible modulo les opérateurs de degré $-\infty$.

Ce calcul est entièrement analogue au calcul pseudodifférentiel usuel, avec la différence que x et ξ sont ici tous deux de degré 1/2 (pour que $\sum dx_j d\xi_j$ soit de degré 1). Il convient donc de diviser les degrés "usuels" par 2. Par exemple l'oscillateur harmonique

(2.6) $\qquad H = (-D+x^2)/2$

se comporte comme un opérateur elliptique de degré 1 (plutôt que 2); et dans ce modèle (où la sphère de l'espace des phases est compacte), la formule de Weyl qui compte asymptotiquement les valeurs propres:

$$N(s) \sim (2\pi)^{-n} \times \text{volume symplectique de la région } \{\sigma(H_0)\le s\}$$

(où N(s) est le nombre de valeurs propres $\le s$) est l'exact analogue de la formule similaire pour un opérateur elliptique d'ordre 1 (et non 2) sur une variété de dimension n. De même dans le mélange $\mathbf{R}_{\text{usuel}} \times \mathbf{R}^n$, l'opérateur $\frac{d}{dt} - iH$ se comporte comme l'opérateur hyperbolique de degré 1 $\frac{d}{dt} - i\sqrt{\Delta}$ sur $\mathbf{R}\times$ variété compacte.

Echelle des espaces de Sobolev

Il y a un analogue de l'échelle des espaces de Sobolev: $\mathcal{H}^s$ est le domaine de n'importe quel opérateur elliptique d'ordre s, par exemple H^s où $H = (-\Delta+x^2)/2$ est l'oscillateur harmonique; l'espace $\mathcal{H}^\infty=\cap\mathcal{H}^s$ est l'analogue de C^∞, et $\mathcal{H}^{-\infty}= \mathcal{H}^s$ est l'analogue de l'espace des distributions. Pour certains calculs il est commode de faire des développements en fonctions de Hermite (fonctions propres de H): rappelons que celles-ci sont indexées par les multi-indices α:

$$h_\alpha = c_\alpha (x - \partial/\partial x)^\alpha e^{-x^2/2}$$

où la constante $c_\alpha > 0$ est telle que $\|h_\alpha\|_{L^2} = 1$ (la constante est $c_\alpha = (\pi^{n/2}\, 2^\alpha \alpha!)^{-1/2}$). On a

$$Hh_\alpha = (|\alpha| + n/2) h_\alpha .$$

Spectre singulier

Il y a enfin un analogue du spectre singulier: si f est une distribution tempérée, $(x_0,\xi_0) \notin SSf$ signifie qu'il existe $P \in OPS^m$ elliptique dans un voisinage conique ouvert de (x_0,ξ_0) tel que $Pf \in \mathcal{H}^{-\infty}$. Le calcul symbolique montre que tous nos opérateurs sont "microlocaux": $SS(Pf) \subset SS(f)$ si $P \in OPS^m_\delta$ si $\delta < 1$.

On peut alors définir l'analogue du faisceau des microfonctions sur le cône symplectique $\mathbf{R}^{2n}$: le germe au point (x,ξ) est l'espace des distributions tempérées f modulo celles telles que $(x,\xi) \notin SSf$. Toute cette situation est microlocalement isomorphe à celle des opérateurs pseudodifférentiels opérant sur le faisceau des microfonctions sur une variété.

Il y a enfin un analogue des opérateurs intégraux de Fourier. Un exemple typique est celui des transformations métaplectiques, qui sont définies par une intégrale de Fourier à phase quadratique, et dont la transformation symplectique sous-jacente est la transformation symplectique linéaire associée à cette phase:

$$\mathcal{F} f(x) = c^{ste} \times \int e^{iq(x,y)}\, f(y)\, dy$$

(la transformation symplectique associe $(x, \partial q/\partial x)$ au point $(y, -\partial q/\partial y)$; ceci suppose que les applications $(x,y) \to (x, \partial q/\partial x)$ ou $(y, -\partial q/\partial y)$ sont bijectives, et il faut modifier un peu la présentation ci-dessus pour les autres cas). Les espaces $\mathcal{H}^s$ sont invariants par transformation intégrale de Fourier, en particulier par la transformation de Fourier et plus généralement par le groupe métaplectique.

1.2 Continuité L^2

On sait qu'un opérateur $P\in OPS^0_\delta$ opère continûment dans L^2 si $\delta\leq 1/2$: cela résulte simplement du calcul symbolique si $\delta<1/2$ (qui montre dans ce cas qu'il existe $B\in OPS^0_\delta$ et $r\in OPS^{-\infty}$ tels que $A^*A+B^*B = Id + r$) ou du théorème de Calderon-Vaillancourt dans le cas $\delta=1/2$. Dans le cas $\delta>1/2$ l'assertion ci-dessus est fausse; en outre le calcul symbolique rappelé ci-dessus ne converge plus du tout; aussi il convient de remplacer l'espace S^0_δ par un espace mieux adapté aux inégalités L^2.

Définition 2.1.- Soit m un nombre réel. On note $OP\Sigma^m_1$ l'espace des opérateurs A réguliers d'ordre m, i.e. tels que A se prolonge en un opérateur continu $\mathcal{H}^s\to\mathcal{H}^{s-m}$ pour tout s. Pour $0\leq\delta\leq 1$ on note $OP\Sigma^m_\delta$ l'espace des $A\in OP\Sigma^m_1$ tels que pour toute suite finie $Q_1,\ldots,Q_N\in OPS^1_0$ d'opérateurs "classiques" de degré 1 on ait

$$[Q_1,[\ldots[Q_N,A]..]]\in OP\Sigma^{m+N\delta}_1.$$

On notera Σ^m_δ l'espace correspondant de symboles (rappelons que a est le symbole de A si $A=a(x,D)$ i.e. si $e^{ix.\xi}a$ est le noyau de Schwartz de $A\mathcal{F}^{-1}$). Il résulte aussitôt de la définition qu'on a $AB\in OP\Sigma^{m+m'}_\delta$ si $A\in OP\Sigma^m_\delta$ et $B\in OP\Sigma^{m'}_\delta$, et aussi $[A,B]\in OP\Sigma^{m+m'-1+\delta+\delta'}_\delta$ si $A\in OP\Sigma^m_\delta$, $B\in\Sigma^{m'}_{\delta'}$ avec $\delta+\delta'\leq 1$.

Pour $\delta<1$ les opérateurs $A\in OP\Sigma^m_\delta$ sont "microlocaux", i.e. ils diminuent le spectre singulier; en effet si $B,C\in OPS_0$ ont des supports disjoints, on a $BAC\sim 0$ (il existe $Q_1,\ldots,Q_N$ de supports disjoints de C, de degré 0, tels que $BQ_1\ldots Q_N\sim B$, on a alors

$$BAC\sim BQ_1\ldots Q_NAC\sim[B[Q_1\ldots[Q_N,A]..]]C$$

qui est de degré $\leq$ deg(BAC)–N(1–δ), aussi petit qu'on veut). En outre si F est un "opérateur intégral de Fourier classique" elliptique, on a $F\,OP\Sigma^m_\delta\,F^{-1}\subset OP\Sigma^m_\delta$ puisque c'est vrai pour $\delta=1$ et qu'on a $F\,OPS^1_0\,F^{-1}\subset OPS^1_0$.

Proposition 2.2.- Pour $\delta\leq 1/2$ on a $OP\Sigma^m_\delta = OPS^m_\delta$.

Examinons d'abord le cas $\delta=1/2$. Il suffit de démontrer la proposition pour m=0. On note $a^{(\alpha)}$ la dérivée $(\partial_{x\xi})^\alpha a$; c'est le symbole du commutateur itéré $A^{(\alpha)}= i^{|\alpha|}(Ad(D,-x))^\alpha A$. Notons E l'espace des distributions a sur $\mathbf{R}^{2n}$ telles que a(x,D) soit continu $L^2\to L^2$. Si $a\in E$, la suite de ses coefficients de Hermite est bornée et par suite $a\in\mathcal{H}^k(\mathbf{R}^{2n})$ pour $k<-n$. Par suite si $a\in\Sigma^0_{1/2}$ on a $a^{(\alpha)}\in E\subset\mathcal{H}^{-n}$ pour tout α donc a est C^∞ et on a une inégalité

$$|a^{(\alpha)}(0,0)| \le c\|A^{(N)}\|_E \text{ pour N assez grand}$$

$(N>3n+2|\alpha|)$ ($A^{(N)}$ désigne la collection des dérivées d'ordre $\le N$).

Pour $(u,v)\in \mathbf{R}^n \times \mathbf{R}^n$, notons T_{uv} l'opérateur $f \to e^{-iv.x} f(x+u)$: les T_{uv} engendrent un groupe de transformations unitaires de $\mathcal{H}^0$ (isomorphe au groupe de Heisenberg) et on a $T_{uv}a(x,D)T_{uv}^{-1}$ $= a(x+u,D+v)$; ceci montre que la norme $\| \|_E$ est invariante par translation en x,ξ: $a(x+u,\xi+v)$ est le noyau de $T_{uv}AT_{uv}^{-1}$. On a donc aussi (avec la même constante)

$$|a^{(\alpha)}(x,\xi)| \le c\,\|(T_{uv}AT\backslash O(_{uv;}^{-1}))^{(N+|\alpha|)}\| = \|T_{uv}A^{(N+|\alpha|)}T\backslash O(_{uv;}^{-1}\|$$

autrement dit toutes les dérivées $a^{(\alpha)}$ sont bornées, ce qui signifie qu'on a $a\in S^0_{1/2}$.

Comme on sait, la réciproque est vraie et résulte du théorème de Calderon-Vaillancourt. Dans le cas $\delta<1/2$, on observe que la condition $a\in \Sigma^m_\delta$ implique $a^{(\alpha)}\in \Sigma^{m-\varepsilon\alpha}_{1/2}$ pour tout α (avec $\varepsilon=1/2-\delta$), et ceci équivaut à $a\in S^m_\delta$.

Dans le cas $\delta>1/2$ il n'y a pas de relation d'inclusion simple entre Σ^m_δ et S^m_δ et l'appartenance à S^m_δ ne se lit pas bien sur le symbole total $a(x,\xi)$ d'un opérateur $a(x,D)$. Néanmoins on a le résultat simple suivant:

Proposition 2.3.- Soit $V=V(x)$ un symbole indépendant de ξ. Alors pour $m\ge 0$, $\delta\ge 1/2$, $V\in S^m_\delta$ implique $V\in \Sigma^m_\delta$.

En effet si $V\in S^m_\delta$, $V^{(\alpha)}$ opère continûment $\mathcal{H}^0 \to \mathcal{H}^{-m-\varepsilon|\alpha|}$ pour tout α ($\varepsilon=1/2-\delta$).

§2. OPÉRATEURS DE TOEPLITZ

Soit Ω la boule unité de $\mathbf{C}^n$, ou plus généralement un espace analytique à bord analytique. On définit alors le noyau de Bergman $B(x,\bar{y})$ de Ω: c'est le noyau du projecteur orthogonal de $L^2(\Omega)$ sur le sous espace des fonctions qui sont holomorphes à l'intérieur de Ω. Ainsi $B(x,\bar{y})$ est une fonction sur $\Omega\times\overline{\Omega}$ holomorphe en x et $\bar{y}$, et on a pour toute fonction f holomorphe, de carré sommable dans Ω:

$$f(x) = \int_\Omega B(x,\bar{y})\, f(y)\, dy$$

Si on veut être intrinsèque, il convient de se placer sur le sous-espace des formes holomorphes dans l'espace des formes différentielles de degré $(n,0)$ sur Ω, i.e. localement de de la forme $\varphi(z)dz_1\ldots dz_n$; ces formes forment de façon naturelle un espace de Hilbert pour le produit hermitien $\langle\omega\,|\,\omega'\rangle = 2^{-n}\, i^{n(n+1)/2}\int \omega\wedge\bar{\omega}$.

Supposons Ω strictement pseudoconvexe, c'est à dire qu'au voisinage de chaque point du bord on peut trouver des coordonnées holomorphes z_j et une fonction u de classe C^∞ telles que Ω soit défini par l'inégalité $u<0$ et que la matrice $(\partial^2 u/\partial z_j\partial\bar{z}_k)$ soit hermitienne $\gg 0$ (dans le cas de la boule: $u=z.\bar{z}-1$ et la matrice des dérivées secondes mixtes est la matrice identité). Dans ce cas la singularité du noyau de Bergman a un aspect remarquable, typique de celle des distributions holonômes simples: B est analytique dans $\Omega\times\bar{\Omega}$ (holomorphe en x et antiholomorphe en y) et se prolonge analytiquement au voisinage de chaque point du bord autre que ceux de la diagonale de $\partial\Omega\times\partial\Omega$. Au voisinage d'un point de cette diagonale, B est de la forme

(2.7) $$B(x,\bar{y}) = f(x,\bar{y})/u(x,\bar{y})^{n+1} + g(x,\bar{y})\mathrm{Log}(x,\bar{y})$$

où f et g sont holomorphes en x, $\bar{y}$ et on a écrit $u = u(z,\bar{z})$ comme fonction holomorphe de z et $\bar{z}$. (Dans le cas où le bord $\partial\Omega$ est seulement C^∞ il y a un résultat asymptotique analogue où, on remplace les fonctions f, g, $u(x,\bar{y})$ par des fonctions "presque holomorphes" i.e. dont les dérivées $(\partial/\partial\bar{x})^\alpha$, $(\partial/\partial y)^\alpha$ s'annulent toutes d'ordre infini sur la diagonale de $\partial\Omega$).

Classiquement un opérateur de Toeplitz est un opérateur linéaire continu A sur l'espace de Hilbert $l^2(N)$ dont la matrice est de la forme (a_{i-j}), i.e. le coefficient d'indice ij ne dépend que de i–j. De façon équivalente, si on interprète l^2 comme l'espace des fonctions holomorphes dans le disque de carré sommable sur le cercle unité de C, A est de la forme $\varphi \to S(f\varphi)$ où f est une fonction bornée sur le cercle, et S est le projecteur orthogonal de L^2(cercle) sur le sous espace des fonctions holomorphes. Cette deuxième formulation se généralise aussitôt à un domaine complexe quelconque: si Ω est un tel domaine et $\mathcal{O}^0(\Omega)$ désigne le sous-espace des fonctions holomorphes de $L^2(\partial\Omega)$, f une fonction continue sur $\partial\Omega$, nous notons T_f l'opérateur linéaire continu sur $\mathcal{O}^0(\Omega)$:

(2.8) $$T_f(\varphi) = S(f\varphi)$$

où B est le projecteur orthogonal $L^2(\partial\Omega)\to\mathcal{O}^0(\Omega)$: T_f est l'opérateur de Toeplitz associé à la fonction f. On pourrait à la place du projecteur de Szegö S (pour la norme de $L^2(\partial\Omega)$) utiliser le projecteur de Bergman B (orthogonal pour la norme de $L^2(\Omega)$; f devant alors être définie dans Ω tout entier); cela ne change pas essentiellement l'algèbre d'opérateurs engendrée.

Nous nous limitons dans la suite au cas où Ω est une variété holomorphe à bord compacte, strictement pseudoconvexe. On généralise encore un peu la définition (2.8) en remplaçant l'opérateur de multiplication par une fonction f par n'importe quel opérateur pseudo-différentiel sur $\partial\Omega$.

Comme au §1 on a une échelle naturelle d'espaces de Hilbert de fonctions holomorphes: $\mathcal{O}^s(\Omega)$ est l'espace hilbertien des fonctions holomorphes φ dans Ω dont la valeur au bord $\varphi \mid \partial\Omega$ (qui est bien définie en tant que distribution) appartient à l'espace de Sobolev $H^s(\Omega)$. Ceci étant:

Définition 2.4.- Un opérateur de Toeplitz de degré m sur Ω est un opérateur linéaire A opérant sur les fonctions holomorphes de Ω de la forme

(2.8)bis $\qquad \varphi \to T_Q\varphi = S(Q\varphi)$

où Q est un opérateur pseudodifférentiel de degré m sur $\partial\Omega$, et S est le projecteur orthogonal $L^2(\partial\Omega) \to \mathcal{O}^0(\Omega)$.

Comme plus haut on a utilisé le projecteur de Szegö S plutôt que celui de Bergman B (plus intrinsèque) parce que $Q\varphi$ n'est pas défini en dehors du bord; en fait les noyaux de Bergman et de Szegö ont des propriétés très similaires, et les classes d'opérateurs qu'ils permettent de définir sont les mêmes.

Parce que les opérateurs de Bergman ou de Szegö sont des "opérateurs intégraux de Fourier" (à phase complexe — c'est ce que montre la formule (2.7)), les opérateurs de Toeplitz se composent bien entre eux; leur loi de composition est décrite par la formule de la phase stationnaire et on a montré dans [BM2] que "localement" ils se comportent exactement comme des opérateurs pseudodifférentiels. Je renvoie à [BM2] pour les détails de la construction et me contenterai ici d'énumérer les principales propriétés de ces opérateurs.

Tout d'abord un opérateur de Toeplitz de degré m opère continûment $\mathcal{O}^s(\Omega) \to \mathcal{O}^{s-m}(\Omega)$ pour tout s; les opérateurs de Toeplitz forment une algèbre, i.e. le composé de deux opérateurs de Toeplitz en est un autre (avec $\deg AB \leq \deg A + \deg B$).

Pour décrire le calcul symbolique, il faut disposer d'un cône symplectique Σ: ici Σ est la moitié du cône caractéristique du système $\bar{\partial}_b$ des équations de Cauchy-Riemann tangentielles qui porte les singularités des valeurs au bord de fonctions holomorphes:

(2.9) $\Sigma \subset \partial\Omega$ est le fibré en demi-droites engendré par $\frac{1}{i}\partial u \mid \partial\Omega$

où u<0 est une inéquation de définition de Ω, ∂u la partie holomorphe de sa différentielle ($\partial u \neq 0$). La condition de stricte pseudoconvexité implique que Σ est symplectique.

Moyennant ceci, si on note T_A l'opérateur de Toeplitz associé à un opérateur pseudodifférentiel A, et qu'on définit le symbole de T_A par

(2.10) $\sigma(T_A) = \sigma(A) \mid \Sigma$ (encore noté $\sigma_m(T_A)$ si on veut mettre en évidence le degré)

on a les formules standard du calcul pseudodifférentiel:

(2.11) $\sigma_m(T_A) = 0$ si et seulement si A est en fait de degré $\leq m-1$

(2.12) $\sigma_{m+m'}(T_A T_B) = \sigma_m(T_A)\, \sigma_{m'}(T_B)$

(2.13) $\sigma_{m+m'-1}([T_A, T_B]) = -i\, \{\sigma_m(T_A), \sigma_{m'}(T_B)\}$

où dans la dernière formule $\{\ ,\ \}$ est le crochet de Poisson de la variété symplectique Σ.

Le cône symplectique Σ n'est pas le fibré cotangent d'une variété en général; en revanche le cône cotangent T^*X d'une variété X est toujours isomorphe à un cône Σ pour un domaine complexe convenable. Par exemple si on suppose X compacte et analytique réelle, plongée comme sous-variété totalement réelle dans une variété complexe $\tilde{X}$ munie d'une métrique hermitienne, on peut prendre $\Omega = X_\varepsilon$, voisinage tubulaire des points de $\tilde{X}$ à distance $<\varepsilon$ de X, pour ε assez petit. Dans ce cas il est d'ailleurs aisé de construire un isomorphisme qui transporte les espaces de Sobolev de X en les espaces $\mathcal{O}^s$ de Ω, et les opérateurs pseudodifférentiels de X en les opérateurs de Toeplitz de Ω.

Si Ω est la boule unité de $\mathbf{C}^n$ le cône Σ n'est pas un fibré cotangent; mais il est de façon presque évidente isomorphe à celui du n°1. Par exemple il est aisé de construire un isomorphisme U de la situation du n°1 (opérateurs pseudodifférentiels polynomiaux) sur celle des opérateurs de Toeplitz de la boule: celui-ci transforme la fonction de Hermite (normalisée) $h_\alpha = c_\alpha (x - \partial/\partial x)^\alpha e^{-x^2/2}$ du n°1 en $c'_\alpha\, z^\alpha$ où on choisit aussi la constante de normalisation $c'_\alpha > 0$.

L'isomorphisme U transporte l'opérateur l'opérateur de création (resp. annihilation) comme suit:

(2.14) $\frac{x_j-\partial/\partial x_j}{\sqrt{2}}$ (création) en l'opérateur $(A+n)^{1/2}z_j$

resp. $\frac{x_j+\partial/\partial x_j}{\sqrt{2}}$ (annihilation) en $\partial/\partial z_j\,(A+n)^{-1/2}$

où A est l'opérateur différentiel $A = \sum z_j \frac{\partial}{\partial z_j}$ ($iA = \frac{\partial}{\partial\theta}$ est le générateur infinitésimal des rotations complexes $z \to e^{i\theta}z$).

§3. STRUCTURES DE CONTACT QUANTIFIÉES

Rappelons qu'un cône symplectique est une variété symplectique Σ (i.e. une variété Σ munie d'une forme symplectique σ) sur laquelle le groupe multiplicatif $\mathbf{R}_+^\times$ des nombres réels positifs opère librement (autrement dit un fibré principal sous $\mathbf{R}_+^\times$), de sorte que pour cette action σ soit homogène de degré 1. La base $\Sigma/\mathbf{R}_+^\times$ de Σ est alors une variété de contact orientée, c'est à dire une variété X munie d'une classe de formes de contact équivalents (deux formes ω et ω' sont équivalentes s'il existe une fonction $\lambda > 0$ qui soit C^∞ et telle que $\omega' = \lambda\,\omega$; la forme ω est de contact si $\omega(d\omega)^{n-1}$ est une forme de volume ou de façon équivalente si le cône Σ engendré par les multiples positifs de ω est un sous-cône symplectique de T^*X). Il revient essentiellement au même de se donner un cône symplectique ou une variété de contact orientée (équivalence de catégorie).

La construction des opérateurs de Toeplitz et les remarques qui terminent le §2 posent naturellement la question suivante: étant donné un cône symplectique Σ, peut-on lui associer naturellement une chaîne d'espaces de Sobolev $\mathcal{O}_\Sigma^s$ et une algèbre filtrée d'opérateurs opérant sur ces espaces, qui donnent lieu au même calcul symbolique que les opérateurs pseudodifférentiels ou de Toeplitz. Nous répondons essentiellement de façon positive à cette question dans [B-G] (voir aussi [BM 3]), du moins en se limitant aux cônes de base compact. Le résultat est le suivant: soit X une variété de contact orientée compacte, $\Sigma \subset T^*X$ le cône symplectique correspondant (engendré par les multiples positifs de la forme symplectique): il existe un opérateur intégral de Fourier S sur X, à phase complexe, de degré 0 qui est un projecteur orthogonal, et dont le spectre singulier est concentré sur Σ, qui se comporte microlocalement exactement comme le noyau de Szegö ou de Bergman décrits au §2. Ceci étant $\mathcal{O}_\Sigma^s$ est l'image par S de l'espace de Sobolev $H^s(X)$, et les opérateurs de Toeplitz relatifs à Σ sont les coefficients des opérateurs pseudodifférentiels sur X, i.e. les opérateurs sur l'échelle des $\mathcal{O}_\Sigma^s$ de la forme

(2.15) $u \to T_Q u = S(Qu)$ pour Q opérateur pseudodifférentiel sur X

Ces opérateurs donnent lieu au même calcul symbolique que les opérateurs de Toeplitz (ou que les opérateurs pseudodifférentiels) i.e. aux formules (2.10) - (2.13) du §2, que nous ne répétons pas.

La construction n'est pas fonctorielle, mais elle jouit tout de même d'une forme d'unicité mod. les sous-espaces de dimension finie: si on a une deuxième telle échelle $\mathcal{O}'^{s}_{\Sigma}$ il existe un quasiisomorphisme F: $\mathcal{O}^{s}_{\Sigma} \to \mathcal{O}'^{s}_{\Sigma}$ qui se comporte comme un opérateur intégral de Fourier elliptique, donc inversible à un opérateur de rang fini (et de degré $-\infty$) près. On peut aussi faire toute la construction de façon équivariante lorsqu'on s'est donné une action de groupe compact. Nous renvoyons à l'appendice d [B-G] ou à [BM 3] pour les détails de ces constructions.

On peut montrer que la théorie des opérateurs de Toeplitz-contact elliptiques donne lieu à une formule d'indice analogue à celle d'Atiyah et Singer, qui s'applique aux matrices carrées de tels opérateurs. La formule de l'indice pour les opérateurs pseudodifférentiels ou de Toeplitz est néanmoins un peu plus générale car elle s'applique aux opérateurs opérants des sections d'un fibré vers celles d'un autre fibré (peut-être pas isomorphe au premier), ce qui n'a guère de sens en vue de l'indice dans cette nouvelle situation où les espaces ne sont définis qu'à un espace de dimension finie près. Dans ce contexte je ne sais pas très bien où est la frontière entre le cas particulier des opérateurs de Toeplitz complexes (ou des opérateurs pseudodifférentiels), et la généralisation ci-dessus; ni d'ailleurs quelles sont les conditions géométriques sur un structure de cône symplectique pour qu'elle provienne d'une structure complexe.

§4 AUTRES MODÈLES

Espace de Bargman

Celui-ci fournit un autre modèle, isomorphe aux deux précédents ("opérateurs pseudo-différentiels polynomiaux", opérateurs de Toeplitz sur la boule). Rappelons que l'espace de Bargman est l'espace $\mathcal{B}$ des fonctions f holomorphes sur $\mathbf{C}^n$ telles que $\int e^{-|z|^2} |f|^2 < \infty$. On prolonge cela en une chaîne d'espaces de Sobolev:

(2.16) $\mathcal{B}^s$ est l'espace des fonctions holomorphes f telles que $\int e^{-|z|^2} |z|^{4s} |f|^2 < \infty$

(on doit compter que, comme pour les opérateurs pseudodifférentiels polynomiaux, z est ici de degré 1/2).

On construit alors l'isomorphisme isométrique V de $L^2(\mathbf{R}^n)$ sur $\mathcal{B}$ qui transporte la fonction de Hermite normalisée h_α sur $c''_\alpha z^\alpha$ où on choisit encore la constante de normalisation positive ($c''_\alpha = (\pi^n \alpha!)^{-1/2}$). Cet isomorphisme a un noyau:

(2.17) $$\text{noyau de } V = c^{ste} \times \exp-(x^2-2\sqrt{2}x.z+z^2)/2$$

(la constante vaut $(2/\pi)^{-n/2}$). L'isomorphisme inverse V^* a pour noyau

$$c^{ste} \times \exp-(2z\bar{z}+x^2-2\sqrt{2}x.\bar{z}+\bar{z}^2)/2.$$

Notons Π le projecteur orthogonal $L^2(\mathbf{C}^n, e^{-|z|^2}) \to \mathcal{B}$. Pour toute fonction a bornée (ou à croissance polynomiale) sur $\mathbf{C}^n$ notons T_a l'opérateur $u \to \Pi au$, par analogie avec la notation des opérateurs de Toeplitz. On montre alors le résultat suivant: les opérateurs sur les $\mathcal{B}^s$ obtenus en transportant les opérateurs pseudodifférentiels polynomiaux de degré m sur $\mathbf{R}^n$ sont les opérateurs de la forme T_a+r où a est un symbole de degré m sur $\mathbf{C}^n = \mathbf{R}^{2n}$ (où on compte les degrés comme au §1), et r est un opérateur de degré $-\infty$ [nous ne donnerons pas de démonstration détaillée de ces faits; ils résultent de ce que les opérateurs $V^{-1}T_aV$ sont visiblement des "opérateurs intégraux de Fourier" dont la phase, résultant des formules (2.6), définit la relation canonique identique des opérateurs pseudodifférentiels]. Le cône symplectique associé à cette nouvelle situation est ici $\mathbf{C}^n-\{0\} = \mathbf{R}^{2n}-\{0\}$ muni de sa forme symplectique canonique; le symbole de T_a est a. On voit directement, ou sur l'expression du noyau de V, comment se transforment les opérateurs de création et d'annihilation:

(2.18) $$V \frac{x_j-\partial/\partial x_j}{\sqrt{2}} V^{-1} \quad \text{est l'opérateur de multiplication par } z_j$$

$$V \frac{x_j+\partial/\partial x_j}{\sqrt{2}} V^{-1} \quad \text{est la dérivation } \partial/\partial z_j.$$

Par suite la transformation canonique associée à l'isomorphisme V est la transformation qui associe z_j à $\frac{x_j-i\xi_j}{\sqrt{2}}$ (resp. ou $\bar{z}_j$ à $\frac{x_j+\xi_j}{\sqrt{2}}$).

On a ainsi construit un troisième modèle équivalent aux précédents. Certains aspects du calcul d'opérateurs sont agréables dans ce modèle; ainsi les opérateurs T_a sont par définition des coefficients d'opérateurs "multiplication par a". Certaines propriétés comme la positivité (T_a est positif si a est positif), les inégalités L^2, l'analogue des inégalités fines de Garding etc... peuvent se lire plus simplement sur ce modèle que pour les opérateurs pseudodifférentiels. Signalons en revanche pour le calcul avec ce modèle une légère différence avec les modèles pseudodifférentiels usuels (phase réelle), en particulier celui du §1: si un opérateur A sur $\mathcal{B}$ s'écrit sous la forme T_a, le symbole a est unique parce que les produits hermitiens $<T_a z^p | z^q>$ déterminent tous les coefficients dans une base de polynômes de Hermite, de même que le symbole d'un opérateur pseudodifférentiel a(x,D) sur $\mathbf{R}^n$ est unique. Dans le modèle pseudodifférentiel réel du §1 où tout opérateur $\mathcal{H}^{\infty} \to \mathcal{H}^{-\infty}$ est de la forme a(x,D) avec a distribution tempérée (a(x,D) est de degré $-\infty$ si et seulement si a est C^{∞} à décroissance rapide); mais il n'y a plus rien de semblable du côté du modèle "espace de Bargman": les opérateurs usuels, en particulier les opérateurs de degré $-\infty$, ne sont pas tous de la forme T_a même si on permet à a d'être une distribution (ce problème ce retrouve chaque fois qu'on travaille avec un phase de partie imaginaire ≥ 0 non nulle). Ceci se reflète aussi sur le transport des opérateurs à symbole moins régulier: si on transporte les opérateurs OPS^m_δ ou $OP\Sigma^m_\delta$: pour $\delta < 1/2$ le développement asymptotique du théorème de la phase stationnaire montre qu'on obtient mod. les opérateurs "régularisant" (i.e. continus $\mathcal{B}^{\infty} \to \mathcal{B}^{-\infty}$) exactement les opérateurs T_a, $a \in S^m_\delta$ et le "calcul symbolique" marche correctement; en revanche pour $\delta \geq 1/2$ le développement de la phase stationnaire diverge et il est faux qu'un opérateur $A \in OP\Sigma^m_\delta$ s'écrive comme somme d'un opérateur de degré $-\infty$ et d'un opérateur T_a avec $a \in S^{m'}_{\delta'}$ ou seulement avec une fonction a un peu régulière.

Fonctions de type exponentiel

Voici une variante du modèle précédent qui a été étudiée par D.Nguon [N].

Soit Ω un ouvert borné strictement convexe de $\mathbf{C}^n$, de frontière C^{∞} (ou analytique si l'on veut rester dans le cadre du §2). La fonction d'appui de Ω est définie par

(2.19) $$h(z) = \sup_{x \in \Omega} \mathrm{Re} <x | z>.$$

C'est une fonction C^{∞} (resp. analytique). On note F la transformation de Fourier-Borel:

(2.20) $$Ff(z) = \int_{\partial\Omega} e^{z.\bar{x}} f(x)\, d\sigma(x)$$

qui est définie lorsque f est holomorphe dans Ω; $d\sigma(x)$ désigne la mesure superficielle du bord $\partial\Omega$. La transformée de Fourier-Borel est entière, à croissance exponentielle, et il est naturel pour l'étudier d'introduire une chaîne d'espaces de Sobolev adaptés:

(2.21) $$\mathcal{E}^s = \{\text{les f holomorphes telles que} \int e^{2h(z)} <z>^{2s-1/2} |f(z)|^2 d^{2n}z < \infty\}$$

(on a posé $<z> = (1+|z|^2)^{1/2}$; le terme $-1/2$ dans l'exposant est là pour que F soit de degré 0 plutôt qu'autre chose, mais n'est pas autrement essentiel).

Le calcul et le formalisme des opérateurs intégraux de Fourier permet de montrer que F est un isomorphisme (non unitaire) de $\mathcal{O}^s(\Omega)$ sur $\mathcal{E}^s$, pour tout s. Lorsqu'on transporte les opérateurs de Toeplitz de Ω au moyen de F, on obtient l'algèbre, analogue à celle du précédent n°, des opérateurs de la forme T_a+r , où T_a est l'opérateur $u \to \Pi au$, Π désignant le projecteur orthogonal sur le sous-espace des fonctions holomorphes dans $L^2\left(e^{-2h(z)} <z>^{-1/2}\right)$, a un symbole sur $\mathbf{C}^n=\mathbf{R}^{2n}$, et r un opérateur de degré $-\infty$ (ici encore parce que les opérateurs transportés $V^{-1}T_aV$ sont des opérateurs intégraux de Fourier à phase complexe, dont la phase, résultant des formules (2.17), est équivalente à celle des opérateurs de Toeplitz). On en déduit un formalisme et un calcul symbolique tout à fait analogues aux précédents, s'appliquant maintenant aux fonctions de type exponentiel. Dans ce contexte (espaces $\mathcal{E}^s$, opérateurs T_a+r) le cône symplectique significatif est encore $\mathbf{C}^n-\{0\}$, muni de la forme symplectique $2i\partial\bar{\partial}h$. Noter aussi que dans ce contexte les opérateurs de multiplication par les z_j sont de degré 1 tandis que les dérivations $\partial/\partial z_j$ sont de degré 0. A part cela ce modèle présente en gros les mêmes avantages et inconvénients que le modèle qui opère sur l'espace de Bargman.

RÉFÉRENCES

[A] M.F.Atiyah. K-theory. Benjamin, Amsterdam.

[A-B-S] M.F.Atiyah, R.Bott, A.Schapiro. Clifford modules. Topology 3, supplément (1964) 3-83.

[A-S] M.F.Atiyah, I.M.Singer. The index of elliptic operators I. Ann. Math. 87 (1968) 484-530; -III, loc. cit. 546-604; -IV, loc. cit. 92 (1970) 119-138.

[B-F-MPh] P.Baum, W.Fulton, R.Mac Pherson. Riemann-Roch and topological K-theory for singular varieties, Acta Math. 143, n°3-4, (1979) 155-192.

[Bea] R.Beals. A general calculus for pseudo-differential operators. Duke Math. J. 42 (1975) 1-42.

[Be] I.N.Bernstein. Modules over a ring of differential operators. An investigation of the fundamental solution of equations with constant coefficients. Funkc. Anal. i Prilozen 5, 2 (1971) 1-16 & Funct. Anal. appl. 5 (1971) 89-101.

[Be-G] I.N.Bernstein, S.I.Gelfand. Meromorphy of the function P^{λ}, Funkc. Anal. i Prilozen 3, (1969) 84-85 & Funct. Anal. appl. 3 (1969) 68-69.

[Bj] J.E.Björk. Rings of Differential Operators. North Holland 1979.

[Bo 1] A.Borel et al. Intersection cohomology. Progress in Math. n°50, Birkhäuser (1984).

[Bo 2] A.Borel et al. Algebraic D-modules. Perspectives in Math. n°2, Academic Press (1987).

[BM 1] L.Boutet de Monvel. Hypoelliptic operators with double characteristics and related pseudo-differential operators. Comm. Pure Appl. Math. 27 (1974) 585-639.

[BM 2] L.Boutet de Monvel. On the index of Toeplitz operators of several complex variables. Invent. Math. 50 (1979) 249-272. Cf. aussi: Opérateurs de Toeplitz. Séminaire EDP 1979, Ecole Polytechnique.

[BM 3] L.Boutet de Monvel. Variétés de contact quantifiées. Séminaire Goulaouic-Schwartz, 1979-80, exposé n°3.

L.Boutet de Monvel. Toeplitz Operators - an asymptotic quantization of symplectic cones. Research Center of Bielefeld-Bochum-Stochastics, University of Bielefeld (FDR) n°215/86 (1986).

[BM 4] L.Boutet de Monvel. Opérateurs pseudodifférentiels polynomiaux, opérateurs de Toeplitz et espace de Bargman. Séminaire EDP, Ecole Polytechnique, 1981-82, exp. n° 3 bis.

[BM 5] L.Boutet de Monvel. Systèmes presqu'elliptiques: une autre démonstration de la formule de l'indice. Astérisque 131 (1985) 201-216.

L.Boutet de Monvel. The index of almost elliptic systems. E. de Giorgi Colloquium, Research notes in Math. 125, Pitman 1985, 17-29.

[BM 6] L.Boutet de Monvel. Opérateurs pseudodifférentiels à bicaractéristiques périodiques. Séminaire Bony-Meyer-Sjöstrand, Ecole Polytechnique, 1984, exposé n°20.

[BM-G] L.Boutet de Monvel, V.Guillemin. The Spectral Theory of Toeplitz Operators. Ann. of Math Studies n°99, Princeton University Press, 1981.

[BM-L-M] L.Boutet de Monvel, M.Lejeune, B.Malgrange. Opérateurs différentiels et pseudo-différentiels. Séminaire, Grenoble 1975-76.

[BM-M] L.Boutet de Monvel, B.Malgrange. Le théorème de l'indice relatif. Ann. Sc. E.N.S. 23 (1990) 151-192.

[BM-Sj] L.Boutet de Monvel, J.Sjöstrand. Sur la singularité des noyaux de Bergman et de Szegö. Astérisque 34-35 (1976) 123-164.

[C1] H.Cartan. Séminaire 1951-52. Fonctions holomorphes de plusieurs variables complexes.

[C2] H.Cartan. Séminaire 1963-64. Théorème d'Atiyah et Singer sur l'indice d'un opérateur différentiel elliptique.

[D-Hö] J.J.Duistermaat, L.Hörmander. Fourier integral operators II. Acta Math. 128 (1972) 183-269.

[Gra] H.Grauert. Ein Theorem der analytischen Garben-theorie und die modulräume komplexe Structuren. IHES Publ. Math. n°5 (1960).

[Gro] A.Grothendieck. SGA V, théorie des intersections et théorème de Riemann-Roch. Lecture Notes in Math. 225, Springer Verlag (1971).

[G-St] V.Guillemin, S.Sternberg. Geometrical asymptotics. Amer. Math. Soc. Surveys 14, Providence RI, 1977.

[Hi] F.Hirzebruch. Neue topologische Methoden in der algebraische Geometrie. Springer Verlag, Berlin.

[Hö 1] L. Hörmander. Pseudodifferential operators. Comm. Pure Appl. Math. 18 (1965) 501-517.

[Hö 2] L. Hörmander. Pseudodifferential operators and hypoelliptic equations. Amer. Math. Soc. Symp. on Singular Integrals (1966) 138-183.

[Hö 3] L. Hörmander. Fourier integral operators I. Acta Math. 127 (1971) 79-183.

[Hö 4] L. Hörmander. The Analysis of Linear Partial Differential Operators, vol. III et IV, Grundlehren der Math. Wiss. 124, Springer Verlag, Berlin.

[H-Sch] Ch.Houzel, P.Schapira. Images directes de modules différentiels, C.R. Acad. Sci. Paris 298 (1984) 461-464.

[K 1] M.Kashiwara. Index theorem for a maximally overdetermined system of linear differential equations, Proc. Jap. Acad. 49-10 (1973) 803-804.

[K 2] M.Kashiwara. b-functions and holonomic systems. Invent. Math. 38 (1976) 33-54.

[K3] M.Kashiwara. Analyse microlocale du noyau de Bergman. Séminaire Goulaouic-Schwartz 1976-77, exp.n°8, Ecole Polytechnique.

[K 4] M.Kashiwara. Introduction to the theory of hyperfunctions. In Sem. on microlocal analysis, Princeton University Press, Princeton N.J. (1979) 3-38.

[K 5] M.Kashiwara. Systems of microdifferential equations. Cours à l'Université Paris Nord, Birkhaüser 1983.

[Ko-N] J.J.Kohn, L.Nirenberg. On the algebra of pseudo-differential operators. Comm. Pure Appl. Math. 18 (1965) 269-305.

[La] G.Laumon. Sur la catégorie dérivée des $\mathcal{D}$-modules filtrés. Thèse, Orsay 1983.

[L] R.N.Levy. Riemann-Roch theorem for complex spaces. Acta Math. 158 (1987) 149-188.

[M1] B.Malgrange. L'involutivité des caractéristiques des systèmes différentiels et microdifférentiels. Séminaire Bourbaki 1977-78, n°552.

[M2] B.Malgrange. Sur les images directes de $\mathcal{D}$-modules. Manuscripta Math. 50 (1985) 49-71.

[Me-Sj] A.Melin, J.Sjöstrand. Fourier Integral operators with complex valued phase functions. Lecture Notes in Math. 459, Springer Verlag (1974) 120-223.

[N] D.Nguon. Modèles d'opérateurs pseudodifférentiels ou de Toeplitz opérant sur des espaces de fonctions de type exponentiel. Thèse, à publier.

[Ph] F.Pham. Singularités des systèmes différentiels de Gauss-Manin. Progress in Math. n°2, Birkhäuser (1980).

[S-K-K] M.Kashiwara, T.Kawai, M.Sato. Microfunctions and pseudo-differential equations, Lecture Notes 287 (1973), Springer-Verlag.

[S 1] M.Sato. Theory of hyperfunctions I. Fac. Sci. Univ. Tokyo I, 8 (1959) 139-193, et II ibid. (1960), 387-437.

[S 2] M.Sato. Hyperfunctions and partial differential equations. Proc. Int. Conf. on Funct. Anal. and rel. topics, 91-94, Tokyo University Press, Tokyo 1969.

[S 3] M.Sato. Regularity of hyperfunction solutions of partial differential equations. Actes Congr. Int. Math. Nice 1970, 2, 785-794.

FOURIER TRANSFORM AND DIFFERENTIAL EQUATIONS

B. MALGRANGE
Institut Fourier
Université de Grenoble 1
BP 74
38402 St Martin d'Hères cedex

SUMMARY. — We study, in the complex domain, the action of the Fourier transform on the solutions of ordinary linear differential equations with polynomial coefficients. In the classical "Laplace method", there are some restrictions; also, some choice of integration contours seem rather unsystematic. We show how to remove these restrictions and how to make these choices in a more systematic way.

ADVERTISEMENT. — These notes do not correspond exactly to the content of the oral lectures. They were devoted to the Fourier transform in the complex domain, and to some applications to linear and non linear equations; especially, they were intended to be an introduction to some of the ideas of J. Ecalle on the "fonctions résurgentes" [E]. However, the material of these lectures is contained in some published paper, namely [Ca], [Ma 1], [Ma 2]. As these papers are already mainly expository papers, it seemed to me unappropriate to rewrite them. Therefore, after having recalled the beginning of [MA 2], I preferred to add a few complements not contained in these papers, on the calculation of "microsolutions" and on the case of non-exponential growth. All this is a part of a more systematic study of Fourier transform and linear differential equations which will be the subject of a forthcoming book [Ma 3].

1. Introduction

Let $p(x, \partial_x) = \sum a_{k,\ell} x^k \partial_x^\ell$ a linear differential operator in one variable $x \in \mathbf{C}$, with $\partial_x = \frac{d}{dx}$. Its "Fourier transform" $q = \mathcal{F}p$ is by definition the operator in ξ, ∂_ξ obtained by the substitution $\partial_x \leftarrow \xi$, $x \leftarrow -\partial_\xi$.

Let us try to write the solutions of $qg = 0$ under the form $g(\xi) = \int_\gamma e^{-x\xi} f(x)dx$, for suitable f and γ. If we differentiate and integrate by parts formally, we have $(qg)(\xi) = \int_\gamma (pf)(x)e^{-x\xi}dx$; therefore, it is natural to try to choose an f verifying $pf = 0$. It is classical ("Laplace method" see f.i.[I]) to do this when q has degree one

A. Boutet de Monvel et al. (eds.), Recent Developments in Quantum Mechanics, 33–48.

in ξ; in that case, p has degree one in ∂_x and the equation $pf = 0$ can be explicitly integrated. If one choose suitably γ , one gets "in general" – but not *always* – all the solutions of $qg = 0$.

Here are a few simple examples (see also [Ma 2]).

EXAMPLE (1.1)

$q = \xi\partial_\xi - \nu$, then $g = a\xi^\nu$ one has $p = -\partial_x x - \nu = -(x\partial_x + \nu + 1)$; here $f = b\cdot x^{-\nu-1}$.

Denote by δ_θ the half-line of end 0 and angle θ , and by γ_θ the following contour : γ_θ goes along δ_θ from ∞ to $\varepsilon e^{i\theta}$ ($\varepsilon > 0$, small), encircle 0 counterclockwise, and goes again along δ_θ from $\varepsilon e^{i\theta}$ to ∞ (see figure 1)

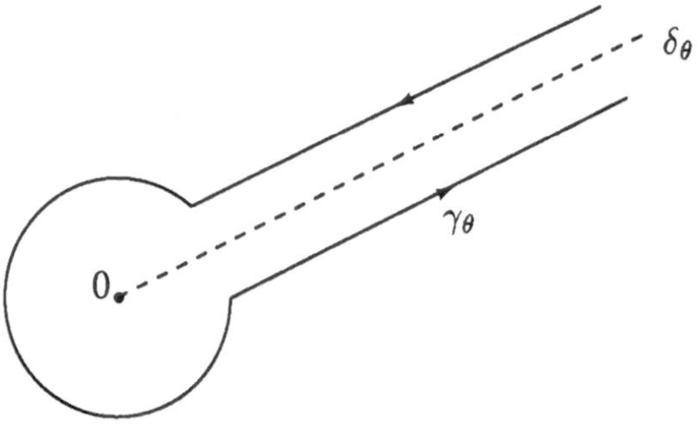

Figure 1

For $\nu + 1 \notin -\mathbf{N}$ ($\mathbf{N}$ = non-negative integers), one has, in $|\arg\xi + \theta| < \pi/2$,

$$\xi^\nu = c\int_{\gamma_\theta} x^{-\nu-1}e^{-x\xi}dx ,$$

[with a suitable choice of the arguments, one has $c = \frac{e^{\pi i\nu}}{2\pi i}\Gamma(1+\nu)$, but it does not matter here].

Therefore, one can represent in the required way the solutions of $qg = 0$; but, for $\nu + 1 \in -\mathbf{N}$, the argument fails since the integral vanishes.

EXAMPLE (1.2)

We consider the Bessel equation ($\nu \in \mathbf{C}$)

$$\frac{d^2h}{dt^2} + \frac{1}{t}\frac{dh}{dt} + (1 - \frac{\nu^2}{t^2})h = 0 .$$

The function g defined by $g(\xi) = \xi^{\nu/2}h(\sqrt{\xi})$ satisfies the equation

(*) $$4\xi g'' - 4(\nu - 1)g' + g = 0 .$$

This equation is the Fourier transform of $4x^2 f' + [4(\nu+1)x+1]f = 0$; a basis of the solutions of this last equation is $f = x^{-\nu-1}e^{1/4x}$.

In the half-plane $|\arg \xi + \theta| < \frac{\pi}{2}$, a basis of the solutions of $(*)$ is given by $g(\xi) = \int_{\gamma_i} f(x)e^{-x\xi}dx$, $(i = 1,2)$ with $\gamma_1 = \gamma_\theta$ (in the notations of the preceding example), and γ_2 is a contour starting from 0 in the direction $\mathbf{R}^-$ and becoming after a turn parallel to δ_θ (figure 2).

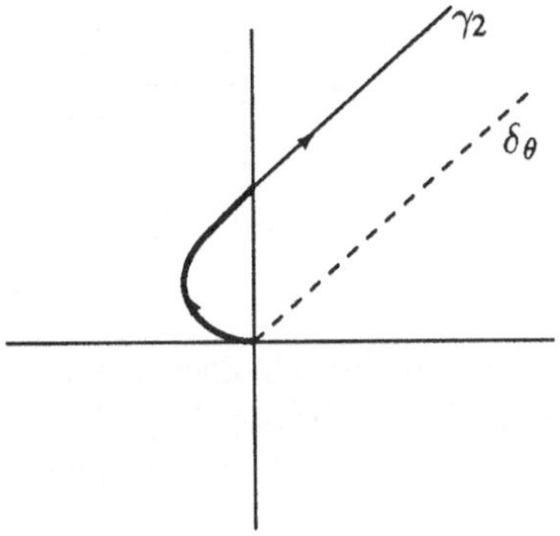

Figure 2

I will verify later that these integrals are linearly independent (for a more precise identification with Bessel functions, I refer to treatises on special functions).

EXAMPLE (1.3)

The equation $f' + x^2 f = 0$ has $e^{-x^3/3}$ as solution; its Fourier transform is

$$(*) \qquad g'' + \xi g = 0 \; .$$

Denote by γ_i , $(i = 1,2,3)$ the contours going from ∞ to 0, and from one domain where $\operatorname{Re} x^3 > 0$ to another one (see figure 3), and put ("Airy integral") $g_i(\xi) = \int_{\gamma_i} e^{-x^3/3 - x\xi}dx$: they are entire functions, and one has obviously $g_1 + g_2 + g_3 = 0$; one can prove that two of these functions, f.i. g_1 and g_2 give a basis of the solutions of $(*)$.

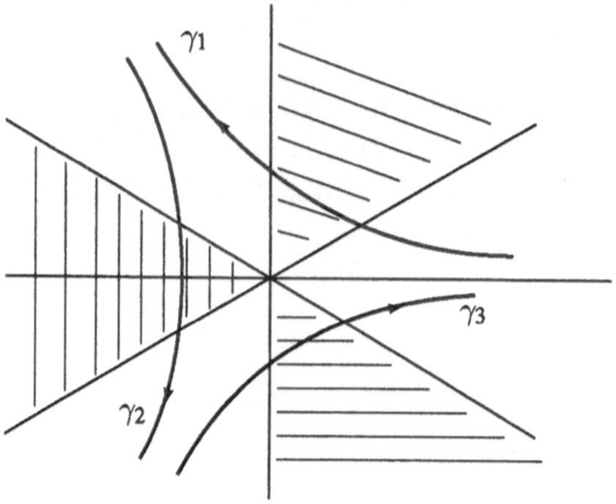

Figure 3

Incidentally, $(*)$ is the equation obtained from the Bessel equation with index $\nu = \frac{1}{3}$ by the transformation $g(\xi) = \xi^{1/2}h(\frac{2}{3}\xi^{3/2})$. I refer again to treatises on special function for the precise identification of the "Airy functions" g_i with Bessel functions.

In the examples 2 and 3, one has obtained a basis of the solutions of $qg = 0$ by Fourier transform; however, a question remains : what is general behind the preceding procedures? The aim of these notes is precisely to discuss this problem.

In the sections 2 and 3, I shall study the case of the equations whose solutions grow at most exponentially at infinity; in the end of these notes, I shall give an idea of the general case. The following theorem will be useful in this context.

THEOREM (1.4). — *The following properties are equivalent*

1)In any sector $|x| \gg 0$, $\alpha < \arg x < \beta$, *the solutions of* $pf = 0$ *grow at most exponentially, e.g. they verify, for some* $A > 0$, $C > 0$ $|f(x)| \leqslant C\ e^{A|x|}$.

2) Denote by m *(resp.* n*) the degree of* p *with respect to* x *(resp.* ∂_x*); then, the coefficient of* $x^m\partial_x^n$ *is* $\neq 0$.

3) $q = \mathcal{F}p$ *verifies 1) or 2).*

2) $\Rightarrow$ 1) Outside of a disc $|x| \leqslant C$, the hypothesis $pf = 0$ implies an inequality $|\partial_x^n f| \leqslant K \sum_{p\leqslant n-1} |\partial_x^k f|$. Then, the results follows from elementary estimates, f.i. from the Gronwall lemma.

1) $\Rightarrow$ 2) This is not so simple; actually, this is a special case of the relationship between the growth of solutions and the so-called "Newton polygon" of the equation. This question can be decomposed in two parts *i)* the relation between the Newton

polygon and the formal solutions, *ii)* the theory of asymptotic expansions, to go from formal to actual solutions. On these points, see [Ra], [Ro], [W]. Finally, the equivalence "p verifies 2)" $\iff$ "q verifies 2)" is obvious, since condition 2) is symmetric in x and ∂_x .

In the next sections, I shall call *equation of exponential type* a differential equation p verifying the conditions of this theorem. For instance, the equations of example 1 and 2 are of exponential type, but not the equations of example 3.

2. Microfunctions

Let me recall briefly the notions of [Ma 1]. Let D be a disc with centre $0 \in \mathbf{C}$, $D^* = D - \{0\}$, and $\widetilde{D}^*$ a universal covering of D^* (determined by the choice of a base-point $a \in D^*$). We denote respectively by $\mathcal{O}(D)$ and $\mathcal{O}(\widetilde{D}^*)$ the space of holomorphic functions on D and $\widetilde{D}^*$; one puts $\mathcal{C}(D) = \mathcal{O}(\widetilde{D}^*)/\mathcal{O}(D)$ and one defines the maps $\mathcal{O}(\widetilde{D}^*) \underset{\text{var}}{\overset{\text{can}}{\rightleftarrows}} \mathcal{C}(D)$ in the following way (Deligne) : "can" is the canonical quotient map; "var" (= the variation) is the unique map such that one has var $\circ$ can $= T - \mathrm{id}$, T the monodromy (= action of a counterclockwise loop of end a on $\mathcal{O}(\widetilde{D}^*)$).

If the radius of D tends to 0 (one moves, f.i., radially the base-point), one obtain spaces denoted respectively $\mathcal{O}$, $\widetilde{\mathcal{O}}$, $\mathcal{C}$ and maps $\widetilde{\mathcal{O}} \underset{\text{var}}{\overset{\text{can}}{\rightleftarrows}} \mathcal{C}$. The space $\mathcal{C}$ is a variant of spaces introduced by Sato, in one or several variables (see f.i. [S.K.K]) under the names of microfunctions or hyperfunctions. Therefore, I call its elements "microfunctions". Its properties are very similar to those of the germs at $0 \in \mathbf{R}$ of distributions or hyperfunctions with support in $\mathbf{R}_+$.

i) One cannot multiply 2 elements of $\mathcal{C}$, but an element of $\mathcal{C}$ and an element of $\mathcal{O}$; on the other hand, the convolution is defined on $\mathcal{C}$; the unit is the "Dirac (micro)function" $\delta = \mathrm{can}\frac{1}{2\pi i x}$.

ii) The action of ∂_x over $\mathcal{C}$ is bijective; one has $\partial_x f = \partial_x \delta * f$, and $\partial_x^{-1} f = Y * f$, $Y = \partial_x^{-1}\delta$ being the Heaviside (micro)function $\mathrm{can}\left(\frac{1}{2\pi i}\log x\right)$.

THEOREM (2.1) (Kashiwara [Ka]). — *Let $p = \sum_0^n a_k \partial_x^k$ a linear differential operator with coefficients in $\mathcal{O}$, and $a_n \neq 0$; denote by $m = v(a_n)$ the order of the zero of a_n in 0. Then $p : \mathcal{C} \to \mathcal{C}$ is surjective, and its kernel has dimension m .*

Consider the exact sequence $0 \longrightarrow \mathcal{O} \longrightarrow \widetilde{\mathcal{O}} \xrightarrow{\text{can}} \mathcal{C} \longrightarrow 0$; applying p one finds an exact sequence

$$
\begin{aligned}
0 \longrightarrow \ker(p,\mathcal{O}) \longrightarrow \ker(p,\widetilde{\mathcal{O}}) \longrightarrow \ker(p,\mathcal{C}) \longrightarrow \operatorname{coker}(p,\mathcal{O}) \\
\longrightarrow \operatorname{coker}(p,\widetilde{\mathcal{O}}) \longrightarrow \operatorname{coker}(p,\mathcal{C}) \longrightarrow 0
\end{aligned}
$$

Now, the usual theorem of existence and uniqueness for differential equations tells us that $\dim \ker(p,\widetilde{\mathcal{O}}) = n$ and that $\operatorname{coker}(p,\widetilde{\mathcal{O}}) = 0$. Therefore, one has also $\operatorname{coker}(p,\mathcal{C}) = 0$.

On the other hand, the "index theorem" for differential equations (proved independently by Kashiwara, Komatsu, and Malgrange, see f.i. [Ka], [Ma 4]) says the following : $\ker(p,\mathcal{O})$ and $\operatorname{coker}(p,\mathcal{O})$ are finite dimensionals, and the difference of their dimensions equals $n-m$. Using this result, and the fact that the alternate sum of dimensions in $(*)$ is zero, one gets the result.

Take now $f \in \mathcal{C}$; suppose that, on the half-line δ_θ (see definition in (1.1)), $\operatorname{var} f$ can be continued analytically in a holomorphic function g , with exponentiel growth at infinity; we will call "hyperfunction in the direction δ_θ" and we will denote by $F = (f,g)$ this pair of datas. One defines the Fourier-Laplace transform of F in the direction θ in the following way : choose $\tilde{f} \in \widetilde{\mathcal{O}}$, with $\operatorname{can} \tilde{f} = f$; choose also $\varepsilon > 0$, small, and denote by γ'_A (resp. γ''_θ) the circular loop around 0 with origine $\varepsilon e^{i\theta}$ (resp. the interval $[\varepsilon e^{i\theta}, +\infty[$ on δ_θ). Put

$$\widehat{F}(\xi) = \overline{\int_{\delta_\theta}} F e^{-x\xi}\, dx = \int_{\gamma'_\theta} \tilde{f}(x) e^{-x\xi}\, dx + \int_{\gamma''_\theta} g(x) e^{-x\xi}\, dx \; .$$

This integral is defined in the half-plane $\operatorname{Re}(\xi e^{i\theta}) > C$, for some $C > 0$ (because g growths exponentially at infinity). It does not depend on $\tilde{f}$ and ε . If $p = \sum a_k \partial_x^k$, $a_k \in \mathbf{C}[x]$, verifies $pf = 0$, one has $pg = 0$, and one verifies at once that one has $q\widehat{F} = 0$, with $q = \mathcal{F}p$.

For $a \in \mathbf{C}$, denote by $\delta_{a,\theta}$ the translated of the half-line δ_θ by the translation a . If we have a "microfunction at the point a" (which means just the translation by a of a microfunction at 0), and if this microfunction has a property of analytic continuation along $\delta_{a,\theta}$ similar to the previous one, this will also define a hyperfunction; its Laplace transform will also be defined and will be holomorphic as before in some half-plane $\operatorname{Re}(\xi e^{i\theta}) \gg 0$.

Consider now a differential operator $p = \sum_0^n a_k \partial_x^k$, $a_k \in \mathbf{C}[x]$, and suppose p of *exponential type*. Let $\alpha_1, \ldots, \alpha_r$ be its singularities, e.g. the zeroes of a_n; if m_k is the multiplicity of α_k, one has $\sum m_k = m$, with $m = \deg a_n$. Let θ be a direction such that the half-lines $\delta_{\alpha_k,\theta}$ are distinct. According to the preceding theorem, there are m_k microfunctions f at a_k linearly independent which verify $pf = 0$ and they can be continued along $\delta_{\alpha_k,\theta}$ into hyperfunctions F which are of exponential growth at infinity. One has the following result :

THEOREM (2.2). — *In some half-plane* $\operatorname{Re}(\xi e^{i\theta}) \gg 0$, *the Laplace transforms of the hyperfunctions F solutions of $pF = 0$ defined along the lines $\delta_{\alpha_k,\theta}$ by the preceding procedure give a basis of the solutions of the equation $qG = 0$ ($q = \mathcal{F}p$).*

As $\sum m_k = m = \deg q/\partial_\xi$, we have "the right number" of solutions. But we have to prove that they are actually linearly independent. The proof will be given in the next section. Let us now take some examples, and especially the examples (1.1) and (1.2).

EXAMPLE (2.3)

We have to look at the microfunctions (at 0) solutions of $(x\partial_x+\nu+1)f=0$. Their space is one-dimensional; for $\nu+1 \notin -\mathbf{N}$, they are given up to a constant by $f=\operatorname{can}(x^{-\nu-1})$; for instance, for $\nu=0$, this gives δ. But, for $\nu+1\in-\mathbf{N}$, they are not of the form can (usual solution); the solution here is $f=\frac{1}{2\pi i}\operatorname{can}(x^{-\nu-1}\log x)=x^{-\nu-1}Y$; one has $\operatorname{var} f=x^{-\nu-1}$, and one has easily

$$\int_{\delta_\theta} \overline{F}e^{-x\xi}\,dx=\int_{\delta_\theta} x^{-\nu-1}e^{-x\xi}\,dx=\Gamma(-\nu)\xi^{\nu}\ .$$

This is of course obvious; but, what we see is the following : generally speaking, the usual procedure of taking usual solutions and contours of the type γ_θ will be sufficient only in the case where

i) the equation has exponential type, and

ii) at all the singular points, the map
can : (usual solutions) $\longrightarrow$ (microsolutions) is surjective.

Another typical example of the same situation is given by the equations with constant coefficients : if $q=\sum_0^n a_k\partial_\xi^k$, $(a_k\in\mathbf{C}\ ,\ a_n\neq 0)$, then $p=\sum a_kx^k$, and there is no usual solution $\neq 0$; the microfunction solutions of $pf=0$ are the "Dirac function" and some of its derivatives at the zeroes of p; by Fourier-Laplace transform, one gets of course the usual exponential $\times$ polynomial solutions .

EXAMPLE (2.4)

Here, one has $pf=4x^2f'+\left[4(\nu+1)x+1\right]f$; the space of solutions has dimension 1 and is generated by $f=x^{-\nu-1}e^{1/4x}$ (with suitable choice of the base point a and of the determination of $\arg x$ at a; I leave those questions to the reader). One microsolution is given by $f_1=\operatorname{can} f$, and it corresponds to the integral along γ_1. But the space of microsolutions has dimension 2, and we need another one. The method used in (1.2) corresponds to the following procedure : we consider on $\mathbf{R}$ the $\mathcal{C}^\infty$ function g equal to $x^{-\nu-1}e^{1/4x}$ for $x<0$ and 0 for $x\gg 0$; by a standard method (f.i. Cauchy integral), we can consider g as difference g_+-g_- , g_+ (resp. g_-) holomorphic in $\operatorname{Im} x>0$ (resp. $\operatorname{Im} x<0$) and $\mathcal{C}^\infty$ up to the boundary. On $\{x>0\}$, g_+ and g_- are analytic continuation of each other; therefore, they define a holomorphic function on $\mathbf{C}-\mathbf{R}_-$, which gives by analytic continuation an element $\tilde{g}$ of $\tilde{\mathcal{O}}$; since $pg=0$ in the sense of $\mathcal{C}^\infty(\mathbf{R})$, pg^+ and pg^- are continuation of each other, or, in other words, $p\tilde{g}$ is holomorphic; therefore, if we put $f_2=\operatorname{can}\tilde{g}$, we have $pf_2=0$ (and, obviously, $\operatorname{var} f_2=f$). The fact that f_2 corresponds to the integral along γ_2 considered in (2.4) can be left to the reader.

The microsolutions f_1 and f_2 are linearly independent; for, suppose that one has $f_1 = \lambda f_2$, $\lambda \in \mathbf{C}$; as $f_1 = \operatorname{can} f$, $f_2 = \operatorname{can} \tilde{g}$, $f - \lambda \tilde{g}$ should be holomorphic at 0; the "principal determination" of $\tilde{g}$ in $\mathbf{C} - \mathbf{R}$ is bounded near 0, then f should have the same property, which is not the case; this proves the result. Now this fact, combined with theorem (2.2) proves that the corresponding integrals $\int_{\gamma_1}$ and $\int_{\gamma_2}$ are linearly independent.

Appendix. On the calculation of the microsolutions

This calculation is based on the exact sequence (see proof of th. 2.1)

$$0 \longrightarrow \ker(p, \mathcal{O}) \longrightarrow \ker(p, \widetilde{\mathcal{O}}) \xrightarrow{\text{can}} \ker(p, \mathcal{C}) \xrightarrow{\alpha} \operatorname{coker}(p, \mathcal{O}) \longrightarrow 0 .$$

To abreviate, we put $\Psi = \ker(p, \widetilde{\mathcal{O}})$, $\Phi = \operatorname{coker}(p, \mathcal{C})$. If we have a basis of Ψ , we can find by usual methods ("variation of constants") solutions $f \in \widetilde{\mathcal{O}}$ of $Df = g$, $g \in \mathcal{O}$; if we have a basis g_i of coker$(p, \mathcal{O})$ [f.i. a suitable collection of monomials x^k], and if f_i is a corresponding solution of $pf_i = g_i$, then $\operatorname{can} \Psi \cup \{\operatorname{can} f_i\}$ generate Φ .

Exercise. Use this method in the preceding example, and compare with the result obtained here.

In this example, I actually used another method, which I will now analyse. First, note the following : let us give a sector $\Sigma : \theta < \arg x < \theta'$, $|x| < \varepsilon$ $(\theta' < \theta + 2\pi)$, and let us give a solution f of $pf = 0$, holomorphic in Σ and decreasing faster than any $|x|^N$ near 0; then the preceding procedure give a function $\tilde{g}$, holomorphic in $\theta < \arg x < \theta' + 2\pi$, with an asymptotic expansion $\hat{g} \in \widehat{\mathcal{O}}(= \mathbf{C}[[x]])$ at the origin, such that the difference of determinations of $\tilde{g}$ is Σ equals f; such a g is unique up to $\mathcal{O}$, and it extends, by the properties of analytic continuations of f , in an element of $\widetilde{\mathcal{O}}$ [but the other determinations may not have the same property of asymptotic expansion]; if we write $\operatorname{can} \tilde{g} = f^\#$, then $f^\#$ is uniquely determined by (f, Σ); one has $pf^\# = 0$, and $\operatorname{var} f^\# = f$.

Denote by $\Phi^\#$ the subspace of Φ generated by the $f^\#$ obtained in this way. We have the following result

Proposition (2.5). — *The sequence* $\Phi^\# \xrightarrow{\alpha} \operatorname{coker}(p, \mathcal{O}) \xrightarrow{\beta} \operatorname{coker}(p, \widehat{\mathcal{O}}) \to 0$ *is exact.*

The fact that β is surjective is well-known (see f.i. [Ma 4]). The rest of the proof is very closely related to the so called "isomorphism of Sibuya and Malgrange", and the fact that, due to the "fundamental theorem on asymptotic expansions", this isomorphism can be restricted to solutions of differential equations (see f.i. [Ma 2], § 7.1).

First, prove that $\beta\alpha = 0$. Let $(f, \Sigma, \tilde{g}, f^\#)$ be as before; one has $\alpha(f^\#) = p\tilde{g}$; as $\tilde{g}$ has an asymptotic expansion $\hat{g} \in \widehat{\mathcal{O}}$ at 0 one has $p\tilde{g} \in \operatorname{Im}(p, \widehat{\mathcal{O}})$; in other words, $\beta(p\tilde{g}) = 0$. Conversely, take $h \in \mathcal{O}$ with $\beta(h) = 0$; there exists $\hat{g} \in \widehat{\mathcal{O}}$ with $p\hat{g} = h$;

by the theorem of asymptotic expansions, in any "sufficiently small" sector Σ , there exists a holomorphic function g in Σ , admitting $\hat{g}$ as asymptotic expansion at 0, and satisfying $pg = h$; choose sections Σ_1 , Σ_n , turning counterclockwise and having only 2 by 2 intersections, such that one can apply the preceding result in these sectors, and let g_i be the functions obtained; put $f_i = g_{i+1} - g_i$ in $\Sigma_i \cap \Sigma_{i+1}$ (one puts $\Sigma_{n+1} = \Sigma_1$, with the angles translated by 2π); one has $pf_i = 0$, and f_i has asymptotic expansion 0, then it has the required property of growth. One verifies easily that $\alpha(\Sigma f_i^{\#}) = h \pmod{p\mathcal{O}}$. This proves the proposition.

Now, as elements of Φ, we have $\operatorname{can}\Psi = \ker\alpha$, and $\Phi^{\#}$; in order to have generators of Φ , we need also to have a section of $\beta \circ \alpha \ : \Phi \longrightarrow \operatorname{coker}(p, \widehat{\mathcal{O}})$. This can be done (in principle!) in the following way

i) Formal level. Denote by $\widehat{\mathcal{O}}^{\sim}$ the series $\sum_0^n a_k(\log x)^k$, with $a_k \in \widehat{\mathcal{O}}[x^{-1}]$, and similarly by $\widehat{\mathcal{C}}$ the space $\widehat{\mathcal{O}}^{\sim}/\widehat{\mathcal{O}}$; the maps "can" and "var" are also defined in the obvious way. We write also $\widehat{\Psi} = \ker(p, \widehat{\mathcal{O}}^{\sim})$ and $\widehat{\Phi} = \ker(p, \widehat{\mathcal{C}})$.

LEMMA (2.6). — *The operator p , acting on $\widehat{\mathcal{O}}^{\sim}$ and $\widehat{\mathcal{C}}$ is surjective.*

We filtrate the differential operators by $\deg x = 1$, $\deg \partial_x = -1$; then $p = \sum_{\ell}^{+\infty} p_k$, p_k of degree k , $p_\ell \neq 0$; this is compatible with the obvious filtration of $\widehat{\mathcal{O}}^{\sim}$ and $\widehat{\mathcal{C}}$: $\deg x = 1$, $\deg(\log x) = 0$; then it is sufficient to prove the lemma for the corresponding graded case, e.g. for p_ℓ; in this case, it can be left to the reader.

From this lemma, one deduces an exact sequence

$$0 \longrightarrow \ker(p, \widehat{\mathcal{O}}) \longrightarrow \widehat{\Psi} \xrightarrow{\text{can}} \widehat{\Phi} \xrightarrow{\hat{\beta}} \operatorname{coker}(p, \widehat{\mathcal{O}}) \longrightarrow 0 .$$

Note also that $\operatorname{coker}(p, \widehat{\mathcal{O}})$ can easily be calculated with the help of the same filtration.

ii) Analytic level. We have to do the following : given $\hat{f} \in \operatorname{coker}(p, \widehat{\mathcal{O}})$ and $\hat{g} \in \widehat{\Phi}$ such that $\hat{\beta}\hat{g} = \hat{f}$, find "as explicitely as possible" $g \in \Phi$ such that $\beta \circ \alpha(g) = \hat{f}$. We will do something more precise : given $\hat{f} \in \widehat{\mathcal{C}}$ and a direction θ at 0 (in the universal covering $\widetilde{D}^*$), we say that $f \in \mathcal{C}$ is a lifting of $\hat{f}$ in the direction θ if one (or any) representative of f in $\widetilde{\mathcal{O}}$ has, in a sector $\theta - \varepsilon < \arg x < \theta + \varepsilon + 2\pi$, $|x| < \varepsilon$ an asymptotic expansion at 0 which is equal $\operatorname{mod} \widehat{\mathcal{O}}$ to $\hat{f}$. The following result is a version for microfunctions of the fundamental theorem on asymptotic expansions.

PROPOSITION (2.7). — *Given $\hat{f} \in \widehat{\Phi}$ and θ, there exists a lifting $f \in \Phi$ of $\hat{f}$ in the direction θ .*

One has $p(\operatorname{var}\hat{f}) = 0$; by the theorem on asymptotic expansions [W], if $\varepsilon < 0$ is small, there exists g holomorphic in $\Sigma = \{\theta - \varepsilon < \arg x < \theta + \varepsilon\}$ having $\operatorname{var}\hat{f} \in \widehat{\mathcal{O}}^{\sim}$ as asymptotic expansion at 0, and verifying $pg = 0$. In order to lift g to Φ , choose a h, holomorphic in $\theta-\varepsilon<\arg x<\theta+2\pi+\varepsilon$, and having, mod $\widehat{\mathcal{O}}$, $\hat{f}$ an asymptotic expansion; such a h exists by a theorem of Borel-Ritt. Put, with obvious notations $f_1=\operatorname{can} h$, $f=f_1+(g-\operatorname{var} f_1)^{\#}$. This is obviously a lifting of $\hat{f}$; therefore pf is a lifting of 0; this fact, plus the equality $\operatorname{var}(pf)=0$ implies easily that $pf=0$. This proves the proposition.

Actually, f is independent of f_1; it depends only on g and $\hat{f}$, and it is even determined by g and the polar part of $\hat{f}$ [e.g. the polar part of the $\hat{a}_k$ in a representation $\hat{f}=\sum\hat{a}_k(\log x)^k \mod \widehat{\mathcal{O}}$]. As in the theory of distributions, f can be considered as a "finite part" of the function g. Also, the corresponding Fourier integral can be considered as a "finite part" integral.

3. Proof of the theorem (2.2)

In order to prove this theorem, I will need some elementary results on the Fourier transform in the complex domain. The following language will be convenient : let $\overline{\mathbf{C}}$ be the compactification of $\mathbf{C}$ (the variable here is ξ) by the circle S of the directions of half-lines; for $\tau\in S$, we denote by $\mathcal{V}(\tau)$ the collection of the open sets $U\subset\mathbf{C}$ such that the closure $\overline{U}\subset\overline{\mathbf{C}}$ is a neighborhood of τ [in the other words, U contains a set of the form $|\xi|>\frac{1}{\varepsilon}$, $|\arg\xi-\tau|<\varepsilon$]. Denote by $\mathcal{A}^{<1}$ the sheaf on S defined in the following way : $\mathcal{A}^{<1}_\tau$ is the space of functions g holomorphic in some $U\in\mathcal{V}(\tau)$ and subexponential at infinity, e.g. verifying the following condition

$$\forall\varepsilon>0\ ,\ \exists c>0 \text{ such that, } \forall\xi\in U\ ,\ \text{ one has } |g(\xi)|\leqslant Ce^{\varepsilon|\xi|}\ .$$

We consider also a class of hyperfunctions slightly different of the class considered in the §2. A hyperfunction F on the half-line δ_θ is defined by the following datas

i) A function f, holomorphic in a sector $\Sigma:|\arg x-\theta|<\varepsilon$

ii) A function $\tilde{f}$, holomorphic in the ramified sector $\{|x|<\varepsilon\ ;\ -\varepsilon<\arg x-\theta<2\pi+\varepsilon\}$, with $\tilde{f}(e^{2\pi i}x)-\tilde{f}(x)=f(x)$ in $\Sigma\cap\{|x|<\varepsilon\}$.

Two such pairs $(f,\tilde{f})$ and $(f_1,\tilde{f}_1)$ define the same hyperfunction F if $f=f_1$ and $\tilde{f}-\tilde{f}_1$ is holomorphic at 0 (possibly, we replace an ε by a smaller one).

Finally we will say that F has *exponential growth* if f has exponential growth in Σ, e.g. if there exists $A>0$, $B>0$ such that, for $x\in\Sigma$, $|x|\gg0$, one has $|f(x)|\leqslant Ae^{B|x|}$.

If one has a hyperfunction F of exponential growth on δ_θ, one defines as in §2 the function $G=\mathcal{F}F=\int_{\delta_\theta}^{-}F(x)e^{-x\xi}\,dx$; G is a function holomorphic is an open set $U=\{\xi\mid \operatorname{Re}(e^{i\theta'}\xi)>C \text{ for } \theta' \text{ in some neighborhood of } \theta\}$, and subexponential at infinity; in other words, if we put $I_\theta=\{\tau\ ;\ |\tau+\theta|\leqslant\frac{\pi}{2}\}$, G is a section of $\mathcal{A}^{<-1}$ on I_θ (or : on some neighborhood of I_θ, which is equivalent by definition). The following proposition is "essentially well-known" :

PROPOSITION (3.1). — *The Fourier transform is a bijection between the space of hyperfunctions of exponential growth on δ_θ and the space $\Gamma(I_\theta, \mathcal{A}^{<1})$.*

The inverse map is defined in the following way : take $G \in \Gamma(I_\theta, \mathcal{A}^{<1})$ and take an open set U as considered, on which G is holomorphic; we choose $a \in U$ and consider the integral $\frac{-1}{2\pi i} \int_{\delta_{a,\tau}} G(\xi) e^{x\xi}\, d\xi$, with τ running in a interval $|\tau+\theta| < \frac{\pi}{2}+\varepsilon$; in this way, one gets a holomorphic function in the ramified sector $-\varepsilon < \arg x - \theta < 2\pi+\varepsilon$, which defines in the obvious way a pair $(\tilde{f}, f)$. The proof that this is the inverse of $\mathcal{F}$ involves permutations of integrations and Cauchy formula; there are no serious problems, and it can be left to the reader.

We can now prove (2.2); let $F_{\alpha_k,j}$ a basis of the hyperfunctions solutions of $pF = 0$ on the lines $\delta_{\alpha_k,\theta}$ $(1 \leqslant 1 \leqslant m_k)$ (in the sense of § 2 or § 3, no matter here); as p is of exponential type, they are of exponential growth; if we denote by $G_{\alpha_k,j}$ their Fourier transform, one has, according to (3.1) : $e^{\alpha_k\xi} G_{\alpha_k,j} \in \Gamma(I_\theta, \mathcal{A}^{<1})$. If k is fixed, the $F_{\alpha_k,j}$ are linearly independent; therefore, according to (3.1) their Fourier transform are also linearly independent. It remains to prove that there is no relation $\Sigma H_k = 0$, H_k a combination of the $G_{\alpha_k,j}$; one has $H_k = e^{-\alpha_k\xi} H_k'$, $H_k' \in \Gamma(I_\theta, \mathcal{A}^{<1})$.

Suppose that we have such a relation; there exists a direction $\tau \in I_\theta$ such that the $\mathrm{Re}(\alpha_k e^{i\tau})$ are distinct; suppose f.i. that one has $\mathrm{Re}(\alpha_1 e^{i\tau}) < \mathrm{Re}(\alpha_k e^{i\tau})$, $k \geqslant 2$. We will prove that $H_1 = 0$.

The preceding inequality is still true for τ' close to τ; therefore, in some sector $U \in \mathcal{V}(\tau)$, H_1' will decrease exponentially. Choose now $a \in U$ such that $\delta_{a,\tau} \in U$; the function $\int_{\delta_{a,\tau}} H_1'(\xi) e^{x\xi}\, dx$ is holomorphic near 0; then, the inversion procedure indicated in (3.1) shows that the corresponding microfunction vanishes; therefore $H_1' = 0$ and $H_1 = 0$, which proves the theorem.

The most interesting application of this theorem is obtained when one varies the direction θ; as long as the half lines $\delta_{\alpha_k,\theta}$ do not meet, one has a decomposition of the space of solutions of $qG = 0$ into $G = \Sigma G_k$, with $e^{\alpha_k\xi} G_k \in \Gamma(I_\theta, \mathcal{A}^{<1})$; but different component are mixed up when θ crosses a direction where some $\delta_{\alpha_k,\theta}$ coincide; this phenomenon, the "Stokes phenomenon" is basic in the study of linear differential equations with irregular singularities. Ecalle [E] has shown that it plays also a basic role in non linear equations, and many other cases. As this is explained in several places, namely in his own papers, and in the references mentioned at the beginning, I will not develop this question here. Instead of that, I will now say a few words on the non-exponential case.

4. Equations of non-exponential type

(4.1)

First, let us introduce some other sheaves on S , e.g. $\mathcal{A}$, $\mathcal{A}^{\leqslant 1}$ and $\mathcal{A}^{<-1}$; they are defined in the following way, for $\tau \in S$:

$\mathcal{A}_\tau$ is the space of functions g holomorphic in some $U \in \mathcal{V}(\tau)$ (no growth condition is required); $\mathcal{A}_\tau^{\leqslant 1} \subset \mathcal{A}_\tau$ is the space of functions increasing at most exponentially at infinity, e.g. verifying the following condition

$$\exists A > 0\ ,\ C > 0 \text{ such that },\ \forall \xi \in U\ ,\ \text{one has } |g(\xi)| \leqslant C\ e^{A|\xi|}\ .$$

Finally $\mathcal{A}_\tau^{\leqslant -1} \subset \mathcal{A}_\tau$ is the space of functions with superexponential decreasing, e.g. verifying the condition

$$\forall A > 0\ ,\ \exists C > 0 \text{ such that },\ \forall \xi \in U\ ,\ \text{one has } |g(\xi)| \leqslant C\ e^{-A|\xi|}\ .$$

The reader could find strange these notations; in fact they become convenient if one has to consider other orders of decreasing or increasing, in which case one would write f.i. $\mathcal{A}^{<r}$, $\mathcal{A}^{<-r}$, *etc.*

I will admit the following basic result.

PROPOSITION (4.1.1). — *Let q be a linear differential operator with polynomial coefficients. Then q , acting on $\mathcal{A}$, $\mathcal{A}^{\leqslant 1}$, $\mathcal{A}^{<1}$, $\mathcal{A}^{<-1}$ is surjective.*

I recall that it means that q is surjective on the *germs* $\mathcal{A}_\tau$, $\mathcal{A}_\tau^{\leqslant 1}$, *etc.* , but of course not necessarily on sections over intervals of S. The first assertion, relative to $\mathcal{A}$, is just the usual existence theorem. The other cases are more delicate; one prove them by arguments close to the arguments used by Wasow [W] in the theory of asymptotic expansions.

In the next sections, I will write $\ker(q, \mathcal{A}) = \mathcal{A}(q)$, $\ker(q, \mathcal{A}^{\leqslant 1}) = \mathcal{A}^{\leqslant 1}(q)$, *etc.* ; note that the preceding proposition implies at once the equalities

(4.1.2) $$\ker(q, \mathcal{A}/\mathcal{A}^{\leqslant 1}) = \mathcal{A}(q)/\mathcal{A}^{\leqslant 1}(q)$$

(4.1.3) $$\ker(q, \mathcal{A}^{\leqslant 1}/\mathcal{A}^{<-1}) = \mathcal{A}^{\leqslant 1}(q)/\mathcal{A}^{<-1}(q)$$

And others similar equalities (but these are the only ones that I will use). I will only explain here how one describes $\mathcal{A}^{<-1}(q)$, $\mathcal{A}^{\leqslant 1}(q)/\mathcal{A}^{<-1}(q)$, and $\mathcal{A}(q)/\mathcal{A}^{\leqslant 1}(q)$ by Fourier transform. This is actually the elementary part of the business. The other points, (glue these spaces together, follow the solutions everywhere and not only near infinity, get an inversion formula, *etc.*), are more delicate; they require a rather heavy machinery of homological algebra which I do not wish to develop here; see [Ma 3].

(4.2) EXPONENTIAL GROWTH

I will use here an idea due to Komatsu [Ko] on the Fourier-Laplace transform of hyperfunctions. Choose a hyperfunction $F = (f, \tilde{f})$ on the half-line δ_θ , as explained in §3; let $\delta_{\theta,n}$ be the interval $\{|x| \leqslant n\}$ on δ_θ, and consider in obvious sense the integral $g_n(\xi) = \int_{\delta_{\theta,n}} F(x) e^{-x\xi}\, dx$; on a secteur $\{|\arg \xi + \theta| < \frac{\pi}{2} - \eta\ ,\ |\xi| \gg A\}$ $(n > 0)$, this gives a collection of functions of subexponential growth which differ, as $n \to +\infty$ by terms decreasing with arbitrary large exponential order; now, an argument of the Mittag-Leffler type shows that one can pass to the $\varprojlim$ and get in this way a section of $\mathcal{A}^{<1}/\mathcal{A}^{<-1}$ on $\{\arg \xi + \theta| < \frac{\pi}{2} - \eta\}$; as η is arbitrarily small, by moving a little bit θ we get finally a $G = \mathcal{F}F \in \Gamma(I_\theta, \mathcal{A}^{<1}/\mathcal{A}^{<-1})$; this is compatible with (3.1) in an obvious sense.

PROPOSITION (4.2.1). — *The Fourier transform which has just been defined is a bijection between the space of hyperfunctions on δ_θ and the space $\Gamma(I_\theta, \mathcal{A}^{<1}/\mathcal{A}^{<-1})$.*

The inverse map *i)* defined in a way similar to (3.1), but a little bit more sophisticated. Note also that this theorem proves the following result : the natural map $\Gamma(I_\theta, \mathcal{A}^{<1}) \to \Gamma(I_\theta, \mathcal{A}^{<1}/\mathcal{A}^{<-1})$ is *injective*, although the corresponding map for the germs is *surjective*; we shall see a more precise result in the next section (remark (4.3.5)).

The application to the differential equations is now a simple generalization of what has been done in §2 and §3. Let $p = \sum_0^n a_k \partial^k$, $a_k \in \mathbf{C}[x]$, and put $m = v(a_n)$; let $\alpha_1, \ldots, \alpha_r$ be the zeroes of a_n, with multiplicity $m_1, \ldots, m_r$ $(m_1 + \cdots + m_r = m)$. Choose a direction θ such that the half-lines $\delta_{\alpha_k,\theta}$ are distinct, and consider for each α_k , the m_k hyperfunctions $F_{\alpha_k,j}$ solutions of $pF = 0$ in the direction θ, and $G_{\alpha_k,j}$ their Fourier transform; as $e^{\alpha_k \xi} G_{\alpha_k,j} \in \Gamma(I_\theta, \mathcal{A}^{<1}/\mathcal{A}^{<-1})$, one has $G_{\alpha_k,j} \in \Gamma(I_\theta, \mathcal{A}^{\leqslant 1}/\mathcal{A}^{<-1})$, and, as these function are solutions of $qG = 0$ $(q = \mathcal{F}p)$, they belong to the space $\Gamma(I_\theta, \mathcal{A}^{\leqslant 1}(q)/\mathcal{A}^{<-1}(q))$ (*cf.* 4.1.3). Then the result is the following

THEOREM (4.2.2).

1) The sheaf $\mathcal{A}^{\leqslant 1}(q)/\mathcal{A}^{<-1}(q)$ is locally constant of rank m .

2) The $G_{\alpha_k,j}$ are a basis of $\Gamma(I_\theta, \mathcal{A}^{\leqslant 1}(q)/\mathcal{A}^{<-1}(q))$.

The first assertion is a generalization of (1.4) and is proved in the same way (see the references given there); when this is proved, *2)* follows by the same arguments as (3.1). Of course, when p is of exponential type, *2)* reduces to (3.1).

(4.3) WILD GROWTH OR WILD DECAY

I will describe here $\mathcal{A}^{<-1}(q)$ and $\mathcal{A}(q)/\mathcal{A}^{\leqslant 1}(q)$ by Fourier transform; one has here also to compactify $\mathbf{C}_x$ by the circle of directions at ∞, which is denoted S_x; if $\tau \in S_\xi$, I put $I_\tau = \{\theta \in S_x \ , \ |\theta + \tau| \leqslant \frac{\pi}{2}\}$ and I denote by I'_τ the symmetric of I_τ with respect to 0. Then, the result is the following

THEOREM (4.3.1). — *There are isomorphisms ("Fourier transform")*

1) $\mathcal{A}^{<-1}(q)_\tau \simeq H^0_c(\mathring{I}_\tau, \mathcal{A}(p)/\mathcal{A}^{\leqslant 1}(p))$.

2) $\mathcal{A}(q)_\tau/\mathcal{A}^{\leqslant 1}(q)_\tau \simeq H^1(I'_\tau, \mathcal{A}^{<-1}(p))$.

(H^0_c means, as usual "sections with compact support").

Hence, here everything "remains at infinity", and the Fourier transform exchanges wild growth and wild decay ("wild" means here just "faster than exponential"). This can of course be precised; actually, growth of order $r > 1$ is exchanged with decay of order r', $\frac{1}{r} + \frac{1}{r'} = 1$. More precise results relating the asymptotic expansions on both side can be obtained, as usual, from the saddle point method.

Example (4.3.2). $pf = f' - xf$; the solutions are $f = C\ e^{x^2/2}$; therefore $\mathcal{A}(p)/\mathcal{A}^{\leqslant 1}(p) = \mathbf{C}$ if $\frac{-\pi}{4} \leqslant \theta \leqslant \frac{\pi}{4}$ or $\frac{3\pi}{4} \leqslant \theta \leqslant \frac{5\pi}{4}$, $= 0$ otherwise; $\mathcal{A}^{<-1}(p) = \mathbf{C}$ on the complementary set, and 0 on the preceding intervals.

One has $pg = g' + \xi g$; the solutions are $g = C\ e^{-\xi^2/2}$; and the corresponding sheaves for q are obtained from those of p after a $\frac{\pi}{2}$ rotation. The reader is invited to verify the isomorphisms *1)* et *2)*, and to examine in the same spirit the example (1.3).

Modulo the formula (4.1.3), the proof of (4.3.1) is an immediate consequence of the similar formulas where "(p)" and "(q)" have been removed. I shall only indicate how are defined the arrows from right to left, and I leave to the reader to guess the definition of the arrows in the opposite direction.

(4.3.3) Isomorphism $\mathcal{F} : H^0_c(\mathring{I}_\tau, \mathcal{A}/\mathcal{A}^{\leqslant 1}) \xrightarrow{\sim} \mathcal{A}^{<-1}_\tau$. An element of the first member is obtained in the following way : suppose for simplicity $\tau = 0$; then $\mathring{I}_\tau$ is the interval $\theta \in]-\frac{\pi}{2}, \frac{\pi}{2}[$; let $\mathcal{J}_i$, $(i = 0, \ldots, n)$ be subintervals $]\alpha_i, \beta_i[$ of $\mathring{I}_\tau$, with $-\frac{\pi}{2} = \alpha_0 < \alpha_1 < \beta_0 < \alpha_2 < \beta_1 < \cdots < \beta_n = \frac{\pi}{2}$; denote by U_i sector $\{\arg x \in \mathcal{J}_i \mid |x| > C\}$ $(C \gg 0)$; finally let f_i be holomorphic in U_i, with the conditions : $f_{i+1} - f_i \in \Gamma(\mathcal{J}_{i+1} \cap \mathcal{J}_i, \mathcal{A}^{\leqslant 1})$, $f_0 \in \Gamma(\mathcal{J}_0, \mathcal{A}^{\leqslant 1})$, $f_n \in \Gamma(\mathcal{J}_n, \mathcal{A}^{\leqslant 1})$; the class of the f_i mod $\mathcal{A}^{\leqslant 1}$ give a section with compact support of $\mathcal{A}/\mathcal{A}^{\leqslant 1}$; its Fourier transform (in the direction $\tau = 0$) is defined in the following way : for $i = 1, \ldots, n-1$, choose a path γ_i starting from ∞ in $U_{i+1} \cap U_i$, and going to infinity in $U_i \cap U_{i-1}$, and consider the integral $\sum \int_{\gamma_i} f_i(x) e^{-x\xi}\, dx$.

[A priori, this is meaningless, but if we cut each γ_i in three pieces, $\gamma_i = \gamma_i' + \gamma_i'' + \gamma_i'''$, $\gamma_i' \in U_i \cap U_{i+1}$, $\gamma_i''' \in U_i \cap U_{i-1}$, γ_i'' compact, and if we arrange that $\gamma_i' = -\gamma_{i+1}'''$, we have a compensation due to the fact that $f_{i+1} - f_i$ has exponential growth, and the integral converges for $|\arg \xi| < \varepsilon$, $|\xi| \gg 0$]

The fact that the integral has a wild decay at infinity is easily obtained by deforming the γ_i to let their support tend to infinity. I leave to the reader to verify that the result depends only on the class defined by the f_i's in $H_c^0(\overset{\circ}{I}_\tau, \mathcal{A}/\mathcal{A}^{\leqslant 1})$.

(4.3.4) Isomorphism $\mathcal{F} : H^1(I'_\tau, \mathcal{A}^{<-1}) \xrightarrow{\sim} \mathcal{A}_\tau/\mathcal{A}_\tau^{\leqslant 1}$. This is simpler to define; let $\{\mathcal{J}_i\}$ be a covering of I'_τ by open intervals; we can suppose that $\mathcal{J}_i \cap \mathcal{J}_j \cap \mathcal{J}_k = \emptyset$ if (i, j, k) are distinct, and that the boundary points of I'_τ are covered only by one of the $\mathcal{J}$'s; then, a collection of $f_{ij} \in \Gamma(\mathcal{J}_i \cap \mathcal{J}_j, \mathcal{A}^{<-1})$ defines an element of the first member; to define its Fourier transform, we have just to take half-lines going to infinity $\gamma_{ij} \subset \{\arg x \in \mathcal{J}_i \cap \mathcal{J}_j \ , \ |x| \gg 0\}$, and to consider the integral $\sum \varepsilon_{ij} \int_{\gamma_{ij}} f_{ij}(x) e^{-x\xi} dx$ ($\varepsilon_{ij} = +1$ if $\mathcal{J}_i$ is before $\mathcal{J}_j$ in the usual orientation of the circle; $\varepsilon_{ij} = -1$ in the opposite case). Modulo $\mathcal{A}_\tau^{\leqslant 1}$, the result is independent of the choice of the γ_{ij}, and of the representative of the cocycle.

Exercise. Consider the integrals of example (1.3); show that, according to the value of τ, they can be considered as well as special case of the procedure (4.3.3) *or* of the procedure (4.3.4)!

Remark (4.3.5). The statements *1)* and *2)* of (4.3.1) (and the same statements with (p) and (q) removed) are not independent; actually denote by *2')* the statement *2)* in which one has exchanged $x \leftrightarrow \xi$, $p \leftrightarrow q$, $I \leftrightarrow I'$; then *1)* and *2')* are equivalent by a purely sheaf theoretic "inversion formula" of Sato ([SKK], chap. 1, prop. 1.4.1).

On the other hand, they are also related to (4.2.1) : in fact, with the notations of (4.2.1), one can prove that one has $H^1(I_\theta, \mathcal{A}^{<1}) = 0$; as we have already seen that $\Gamma(I_\theta, \mathcal{A}^{<1}) \longrightarrow \Gamma(I_\theta, \mathcal{A}^{<1}/\mathcal{A}^{<-1})$ is injective, the exact sequence of cohomology gives an exact sequence

$$0 \longrightarrow \Gamma(I_\theta, \mathcal{A}^{<1}) \longrightarrow \Gamma(I_\theta, \mathcal{A}^{<1}/\mathcal{A}^{<-1}) \longrightarrow H^1(I_\theta, \mathcal{A}^{<-1}) \longrightarrow 0 .$$

Therefore, according to (3.1) and (4.2.1), $H^1(I_\theta, \mathcal{A}^{<-1})$ is isomorphic to the space (hyperfunctions on δ_θ) / (hyperfunctions of exponential growth on δ_θ); now, it is not difficult to prove that this space is just isomorphic with $\mathcal{A}_\theta/\mathcal{A}_\theta^{\leqslant 1}$. One verifies that the isomorphism obtained in this way $H^1(I_\theta, \mathcal{A}^{<-1}) \simeq \mathcal{A}_\theta/\mathcal{A}_\theta^{\leqslant 1}$ coincides with *2')* up to a factor $\pm 2\pi i$.

Litterature

[Ca] CANDELPERGHER B. — *Une introduction à la résurgence*, Gazette des Mathématiciens, Soc. Math. Fr., **42** (1989), 36–64.

[E] ECALLE J. — *Les fonctions résurgentes*, vol. I to III, Publications mathématiques d'Orsay, 1981–85.

[I] INCE E.L.. — *Ordinary differential equations*, Dover, New-York, 1956.

[Ka] KASHIWARA M. — *Systems of microdifferential equations*, Progress in math., Birkhaüser, 1983.

[Ko] KOMATSU H. — *Operational calculus, hyperfunctions, and ultradistributions*, Algebraic analysis (papers dedicated to M. Sato), Academic Press (1988), 357–372.

[Ma 1] MALGRANGE B. — *Introduction aux travaux de J. Ecalle*, l'Enseignement mathématique, **31** (1985), 261–282.

[Ma 2] MALGRANGE B. — *Equations différentielles linéaires et transformation de Fourier*, Ensaios Matemáticos, vol. 1, Soc. Brasil de Matemática, 1989.

[Ma 3] MALGRANGE B. — *Systèmes holonomes à une variable*, (book, to be published).

[Ma 4] MALGRANGE B. — *Sur les points singuliers des équations différentielles*, l'Enseignement mathématique, **20**, fasc.1-2 (1974), 147–176.

[Ra] RAMIS J.-P. — *Dévissage Gevrey*, Astérisque, **59–60** (1978), 173–204.

[Ro] ROLBA P. — *Lemme de Hensel pour des opérateurs différentiels*, l'Enseignement mathématique, **26**, fasc. 3-4 (1980), 279–311.

[S-K-K] SATO M., KAWAI T., KASHIWARA M. — *Hyperfunctions and pseudodifferential equations*, Lect. Notes in Math., **287** (1973), 265–529, Springer-Verlag.

[W] WASOW W. — *Asymptotic expansions for ordinary differential equations*, Interscience publishers, 1965.

EXCURSIONS AND ITÔ CALCULUS IN NELSON'S STOCHASTIC MECHANICS

Aubrey Truman
Department of Mathematics
and Computer Science
University College Swansea
Singleton Park
Swansea SA2 8PP

David Williams
Department of Pure Mathematics
and Mathematical Statistics
University of Cambridge
16 Mill Lane
Cambridge CB2 1SB

ABSTRACT. Using a simple-minded approach, we give a more or less self-contained account of Itô calculus and excursion theory in stochastic mechanics. We present some new results on Poisson-Lévy excursion measures for radial ground-state Nelson diffusions in Coulomb-type potentials: for these diffusions we consider excursions from a spherical shell of radius a. Let $\#^{\pm}(s,t)$ be the number of $\begin{smallmatrix}\text{outward}\\ \text{inward}\end{smallmatrix}$ excursions of duration s upto the local time at a equals t for our diffusion X corresponding to the radial ground-state wave-function f_E. Then, for N = 0,1,2, ...,

$$\mathbb{P}(\#^{\pm}(s,t) = N) = e^{-t\,d\upsilon^{\pm}(s)} \frac{(t\,d\upsilon^{\pm}(s))^N}{N!},$$

where

$$\frac{d\upsilon^{\pm}(s)}{ds} = f_E^{-2}(a)(f_E,(H_{\pm} - E)^2 \exp(-s(H_{\pm} - E))f_E)_{L^2},$$

$H_{\pm}$ being a Dirichlet Hamiltonian for the Coulomb-type potential, with Dirichlet boundary conditions $\begin{smallmatrix}\text{inside}\\ \text{outside}\end{smallmatrix}$ the sphere of radius a.

1. Introduction

In this paper we give an expository account of Itô calculus and excursion theory in Nelson's stochastic mechanics. Within the limitations of space, we have striven to make our paper accessible to non-specialists and to make our presentation more or less self-contained. To this end, with mathematical physicists in mind, we give an elementary (partly novel) account of stochastic integrals with respect to Brownian motion and derive the basic results of the Itô calculus. Some of the ideas here are our own but our treatment owes a great deal to the earlier work of Henry McKean and Barry Simon (see Refs. (13),(18) and (7)). We urge the interested reader to consult these authors for further background material. Similar remarks apply to our treatment of Nelson's stochastic mechanics which is necessarily very brief. Here the reader is encouraged to consult the original accounts of Ed Nelson, Francesco Guerra (especially on variational principles for stochastic mechanics) and the recent work of Eric Carlen (see Refs.

A. Boutet de Monvel et al. (eds.), Recent Developments in Quantum Mechanics, 49–83.

(4),(5),(11),(14) and (15)). We also present here some new results on the application of excursion theory in Nelson's stochastic mechanics. We have given only a minimal account of excursion theory. Indeed we have used Nelson's stochastic mechanics to develop some of our excursion theory results. In the last few months a very readable account of excursion theory has been given by Chris Rogers (Ref. (16)). A much more detailed account can be found in Rogers and Williams (Ref. (17)). Further results applying stochastic mechanics to excursion theory are given in Truman and Williams (Ref. (19)). Our account of first hitting times in Nelson's stochastic mechanics is not complete. Additional results for Coulomb potentials are discussed in Batchelor and Truman (see Refs. (1),(2),(3)). Our treatment uses the Feynman-Kac formula for Schrödinger operators in an intrinsic way. A good review of the applications of the Feynman-Kac formula for Schrödinger operators is given in the article written by Michael Demuth in the present volume (Ref. (6)).

2. Preliminaries on Brownian Motion

The Wiener process $B(t)$, starting at the origin at time zero, is a Gaussian stochastic process, with independent increments, which is temporally homogeneous. That is $B(t)$ satisfies $B(0) = 0$ and $\{B(t+h) - B(h)\}$ has the same distribution as $B(t)$ for $t,h > 0$, this distribution being given by

$$\mathbb{P}(B(t)\ \varepsilon(a,b)) = (2\pi t)^{-1/2}\int_a^b \exp\left(-\frac{u^2}{2t}\right) du,$$

$\{B(t+h) - B(t)\}$ being independent of $B(s)$ for $0 \leq s \leq t$.

We shall see that it is very easy to calculate with the Wiener process. The difficult thing to establish for the Wiener process is the existence result - the fact that a process with the above properties exists with continuous sample paths. Let us spell out in more detail what this existence result means.

To be explicit $B(t)$ is a real-valued function $B(t) = B(t)(\omega)$ for $\omega \in \Omega$, the so-called probability space, with associated Wiener measure μ, so for $t \geq 0$ $B(t) : \Omega \to \mathbb{R}$. Inspired by Einstein's work on Brownian motion, Wiener showed that we can take $\Omega = C_0$, the space of continuous functions $\omega : [0,\infty) \to \mathbb{R}$ with $\omega(0) = 0$, and for $t \geq 0$

$$B(t)(\omega) = \omega(t).$$

C_0 is equipped with a distinguished family of subsets, $\mathbf{C}$, the measurable sets in C_0.

$\mathbf{C}$ = the smallest σ-algebra which makes all coordinate maps $\omega \to \omega(t)$ measurable.

$\mathbf{C}$ is the family of sets generated by ϕ, C_0 and $\{\omega : \omega(t) \in B$, Borel $B \subset \mathbb{R}\}$ for $t \in [0,\infty)$ i.e. sets of the form ϕ, C_0, $\omega(t)^{-1}B$ and these obtained by taking countable unions and complements. Let $\mathbb{Q}[0,1]$ be the rationals in $[0,1]$. Then, because each ω is continuous, we see that

$$\bigcap_{t\in\mathbb{Q}[0,1]} \{\omega \in C_0 : \omega(t) \in [-\varepsilon_n,\varepsilon_n]\} = \{\omega \in C_0 : \|\omega\|_\infty \leq \varepsilon_n\},$$

where $\|\omega\|_\infty = \sup_{t\in[0,1]} |\omega(t)|$. Setting $\varepsilon_n = \varepsilon - \frac{1}{n}$ for $n = 1,2, ...,$ since

$$\bigcup_n \{\omega : ||\omega||_\infty \leq \varepsilon_n\} = \{\omega : ||\omega||_\infty < \varepsilon\},$$

we can deduce that **C** contains the Borel sets in the topology of uniform convergence on compact subsets of $\mathbb{R}$.

The sets are measurable in the sense that μ is a countably additive function so that, for each $A_i \in C$, $\mu(A_i) \geq 0$ and for disjoint A_i

$$\mu\left(\bigcup_{j=1}^{\infty} A_j\right) = \sum_{j=1}^{\infty} \mu(A_j),$$

with, needless to say, $\mu(\phi) = 0$ and $\mu(C_o) = 1$ and

$$\mu\{\omega : \omega(t) \in (a,b)\} = \mathbb{P}(B(t) \in (a,b)).$$

Integrals with respect to μ are denoted by $\mathbb{E}$. Observe first that

$$\mathbb{E}\{B(t)B(s)\} = t \wedge s,$$

where $(t \wedge s)$ is the minimum of t and s. This follows because, by independence if $t \geq s \geq 0$

$$\mathbb{E}\{B(t)B(s)\} = \mathbb{E}\{(B(t)-B(s))B(s)\} + \mathbb{E}\{B^2(s)\} = \mathbb{E}\{B(t-s)\}\mathbb{E}\{B(s)\} + \mathbb{E}\{B^2(s)\} = \mathbb{E}\{B^2(s)\},$$

since $\mathbb{E}\{B(s)\} = (2\pi s)^{-1\backslash 2} \int_{-\infty}^{\infty} xe^{-x^2/2s}\,dx = 0$, and integrating by parts

$$\mathbb{E}\{B^2(s)\} = (2\pi s)^{-1\backslash 2} \int_{-\infty}^{\infty} x^2 e^{-x^2/2s}\,dx = s(2\pi s)^{-1\backslash 2} \int_{-\infty}^{\infty} e^{-x^2/2s}\,dx = s. \quad (s \geq 0).$$

This leads to:

PROPOSITION 2.1

Given the intervals $I = [a,b]$, $J = [c,d]$, for $a,b,c,d \geq 0$, and defining $\Delta B(I) = B(b) - B(a)$, $\Delta B(J) = B(d) - B(c)$,

$$\mathbb{E}\{\Delta B(I)\Delta B(J)\} = \lambda(I \cap J),$$

where λ is the Lebesgue measure.
We shall also need:

PROPOSITION 2.2

For $\alpha_j \in \mathbb{R}$ and $t_j \geq 0$, for $j = 1,2, \ldots, n$,

$$\mathbb{E}\{\exp(i \sum_{j=1}^{n} \alpha_j B(t_j))\} = \exp\{-\tfrac{1}{2} \sum_{j,k=1}^{n} \alpha_j \alpha_k (t_j \wedge t_k)\}.$$

PROOF

The proof proceeds by induction and the independent increments property. We can assume without loss of generality that $0 \leq t_1 \leq t_2 \ldots \leq t_n$. It is a routine matter to establish the result when $n = 1$. Assume now that the result is valid when $n = k - 1$ and write $B(t_k) = (B(t_k) - B(t_{k-1})) + B(t_{k-1})$ and use the independent increments property. The desired result for $n = k$ then follows, because

$$\left\{\sum_{j<k-1}\left(\frac{2\alpha_{k-1}\alpha_j t_j}{2}\right)+\frac{\alpha_{k-1}^2 t_{k-1}}{2}\right\}_{\alpha_{k-1}\to\alpha_k+\alpha_{k-1}}+\frac{\alpha_k^2}{2}(t_k-t_{k-1})$$

$$=\sum_{j<k}\frac{2\alpha_j\alpha_k t_j}{2}+\sum_{j<k-1}\frac{2\alpha_j\alpha_{k-1}t_j}{2}+\frac{\alpha_{k-1}^2 t_{k-1}}{2}+\frac{\alpha_k^2 t_k}{2}. \quad //$$

Taking Fourier transforms one can show that:

PROPOSITION 2.3

Let $A=\{\omega:\omega(t_1)\in B_1,\ \omega(t_2)\in B_2,\ldots,\omega(t_n)\in B_n\}\in \mathbf{C}$, for Borel $B_i\subset\mathbb{R}$. Then for $0<t_1<t_2<\ldots<t_n$

$$\mu(A)=(2\pi t_1)^{-1/2}(2\pi(t_2-t_1))^{-1/2}\ldots(2\pi(t_n-t_{n-1}))^{-1/2}\int_{B_1}dx_1\int_{B_2}dx_2$$

$$\ldots\int_{B_n}dx_n\ e^{-x_1^2/2t_1}\ e^{-(x_2-x_1)^2/2(t_2-t_1)}\ldots e^{-(x_n-x_{n-1})^2/2(t_n-t_{n-1})}.$$

This result embodies the idea of a particle diffusing in $\mathbb{R}$ with a transition density

$$p_t(x_1,x_2)=(2\pi t)^{-1/2}\,e^{-(x_2-x_1)^2/2t}.$$

Another consequence of the penultimate result is:

PROPOSITION 2.4 (Wick's Theorem)

For $t_j\geq 0$, $j=1,2,\ldots,2n+1$,

$$\mathbb{E}(B(t_1)B(t_2)\ldots B(t_{2n+1}))=0,$$

$$\mathbb{E}(B(t_1)B(t_2)\ldots B(t_{2n}))=\sum(t_{i_1}\wedge t_{j_1})\ldots(t_{i_n}\wedge t_{j_n}),$$

where the sum is taken over distinct pairings $(t_{i_1},t_{j_1})\ldots(t_{i_n},t_{j_n})$ of $t_1,t_2,\ldots,t_{2n}$.

PROOF

We merely equate coefficients of powers of α in above. Let $C(\alpha_1,\ldots\alpha_{2n})$ be the coefficient of $\alpha_1\alpha_2\ldots\alpha_{2n}$ in $\exp\{-\frac{1}{2}\sum\alpha_j\alpha_k(t_j\wedge t_j)\}$, say. Then

$C(\alpha_1,\alpha_2\ldots\alpha_{2n})$

$=$ coeff of $(\alpha_1\alpha_2\ldots\alpha_{2n})$

$$\text{in }\{(e^{-\alpha_1\alpha_2(t_1\wedge t_2)}\,e^{-\alpha_1\alpha_3(t_1\wedge t_3)}\ldots e^{-\alpha_1\alpha_{2n}(t_1\wedge t_{2n})}\ldots e^{-\alpha_{2n-1}\alpha_{2n}(t_{2n-1}\wedge t_{2n})}\}$$

$=$ coeff of $(\alpha_1\alpha_2\ldots\alpha_{2n})$

$$\text{in } \{(1-\alpha_1\alpha_2(t_1\wedge t_2))(1-\alpha_1\alpha_3(t_1\wedge t_3)) \dots (1-\alpha_1\alpha_{2n}(t_1\wedge t_{2n}))$$
$$(1-\alpha_2\alpha_3(t_2\wedge t_3))(1-\alpha_2\alpha_4(t_1\wedge t_4)) \dots$$
$$\dots\dots\dots\dots\dots$$
$$(1-\alpha_{2n-1}\alpha_{2n}(t_{2n-1}\wedge t_{2n}))\} .$$

There are $(2n-1)$ terms in first line, $(2n-2)$ terms in second line, $(2n-3)$ in third line, and so on. We must choose one factor from the first line, say (t_1, t_{i_1}). If $i_1 = 2$ we must make the next choice from the third line, if not we can choose any factor from second line save for (t_2, t_{i_1}) term, so there are $(2n-3)$ choices for second term either way. Thus, there are $(2n-1)(2n-3) \dots 3.1$ distinct pairings giving the above result. //

COROLLARY 2.5

$$\mathbb{E}\{B^{2n-1}(t)\} = 0, \qquad \mathbb{E}\{B^{2n}(t)\} = (2n-1)(2n-3)\dots 3.1t^n.$$

A simple calculation now gives

$$\mathbb{E}\left\{\left(\frac{B(t+h)-B(t)}{h}\right)^2\right\} = h^{-1/2} \to \infty,$$

as $h \to 0^+$. In fact, Paley, Wiener and Zygmund showed that

$$\mathbb{P}\{\omega : t \to B_t(\omega) \text{ is differentiable at some } t\} = 0.$$

This means that we cannot define $\int_a^b f(t)\, dB(t)$ by $\int_a^b f(t) \frac{dB(t)}{dt}\, dt$. We address this problem in the next section. Before leaving this section the reader should note that trivially from the above there exist positive constants, C, r, s with $\mathbb{E}|B(t+h)-B(t)|^r \le Ch^{1+s}$ and, according to a theorem of Kolmogorov (see Ref. (10)) this ensures that Ω can be taken to be a space of continuous functions.

3. Stochastic Integrals with respect to Brownian motion and S.D.E.s

First we treat stochastic integrals and then consider the ramifications of our results for stochastic differential equations.

3.1 STOCHASTIC INTEGRALS

We begin by giving what appears to be an unnatural definition of $\int_a^b .dB$.

Definition

Let f be any C^2 function $f : \mathbb{R} \to \mathbb{R}$. Then define for each $b \ge a \ge 0$

$$\int_a^b f'(B(s))\, dB(s) = [f(B(b)) - f(B(a))] - 2^{-1}\int_a^b f''(B(s))\, ds.$$

Remark

The final term on the right hand side of the above equation is called the Itô correction. One remembers the above equation by rewriting it as

$$df(B(s)) = f'(B(s))\, dB(s) + 2^{-1}f''(B(s))\, dB(s)\, dB(s)$$

and using McKean's multiplication $dB(s)\, dB(s) = ds$.

PROPOSITION 3.1.1

Let f be any C^2 function $f : \mathbb{R} \to \mathbb{R}$ bounded together with its first two derivatives. Then

$$\mathbb{E}\left\{\int_a^b f'(B(s))\, dB(s)\right\} = 0.$$

PROOF

For $f(B(t)) = B^{2n}(t)$, for each integer $n \geq 0$, our definition gives

$$\mathbb{E}\left\{2n\int_a^b B^{2n-1}(t)\, dB(t)\right\} = \mathbb{E}\left\{B^{2n}(b) - B^{2n}(a)\right\} - 2^{-1}\int_a^b 2n\,(2n-1)\mathbb{E}(B^{2n-2}(t))\, dt$$

and

$$\text{r.h.s.} = (2n-1)(2n-3)\ \ldots\ 3.1\left\{b^n - a^n - \int_a^b nt^{n-1}\, dt\right\} = 0.$$

The result for $f(B(t)) = B^{2n+1}(t)$ is even simpler to establish as all this requires is $\mathbb{E}\left\{B^{2n-1}(t)\right\} = 0$ for each integer $n \geq 1$. It follows that for any polynomial p

$$\mathbb{E}\left\{\int_a^b p(B(t))\, dB(t)\right\} = 0.$$

Since any bounded C^2 function f can be uniformly approximated on compacts together with its first two derivatives by polynomials and their derivatives the result follows by the dominated convergence theorem. //

There is a second important consequence of our definition.

PROPOSITION 3.1.2

Let $f,g \in \mathcal{S}(\mathbb{R}) = \left\{h \in C^\infty(\mathbb{R}) : \sup_x \left| x^r \frac{d^s h(x)}{dx^s} \right| < \infty \text{ for } r,s = 0,1,2,\ldots\right\}$. Then

$$\mathbb{E}\left\{\int_0^t f(B(s))\, dB(s) \int_0^t g(B(s))\, dB(s)\right\} = \int_0^t \mathbb{E}\{(fg)(B(s))\}\, ds.$$

PROOF

All we need to show is that for each $\alpha,\beta \in \mathbb{R}$

$$\mathbb{E}\left\{\int_0^t e^{i\alpha B(u)} dB(u) \int_0^t e^{i\beta B(v)} dB(v)\right\} = \int_0^t \mathbb{E}\left\{e^{i(\alpha+\beta)B(u)}\right\} du,$$

the desired result will then follow by multiplying by $\tilde{f}(\alpha)\tilde{g}(\beta)$ and integrating with respect to α and β. To prove the last identity one merely notes that according to our definition

$$i\alpha\int_0^t e^{i\alpha B(u)}dB(u) = (e^{i\alpha B(t)} - 1) + \frac{\alpha^2}{2}\int_0^t e^{i\alpha B(u)}du.$$

A simple calculation using PROPOSITION 2.2 and the independent increments property gives

$$\mathbb{E}\{(e^{i\alpha B(t)} - 1)(e^{i\beta B(t)} - 1)\} = e^{-\frac{(\alpha+\beta)^2 t}{2}} - e^{-\frac{\alpha^2 t}{2}} - e^{-\frac{\beta^2 t}{2}} + 1,$$

$$\mathbb{E}\left\{\int_0^t e^{i\alpha B(u)}du\int_0^u e^{i\beta B(v)}dv\right\} = \left[\frac{(e^{-\frac{(\alpha+\beta)^2 t}{2}} - 1)}{\frac{(\alpha+\beta)^2}{2}} - \frac{(e^{-\frac{\alpha^2 t}{2}} - 1)}{\frac{\alpha^2}{2}}\right]\frac{1}{\left(\frac{\beta^2}{2} + \alpha\beta\right)}$$

and

$$\mathbb{E}\left\{(e^{i\beta B(t)} - 1)\int_0^t e^{i\alpha B(u)}du\right\} = \frac{(e^{-\frac{\beta^2 t}{2}} - e^{-\frac{(\alpha+\beta)^2 t}{2}})}{(\frac{\alpha^2}{2} + \alpha\beta)} + \frac{(e^{-\frac{\alpha^2 t}{2}} - 1)}{\frac{\alpha^2 t}{2}}.$$

Combining these we arrive at

$$\mathbb{E}\left\{\int_0^t e^{i\alpha B(u)}dB(u)\int_0^t e^{i\beta B(v)}dB(v)\right\} = (e^{-\frac{(\alpha+\beta)^2 t}{2}} - 1)\frac{-2}{(\alpha+\beta)^2},$$

which is the desired result. //

The above ansatz for $\int dB$ can be obtained by insisting that dB(t) be made to point into the future and using the normal limiting procedure for integrals. To see this we define for $\int_0^1 dB$, say,

$$\Sigma_n(f) = \sum_{m=1}^{2^n} f\left(B\left(\frac{m-1}{2^n}\right)\right)\left(B\left(\frac{m}{2^n}\right) - B\left(\frac{m-1}{2^n}\right)\right).$$

Then, because

$$f\left(B\left(\tfrac{2m-2}{2^{n+1}}\right)\right)\left(B\left(\tfrac{2m-1}{2^{n+1}}\right) - B\left(\tfrac{2m-2}{2^{n+1}}\right)\right) + f\left(B\left(\tfrac{2m-1}{2^{n+1}}\right)\right)\left(B\left(\tfrac{2m}{2^{n+1}}\right) - B\left(\tfrac{2m-1}{2^{n+1}}\right)\right) - f\left(B\left(\tfrac{m-1}{2^n}\right)\right)\left(B\left(\tfrac{m}{2^n}\right) - B\left(\tfrac{m-1}{2^n}\right)\right)$$
$$= \left(f\left(B\left(\tfrac{2m-1}{2^{n+1}}\right)\right) - f\left(B\left(\tfrac{2m-2}{2^{n+1}}\right)\right)\right)\left(B\left(\tfrac{2m}{2^{n+1}}\right) - B\left(\tfrac{2m-1}{2^{n+1}}\right)\right),$$

we obtain

$$\Sigma_{n+1}(f) - \Sigma_n(f) = \sum_{m=1}^{2^n} \left(f\left(B\left(\tfrac{2m-1}{2^{n+1}}\right)\right) - f\left(B\left(\tfrac{2m-2}{2^{n+1}}\right)\right)\right)\left(B\left(\tfrac{2m}{2^{n+1}}\right) - B\left(\tfrac{2m-1}{2^{n+1}}\right)\right).$$

Therefore, since $|f(x) - f(y)| \leq |x - y| \sup_{x \in \mathbb{R}} |f'(x)|$, we obtain by the independent increments property and the fact that $\mathbb{E}\left\{B\left(\tfrac{2m}{2^{n+1}}\right) - B\left(\tfrac{2m-1}{2^{n+1}}\right)\right\} = 0$ (see above)

$$\mathbb{E}\left(\Sigma_{n+1}(f) - \Sigma_n(f)\right)^2 = \sum_{m=1}^{2^n} \mathbb{E}\left\{\left(f\left(B\left(\tfrac{2m-1}{2^{n+1}}\right)\right) - f\left(B\left(\tfrac{2m-2}{2^{n+1}}\right)\right)\right)^2\right\} \frac{1}{2^{n+1}}$$

$$\leq \left\{\sup_{x \in \mathbb{R}} |f'(x)|\right\}^2 \sum_{m=1}^{2^n} \frac{1}{2^{n+1}} \frac{1}{2^{n+1}} \leq \left\{\sup_{x \in \mathbb{R}} |f'(x)|\right\}^2 / 2^{n+2},$$

so that $\Sigma_n(f)$ is a Cauchy sequence in $L^2(C_o, d\mu)$. This is the key to:-

PROPOSITION 3.1.3

Let $f : \mathbb{R} \to \mathbb{R}$ be a C^2 function bounded together with its first two derivatives. Then

$$\Sigma_n(f') \xrightarrow{L^2(C_o, d\mu)} [f(B(1)) - f(B(0))] - 2^{-1}\int_0^1 f''(B(s))\, ds$$

as $n \to \infty$.

The last result and the independent increments property explain the origin of PROPOSITIONS 3.1.1 and 3.1.2. The class of Itô integrable functionals can be enlarged by taking limits of simple non-anticipating functions.

Definition

For $t \geq 0$, $f(t) : \Omega \to \mathbb{R}$, is said to be non-anticipating if $f(t)$ is $\mathcal{F}_t$-measurable for each $t \geq 0$, where $\mathcal{F}_t$ is the smallest σ-algebra generated by $\{B(s)^{-1}B$, for Borel $B \subset \mathbb{R}$ and $s \leq t\}$. The time-dependent random variable $f(t)$ is said to be a simple non-anticipating function if for $0 \leq t_1 < t_2 < \ldots < t_m = b < \infty$

$$f(t) = \sum_{k=0}^{m-1} f(t_k)\, \chi_{[t_k, t_{k+1})}(t),$$

where each random variable $f(t_k)$ is $\mathcal{F}_{t_k}$-measurable. Imitating the above, we define (almost surely)

$$\int_0^T f(t)\, dB(t) = \sum_{k=0}^{m'-1} f(t_k)[B(t_{k+1}) - B(t_k)],$$

where $t'_m = T$ and $(m'-1)$ is the largest integer r with $t_r < T$. This makes $T \to \int_0^T f(t)\, dB(t)$ continuous almost surely.

Following McKean we now introduce:

<u>Definition</u>

$H_2[0,T]$ is the class of all non-anticipating functions $f(t)$ such that $\int_0^T \mathbb{E}\{f^2(t)\}\, dt < \infty$.

McKean proved the following powerful approximation result. (See Ref. (13))

<u>PROPOSITION 3.1.4</u>

Corresponding to any function $g(\cdot) \in H_2[0.T]$ there exists a sequence $\{g_n(\cdot)\}$ of simple non-anticipating functions such that, almost surely with respect to μ,

$$\int_0^u |g(t) - g_n(t)|^2\, dt \to 0$$

and as $n \to \infty$

$$\int_0^u g_n(s)\, dB(s) \to \text{limit} \stackrel{\text{def}}{=} \int_0^u g(t)\, dB(t),$$

uniformly for $u \in [0,T]$, the limit being independent of the sequence $\{g_n\}$.

The proof of this result is quite hard. It uses McKean's exponential martingale inequality:

<u>LEMMA</u> (Exponential Martingale Inequality)

Let f be a simple function $f \in H_2[0,T]$. Then for each $\alpha > 0$

$$\mathbb{P}\,[\max_{t \le 1} (\int_0^t f(s)\, dB(s) - \frac{\alpha}{2} \int_0^t f^2(s)\, ds) > \beta] \le e^{-\alpha\beta}$$

$\beta \in \mathbb{R}$.

<u>Remark</u>

Since $u \to \int_0^u g_n(t)\, dB(t)$ is continuous for simple non-anticipating $g_n \in H_2[0,T]$, the uniformity of convergence in McKean's approximation result implies that $u \to \int_0^u g(t)\, dB(t)$ is continuous for $g \in H_2[0,T]$, $u \le T$. The last lemma is a consequence of Doob's martingale inequality in the next section.

3.2 <u>STOCHASTIC DIFFERENTIAL EQUATIONS</u>

We need the concept of martingale.

Definition

The family of random variables $\{X_t\}_{t\in[0,T]}$, with $\mathbb{E}(|X_t|) < \infty$ for $t \in [0,T]$, is a submartingale / martingale if

$$\mathbb{E}(X_t \mid \{X_s : s \le u\}) \geqq X_u, \quad \text{for } u \le t \le T.$$

Doob's Martingale Inequality (See Ref. (13))
For submartingales we have Doob's powerful inequality : for $\{X_t\}$ a submartingale with continuous sample paths, $t \to X_t$, for each $\lambda > 0$,

$$\mathbb{E}(\max_{0\le s\le t} X_s \ge \lambda) \le \lambda^{-1}\, \mathbb{E}(X_t^{\vee}0),$$

for $(X_t^{\vee}0)$ the maximum of X_t and 0.

By the independent increments property and the above remark $X_u = \int_0^u g(t)\, dB(t)$ is a martingale for $g \in H_2[0,T]$, for $u \in [0,T]$, with continuous sample paths. Therefore, using $-|X_t| \le X_t \le |X_t|$,

$$-\mathbb{E}\{|X_t| | \{X_s : s \le u\}\} \le \mathbb{E}\{X_t | \{X_s : s \le u\}\} \le \mathbb{E}\{|X_t| | \{X_s : s \le u\}\}.$$

Taking conditional expectations with respect to the sub-σ-algebra $\{|X_s| : s \le u\}$, we see that $\{|X_t|\}_{t\in[0,T]}$ is a positive submartingale with continuous sample paths. For such positive submartingales we have:

LEMMA 3.2.1 (Doob's L^p submartingale inequality)
For $p > 1$ and for any positive submartingale $\{X_t\}_{t\in[0,T]}$, setting $\bar{X}_t = \max_{0\le s\le t} X_s$, $t \in [0,T]$,

$$\mathbb{E}\left[\bar{X}_t^p\right] \le \left(\frac{p}{p-1}\right)^p \mathbb{E}\left[(X_t)^p\right].$$

PROOF

$$\mathbb{E}\left[\bar{X}_t^p\right] = -\int_0^\infty \lambda^p d_\lambda\, \mathbb{P}\{\bar{X}_t > \lambda\} = \int_0^\infty p\lambda^{p-1} d\lambda\, \mathbb{P}\{\bar{X}_t > \lambda\} - \lim_{\lambda\uparrow\infty} \lambda^p\, \mathbb{P}\{\bar{X}_t > \lambda\}.$$

Therefore, from Fubini's theorem and Doob's inequality

$$\mathbb{E}\left[\bar{X}_t^p\right] \le \int_0^\infty p\lambda^{p-1}\, \mathbb{P}\{\bar{X}_t > \lambda\}\, d\lambda \le \int_0^\infty p\lambda^{p-1}\left(\lambda^{-1}\int_{\bar{X}_t \ge \lambda} X_t\, d\mu\right) d\lambda.$$

Therefore $\mathbb{E}\left[\bar{X}_t^p\right] \leq \int_\Omega X_t \left(\int_0^{\bar{X}_t} p\lambda^{p-2} d\lambda\right) d\mu = \frac{p}{(p-1)} \int_\Omega (\bar{X}_t)^{p-1} X_t \, d\mu.$

If we now let $\frac{p}{(p-1)}$ be the exponent conjugate to p and apply Hölder's inequality with $X = X_t$, $Y = (\bar{X}_t)^{p-1}$, we have

$$\mathbb{E}[\,|XY|\,] \leq \mathbb{E}[\,|X|^p]^{1/p}\, \mathbb{E}[\,|Y|^q]^{1/q}.$$

We obtain therefore

$$\mathbb{E}\left[\bar{X}_t^p\right] \leq q\left(\mathbb{E}(|X_t|^p)\right)^{1/p}\left(\mathbb{E}(\,|\bar{X}_t|^p)\right)^{1/q}.$$

If $\mathbb{E}\left[\bar{X}_t^p\right] < \infty$ the result follows after dividing by $\left(\mathbb{E}(\,|\bar{X}_t|^p)\right)^{1/q}$. To avoid the last assumption observe that $\mathbb{P}\{(\bar{X}_t \wedge N) > \lambda\} \leq \mathbb{P}\{\bar{X}_t > \lambda\}$ and so from above

$$\left[\mathbb{E}\{(\bar{X}_t \wedge N)^p\}\right]^{1/p} \leq q\left[\mathbb{E}\{\,|X_t|^p\}\right]^{1/q}.$$

Letting $N \uparrow \infty$ the result follows. //

This enables one to establish Itô's beautiful result. (We write $dB(t)$ as dB_t.)

THEOREM 3.2.2

If for all x and y $|\sigma(x) - \sigma(y)| \leq K|x - y|$ and $|b(x) - b(y)| \leq K|x - y|$, then the stochastic differential equation

$$X_t = x + \int_0^t \sigma(X_s)\, dB_s + \int_0^t b(X_s)\, ds$$

has a unique non-anticipating solution, with $X_o = x$.

PROOF

The proof is based on the Picard theorem for ordinary differential equations. Set $X_t^o \equiv x$ and define recursively

$$X_t^n = x + \int_0^t \sigma(X_s^{n-1})\, dB_s + \int_0^t b(X_s^{n-1})\, ds, \quad \text{for } n \geq 1.$$

Let $D_n(t) = \mathbb{E}\left\{\sup_{0 \leq s \leq t} |X_s^n - X_s^{n-1}|^2\right\}$. Observing $B(s) \cong t^{1/2} B(\frac{s}{t})$, it is a simple matter to show that $D_1(t) \leq C(t + t^2)$ for a constant C.

Using $|a + b|^2 \leq 2a^2 + 2b^2$, one obtains

$$D_{n+1}(T) \le 2\mathbb{E} \sup_{0\le t\le T} \left|\int_0^t (\sigma(X_s^n)-\sigma(X_s^{n-1}))\, dB_s\right|^2 + 2\mathbb{E} \sup_{0\le t\le T} \left|\int_0^t (b(X_s^n)-b(X_s^{n-1}))\, ds\right|^2.$$

From the Cauchy-Schwarz inequality

$$2\mathbb{E} \sup_{0\le t\le T} \left|\int_0^t (b(X_s^n) - b(X_s^{n-1}))\, ds\right|^2 \le 2T\mathbb{E} \int_0^T |(b(X_s^n) - b(X_s^{n-1})|^2\, ds$$

$$\le 2TK^2\mathbb{E}\left[\int_0^T |X_s^n - X_s^{n-1}|^2 ds\right].$$

Because by hypothesis σ is continuous one can prove inductively that $X_t^n \in H_2[0,T]$ and therefore $\sigma(X_s^n) \in H_2[0,T]$, for each integer $n \ge 0$. It follows from Doob's L^2-inequality that

$$2\mathbb{E} \sup_{0\le t\le T} \left|\int_0^t (\sigma(X_s^n) - \sigma(X_s^{n-1}))\, dB_s\right|^2 \le 8\mathbb{E} \int_0^T |\sigma(X_s^n) - \sigma(X_s^{n-1})|^2 ds$$

$$\le 8K^2\mathbb{E} \int_0^T |X_s^n - X_s^{n-1}|^2 ds.$$

Combining the last two inequalities $D_{n+1}(T) \le B \int_0^T D_n(s)\, ds$, for $B = 2TK^2 + 8K^2$. Arguing inductively gives $D_n(T) \le B^{n-1}C\left(\frac{T^n}{n!} + \frac{2T^{n+1}}{(n+1)!}\right)$. Chebyshev's inequality gives

$$\mathbb{P}\left(\sup_{0\le t\le T} |X_t^n - X_t^{n-1}| > 2^{-n}\right) \le 2^{2n} D_n(T)$$

and $\sum_n 2^{2n} D_n(T) < \infty$, so by the Borel-Cantelli lemma

$$\mathbb{P}\left(\sup_{0\le t\le T} |X_t^n - X_t^{n-1}| > 2^{-n} \text{ i.o.}\right) = 0$$

i.e. with probability one $X_t^n \to X_t^\infty$ uniformly on $[0,T]$. We show next that $X_t \overset{\text{def}}{=} X_t^\infty$ satisfies

$$X_t = x + \int_0^t \sigma(X_s)\, dB_s + \int_0^t b(X_s)\, ds.$$

From the triangle inequality for $\|\cdot\|_2 = \{\mathbb{E}(|\cdot|^2)\}^{1/2}$, the usual L^2 norm, for $n > m$

$$\|\sup_{0\le s\le T}|X_s^m - X_s^n|\|_2 \le \sum_{k=m+1}^{n} \|\sup_{0\le s\le T}|X_s^k - X_s^{k-1}|\|_2 = \sum_{k=m+1}^{n} D_k(T)^{1/2},$$

so letting $n \uparrow \infty$, for $m \le n \le \infty$,

$$\mathbb{E}[\sup_{0\le s\le T}|X_s^m - X_s^n|^2] \le \left(\sum_{k=m+1}^{n} D_k^{1/2}(T)\right)^2.$$

Since by induction $X_t^n \in H_2[0,T]$, we see that $X_t \in H_2[0,T]$ as well and so $\sigma(X_t) \in H_2[0,T]$. Therefore, defining $\tilde{Y}_t$ by

$$\tilde{Y}_t = x + \int_0^t \sigma(Y_s)\,dB_s + \int_0^t b(Y_s)\,ds,$$

and $\tilde{Z}_t$ similarly, for $Y_s, Z_s \in H_2[0,T]$, arguing as above,

$$\mathbb{E}\{\sup_{0\le t\le T}|\tilde{Y}_t - \tilde{Z}_t|^2\} \le B\,\mathbb{E}\int_0^T |Y_s - Z_s|^2 ds \le BT\,\mathbb{E}[\sup_{0\le s\le T}|Y_s - Z_s|^2].$$

Setting $Y_t = X_t^n$ and $Z_t = X_t$, shows that

$$\mathbb{E}\{\sup_{0\le t\le T}|X_t^{(n+1)} - \tilde{X}_t|^2\} \le BT\,\mathbb{E}\{\sup_{0\le s\le T}|X_s^n - X_s|^2\} \to 0,$$

as $n \to \infty$ by above, so that $\tilde{X}_t = \lim_{n\uparrow\infty} X_t^{(n+1)} = X_t$ and X_t is a solution. We leave the proof of uniqueness as an exercise for the reader. //

Remarks

(1) The continuity of $t \to X_t^n$ and the uniform convergence of X^n to X guarantee that almost surely $t \to X_t$ is continuous.

(2) In the last proposition and theorem we can allow time-dependent integrands in the stochastic integrals.

(3) The above generalises to matrix-valued $\underset{\sim}{\sigma} = (\sigma_{ij})_{i=1,2,\ldots,d;j=1,2,\ldots,m}$, column-vector-valued $\underset{\sim}{b} = (b_i)_{i=1,2,\ldots,d}$, and $\underset{\sim}{X} = (X_i)_{i=1,2,\ldots,d}$, $\underset{\sim}{B} = (B_i)_{i=1,2,\ldots,m}$ being a $BM(\mathbb{R}^m)$ process, consisting of a column vector of independent $BM(\mathbb{R})$ processes $B_1, B_2, \ldots, B_m$. We only need to demand Lipschitz entries for $\underset{\sim}{\sigma}$ and $\underset{\sim}{b}$.

4. Itô Calculus

We begin with a discussion of Itô's formula, move on to applications and generalisations and conclude with stopping times.

4.1 ITÔ'S FORMULA IN ONE-DIMENSION

The most powerful tool in stochastic calculus is Itô's formula.

THEOREM 4.1

Let $f \in C_0^\infty(\mathbf{R} \times \mathbf{R}^+)$ and set $f = f(x,t)$. If $X(t) = X_t$ is the H_2-solution of the above stochastic differential equation, for bounded σ,

$$dX(t) = b(X(t))\,dt + \sigma(X(t))\,dB(t),$$

b and σ being Lipschitz, then

$$f(X(t),t) - f(X(0),0) =$$

$$\int_0^t \left[\frac{\partial f}{\partial s}(X(s),s) + b(X(s))\frac{\partial f}{\partial x}(X(s),s;) + 2^{-1}\sigma^2(X(s))\frac{\partial^2 f}{\partial x^2}(X(s),s)\right] ds$$

$$+ \int_0^t \frac{\partial f}{\partial x}(X(s),s)\,\sigma(X(s))\,dB(s).$$

PROOF

A slight extension of the result in PROPOSITION 3.1.3 gives the above result in the special case when $b \equiv 0$, $\sigma \equiv 1$. Thus,

$$df(B(t),t) = \left(\frac{\partial f}{\partial t}(B(t),t) + 2^{-1}\frac{\partial^2 f}{\partial x^2}(B(t),t)\right) dt + \frac{\partial f}{\partial x}(B(t),t)\,dB(t).$$

We must now replace $B(t)$ by $X(t)$ the above stochastic process. Firstly let

$$x(t) = x_0 + b_0 t + \sigma_0 B(t)$$

for x_0, b_0 and σ_0 constants. Then

$$f(x(t),t) = f(x_0 + b_0 t + \sigma_0 B(t),t) = \phi(B(t),t),$$

for $\phi(x,t) = f(x_0 + b_0 t + \sigma_0 x,t)$. Then

$$\frac{\partial \phi}{\partial t} = \frac{\partial f}{\partial t} + b_0 \frac{\partial f}{\partial x}, \quad \frac{\partial \phi}{\partial x} = \sigma_0 \frac{\partial f}{\partial x}, \quad \frac{\partial^2 \phi}{\partial x^2} = \sigma_0^2 \frac{\partial^2 f}{\partial x^2}.$$

Therefore, we obtain

$$df = d\phi(B(t),t) = \frac{\partial \phi}{\partial x}(B(t),t)\,dB(t) + \left\{2^{-1}\frac{\partial^2 \phi}{\partial x^2}(B(t),t) + \frac{\partial \phi}{\partial t}(B(t),t)\right\} dt$$

$$= \left[\frac{\partial f}{\partial t}(x(t),t) + b_o \frac{\partial f}{\partial x}(x(t),t) + 2^{-1}\sigma_o^2 \frac{\partial^2 f}{\partial x^2}(x(t),t)\right] dt + \sigma_o \frac{\partial f}{\partial x}(x(t),t)\, dB(t).$$

The above result is still valid when $b_o = b_o(\cdot)$, $\sigma_o = \sigma_o(\cdot)$ are simple (step) functions. From the above the H_2-functions $X(s)$, $b(X(s))$ and $\sigma(X(s))$ can be uniformly approximated by simple step functions with stochastic integrals whose limits define stochastic integrals of the corresponding functions. It only remains to take limits. The stochastic integral term is handled by observing that from the hypotheses on σ and f, for constants M, M′,

$$\Big|\Big|\int_0^t (\sigma_n(s) \frac{\partial f}{\partial x}(X_n(s),s) - \sigma(X(s)) \frac{\partial f}{\partial x}(X(s),s))\, dB(s)\Big|\Big|_2^2$$

$$\le M \Big|\Big|\int_0^t (\sigma_n(s) - \sigma(X(s)))\, dB(s)\Big|\Big|_2^2 + M' \Big|\Big|\int_0^t (X(s) - X_n(s))\, dB(s)\Big|\Big|_2^2 \to 0, \text{ as } n\to\infty. \;//$$

Remarks

(1) The above result can be remembered by writing

$df = \frac{\partial f}{\partial t} dt + \frac{\partial f}{\partial X} dX + 2^{-1}\frac{\partial^2 f}{\partial x^2} dX\, dX$, $dX = b\, dt + dB(t)$, using McKean's multiplication table:

$$dB(t)\, dB(t) = dt, \quad dt^2 = dt\, dB(t) = 0.$$

(2) Because of the hypotheses on σ the integrand in the stochastic integral term is an H_2-function.

4.2 THE FORWARD KOLMOGOROV EQUATION AND THE GCM THEOREM

Itô's formula has many applications. We give two important applications here.

COROLLARY 4.2.1

Let $X(t)$ satisfy the stochastic differential equation

$$dX(t) = b(X(t))\, dt + \sigma(X(t)\, dB(t),$$

for Lipschitz $\sigma, b \in C^\infty(\mathbb{R})$. Then if $\rho_t(x,y)$ is smooth where

$$\mathbb{P}(X(t) \in A \mid X(0) = x) = \int_{y\in A} \rho_t(x,y)\, dy,$$

each Borel $A \subset \mathbb{R}$, $\rho = \rho_t(x,y)$ satisfies the forward Kolmogorov equation

$$\frac{\partial \rho_t}{\partial t} = L_y^* \rho_t, \qquad \lim_{t\downarrow 0} \rho_t(x,y) = \delta_x(y),$$

where

$$L_y^* \cdot = \frac{\partial^2}{\partial y^2}\{\frac{\sigma^2(y)}{2}\cdot\} - \frac{\partial}{\partial y}\{b(y)\cdot\}$$

is the L^2-adjoint of $L_y = \frac{\sigma^2(y)}{2}\frac{\partial^2}{\partial y^2} + b(y)\frac{\partial}{\partial y}$.

PROOF

Let $\omega \in C_0^\infty(\mathbb{R}\times\mathbb{R}^+)$ have support in the time-slice $t = a$, $t = b$. By Itô's formula

$$0 = \omega(X(b),b) - \omega(X(a),a) = \int_a^b \left(\frac{\partial\omega}{\partial t}dt + \frac{\partial\,\omega}{\partial X}dX + 2^{-1}\frac{\partial^2\omega}{\partial X^2}dX\,dX\right).$$

By the hypotheses on σ, $\frac{\partial\omega}{\partial X}(X(s),s)\sigma(X(s)) \in H_2[0,T]$ for each $T > 0$ since $\omega \in C_0^\infty$. Therefore, $\mathbb{E}\left\{\int_a^b \frac{\partial\,\omega}{\partial X}\sigma\, dB\right\} = 0$, giving

$$\mathbb{E}\left\{\int_a^b\left(\frac{\partial\omega}{\partial t}(X(t),t) + b(X(t))\frac{\partial\omega}{\partial X}(X(t),t) + 2^{-1}\sigma^2(X(t))\frac{\partial^2\omega}{\partial X^2}(X(t),t)\right)dt\right\} = 0.$$

The defining property of ρ gives

$$\int_0^\infty dt\int_{\mathbb{R}} dy\,\rho(t,x,y)\left(\frac{\partial\omega}{\partial t}(y,t) + b(y)\frac{\partial\omega}{\partial y}(y,t) + 2^{-1}\sigma^2(y)\frac{\partial^2\omega}{\partial y^2}(y,t)\right) = 0$$

for all $\omega \in C_0^\infty(\mathbb{R}\times\mathbb{R}^+)$. The desired result follows by partial integration. //

Remarks

(1) According to a lemma of Weyl there always exists a smooth ρ above if σ and b are smooth. (See Ref. (13)). Also this Weyl lemma extends to higher dimension.

COROLLARY 4.2.2 (Girsanov-Cameron-Martin Theorem)
Let $X(t)$ be the solution of the stochastic differential equation

$$dX(t) = b(X(t))\,dt + dB(t),$$

b being Lipschitz and bounded. Set

$$M_t = \exp\left\{-\frac{1}{2}\int_0^t b^2(B(s))\,ds + \int_0^t b(B(s))\,dB(s)\right\}.$$

Then for $0 \le t_1 \le t_2 \dots \le t_k \le t$ and for $f_i \in C_0^\infty(\mathbb{R})$, $i = 1,2,\dots,k$,

$$\mathbb{E}_x[M_t f_1(B(t_1)\dots f_k(B(t_k)] = \mathbb{E}_x[f_1(X(t_1)\dots X(t_k)].$$

PROOF

Here we only prove the result when $k = 1$. A straight-forward application of Itô's formula gives for $t > u$

$$\mathbb{E}_x[f(X(t))] = \mathbb{E}_x[f(X(u))] + \int_u^t \mathbb{E}_x[(Lf)(X(s))]\,ds,$$

L being the generator of the diffusion $L = 2^{-1}\frac{d^2}{dx^2} + b(x)\frac{d}{dx}$. We next observe that M_t is a supermartingale. In fact, setting $Z(t) = -\frac{1}{2}\int_0^t b^2(B(s))\,ds + \int_0^t b(B(s))\,dB(s)$, $M_t = \exp(Z(t))$ so

$$dM_t = (dZ(t) + 2^{-1}dZ(t)\,dZ(t))\exp(Z(t)) = b(B(t))\,dB(t)\,M_t.$$

Therefore, using Itô's formula again,

$$\begin{aligned} d(M_t f(B(t)) &= (dM_t)f(B(t)) + M_t df(B(t)) + dM_t df(B(t)) \\ &= b(B(t))\,dB(t)M_t f(B(t)) + M_t(f'(B(t))\,dB(t) + 2^{-1}f''(B(t))\,dt) + b(B(t))f'(B(t))M_t\,dt. \end{aligned}$$

Taking expectations

$$\mathbb{E}_x[M_t f(B(t))] = \mathbb{E}_x[M_u f(B(u))] + \int_u^t \mathbb{E}_x[M_s(Lf)(B(s))]\,ds.$$

The desired result follows from uniqueness of solutions to the equation

$$\frac{d}{dt}(P_t f) = P_t(Lf).$$

(See e.g. Ref. (17) for more detailed proof.) //

4.3 GENERALIZATIONS TO d-DIMENSIONS

Let $\underset{\sim}{B}^T = (B_1(t), B_2(t), \ldots, B_m(t))$ be a $BM(\mathbb{R}^m)$ valued process, where $B_i(\cdot)$ are independent $BM(\mathbb{R})$ processes, with

$$\mathbb{E}\{B_i(t)B_j(s)\} = \delta_{ij}(t \wedge s), \quad \text{for } i,j = 1,2,\ldots,m.$$

From the above we can deduce that for each integer $n \geq 1$ and each $t_j \in \mathbb{R}^+$

$$\mathbb{E}\left\{\exp\left(i\sum_1^n \underset{\sim}{\alpha}_j.\underset{\sim}{B}(t_j)\right)\right\} = \exp\left\{-\frac{1}{2}\sum_{i,k=1}^n \underset{\sim}{\alpha}_j.\underset{\sim}{\alpha}_k\,(t_j \wedge t_k)\right\},$$

where each $\underset{\sim}{\alpha}_j \in \mathbb{R}^m$ and $(\underset{\sim}{\alpha}_j.\underset{\sim}{\alpha}_k)$ denotes the Euclidean scalar product in $\mathbb{R}^m$.

It follows that, if we define for a C^2 $f : \mathbb{R}^m \to \mathbb{R}$, with bounded first and second order partial derivatives,

$$\int_0^t \underset{\sim}{\nabla} f(\underset{\sim}{B}(u)).d\underset{\sim}{B}(u) = [f(\underset{\sim}{B}(t)) - f(\underset{\sim}{B}(0))] - 2^{-1}\int_0^t \Delta f(\underset{\sim}{B}(u))\,du,$$

so that

$$i\int_0^t e^{i\underset{\sim}{\alpha}.\underset{\sim}{B}(s)}\underset{\sim}{\alpha}.d\underset{\sim}{B}(s) = (e^{i\underset{\sim}{\alpha}.\underset{\sim}{B}(t)} - 1) + \frac{|\underset{\sim}{\alpha}|^2}{2}\int_0^t e^{i\underset{\sim}{\alpha}.\underset{\sim}{B}(u)}\,du,$$

for $|\underset{\sim}{\alpha}|^2 = \underset{\sim}{\alpha}.\underset{\sim}{\alpha}$, then necessarily for $f,g \in \mathcal{S}(\mathbb{R}^m)$

$$\mathbb{E}\left\{\int_0^t (\underset{\sim}{\nabla} f)(\underset{\sim}{B}(u)).d\underset{\sim}{B}(u)\right\} = 0$$

and

$$\mathbb{E}\left\{\int_0^t (\underset{\sim}{\nabla} f)(\underset{\sim}{B}(u)).d\underset{\sim}{B}(u)\int_0^t (\underset{\sim}{\nabla} g)(\underset{\sim}{B}(v)).d\underset{\sim}{B}(v)\right\} = \int_0^t \mathbb{E}\{(\underset{\sim}{\nabla} f).(\underset{\sim}{\nabla} g)(\underset{\sim}{B}(u))\}\,du.$$

The first of these identities is obvious the second follows by proving that from the above analogue of PROPOSITION 2.2.

$$\mathbb{E}\left\{\int_0^t e^{i\underset{\sim}{\alpha}.\underset{\sim}{B}(u)}\underset{\sim}{\alpha}.d\underset{\sim}{B}(u)\int_0^t e^{i\underset{\sim}{\beta}.\underset{\sim}{B}(v)}\underset{\sim}{\beta}.d\underset{\sim}{B}(v)\right\}du = \int_0^t \underset{\sim}{\alpha}.\underset{\sim}{\beta}\;\mathbb{E}\left\{e^{i(\underset{\sim}{\alpha}+\underset{\sim}{\beta}).\underset{\sim}{B}(u)}\right\}du.$$

Moreover, making $d\underset{\sim}{B}$ stick into the future, $\int_0^t \underset{\sim}{\nabla} f(\underset{\sim}{B}(u).d\underset{\sim}{B}(u)$ can be obtained as a limit of Itô sums. This yields a d-dimensional Itô formula. Let $\underset{\sim}{X} = (X_1,X_2,\ldots,X_d)^T$, where $\underset{\sim}{X}$ satisfies the stochastic differential equation

$$d\underset{\sim}{X}(t) = \underset{\sim}{b}(\underset{\sim}{X}(t))\,dt + \underset{\sim}{\sigma}(\underset{\sim}{X}(t))\,d\underset{\sim}{B}(t),$$

$\underset{\sim}{b}$ and $\underset{\sim}{\sigma}$ having globally Lipschitz entries, $\underset{\sim}{b} = (b_1,b_2,\ldots,b_d)^T$, $\underset{\sim}{\sigma} = (\sigma_{ij})_{i=1,2,\ldots,d;j=1,2,\ldots,m}$, $\underset{\sim}{B} = (B_1,\ldots,B_m)^T$ being a $BM(\mathbb{R}^m)$ process. Then let $f \in C_0^\infty(\mathbb{R}^d \times \mathbb{R}^+)$ and consider $f(\underset{\sim}{X}(t),t)$. One can then prove

$$df(\underset{\sim}{X}(t),t) = \left[\frac{\partial f}{\partial t}(\underset{\sim}{X}(t),t) + \underset{\sim}{b}(\underset{\sim}{X}(t)).\underset{\sim}{\nabla} f(\underset{\sim}{X}(t),t) + 2^{-1}\sum_{i,j=1}^{d} A_{ij}(\underset{\sim}{X}(t))\frac{\partial^2 f}{\partial x_i \partial x_j}(\underset{\sim}{X}(t),t)\right]dt$$

$$+\left[\sum_{i=1}^{d}\sum_{j=1}^{m}\sigma_{ij}(\underset{\sim}{X}(t))\frac{\partial f}{\partial x_i}(\underset{\sim}{X}(t),t)\,dB_j(t)\right], \qquad A_{ij} = (\sigma\sigma^T)_{ij},$$

where we have written $f = f(\underset{\sim}{x},t)$, $\underset{\sim}{x} = (x_1,x_2,\ldots,x_d)$ and the last term is the stochastic integral $(\underset{\sim}{\sigma}\,\underset{\sim}{\nabla} f).d\underset{\sim}{B}$. Needless to say this can be obtained from Taylor's formula, retaining the terms of quadratic variation and using McKean's multiplication rules:

$dB_i(t)\, dB_j(t) = \delta_{ij}\, dt$, $(dt)^2 = dt\, dB_i(t) = 0$, $i,j = 1,2,\dots,m$. If we take $m = d$ and σ to be the $(d \times d)$ identity we obtain:

COROLLARY 4.3

Let $\underset{\sim}{X}(t)$ satisfy the above stochastic differential equation, for Lipschitz smooth $\underset{\sim}{\sigma}$ and $\underset{\sim}{b}$. Then the transition density $\rho_t(\underset{\sim}{x},\underset{\sim}{y})$ with

$$\mathbf{P}(\underset{\sim}{X}(t) \in A \,|\, \underset{\sim}{X}(0) = x) = \int_{\underset{\sim}{y} \in A} \rho_t(\underset{\sim}{x},\underset{\sim}{y})\, dy$$

each Borel $A \subset \mathbf{R}^d$, satisfies

$$\frac{\partial \rho_t}{\partial t} = L_y^* \rho_t, \qquad \lim_{t \downarrow 0} \rho_t(\underset{\sim}{x},\underset{\sim}{y}) = \delta_{\underset{\sim}{x}}(\underset{\sim}{y}),$$

for $L_y^* \cdot = \mathrm{div}_y\{2^{-1}\underset{\sim}{\nabla}_y \cdot - \underset{\sim}{b}(\underset{\sim}{y})\cdot\}$, the L^2-adjoint of $L_y = 2^{-1}\Delta_y + \underset{\sim}{b}(\underset{\sim}{y}).\underset{\sim}{\nabla}$.

4.4 STOPPING TIMES

Itô's formula comes into its own when used in conjunction with stopping times.

Definition

A Brownian stopping time τ is a non-negative random variable with the property that $\{\omega : \tau(\omega) < t\} \in \mathcal{F}_t$, for each $t > 0$ and so that $\mathbf{P}(\tau < \infty) > 0$.

If τ is a Brownian stopping time, defining $\int_0^\tau \cdot\, dB(s) = \int_0^\infty \cdot\, \chi_{\{<\tau\}}(s)\, dB(s)$, for $\chi_{\{<\tau\}}$ the characteristic function of $\{u : u \le \tau\}$, leads to stochastic integrals with stopping times as the limits of integration and so to Itô's formulae with stopping time arguments.

Caution

First hitting times are stopping times but last exist times are not.

Definition

For a stopping time τ define $\mathcal{F}_{\tau^+} = \{B \in \mathbf{C} \,|\, B \cap (\tau < t) \in \mathcal{F}_t$, for all $t\}$.

This definition makes τ into an $\mathcal{F}_{\tau^+}$-measurable random variable. For the stopping time τ we have the strong Markov property of Dynkin and Hunt: for all bounded measurable F

$$\mathbb{E}\,[(F(\theta_\tau B(t))\, \chi_{\{\tau<\infty\}} \,|\, \mathcal{F}_{\tau^+}] = \chi_{\{\tau<\infty\}} \mathbb{E}\,[F(B(t))],$$

where $\theta_\tau B(t) = B(t + \tau) - B(\tau)$, $t \ge 0$. (For a proof of this result see Simon Ref. (18) p.68).

5. Nelson's Stochastic Mechanics

Here we present a brief account of stochastic kinematics and dynamics after Ed Nelson (see Refs. (14) and (15)). Consider the equation for a unit mass particle diffusing in d-dimensional Euclidean space, $\mathbb{R}^d$, the particle being at the point with position vector $\underset{\sim}{x} \in \mathbb{R}^d$, with probability $\rho_t(\underset{\sim}{x})$, at time t, the corresponding diffusion equation being

$$d\underset{\sim}{X}(t) = \underset{\sim}{b}(\underset{\sim}{X}(t),t)\,dt + d\underset{\sim}{B}(t),$$

$\underset{\sim}{B}$ being a $BM(\mathbb{R}^d)$ process with the above covariance.

The sample paths $t \to \underset{\sim}{X}_t$ are, with probability one, nowhere differentiable. To overcome this problem Nelson introduced the mean forward and backward time derivatives

$$D_{\pm}f(\underset{\sim}{X}(t),t) = \lim_{h\downarrow 0} \mathbb{E}\left\{\frac{f(\underset{\sim}{X}(t \pm h),t \pm h) - f(\underset{\sim}{X}(t),t)}{\pm h} \,\Big|\, \underset{\sim}{X}(t)\right\},$$

so that for sufficiently regular $\underset{\sim}{b}$, from above

$$D_+\underset{\sim}{X}(t) = \underset{\sim}{b}(\underset{\sim}{X}(t),t) \overset{\text{def}}{=} \underset{\sim}{b}_+(\underset{\sim}{X}(t),t).$$

It follows from Itô's formula that for $f \in C_0^\infty(\mathbb{R}^d \times \mathbb{R}^+)$, say $f = f(\underset{\sim}{x},t)$,

$$D_{\pm}f(\underset{\sim}{X}(t),t) = \left(\frac{\partial}{\partial t} + D_{\pm}\underset{\sim}{X}.\underset{\sim}{\nabla}_x \pm 2^{-1}\Delta_x\right) f(\underset{\sim}{X}(t),t).$$

Nelson proved that for sufficiently well-behaved f,g

$$\frac{d}{dt}\mathbb{E}(fg) = \mathbb{E}\{fD_+g + gD_-f\},$$

so that in particular it is necessary that

$$\mathbb{E}\{D_+(fg)\} = \mathbb{E}\{fD_+g + gD_-f\}.$$

When f and g are independent of the time, we see that in particular

$$\mathbb{E}\{f(D_+\underset{\sim}{X}.\underset{\sim}{\nabla}g) + g(D_+\underset{\sim}{X}.\underset{\sim}{\nabla}f) + 2^{-1}\Delta(fg)\}$$

$$= \mathbb{E}\{f(D_+\underset{\sim}{X}.\underset{\sim}{\nabla}g) + g(D_-\underset{\sim}{X}.\underset{\sim}{\nabla}f) + 2^{-1}f\Delta g - 2^{-1}g\Delta f\}.$$

Using $\Delta(fg) = f\Delta g + g\Delta f + 2\underset{\sim}{\nabla}f.\underset{\sim}{\nabla}g$, we therefore obtain

$$\mathbb{E}\{\underset{\sim}{\nabla}f.\underset{\sim}{\nabla}g + g\Delta f + g(D_+\underset{\sim}{X}.\underset{\sim}{\nabla}f)\} = \mathbb{E}\{g(D_-\underset{\sim}{X}.\underset{\sim}{\nabla}f)\}.$$

Integration by parts gives in terms of the density ρ

$$\mathbb{E}\{\underset{\sim}{\nabla}f.\underset{\sim}{\nabla}g\} = -\mathbb{E}\{\rho^{-1}g\,\mathrm{div}(\rho\underset{\sim}{\nabla}f)\} = -\mathbb{E}\{g\rho^{-1}\underset{\sim}{\nabla}\rho.\underset{\sim}{\nabla}f + g\Delta f\},$$

so we arrive at

$$\mathbb{E}\{g(D_+\underset{\sim}{X} - \underset{\sim}{\nabla}\ln\rho).\underset{\sim}{\nabla}f\} = \mathbb{E}\{gD_-\underset{\sim}{X}.\underset{\sim}{\nabla}f\},$$

for all $f,g \in C_0^\infty(\mathbb{R}^d)$. Therefore, we arrive at Nelson's result

$$D_+\underset{\sim}{X} - \underset{\sim}{\nabla} \ln \rho = D_-\underset{\sim}{X},$$

so that

$$D_-\underset{\sim}{X}(t) = \underset{\sim}{b}(\underset{\sim}{X}(t),t) - \underset{\sim}{\nabla} \ln \rho(\underset{\sim}{X}(t),t) \overset{\text{def}}{=} \underset{\sim}{b}_-(\underset{\sim}{X}(t),t).$$

Following Nelson, we now write $\rho = e^{2R}$ and assume that $\underset{\sim}{b}$ is a gradient, defining S by

$$\underset{\sim}{b} = \underset{\sim}{\nabla}(S + R) = \underset{\sim}{b}_+$$

and so

$$\underset{\sim}{b}_- = \underset{\sim}{\nabla}(S - R).$$

Nelson's stochastic acceleration, $\underset{\sim}{a} = 2^{-1}(D_+D_- + D_-D_+)\underset{\sim}{X}(t)$, can then be calculated in terms of R and S. A tedious calculation gives

$$\underset{\sim}{a} = 2^{-1}(D_+\underset{\sim}{b}_- + D_-\underset{\sim}{b}_+) = \underset{\sim}{\nabla}\left(\frac{\partial S}{\partial t} - 2^{-1}(|\underset{\sim}{\nabla} R|^2 - |\underset{\sim}{\nabla} S|^2 + \Delta R)\right),$$

so the stochastic acceleration is also a gradient. If the unit mass particle at $\underset{\sim}{X}(t)$ is subject to a force $-\underset{\sim}{\nabla} V(\underset{\sim}{X}(t))$ Nelson postulated that the stochastic acceleration $\underset{\sim}{a}(\underset{\sim}{X}(t),t)$ obey a dynamical law, namely the Nelson-Newton law

$$\underset{\sim}{a}(\underset{\sim}{X}(t),t) = -\underset{\sim}{\nabla} V(\underset{\sim}{X}(t)).$$

This is the analogue of Newton's second law of motion for a unit mass diffusing particle. It leads to the identity

$$\frac{\partial S}{\partial t} - 2^{-1}(|\underset{\sim}{\nabla} R|^2 - |\underset{\sim}{\nabla} S|^2 + \Delta R) = -V + \varphi(t),$$

for some function $\varphi(t)$. Nelson showed that without loss of generality we can take $\varphi \equiv 0$. Therefore, we obtain

$$\frac{\partial S}{\partial t} - 2^{-1}(|\underset{\sim}{\nabla} R|^2 - |\underset{\sim}{\nabla} S|^2 + \Delta R) = -V.$$

Also $\rho = e^{2R}$ satisfies the forward Kolmogorov equation for drift $\underset{\sim}{b} = \underset{\sim}{\nabla}(R + S)$ giving

$$\frac{\partial R}{\partial t} + \underset{\sim}{\nabla} S \,.\, \underset{\sim}{\nabla} R + 2^{-1}\Delta S = 0.$$

The last pair of equations give the necessary and sufficient conditions for the diffusion $\underset{\sim}{X}$ to obey the Nelson-Newton law as a pair of nonlinear equations for the forward and backward drifts $\underset{\sim}{b}_\pm = \underset{\sim}{\nabla}(S \pm R)$. The amazing fact discovered by Nelson is that this pair of equations can be linearized by writing $\psi = e^{R+iS}$ for the above real-valued R and S, the linearized equation for ψ being nothing other than Schrödinger's equation for $\psi = \psi(x,t)$

$$i\frac{\partial\psi}{\partial t} = -2^{-1}\Delta_x\psi + V(x)\psi.$$

This suggests that the sample paths of the above diffusion are of some physical significance in studying the quantum mechanics of the corresponding Schrödinger equation and that traditional quantum mechanical observables play a roll in determining the statistical properties of our diffusion. These are the two main themes in the present paper.

6. Nelson diffusions corresponding to ground-states in spherically symmetric potentials

We consider ground-state Nelson diffusions corresponding to a quantum mechanical Hamiltonian $H = -2^{-1}\Delta + V$, a self-adjoint linear operator on some domain in $L^2(\mathbb{R}^d)$. We specialize to $d = 3$ and to spherically symmetric potentials, $V(\underset{\sim}{x}) = V(|\underset{\sim}{x}|)$, $|\underset{\sim}{x}|$ being the Euclidean norm of $\underset{\sim}{x} \in \mathbb{R}^3$.

For $E = \inf\operatorname{spec}(H)$, let $\psi_E(\underset{\sim}{x},t)$ be the ground-state with $\psi_E(\cdot,t) \in L^2(\mathbb{R}^3)$ given by

$$\psi_E(\underset{\sim}{x},t) = |\underset{\sim}{x}|^{-1} f_E(|\underset{\sim}{x}|)e^{-iEt}.$$

In order for ψ_E to satisfy the Schrödinger equation, because ψ_E must be a state with zero angular momentum, it is necessary that f_E satisfies $(H_r - E)f_E = 0$, with $H_r = -2^{-1}\frac{d^2}{dx^2} + V(x)$, the radial Hamiltonian, so that

$$(-2^{-1}\frac{d^2}{dx^2} + V(x) - E)f_E(x) = 0.$$

We need to assume that V is continuous on $(0,\infty)$, $xV(x)$ analytic in a neighbour of the origin, so that f_E is C^2 and $f_E > 0$ on $(0,\infty)$, with $f_E(0) = 0$, $f_E \in L^2(\mathbb{R}^+)$.

6.1 TRANSITION DENSITIES

LEMMA 6.1.1

Let $\hat{\underset{\sim}{X}}(s)$ be a non-anticipating unit vector. Then $\beta(t) = \int_0^t \hat{\underset{\sim}{X}}(s).d\underset{\sim}{B}(s)$ is a $BM(\mathbb{R})$ process.

PROOF

Let $v(t) = \int_0^t \sum_1^n \alpha_i\chi_{[0,t_i)}(s)\, \hat{\underset{\sim}{X}}(s).d\underset{\sim}{B}(s) = \sum_1^n \alpha_i\beta(t_i)$, for fixed $(\alpha_1,\alpha_2,\ldots,\alpha_n) \in \mathbb{R}^n$. Itô's formula gives

$$e^{iv(t)} - 1 = i\int_0^t e^{iv(s)} dv(s) - 2^{-1}\int_0^t e^{iv(s)} \sum_{i,j} \alpha_i\alpha_j \chi_{[0,t_i\wedge t_j)}(s)\, ds.$$

Taking expectations

$$\mathbb{E}(e^{iv(t)}) = \exp\left\{-\frac{1}{2}\sum_{i,j}\alpha_i\alpha_j(t_i\wedge t_j)\right\}. \qquad //$$

Now let $\underset{\sim}{X}$ be the ground-state Nelson diffusion corresponding to the state ψ_E. We consider its radial component $|\underset{\sim}{X}|$. Using $d|\underset{\sim}{X}| = d\underset{\sim}{X}.\underset{\sim}{\nabla}(|\underset{\sim}{X}|) + 2^{-1}\Delta|\underset{\sim}{X}|\,dt$ and $d\underset{\sim}{X}(s) = \underset{\sim}{\nabla}\ln\left(|\underset{\sim}{X}|^{-1} f_E|\underset{\sim}{X}|\right)ds + d\underset{\sim}{B}(s)$, we obtain from the last lemma

$$d|\underset{\sim}{X}|(t) = \frac{d}{d|\underset{\sim}{X}|}\ln f_E(|\underset{\sim}{X}|(t))dt + d\beta(t),$$

β being a BM($\mathbb{R}$) process. Thus, $|\underset{\sim}{X}|$ is a one-dimensional time-homogeneous process with generator

$$L_x = 2^{-1}\frac{d^2}{dx^2} + \frac{d}{dx}\ln f_E(x)\frac{d}{dx}.$$

Because f_E satisfies a one-dimensional Schrödinger equation, $|\underset{\sim}{X}|$ satisfies a one-dimensional Nelson-Newton law

$$2^{-1}(D_+D_- + D_-D_+)|\underset{\sim}{X}|(t) = -\frac{d}{d|\underset{\sim}{X}|}V(|\underset{\sim}{X}|)(t)).$$

We set $X = |\underset{\sim}{X}|$.

We also need some assumptions on the rate of growth of V. Weaker assumptions would probably do, but here we assume that $\lim_{x\to\infty} V(x)$ exists. This ensures that the drift $b = f_E'/f_E$ is bounded away from the origin O. We also need to assume that, almost surely, $\tau_x(n^{-1}) \to \infty$ as $n\to\infty$, where

$$\tau_x(a) = \inf\{s>0 : X(s) = a\,|\,X(0) = x\}, \qquad \text{each } x > 0.$$

A sufficient condition ensuring that this is so is: $\int^y f_E^{-2}(u)\,du$ diverges as $y\to 0$, as the next lemma shows.

LEMMA 6.1.2

Set $S(x) = \int^x f_E^{-2}(u)\,du$ (the scale function) and for each $x\in[a,b]\subset\mathbf{R}^+$ define

$$U(x) = (S(x) - S(a))/(S(b) - S(a)).$$

Then

$$U(x) = \mathbb{P}(\tau_x(b) < \tau_x(a)).$$

PROOF

By inspection $S(x) = \int^x f_E^{-2}(u)\,du$ satisfies the ordinary differential equation

$$2^{-1}y''(x) + b(x)y'(x) = 0, \qquad b = f_E'/f_E,$$

therefore so does $U(x)$ with $U(a) = 0$, $U(b) = 1$. Let $X_x(t)$ be the solution of

$$dX(t) = b(X(t))dt + dB(t),$$

for $b = f_E'/f_E$, with $X_x(0) = x$. Itô's formula then gives

$$U(X_x(t)) - U(x) = \int_0^t U'(X_x(s))dX_x(s) + 2^{-1}\int_0^t U''(X_x(s))\,dX(s)\,dX(s)$$

$$= \int_0^t U'(X_x(s))\{b(X_x(s))ds + dB(s)\} + 2^{-1}\int_0^t U''(X_x(s))\,ds.$$

Let $\tau_x(a,b) = \tau_x(a) \wedge \tau_x(b)$. We can now set $t = \tau_N = N \wedge \tau_x(a,b)$. Taking expectations, gives

$$\mathbb{E}\{U(X_x(\tau_N))\} = U(x),$$

for each fixed integer $N > 0$. Letting $N \uparrow \infty$ and using monotone convergence, we obtain

$$\mathbb{E}\{U(\tau_x(a,b))\} = U(x).$$

But

$$\mathbb{E}\{U(\tau_x(a,b))\} = U(a)\,\mathbb{P}(\tau_x(a) < \tau_x(b)) + U(b)\,\mathbb{P}(\tau_x(b) < \tau_x(a))$$

and, since $U(a) = 0$, $U(b) = 1$,

$$\mathbb{P}(\tau_x(b) < \tau_x(a)) = U(x),$$

as asserted. //

Therefore, for $S(x) = \int^x f_E^{-2}(u)\,du$, and for $x \in [a,b] \subset \mathbb{R}^+$,

$$\mathbb{P}(\tau_x(b) < \tau_x(a)) = (S(x) - S(a))/(S(b) - S(a)).$$

Therefore, if $S(a)$ diverges as $a \to 0+$, for each $x \in (0,b]$,

$$\lim_{a \downarrow 0} \mathbb{P}(\tau_x(b) < \tau_x(a)) = 1$$

i.e. $\tau_x(a) \uparrow \infty$ as $a \downarrow 0$. If $xV(x)$ is analytic in a neighbourhood of the origin O then $\int^x f_E^{-2}(u)\,du$ will diverge like $\int^x u^{-2}du$ for x in a neighbourhood of the origin. Of course much weaker assumptions on V will do, but, bearing in mind that the physically important potential is the Coulomb potential, we assume that $xV(x)$ is analytic in a neighbourhood of the origin.

PROPOSITION 6.1.3

Let the real-valued potential $V \in C(\mathbb{R}^+)$ be such that $\lim_{x\to\infty} V(x)$ exists and assume that $xV(x)$ is analytic in a neighbourhood of the origin O. Then $\rho_t(x,y)$ the transition density corresponding to the radial ground-state Nelson diffusion X associated with the Hamiltonian $H = -2^{-1}\Delta_{\underset{\sim}{x}} + V(|\underset{\sim}{x}|)$ for $E = \inf \operatorname{spec}(H)$ is given by

$$\rho_t(x,y) = f_E^{-1}(x) \exp\{-t(H_0 - E)\}(x,y) f_E(y), \qquad (x,y > 0)$$

H_0 being the one-dimensional Dirichlet Hamiltonian (densely defined in $L^2(\mathbb{R})$ by a Feynman-Kac formula)

$$H_0 = \lim_{\lambda\uparrow\infty} \left(-2^{-1}\frac{d^2}{dx^2} + V(x) + \lambda\chi\_(x)\right),$$

$\chi\_$ being the characteristic function of $\mathbb{R}^-$, f_E the radial eigenfunction with

$$\left(-2^{-1}\frac{d^2}{dx^2} + V(x) - E\right) f_E(x) = 0,$$

$f_E > 0$ on $(0,\infty)$ and $f_E(0) = 0$. The corresponding invariant measure for the process X is $f_E^2(x)\,dx$.

PROOF

Let f be bounded and measurable. Then for each $x > 0$

$$\mathbb{E}_x\{f(X(t))\} = \lim_{n\uparrow\infty} \mathbb{E}\{\chi_{\{\tau_x(n^{-1})>t\}} f(X_x(t))\},$$

X_x being the radial Nelson diffusion with $X_x(0) = x$. Because the drift b is bounded away from the origin, we obtain from the Girsanov-Cameron-Martin theorem

$\mathbb{E}_x\{f(X(t))\} =$

$$\lim_{n\uparrow\infty} \mathbb{E}\left\{\exp\{-\tfrac{1}{2}\int_0^t b^2(x+B(s))\,ds + \int_0^t b(x+B(s))\,dB(s)\}\ f(x+B(t))\ \chi_{\{\tau_x(n^{-1})>t\}}\right\},$$

the function χ now referring to the Brownian motion.

Using

$$d \ln f_E = (f_E'/f_E)\,dB(s) + 2^{-1}[(f_E f_E'') - f_E'^2)/f_E^2]\,ds = b\,dB(s) - 2^{-1}b^2 ds + 2^{-1} f_E''/f_E\,ds,$$

and $-2^{-1}f_E'' + (V - E)f_E = 0$, we obtain

$$\mathbb{E}_x\{f(X(t))\} = e^{Et} f_E^{-1}(x) \lim_{n\uparrow\infty} \mathbb{E}\left\{\chi_{\{\tau_x(n^{-1})>t\}} \exp\left\{-\int_0^t V(x+B(s))\,ds\right\} (f_E f)(x+B(t))\right\}.$$

Therefore, we arrive at for each $x>0$

$$\mathbb{E}_x\{f(X(t))\} = e^{Et} f_E^{-1}(x)\, \mathbb{E}\left\{\chi_{\{\tau_x(0)>t\}} \exp\left\{-\int_0^t V(x+B(s))\,ds\right\} (f_E f)(x+B(t))\right\}.$$

The result for a.e.y. follows from the Feynman-Kac formula for H_0. Continuity in y completes the argument. Evidently f_E is the ground-state for H_0 with eigenvalue E. //

6.2 FIRST HITTING TIMES

We next prove:

PROPOSITION 6.2

Let V satisfy the hypotheses in the last proposition. Then for each $\lambda>0$ the equation (*),

$$-2^{-1}\frac{d^2y}{dx^2} + (V(x)+\lambda-E)y = 0 \qquad (x>0), \tag{*}$$

has a unique solution $f_{E-\lambda}$ satisfying $f_{E-\lambda}(0+)=0$, $f'_{E-\lambda}(0+)=1$ and a unique solution $h_{E-\lambda}$ exponentially decreasing at infinity. The first hitting time $\tau_x(a)$ for the radial ground-state Nelson diffusion X satisfies

$$\mathbb{E}\{\exp\{-\lambda\tau_x(a)\}\} = f_{E-\lambda}(x)f_E(a)/f_{E-\lambda}(a)f_E(x) \qquad (x<a),$$

and

$$\mathbb{E}\{\exp\{-\lambda\tau_x(a)\}\} = h_{E-\lambda}(x)f_E(a)/h_{E-\lambda}(a)f_E(x) \qquad (x>a).$$

PROOF

The diffusion X is a one-dimensional time-homogeneous diffusion. For such a diffusion to have gone from x to y in time t the first hitting time of any intermediate point a must be less than t. Since the process starts afresh from a, for each fixed x,y and any intermediate a,

$$p_t(x,y) = \int_0^t \mathbf{P}(\tau_x(a)\in du)\,p_{t-u}(a,y)\,du.$$

Taking Laplace transforms, for each $\lambda>0$

$$\tilde{p}_\lambda(x,y) = \mathbb{E}\{e^{-\lambda\tau_x(a)}\}\tilde{p}_\lambda(a,y),$$

where $\tilde{p}_\lambda(x,y) = \int_0^\infty e^{-\lambda t}p_t(x,y)dt$. Since $\tilde{p}_\lambda(x,\cdot)$ is continuous, letting $y\to a$, we arrive at

$$\mathbb{E}\{e^{-\lambda\tau_x(a)}\} = \tilde{p}_\lambda(x,a)/\tilde{p}_\lambda(a,a).$$

From the last proposition therefore for $\lambda > 0$

$$\mathbb{E}\{e^{-\lambda\tau_x(a)}\} = f_E^{-1}(x)\,(H_o + \lambda - E)^{-1}(x,a)\, f_E(a)/(H_o + \lambda - E)^{-1}(a,a),$$

$(H_o + \lambda - E)^{-1}(\cdot,\cdot)$ being the resolvent kernel of H_o. The desired result follows by standard theory of o.d.e.s. (See Ref. (12)). //

Example (Coulomb Problem)
Let $V(\underset{\sim}{x}) = -1/|\underset{\sim}{x}|$. The ground-state energy is $E_o = -\frac{1}{2}$ and the corresponding ground-state wave-function is $f_{E_o}(x) = x\exp(-x)$. In this case

$$\mathbb{E}\{-\lambda\tau_x(a))\} = \frac{a}{x}\exp(x-a)\frac{G_{\frac{1}{k},\frac{1}{2}}(2kx)}{G_{\frac{1}{k},\frac{1}{2}}(2ka)},$$

$k = (1+2\lambda)^{1/2}$, $G \equiv W$ for $x \geq a$ and $G \equiv M$ for $x \leq a$, W and M being Whittaker functions. (See Refs. (1),(2),(3)).

Remark
When the radial Hamiltonian H is limit point at 0 and ∞, $f_{E-\lambda}$ and $h_{E-\lambda}$ in the last proposition have to be taken to be the unique L^2-eigenfunctions of H corresponding to eigenvalue $(E-\lambda)$, because $\mathbb{E}(e^{-\lambda\tau_x(a)}) < 1$ and $f_E \in L^2(\mathbb{R}^+)$.

6.3 A GENERALISED ARC-SINE LAW

We conclude this section with a generalised arc-sine law. For the above radial Nelson diffusion process X, $a > 0$, set $L^{\pm}(t) = \mathrm{Leb}\{s \in [0,t] : X(s) \gtrless a\}$. Then we have:

PROPOSITION 6.3
Assume that the potential V satisfies the hypotheses of the last proposition. Then, for all choices of constants $\alpha, \lambda_+, \lambda_- > 0$,

$$\int_0^\infty e^{-\alpha t}\,\mathbb{E}_a\left\{e^{-\lambda_+ L^+(t) - \lambda_- L^-(t)}\right\} dt$$

$$= \frac{\mathrm{Disc}\big|_{x=a}(\alpha+\Lambda)^{-1}\frac{d}{dx}\ln\left\{f_E^{-1}(x)(H_o+\alpha+\Lambda-E)^{-1}(x,a)\right\}}{\mathrm{Disc}\big|_{x=a}\frac{d}{dx}\ln\left\{f_E^{-1}(x)(H_o+\alpha+\Lambda-E)^{-1}(x,a)\right\}},$$

where $\Lambda = \Lambda(x) = \lambda_{\pm}$ for $x \gtrless a$, respectively.

PROOF

Set

$$u(x)=\int_0^\infty e^{-\alpha t}\mathbb{E}_x\{\exp(-\lambda_-L^-(t)-\lambda_+L^+(t))\}\,dt=\int_0^\infty e^{-\alpha t}\mathbb{E}_x\{\exp\{-\int_0^t\Lambda(X(s))ds\}\}\,dt,$$

where $\alpha, \lambda_+, \lambda_-$ are constants and $\Lambda(x)=\lambda_-\chi_-(x)+\lambda_+\chi_+(x)$, $\chi_\pm$ being the characteristic functions of $\{x: x \gtrless a\}$. Therefore, because $\tau_x(n^{-1})\uparrow\infty$ almost surely as $n\uparrow\infty$,

$$u(x)=\lim_{n\uparrow\infty}\int_0^\infty e^{-\alpha t}dt\,\mathbb{E}_x\{\chi_{\{\tau(n^{-1})>t\}}\exp(-\int_0^t\Lambda(X(s))ds)\}.$$

Applying the Girsanov-Cameron-Martin theorem as in PROPOSITION 6.1.3, gives

$$u(x)=f_E^{-1}(x)\lim_{n\uparrow\infty}\int_0^\infty e^{-\alpha t}e^{Et}\,\mathbb{E}_x\{\chi_{\{\tau(n^{-1})>t\}}\exp(-\int_0^t(V+\Lambda)(B(s))ds)\,f_E(B(t))\}\,dt.$$

Therefore, the Feynman-Kac formula gives

$$u(x)=f_E^{-1}(x)\{\exp\{-t(H_0+\Lambda-E)\}f_E(x)\},$$

where $H_0=\lim_{\lambda\uparrow\infty}\left(-2^{-1}\frac{d^2}{dx^2}+V(x)+\lambda\chi_{\mathbb{R}^-}(x)\right)$, $\chi_{\mathbb{R}^-}$ being the characteristic function of $\{x: x<0\}$. As in the proof of Kac's theorem (see Ref. (9)), $u(x)$ is C^2 and satisfies

$$(L_x-(\Lambda(x)+\alpha))u(x)=-1,$$

with $u(a+)=u(a-)$ and $u'(a+)=u'(a-)$, L_x being the generator of the diffusion $L_x=2^{-1}\frac{d^2}{dx^2}+\frac{f_E'(x)}{f_E(x)}\frac{d}{dx}$. However, the only solutions of

$$L_x y(x)=\gamma y(x)\qquad(\gamma>0)$$

bounded in a neighbourhood of $x=0$ and $x=\infty$ are $\mathbb{E}\{e^{-\gamma\tau_x(a)}\}$. Setting

$$u(x)=\frac{1}{(\alpha+\lambda_-)}+A\,\mathbb{E}\{e^{-(\alpha+\lambda_-)\tau_x(a)}\},\qquad x<a,$$

and

$$u(x)=\frac{1}{(\alpha+\lambda_+)}+B\,\mathbb{E}\{e^{-(\alpha+\lambda_+)\tau_x(a)}\},\qquad x>a,$$

for constants A,B and satisfying the above boundary conditions gives

$$u(a)=\frac{\mathrm{Disc}\,|_{x=a}(\alpha+\Lambda(x))^{-1}\frac{d}{dx}\mathbb{E}\{e^{-(\alpha+\Lambda(x))\tau_x(a)}\}}{\mathrm{Disc}\,|_{x=a}\frac{d}{dx}\mathbb{E}\{e^{-(\alpha+\Lambda(x))\tau_x(a)}\}},$$

for $\mathrm{Disc}\big|_{x=a} f(x) = f(a+) - f(a-)$. The desired result now follows from the last proposition. //

The above is a slight extension of the Batchelor-Truman formula. It can be viewed as a generalised arc-sine law (see Ref. (19)).

Example (Coulomb Problem)

For the ground-state radial Nelson diffusion process, associated with the Coulomb problem for the potential $V(\underset{\sim}{x}) = -1/|\underset{\sim}{x}|$, for all possible choices of constants $\alpha, \lambda_+, \lambda_- > 0$,

$$\int_0^\infty e^{-\alpha t}\,\mathbb{E}_a\{e^{-\lambda_- L^-(t)-\lambda_+ L^+(t)}\}\,dt = \frac{\mathrm{Disc}\big|_{x=a}(\alpha+\Lambda(x))^{-1}\frac{d}{dx}x^{-1}e^x\,G_{\frac{1}{k},\frac{1}{2}}(2kx)}{\mathrm{Disc}\big|_{x=a}\frac{d}{dx}x^{-1}e^x\,G_{\frac{1}{k},\frac{1}{2}}(2kx)},$$

where $k = k(x) = (1 + 2\alpha + 2\Lambda(x))^{1/2}$, $\Lambda(x) = \lambda_\pm$ for $x \gtrless a$; $G \equiv W$, $x > a$, $G \equiv M$, $x < a$, W and M being Whittaker functions.

7. Excursions in Nelson's Stochastic Mechanics

Let $a(>0)$ be a regular point for the diffusion X so that

$$\mathbb{E}\{\tau_a(a) = 0\} = 1.$$

We shall consider excursions away from the point a for the above diffusion X with $X(0) = a$.

7.1 LOCAL TIME L^a (See Rogers and Williams Ref. (17)).

The local time at a upto time t, $L^a(t) = \int_0^t \delta(X(s) - a)\,ds$, is defined through Trotter's theorem and the identity

$$\int_0^t f(X(s))\,ds = \int_{-\infty}^\infty f(a)L^a(t)\,da,$$

for all bounded measurable functions f. The points of increase of $L^a(\cdot)$ are therefore the zeros of $(X(\cdot) - a)$.

Firstly, observe that: $\mathrm{Leb}\{t > 0 : X(t) = a\} = 0$, almost surely. For

$$\mathbb{E}\{\mathrm{Leb}\{t > 0 : X(t) = a\}\} = \mathbb{E}\{\int_0^\infty \chi_{\{a\}}(X(t))\,dt\} = \int_0^\infty \mathbb{P}(X(t) = a)\,dt = 0.$$

Since $X(\cdot)$ is continuous if $Z = \{t > 0 : X(t) = a\}$, $(\mathbb{R}^+\backslash Z)$ is almost surely open and can be decomposed into open components - the *excursion intervals* for X away from a.

Secondly, denoting by ρ_t the transition density for the process X,

$$\mathbb{E}_a(L^a(t)) = \mathbb{E}_a\{\int_0^t \delta(X(s) - a)\,ds\} = \int_0^t \rho_s(a,a)\,ds.$$

To understand the statistics of excursions better we have to consider how long one has to wait before the local time at a equals t.

7.2 SUBORDINATOR γ^a

Define γ^a (the right inverse of L^a) by

$$\gamma^a(t) = \inf\{u > 0 : L^a(u) > t\},$$

so that $L^a(\gamma^a(t)) = t$. Then each $\gamma^a(t)$ is a *stopping time* with $X(\gamma^a(t)) = a$. The jumps in γ^a are the excursions for X.

The all-important property for γ^a is that $\gamma^a(t)$ is a *subordinator* in the sense that

$$\gamma^a(t+s)(\omega) - \gamma^a(t)(\omega) = \gamma^a(s)(\theta_{\gamma^a(t)})(\omega), \qquad s,t > 0,$$

where $\theta_{\gamma^a(t)}$ is the usual shift

$$\theta_{\gamma^a(t)}(\omega)(s) = \omega(s + \gamma^a(t)) - \omega(\gamma^a(t)).$$

This gives for some constant $\Psi(\lambda)$

$$\mathbb{E}_a(e^{-\lambda\gamma^a(t)}) = \exp(-t\Psi(\lambda)), \qquad \text{each } \lambda > 0, \text{ each } t > 0.$$

To see this observe that

$$\mathbb{E}_a(e^{-\lambda\gamma^a(t+s)}) = \mathbb{E}_a\left(e^{-\lambda\gamma^a(t)}\mathbb{E}\left(e^{-\lambda(\gamma^a(t+s)-\gamma^a(t))} \,|\, \mathcal{F}_{\gamma^a(t)+}\right)\right),$$

and by the strong Markov property

$$\mathbb{E}\left(e^{-\lambda(\gamma^a(t+s)-\gamma^a(t))} \,|\, \mathcal{F}_{\gamma^a(t)+}\right) = \mathbb{E}_a(e^{-\lambda\gamma^a(s)}),$$

proving the existence of $\Psi(\lambda)$.

Using the inverse property of γ^a, $\int_0^\infty e^{-\lambda\gamma^a(t)}dt = \int_0^\infty e^{-\lambda s}dL^a(s)$, almost surely, giving

$$\Psi(\lambda)^{-1} = \int_0^\infty \mathbb{E}_a(e^{-\lambda\gamma^a(t)})\,dt = \mathbb{E}_a\left\{\int_0^\infty e^{-\lambda s}dL^a(s)\right\} = \int_0^\infty e^{-\lambda s}d(\mathbb{E}_a(L^a(s))) = \tilde{\rho}_\lambda(a,a),$$

with $\tilde{\rho}_\lambda(a,a) = \int_0^\infty e^{-\lambda s}\rho_s(a,a)\,ds$. Therefore, we have shown that for our diffusion X

$$\mathbb{E}_a(e^{-\lambda\gamma^a(t)}) = \exp\{-t/\tilde{\rho}_\lambda(a,a)\},$$

$\tilde{\rho}_\lambda$ being the Laplace transform of ρ_s, the transition density associated with X.

LEMMA 7.2.1

For the above radial ground-state Nelson diffusion X, associated with the radial eigenfunction f_E in a spherically symmetric potential V, satisfying the hypotheses of the last proposition, for each $t,\lambda > 0$,

$$\mathbb{E}_a\left(e^{-\lambda\gamma^a(t)}\right) = \exp\left\{-t\lambda\int_0^\infty f_E^2(x)\, f_E^{-2}(a)\, \mathbb{E}\left(e^{-\lambda\tau_x(a)}\right) dx\right\} \qquad (a > 0),$$

$\tau_x(a)$ being the first hitting time of a for the process starting from x.

PROOF
Recall that from

$$\mathbb{E}\{e^{-\lambda\tau_x(a)}\} = \int_0^\infty e^{-\lambda s}\rho_s(x,a)\, ds/\tilde{\rho}_\lambda(a,a).$$

Multiplying both sides by $f_E^2(x)$ and integrating by Fubini's theorem

$$\int_0^\infty f_E^2(x)\, \mathbb{E}\{e^{-\lambda\tau_x(a)}\}\, dx = \int_0^\infty e^{-\lambda s} ds \int_0^\infty f_E^2(x)\, \rho_s(x,a)\, dx/\tilde{\rho}_\lambda(a,a)$$

$$= \lambda^{-1} f_E^2(a)/\tilde{\rho}_\lambda(a,a). \qquad //$$

COROLLARY 7.2.2
Since $f_E \in L^2(\mathbb{R}^+)$, $\mathbb{P}(\gamma^a(t) < \infty) = \mathbb{P}(L^a(\infty) > t) = 1$.

PROOF

$$\lim_{\lambda\downarrow 0} \mathbb{E}\left(e^{-\lambda\gamma^a(t)}\right) = \lim_{\lambda\downarrow 0} \exp\left\{-t\lambda f_E^{-2}(a)\int_0^\infty f_E^2(x)\, \mathbb{E}\left(e^{-\lambda\tau_x(a)}\right) dx\right\} = 1. \qquad //$$

7.3 THE POISSON-LÉVY EXCURSION MEASURES

We now give the main result in this section:

PROPOSITION 7.3.1
For the above diffusion X, associated with the radial ground-state f_E in the spherically symmetric potential V, satisfying above hypotheses,

$$\mathbb{E}_a\left(e^{-\lambda\gamma^a(t)}\right) = \exp\left\{-t\int_0^\infty (1 - e^{-\lambda s})\, d\upsilon(s)\right\}, \qquad \text{each } \lambda, t > 0,$$

where

$$\frac{d\upsilon(s)}{ds} = f_E^{-2}(a)\Big\{\left(f_E, (H_+ - E)^2 \exp(-s(H_+ - E)) f_E\right)_{L^2}$$

$$+ \left(f_E, (H_- - E)^2 \exp(-s(H_- - E)) f_E\right)_{L^2}\Big\},$$

for

$$H_{\pm} = \lim_{\lambda\downarrow\infty} (H_0 + \lambda\chi_{\mp}),$$

$\chi_{\mp}$ being the characteristic function of $\{x : x \gtrless a\}$, H_0 being defined as above.

PROOF

The proof depends upon:

LEMMA 7.3.2

For the above diffusion X, for $x \gtrless a$,

$$\mathbb{P}(\tau_x(a) > t) = f_E^{-1}(x) \{\exp(-t(H_{\pm} - E))f_E(x)\},$$

respectively.

PROOF

For definiteness sake assume that $x > a$. Since $\int_0^t \chi_-(X_x(s))\,ds > 0$ if $X_x(s_0) < a$ for some $s_0 \in (0,t)$,

$$\mathbb{P}(\tau_x(a) > t) = \lim_{\lambda\uparrow\infty} \mathbb{E}\left\{\exp(-\lambda\int_0^t \chi_-(X_x(s))\,ds)\right\}.$$

Arguing as in PROPOSITION 6.3 the Girsanov-Cameron-Martin theorem gives the desired result. //

As a consequence we obtain for the ground-state f_E with energy E

$$\int_0^\infty f_E^2(x)\,\mathbb{P}(\tau_x(a) \in ds)\,dx = \int_0^\infty f_E^2(x) - \frac{d}{ds}\mathbb{P}(\tau_x(a) > s)\,ds\,dx$$

$$= \{(f_E,(H_+ - E)\exp(-s(H_+ - E))f_E)_{L^2} + (f_E,(H_- - E)\exp(-s(H_- - E))f_E)_{L^2}\}\,ds\,.$$

The desired result follows after an integration by parts in the last but one lemma. //

Identifying the jumps in γ^a with the excursions gives:

COROLLARY 7.3.3

Define # by

#(s,t) = the number of excursions of duration s of X away from a upto the local time at a equals t.

Then

$$\mathbb{P}(\#(s,t) = N) = e^{-t\,d\upsilon(s)} \frac{(t\,d\upsilon(s))^N}{N!}, \qquad \text{for } N = 0,1,2,\ldots,$$

υ being given as in the last proposition.

PROOF

The point is that, for $t = \sum_j \delta t_j$ and $\int_0^\infty \upsilon(ds) = \sum_i \Delta_i = \sum_i \upsilon(ds_i)$,

$$\mathbb{E}_a(e^{-\lambda\gamma^a(t)}) = \prod_{\Delta_i,\delta t_j} e^{-\delta t_j \Delta_i} e^{\delta t_j e^{-\lambda s_i}\Delta_i},$$

showing independent contributions from disjoint $(\delta t_j \times \Delta_i)$ to $\mathbb{E}_a(e^{-\lambda\gamma^a(t)})$. Therefore,

$$\mathbb{E}_a(e^{-\lambda\gamma^a(t)}) = \prod_{\Delta_i} e^{-t\Delta_i} e^{te^{-\lambda s_i}\Delta_i},$$

and

$$e^{-t\Delta} e^{te^{-\lambda s}\Delta} = e^{-t\Delta} \sum_{n=0}^{\infty} e^{-\lambda sn} \frac{(t\Delta)^n}{n!} = \sum_{n=0}^{\infty} e^{-\lambda sn} \left(e^{-t\Delta} \frac{(t\Delta)^n}{n!} \right).$$

This gives the distribution of the jumps in γ^a or the excursions for X away from a. //

7.4 THE INS AND OUTS OF EXCURSIONS

Because $t \to X(t)$ is almost surely continuous $\{t > 0 : X(t) > a\}$ is almost surely open and can be decomposed into components - the outward excursion intervals. Similarly by considering $\{t > 0 : X(t) < a\}$ we obtain the inward excursions into the sphere of radius a. The statistics of the inward and outward excursions is controlled by the *independent subordinators* $L^{\pm}(\gamma^a(t))$, with

$$L^+(\gamma^a(t)) + L^-(\gamma^a(t)) = \gamma^a(t),$$

almost surely; $L^{\pm}(t) = \text{Leb}\,\{s \in [0,t] : X(s) \gtrless a\}$, for each $t > 0$. We conclude by stating without proof one of the consequences of PROPOSITION 7.3.1 and the last corollary (see Ref. (19)).

PROPOSITION 7.4.1

For the ground-state radial Nelson diffusions associated with the radial ground-state f_E in the spherically symmetric potential V, satisfying the above hypotheses,

$$\mathbb{E}\,\{\exp(-\lambda L^{\pm}(\gamma^a(t)))\} = \exp\left\{-t \int_0^\infty (1 - e^{-\lambda s})\, d\upsilon^{\pm}(s)\right\}, \qquad \text{each } \lambda, t > 0,$$

with

$$\frac{d\upsilon^{\pm}(s)}{ds} = f_E^{-2}(a)(f_E, (H_{\pm} - E)^2 \exp(-s(H_{\pm} - E)) f_E)_{L^2},$$

E being the ground-state energy, $H_{\pm}$ being defined as in the last proposition. An alternative expression for the above is

$$\mathbb{E}\{\exp(-\lambda L^{\pm}(\gamma^{a}(t)))\} = \exp\left\{\pm\,\frac{t}{2}\,\frac{d}{dx}\Big|_{x=a\pm} \ln\{f_E^{-1}(x)(H_o+\lambda-E)^{-1}(x,a)\}\right\}.$$

$(H_o+\lambda-E)^{-1}(x,a)$ being the resolvent kernel of the Hamiltonian H_o defined above.

COROLLARY 7.4.2

Define $\#^{\pm}$ by

$\#^{\pm}(s,t)$ = the number of $\begin{smallmatrix}\text{outward}\\\text{inward}\end{smallmatrix}$ excursions of duration s of X away from a upto the local time at a equals t.

Then

$$\mathbb{P}(\#^{\pm}(s,t)=N) = e^{-t\,d\upsilon^{\pm}(s)}\,\frac{(t\,d\upsilon^{\pm}(s))^{N}}{N!}, \qquad \text{for } N=0,1,2,\ldots$$

$\upsilon^{\pm}$ being defined as above.

The above are some new variants of Lévy's formulae. (See Refs. (8) and (19)).

Example (Coulomb Problem)

In this case $V(x)=-\frac{1}{x}$, $f_E(x)=xe^{-x}$, $E=-\frac{1}{2}$ and setting $k=\sqrt{(1+2\lambda)}$, gives for $\lambda>0$

$$(H_o+\lambda-E)^{-1}(x,a) = \frac{\Gamma(1-\frac{1}{k})}{k}\, M_{\frac{1}{k},\frac{1}{2}}(2kx\wedge a)\, W_{\frac{1}{k},\frac{1}{2}}(2kx\vee a),$$

W and M being Whittaker functions.

In the last example finding $\upsilon^{\pm}$ explicitly seems to be a very difficult task. We give a final example.

Example (Extra-Nuclear Excursions)

Let the potential V have compact support in the ball $\{\underset{\sim}{x} : |\underset{\sim}{x}| \le a_o\}$, V satisfying above hypotheses. Consider the radial Nelson diffusion, with energy $E(<0)$. Then for $x > a \ge a_o$

$$\mathbb{E}(e^{-\lambda\tau_x(a)}) = \exp\{(\sqrt{-2E}-\sqrt{-2(E-\lambda)})\,(x-a)\} \qquad (E<0)$$

In this case there is an explicit formula for excursions away from a

$$\frac{d\upsilon^{+}(s)}{ds} = \frac{e^{Es}ds}{(2\pi s^{3})^{1/2}},$$

so υ is independent of the value of $a\ (\ge a_o)$

$$\mathbb{P}(\#^{+}(s,t)=N) = e^{-t\,d\upsilon^{+}(s)}\,\frac{(t\,d\upsilon^{+}(s))^{N}}{N!},$$

for $N=0,1,2,\ldots$

8. Acknowledgement

One of us, AT, would like to thank the conference organisers for inviting him to present our results and for giving him an opportunity to attend so many stimulating lectures.

References

1. A. Batchelor and A. Truman, 'On capture times and hitting times in stochastic mechanics', in Stochastic Methods in Mathematics and Physics, XXIV Karpacz Winter School on Theoretical Physics, Jan. 1988, editors R. Gielerak and W. Karwowski, World Scientific Press, Singapore, 1989.
2. A. Batchelor and A. Truman, 'On first hitting times in stochastic mechanics' in Stochastic Mechanics and Stochastic Processes, Proceedings, Swansea 1986, editors A. Truman and I.M. Davies, Springer Lecture Notes in Mathematics, Volume 1325, Berlin 1988.
3. A. Batchelor and A. Truman, 'Hitting, killing and capturing in Nelson's stochastic mechanics', in Proceedings of the International Conference on Stochastic Processes - Geometry and Physics, Ascona-Locarno, Summer 1988, edited by S. Albeverio and G. Casati et al, World Scientific Press, Singapore, 1989.
4. E.A. Carlen, 'Conservative Diffusions', Com. Math. Phys. t 94, 273-296, 1984.
5. E.A. Carlen, 'Progress and Problems in Stochastic Mechanics', in Stochastic Methods in Mathematics and Physics, XXIV Karpacz Winter School on Theoretical Physics, Jan. 1988, editors R. Gielerak and W. Karwowski, World Scientific Press, Singapore, 1989.
6. M. Demuth, 'On topics in spectral and stochastic analysis for Schrôdinger operators', this volume.
7. R. Durrett, Brownian Motion and Martingales in Analysis, Wadsworth, California, 1984.
8. K. Itô and H.P. McKean, Diffusion Processes and Their Sample Paths, Springer, Berlin, 1965.
9. M. Kac, 'On some connections between probability theory and differential and integral equations', in Proceedings of Second Berkeley Symposium on Math. Stat. and Probability 189-215, University of California Press, Berkeley, 1951.
10. A. Kolmogorov, Grundbegriffe der Wahrscheinlichkeitsrechnung, Springer, Berlin, 1933.
11. F. Guerra and L. Morato, Quantization of Dynamical Systems and Stochastic Control Theory, Phys. Rev. D27, 1774-1786, 1983.
12. P. Mandl, Analytical Treatment of One-dimensional Markov processes, Springer, Berlin-Heidelberg-New York, 1968.
13. H.P. McKean, Stochastic Integrals, Academic Press, New York, 1969.
14. E. Nelson, Dynamical Theories of Brownian Motion, Princeton University Press, Princeton, 1967.
15. E. Nelson, Quantum Fluctuations, Princeton University Press, Princeton, 1984.
16. L.C.G. Rogers, 'A guided tour through excursions', Bull. L.M.S. #91, Volume 21, Part 4, July 1989.
17. L.C.G. Rogers and D. Williams, Diffusions, Markov Processes and Martingales, Vol. 2 Itô Calculus, John Wiley, Chichester, 1987.
18. B. Simon, Functional Integration and Quantum Physics, Academic Press, New York, 1979.
19. A. Truman and D. Williams, 'A generalised arc-sine law and Nelson's stochastic mechanics of one-dimensional time-homogeneous diffusions', Swansea preprint 1990.

STARK - WANNIER RESONANT STATES

F. BENTOSELA
Centre de Physique Théorique
CNRS Luminy - Case 907
13288 Marseille Cedex 9 France

ABSTRACT.The resonance states which appear in a crystal submitted to an external constant electric field are shown to be close to the eigenstates of the Schrödinger operator obtained neglecting the interband terms .

1. Introduction

Schrödinger equation involving a linear potential plus a periodic potential has attracted much attention, as it is a first step in the study of conductivity in crystalline solids [11]. In this school we had the opportunity to hear about two other approaches, the one by Buslaev and Dmitrieva [6], the second by Combes and Hislop [7] .

Ohm's law says that current is proportional to external applied electric field. This fact shows that the behavior of electrons in solids is not the same as in the vacuum, where electrons are uniformly accelerated when submitted to a constant electric field. In solids it seems, that after a transient regime, their velocity reach a constant value.

To explain this fact, in most of the books, the authors consider the conduction electrons as free,except they are submitted to collisions by impurities and phonons; electrons are accelerated by the electric field in between two collisions. There is a transfert energy from the electrons to the lattice : phonons (i.e. lattice vibrations) are created and the electron velocity, on large scales, will remain almost constant. At the contrary when considering the periodic potential due to the ions in a crystal and neglecting the effect of the impurities and lattice vibrations, at least in one dimensional models the electron motion is not similar to the free one.Even if, the electron will finally escape in the direction opposite to the electric field, it remains for a long time in a space region ,going back and forth in it with the quasi-period, $T = \frac{h}{|e|Ea}$ (where a is the lattice period e the electric charge, E the constant electric field, h the Planck constant).

A. Boutet de Monvel et al. (eds.), Recent Developments in Quantum Mechanics, 85–95.

These resonance states, as it will be shown in this lecture, live in regions whose lenght is inversely proportional to the external field and to the spectral bandwidths of the Bloch Hamiltonian To have a chance to observe them it is necessary to create large external electric fields.Some experiments where done using the large electric field created at the junction of two semiconductors of different type: p and n. Another favourable condition for their observation is the ocurrence of small energy bands near the Fermi energy; such a situation occurs in superlattices, which are man-made crystals in which layers of two distinct semi conductors alternate ; the period in the perpendicular direction to the layers can be of the order of hundreds of normal lattice period.This structure creates conduction minibands of the order of some meV. In any case, as the mean free path of the electron has to be larger than the space which is needed for the resonance states, the samples have to be extremely pure and the temperature has to be low to decrease the number of phonons; this probably explain why, they were so difficult to observe and why no everybody was convinced by the claims they have been observed.[14] Only recently their effect appeared in the electro-optical properties of semiconductor superlattices [4],[12] and came the mathematical proof of their existence [1] .

In [3] can be found the proofs of the main theorems, here I only sketch some of them, giving the main ideas. I also prove that the resonances states are very near the Wannier states, so they will be called Stark-Wannier resonance states to emphasize the fact that G. Wannier did the pioneer work in the subject [13] .

In chapter 1 is given the resonance definition and indicated the space deformation we use to construct a family of non self-adjoint operators, the eigenvalues of which are the resonances.

In chapter 2 is explained the main idea which uses the tilted band picture introduced by Zener and the techniques borrowed from the Briet, Combes, Duclos[5] and Helffer Sjöstrand [9] papers on multiple wells resonances.

In chapter 3 we study the spectral properties of operators introduced in chapter 2, in particular the behavior of their associated Green functions.Exponential decrease of these allow us to prove that resonance widths are exponentially small with respect to the electric field.

In chapter 4, I prove that Stark-Wannier states are good approximations for the resonance states, using and improving a technique given by A. Nenciu and G.Nenciu in [10].

2. Definitions - The local space deformation

The operator, the spectral properties we want to study is: $H = -\frac{d^2}{dx^2} + V_P(x) + Fx$, acting on $L^2(\mathbb{R})$,where V_P is a real periodic function : $V_p(x+a) = V_p(x)$ and F is the product of the electron charge by the electric field; $V_P(z)$ is chosen to be, analytic in the strip $|\mathrm{Im}\, z| < A$.

For every $E \in \mathbb{C}$ such that $\mathrm{Im}\, E \neq 0$ one can define the Green function $G(x, y ; E)$; it is analytic with respect to E in the upper and lower complex half planes. The resonances can be defined as the poles of the analytic continuation of the Green function through the real axis.Resonances are also eigenvalues for an analytic family of operators H(b) we construct now, using the space transform:

$$s : x \in \mathbb{R} \rightarrow s(x) = x + ibf(x) \ ; \quad 0 < b < A$$

where f is a real C^3 function whose graph is represented below:

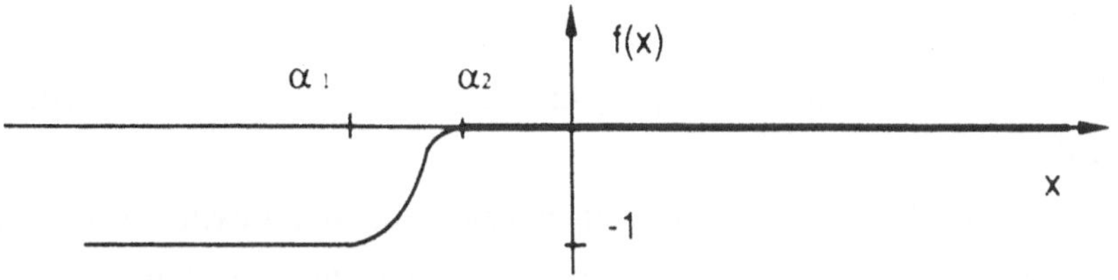

Fig.1

f is constant outside interval $[\alpha_1, \alpha_2]$ which will be precised later.

We define a transformation U_b on L^2 (R) by

$$U_b : g(x) \rightarrow (U_b g)(x) = \sqrt{1 + ibf'(x)} \ g(x + ibf(x))$$

Under this transformation our hamiltonian, $H = -\frac{d^2}{dx^2} + V_p(x) + Fx$ becomes

$$H(b) := U_b H U_b^{-1} = \frac{1}{1 + ibf'(x)} \left(-\frac{d^2}{dx^2}\right) \frac{1}{1 + ibf'(x)} + V_p(x + ibf(x)) + F(x + ibf(x))$$
$$+ \frac{1}{1 + ibf'(x)} \{s, x\} \frac{1}{1 + ibf'(x)}$$

where $\{s, x\}$ is the Schwarzian :

$$\{s, x\} = \frac{1}{2} \left[\frac{s'''}{s'} - \frac{3}{2} \left(\frac{s''}{s'}\right)^2\right]$$

Remark : if the support of ϕ is included in $[\alpha_2, +\infty)$ then

$$(H(b)\, \phi)(x) = -\frac{d^2}{dx^2} \phi(x) + V_p(x)\, \phi(x) + Fx\, \phi(x)$$

If $b \neq 0$, H(b) may have eigenvalues in the strip $\{E \in \mathbb{C} \mid -b < \mathrm{Im}\, E < 0\}$. It can be proven that these eigenvalues are stable, i.e., if b'>b , eigenvalues of H(b) and eigenvalues of H(b') coincide in the strip $\{E \in \mathbb{C} \mid -b < \mathrm{Im}\, E < 0\}$. Eigenvalues of H(b) are the resonances for H.

3. The tilted band picture and the partition

If F is small , near $x = x_0$ solutions for:

$$\left(-\frac{d^2}{dx^2} + V_P(x) + Fx\right)\varphi = E\,\varphi$$

are near the solutions for :

$$\left(-\frac{d^2}{dx^2} + V_P(x)\right)\varphi_0 = (E - Fx_0)\,\varphi_0$$

φ_0 are linear combinations of Bloch waves ψ which satisfy:

$$\psi(x+a) = e^{ik(E - Fx_0).a}\,\psi(x) \text{ or } \psi(x) = e^{ik(E - Fx_0).x}\,u_k(x) \text{ with } u_k(x+a) = u_k(x)\ ;$$

$k(E - Fx_0)$ are real or complex depending on the value taken by the " effective energy ", $E - Fx_0$.

The spectrum of $-\frac{d^2}{dx^2} + V_P(x)$ is constituted by intervals, named bands, separated by intervals named gaps. If $E - Fx_0$ belongs to a band, $k(E - Fx_0)$ is real and the solution ψ oscillates; if $E - Fx_0$ belongs to a gap, $\text{Im}\, k(E - Fx_0) \neq 0$ and the solution behaves exponentially.

One can draw the so-called Zener tilted bands, adding for each x the quantity Fx to the energy band spectrum. Drawing the horizontal line going through E and looking at its intersections with the tilted bands enables us to visualize better the regions $(\mu_i a, \mu'_{i-1} a)$ where the solution oscillate and the ones $(\mu'_i a, \mu_i a)$, where it behaves exponentially,[fig.2].

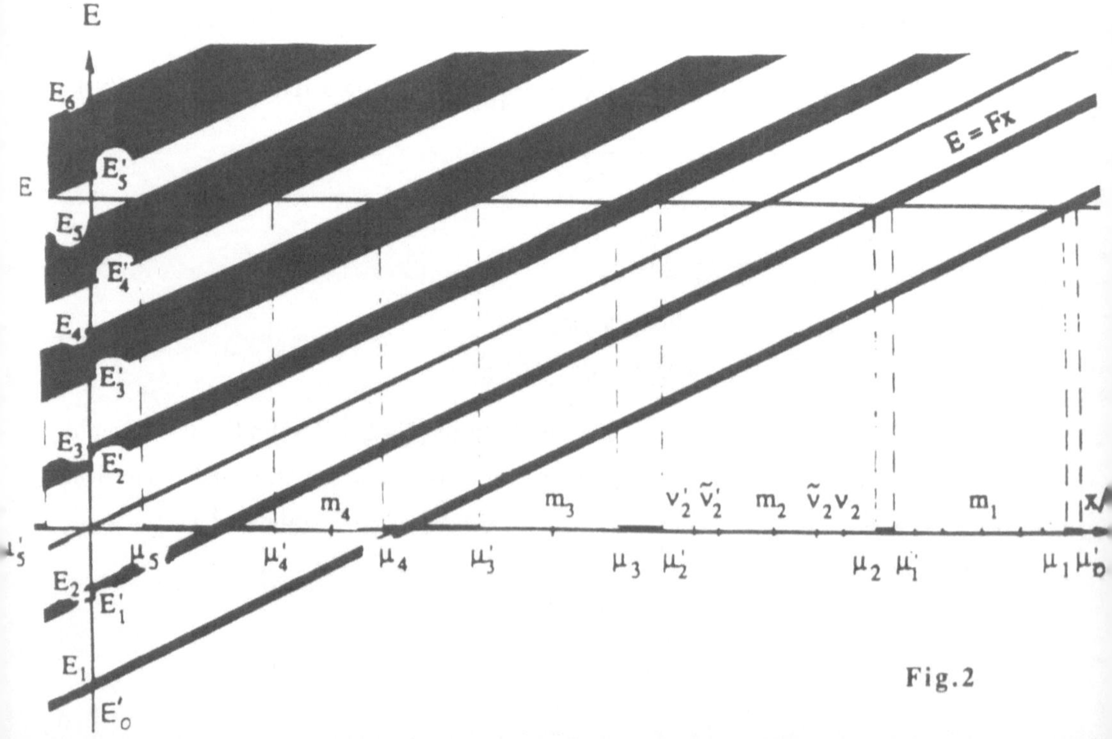

Fig.2

The regions where the solution oscillates are analogous to the the wells in the multiple well problem (MPW). In the regions where the solution behave exponentially we get the same situation as in the classically forbidden regions in MWP.

In the MWP the aim was to relate the resonances to the eigenvalues of operators constructed selecting a well and defining the new potential in the following manner : it takes the same value as the former inside the well and large constant value outside.

In our case to define the "allowed" and"forbidden regions", we will choose an energy, E, inside the first gap which is just above the maximum of the periodic potential.

We select an allowed region $(\mu_i a, \mu'_{i-1}a)$ and construct a continuous potential which is equal to $V_P(x) + Fx$ is an interval sligthly larger $(\nu'_i a, \nu_{i-1}a)$ and take the value $V_p(x)$, up to a constant, outside this interval.

The precise values for the μ's, m's, ν's and $\tilde{\nu}$'s are given in [2]. If there are N bands below E we get N allowed regions on $\mathbb{R}^+$ and so define N operators H_i : i = 1, ...N

$$H_i = -\frac{d^2}{dx^2} + V_p(x) + \chi(-\infty, \nu'_i a)\, F\nu'_i\, a + \chi(\nu'_i a, \nu_{i-1}\, a)\, Fx + \chi(\nu_{i-1}\, a, +\infty)\, F\nu_{i-1}\, a$$

We also define two special operators H_0 and H_{N+1} in the following manner :

$$H_0 = -\frac{d^2}{dx^2} + V_p(x) + \chi(-\infty, \nu'_0 a)\, Fb + \chi(\nu'_0 a, +\infty).\, Fx$$

$$H_{N+1} = \frac{1}{1+ibf'(x)}\left(-\frac{d^2}{dx^2}\right)\frac{1}{1+ibf'(x)} + \frac{1}{1+ibf'(x)}\{s, x\}\frac{1}{1+ibf'(x)} +$$
$$+ V_p(x + ibf(x)) + \chi(-\infty, \tilde{\nu}_N a)\, F(x + ibf(x)) + \chi(\tilde{\nu}_N a, +\infty)\, F\tilde{\nu}_N a$$

The resolvent $(H(b)-z)^{-1}$ can be linked to the resolvents $R_i = (H_i - z)^{-1}$ by the formula:

$$(H(b)-z)^{-1} = \left(\sum_{i=0}^{N+1} (H_i - z)^{-1}\right)\left(1 + \sum_{i=0}^{N+1} K_i\right)^{-1} ,$$

where: $K_i = \left[-\frac{d^2}{dx^2}, J_i\right] R_i \tilde{J}_i$; $\tilde{J}_i$ is the caracteristic function of the $[m_i, m_{i-1}]$ interval ; J_i are defined in the following manner : the support of J_i is $[\nu'_i a, \nu_{i-1}\, a]$ for i=1 to N, J_i is a C_0^∞ function which takes the value 1 on $[\tilde{\nu}'_i\, a, \tilde{\nu}_{i-1}\, a]$.

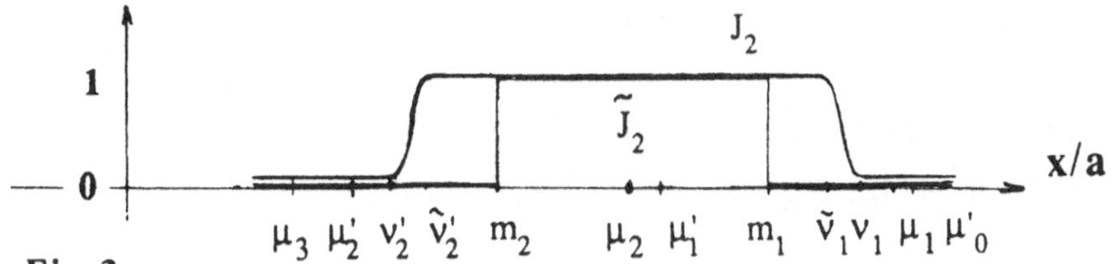

Fig.3

4. Study of the H_i

The aim is to prove that H_i, i = 1, ...N have eigenvalues in a neighborhood of E : secondly to prove that $\sum_{i=0}^{N+1} ||K_i(z)||$ is smaller than 1 if z is at an exponentially small distance of an eigenvalue of some H_i .

In [3] we prove :

- H_i , i = 1, ...N have eigenvalues in a neighborhood of E, separated by Fa $+ O(F^2)$

- If x and y belong to a "forbidden region" the Green function associated to H_i has the following behavior:

$$|G_i(x, y ; E)| \leq c_i \frac{e^{-a \sum_{i=[x]}^{[y]} \kappa(E-Fai)}}{\text{dist}(E,\sigma(H_i))}$$

where κ(E-Fai) is the imaginary part of the crystal momentum k(E - Fai), [x] is the entire part of $\frac{x}{a}$ and c_i is a constant.

As F becomes smaller the supports of $[-\frac{d^2}{dx^2}, J_i]$ and $\tilde{J}_i$ are more distant and $K_i = [-\frac{d^2}{dx^2}, J_i]R_i\tilde{J}_i$ becomes smaller because as x and y are further apart the number of terms in $\sum_{i=[x]}^{[y]} \kappa(E\text{-}Fai)$ increase. So we get $||K_i(E)|| \leq \frac{C_i}{\text{dist}(E,\sigma(H_i))} e^{-\alpha_i/F}$.

Operator H_{N+1} plays a special role, it is the only one affected by the spatial deformation. The easiest way to control K_{N+1} is to prove that H_{N+1} has no eigenvalues in a neighborhood of E which includes some of the H_i eigenvalues. In order to prove this result we need to introduce a parameter ε in the definition of H

$$H(\varepsilon) = -\varepsilon^2 \frac{h_0^2}{2m} \frac{d^2}{dx^2} + V_p(x) + \varepsilon F_0 x$$

where h_0 and F_0 are fixed quantities and $\varepsilon \rightarrow 0$. That is, we take simultaneously the limits: $h \rightarrow 0$ and $F \rightarrow 0$.In [3] we prove the following lemma.

LEMMA

It exists: an interval included in $[E_N(\varepsilon), E'_N(\varepsilon)]$, called $J_N \cap K_N$ in [3] ; $b_0 \in \mathbb{R}$ and ε_0 such that : $D_N = \{z \in \mathbb{C} \mid \text{Re } z \in J_N \cap K_N ; |\text{Im } z| < \varepsilon b_0\}$ is in the $H_{N+1}(\varepsilon)$ resolvent set, if $\varepsilon < \varepsilon_0$.

Main steps of the proof

We construct an operator $\hat{H}_{N+1}(\varepsilon) = H_{N+1}(\varepsilon) + Q(x)$ whose Green function can be explicitly calculated because solutions of $\hat{H}_{N+1}(\varepsilon)\,\phi = E\phi$ are explicit. They are the WKB solutions for $H_{N+1}(\varepsilon)$.

Studying the Wronskian which enters in the Green function we deduce that D_N is in the resolvent set for $\hat{H}_{N+1}(\varepsilon)$.

Using the fact Q(x) decreases at $-\infty$: $\| Q(\hat{H}_{N+1} - E)^{-1} \| \underset{\varepsilon \to 0}{\to} 0$, so that for sufficiently small ε, D_N is in the resolvent set of $H_{N+1}(\varepsilon)$.

Collecting the sparse results we get the following

THEOREM

If F_0 satisfies $F_0 < \frac{\Gamma_N}{a}$ (where Γ_N is the Nth gap width), it exists ε_0 such that if $\varepsilon < \varepsilon_0$ there are at least $N(\varepsilon)$ ladders of resonances for $-\varepsilon^2 \frac{h_0^2}{2m} \frac{d^2}{dx^2} + V_p(x) + \varepsilon F_0 x$; their width is exponentialty small with respect to F_0, being smaller than $\varepsilon^2 F_0\, a\, e^{-\Gamma_N/F_0 a}$.

5. Stark - Wannier resonance states

In this paragraph we show how the resonance states are linked with the so-called Stark- Wannier states.

Let us first recall the Stark-Wannier states definition. If P_n is the spectral projection of H_B on the nth band , operator $H = H_B + \varepsilon F_0 x$ can be written in the form:

$$H = \sum_n P_n (H_B + \varepsilon F_0 x)\, P_n + \varepsilon F_0 \sum_n P_n\, x\, (1-P_n)$$

In the crystal momentum representation (CMR) the Hilbert space is :

$L^2\,(\mathbf{B}\mathrm{x}N) = \left\{ a_n(k) \mid \sum_n \int_{\mathbf{B}} |a_n(k)|^2\, dk < \infty \right\}$ where **B** is the Brillouin zone,

$P_n H_B P_n$ is just the multiplication by $E_n(k)$ and $P_n\, x\, P_n$ is the derivation : $-i\frac{d}{dk}$.

The Stark-Wannier states are the eigenvectors of $P_n\,(H_B + Fx)\,P_n$ .In the CMR their expression is :

$$a^{(w)}_{nj}(k) = e^{\frac{i}{\varepsilon F_0} \int_0^k [E_{n,j} - E_n(k')]dk'}$$

where $E_{n,j}$ are the corresponding eigenvalues :

$$E_{n,j} = \frac{a}{2\pi} \int_0^{\frac{2\pi}{a}} E_n(k)dk + j\varepsilon F_0 a = \langle E_n \rangle + j\varepsilon F_0 a \; ; j \in \mathbb{Z}.$$

So what is called the intraband term : $\sum P_n (H_B + \varepsilon F_0 x) P_n$ has a discrete spectrum constituted , in general, by an infinite number of ladders. These eigenvalues accumulate at least once on an interval of length Fa, and even if $P_n x (1 - P_n)$ is a bounded operator , to get the resonances,starting from the Wannier states by naive perturbation theory , failed.

One way to attack the problem was to insert part of the linear potential in the periodic one [2], this was systematized by A.Nenciu and G.Nenciu [10].We rewrite here their procedure with some slight modifications :

$$H = H_B + \varepsilon F_0[(1 - P_n) x P_n + P_n x (1 - P_n)] + \varepsilon F_0 P_n x P_n + \varepsilon F_0(1 - P_n) x (1 - P_n)$$

Call : $X^{(1)} = P_n x P_n + (1 - P_n) x (1 - P_n)$

Notice that , $(1 - P_n) x P_n + P_n x (1 - P_n) = (1 - 2P_n) [x, P_n]$

and $[x, P_n] = \int_{C_n} [x, \frac{1}{H_B - \zeta}] d\zeta = \int_{C_n} \frac{1}{H_B - \zeta}[x, H_B] \frac{1}{H_B - \zeta} d\zeta = \int_{C_n} \frac{1}{H_B - \zeta} p \frac{1}{H_B - \zeta} d\zeta$

where C_n is a contour in the complex plane around the nth band.

So it is clear that $(1 - 2P_n) [x, P_n]$ commutes with $T_a = e^{i\hat{p}a}$.

As $H_B^{(1)} = H_B + \varepsilon F_0[(1 - P_n) x P_n + P_n x (1 - P_n)]$ commutes with T_a , it is a new Bloch hamiltonian whose energy bands, since $[x,P_n]$ is a bounded operator,are very close to the energy bands of H_B when ε is small. Let us call : $P_n^{(1)}$ the spectral projection of $H_B^{(1)}$.

Now we can repeate the procedure starting from :

$$H = H_B^{(1)} + \varepsilon F_0 X^{(1)} = H_B^{(1)} + \varepsilon F_0\left((1 - P_n^{(1)})X^{(1)} P_n^{(1)} + P_n^{(1)} X_1 (1 - P_n^{(1)})\right) +$$

$$+ \varepsilon F_0\left(P_n^{(1)} X^{(1)} P_n^{(1)}) + (1 - P_n^{(1)}) X^{(1)} (1 - P_n^{(1)})\right)$$

$$= H_B^{(2)} + \varepsilon F_0 X^{(2)}$$

Let us notice that $(1-P_n^{(1)})\,X_1\,P_n^{(1)} = [1-P_n^{(1)}]\,[P_n\,x\,P_n + (1-P_n)\,x\,(1-P_n)]\,P_n^{(1)}$ is $O(\varepsilon)$ because $(1-P_n^{(1)})\,P_n$ is $O(\varepsilon)$. Then in $H_B^{(2)}$, the interband is $O(\varepsilon^2)$.

As, it can be proven that $P_n^{(m)}\,X^{(m)}\,P_n^{(m)}$ is the derivation in the corresponding CMR, at each stage we can define Stark - Wannier states. So to each energy band $E_n^{(m)}(p)$ correspond a ladder of eigenvalues. What we gain, is the fact, the interband term $P_n^{(m)}\,X^{(m)}(1-P_n^{(m)})$ is $O(\varepsilon^{m+1})$.

Now we want to show that eigenvectors of H_i can be well approximated by Stark Wannier states.

First let us calculate $\|(H_i - E_{nj}^{(m)})\,\phi_{n,j}^{(w)(m)}\|$

$$\begin{aligned} H_i &= H_B + \varepsilon F_0 x + \chi(-\infty,\upsilon'_i a)\,\varepsilon F_0(\upsilon'_i a - x) + \chi(\upsilon_{i-1}a, +\infty)\,\varepsilon F_0\cdot(\upsilon_{i-1}a - x) \\ &= H_B^{(m)} + \varepsilon F_0\,X^{(m)} + \chi(-\infty,\upsilon'_i a)\,\varepsilon F_0(\upsilon'_i a - x) + \chi(\upsilon_{i-1}a, +\infty)\varepsilon F_0\cdot(\upsilon_{i-1}a - x) \\ &= P_n^{(m)}\,H_B^{(m)}\,P_n^{(m)} + (1-P_n^{(m)})\,H_B^{(m)}\,(1-P_n^{(m)}) + \varepsilon F_0 P_n^{(m)}\,X^{(m)}\,P_n^{(m)} + \\ &\quad + \varepsilon F_0 P_n^{(m)}\,X^{(m)}(1-P_n^{(m)}) + \varepsilon F_0(1-P_n^{(m)})X^{(m)}\,P_n^{(m)} + \varepsilon F_0(1-P_n^{(m)})X^{(m)}(1-P_n^{(m)}) \\ &\quad + \chi(-\infty,\upsilon'_i a)\,\varepsilon F_0\cdot(\upsilon'_i a - x) + \chi(\upsilon_{i-1}a, +\infty)\,\varepsilon F_0\cdot(\upsilon_{i-1}a - x) \end{aligned}$$

So,
$$\begin{aligned} (H_i - E_{nj}^{(m)})\phi_{n,j}^{(w),(m)} &= \varepsilon F_0(1-P_n^{(m)})X^{(m)}\,P_n^{(m)}\phi_{n,j}^{(w),(m)} + \chi(-\infty,\upsilon'_i a)\,\varepsilon F_0\cdot(\upsilon'_i a - x)\,\phi_{n,j}^{(w),(m)} + \\ &\quad + \chi(\upsilon_{i-1}a, +\infty)\,\varepsilon F_0\cdot(\upsilon_{i-1}a - x)\,\phi_{n,j}^{(w),(m)} \end{aligned}$$

The first term has a norm which is $O(\varepsilon^{m+1})$. Now we have to control the behavior of $\phi_{n,j}^{(w),(m)}$ outside the interval $[\upsilon'_i a, \upsilon_{i-1}a]$.

Calling $w_n^{(m)}(x)$ the nth Wannier function corresponding to $H_B^{(m)}$, the Stark-Wannier state in the space-representation is:

$$\phi_{n,j}^{(SW),(m)} = \sum_l a_n^{(m)}(l)\;w_n^{(m)}(x - la - ja) \quad , \text{ where:}$$

$$a_n^{(m)}(l) = \frac{a}{2\pi}\int_0^{\frac{2\pi}{a}} \left(e^{-ikla - \frac{i}{\varepsilon F_0}\int_0^k E_n^{(m)}(k') - \langle E_n^{(m)}\rangle dk'}\right) dk$$

The Wannier functions are localized, so to study the localization of $\phi_{n,j}^{(SW),(m)}$ it is sufficient to study the decrease of $a_n^{(m)}(l)$ as $|l| \to +\infty$. Write:

$$a_n^{(m)}(l) = \frac{i\varepsilon F_0 a}{2\pi}\int_0^{\frac{2\pi}{a}} \frac{1}{(\varepsilon F_0 la + E_n^{(m)}(k) - \langle E_n\rangle)} \frac{d}{dk}\left(e^{-ikla - \frac{i}{\varepsilon F_0}\int_0^k (E_n^{(m)}(k') - \langle E_n^{(m)}\rangle)dk'}\right)dk$$

and integrate by parts

$$a_n^{(m)}(l) = \frac{i\varepsilon F_0 a}{2\pi}\int_0^{\frac{2\pi}{a}} e^{-ikla - \frac{i}{\varepsilon F_0}\int_0^k (E_n^{(m)}(k') - \langle E_n^{(m)}\rangle)\, dk'} \frac{dE_n^{(m)}}{dk} \frac{1}{[Fla + E_n^{(m)}(k) - \langle E_n\rangle]^2}\, dk$$

Notice that if l is such that $Fla + E_n^{(m)}(k) - \langle En\rangle \neq 0$ there are no singularities in the integrand.

Performing the same trick several times enable us to conclude that $a_n^{(m)}(l)$ decreases as $l^{-N}, \forall N$. So it is clear that terms, $\chi(-\infty, \upsilon'_i a)\, F(\upsilon'_i a - x)\, \phi_{n,j}^{(SW),(m)}$ and $\chi(\upsilon_{i-1} a, +\infty)$. $F(\upsilon_{i-1} a - x)\, \phi_{n,j}^{(SW),(m)}$ are L^2- functions whose norm decrease with ε.

Using the fact , we know, that eigenvalues of H_i are distant by $\varepsilon_0 Fa$ and a variant of the Eckart bound (see in [8],corollary (1.17) and proposition (1.18)), finally we get that the trial function $\phi_{n,j}^{(SW),(m)}$ approach the eigenfunction of H_i as $\varepsilon \to 0$.

REFERENCES

(1) Agler J. and Froese R.(1985),'Existence of Stark ladder resonances' Commun. Math. Phys. 100, 161-171 .

(2) Bentosela F.(1979) 'Bloch electrons in constant electric field' Commun.Math.Phys. 68, 173-182

(3) Bentosela F., Grecchi V.(1989) 'Stark-Wannier ladders' (preprint n°2117,CPT - Marseille)

(4) Bleuse J, Bastard G,Voisin P.(1988) 'Electric field induced localization and oscillatory electro-optical properties of semiconductor superlattices Phys Rev.Lett.Vol.60, N° 3, 220-223.

(5) Briet P., Combes J.M. and Duclos P(1988). 'Spectral stability under tunneling for Schrôdinger operators' .Proceedings of the Holzhau conference on Partial Differential Equations. Teuber -Texte zur Mathematik .

(6) Buslaev V.S., Dmitrieva L.A.(1989) 'Bloch electrons in an external electric field' (preprint Leningrad)

(7) Combes J.M. and Hislop P.(1989), Stark ladder resonances for small electric fields (preprint ,Univ.Kentucky)

(8) Harrell E.(1980) 'Double wells'Commun.Math. Phys. 75, 239-261

(9) Helffer B. and Sjöstrand J.(1985)'Puits multiples en limite semi-classique II' Annales de l'Institut Henri-Poincaré, Vol. 42 , N°2, 127-212.

(10) Nenciu A. and Nenciu G.(1981),'Dynamics of Bloch electrons in external electric fields: I. Bounds for interband transitions and effective Wannier hamiltonians'. J. Phys. A., Math. Gen.14, 2817
II.The existence of Stark-Wannier ladder resonances
J. Phys. A., Math. Gen.15, 3313

(11) Nenciu G.Proceedings of the Poiana-Brasov school (1989)

(12) Voisin P., Bleuse J., Bouche C., Gaillard S., Alibert C., Regreny A.(1988) 'Observation of the Wannier-Stark quantization in a semiconductor superlattice' Phys.Rev.Lett.Vol.61 N° 14, 1639-1642.

(13) Wannier G. (1960) 'Wave functions and effective hamiltonian for Bloch electrons in an electric field' Phys. Rev.117, 432-439 .
(1969) 'Stark ladders in solids?A reply.'Phys. Rev. 181, 1364-1365.

(14) Zak J.(1968)'Stark ladders in solids?' Phys Rev.Lett.Vol.20,1477-1481.
(1969)'Stark ladders in solids? A reply to a reply' Phys. Rev.181, 1366-1367.

SPECTRAL PROPERTIES OF ADIABATICALLY PERTURBED DIFFERENTIAL OPERATORS WITH THE PERIODIC COEFFICIENTS

V. S. BUSLAEV
Institute of Physics
Leningrad University
Petrodworetz, Leningrad, 198904
USSR

ABSTRACT. We describe the asymptotic structure of the spectrum and the asymptotic behavior of the eigenfunctions of the operators

$$Hz = -z_{xx} + p(x)z + v(\varepsilon x)z,$$

p is periodic, $0 < \varepsilon << 1$.

1. INTRODUCTION

In recent years there have appeared a series of works devoted to the asymptotic investigation of the differential equations of the form

$$L(\varepsilon x, x, -id_x)z = 0, \ x \in R^d, \ 0 < \varepsilon << 1. \tag{1.1}$$

The symbol L was considered as a periodic function of the second variable x [1-4]. Such equations have various applications and arise quite naturally in solid state physics. From a mathematical point of view they can be considered as generalizations of the usual semi-classical equations

$$L(\varepsilon x, x, -id_x)z = 0. \tag{1.2}$$

It has been clear that all main constructions, known in the standard semiclassical approach, can be generalized to the equations (1.1). However, the new constructions are not simple repetitions of the standard ones, they are real generalizations of the usual constructions and generate new questions.

In this report the general ideas, worked out for the equation (1.1), will be applied to some concrete spectral problems for ordinary differential equations of the following form

$$-z_{xx} + p(x)z + v(\varepsilon x)z = Ez, \ x \in R, \tag{1.3}$$

A. Boutet de Monvel et al. (eds.), Recent Developments in Quantum Mechanics, 97–112.

the potentials $p(y)$ and $v(x)$ here are supposed to be smooth functions, the function p in addition is supposed periodic, $p(x + a) = p(x)$, $a > 0$ is the period.

In quantum physics the equations of this type can appear when we describe the motion of the Bloch electron in a crystal with an internal periodic field $p(x)$, the crystal assumed to be in an external field $v(\varepsilon x)$.

The spectral properties of the operator H,

$$Hz = -z_{xx} + p(x)z + v(\varepsilon x)z, \qquad (1.4)$$

considered on $L_2(R)$, depend on the behavior of the function $v(y)$ as y tends to infinity. We shall consider three cases,

A. $v(y) \to +\infty$, $y \to \infty$,

B. $v(z) \to 0$, $y \to \infty$,

C. $v(y) = - y$.

We are interested in the asymptotic character of the spectrum of the operator H and in the asymptotic behavior of its eigenfunctions.

In the case A the operator H has the purely discrete simple spectrum the same way as the operator H_1,

$$H_1z = -z_{xx} + v(\varepsilon x)z. \qquad (1.5)$$

The periodic term $p(x)$ has a strong influence on the asymptotic structure of this spectrum. Such operator H can arise in the theory of wave propagation.

In a physical sense the case B corresponds to a local disturbance of the regular structure of the cristal, for example by impurities. The structures of the continuous spectra of the operators H and H_0,

$$H_0z = - z_{xx} + p(x)z, \qquad (1.6)$$

are identical. They are continuous of multiplicity two consisting of some intervals bounded from below. The main question of interest is the structure of the discrete spectrum of the operator H which lies on the gaps. Besides, the operator H has complex resonances and we will discuss their asymptotic behavior.

The case C corresponds to the crystal being plased in the homogenious external field. This problem can be considered as a classical one, it was the subject of numerous investigations. We shall discuss the case C very briefly because now there exists the detailed paper [5], devoted to this case. Other references also can be found in [5].

In the case C the structure of the spectrum of the operator H coincides with the structure of the spectrum of the operator H_1. It means that there is the essential difference between the spectra of the operator H and the operator H_0, which arises if ε is supposed to be 0 in H. In this situation we have to expect the appearence of the resonances of the operator H with nontrivial properties. In fact our aim is the rigorous investigation of these resonances and the corresponding resonance states. In contrast with the previous works , devoted to this subject, we are able to obtain not only some qualitative facts and estimates, but also simple asymptotic formulas for the resonances and the resonance states.

The outline of the report is as follows. In Section 2 we will shortly describe some asymptotic properties of the solutions of the Eq. (1.3). In Sections 3,4,5 we will consider the operator H in the cases A,B,C respectively. Since we cannot present in this report complete proofs we restrict ourselves to description of the results in general terms, their relations and motivations.

I want to express my gratitude to my student L.Dmitrieva who was very helpful during the preparation of the lecture at the Brasov school. Also I would like to thank the organizers of the school for the kind invitation.

2. ON THE ASYMPTOTIC BEHAVIOR OF THE SOLUTIONS OF EQ. (1.3)

2.1. The equations with the periodic potentials

We will reproduce here in short some results of the papers [1-3] which will be used in the following.

First of all we need some general facts about the solutions of the equation

$$-\Phi_{xx} + p(x)\Phi = E\Phi \qquad (2.1)$$

with the purely periodic potential p. This equation admits solutions of the following form:

$$\Phi(x,k) = \exp(ikx)\phi(x,k),\ \phi(x + a,k) = \phi(x,k),\ E = E(k). \qquad (2.2)$$

The solutions Φ are called Bloch solutions, the function E is called dispersion function, the parameter k is called quasimomentum. The latter is a complex number, more precisely, it is a point of some single sheet's Riemann surface K. Positions of branching points b_l, b_l^*, $b_l = k_l + ih_l$, , $k_l = l\pi/a$, $h_l \geq 0$, $l \in Z$, on the surface depend on the potential p. The branching points b_l and b_l^* have to be connected by the cut $c_l = k_l + it$, $-h_l \leq t \leq h_l$. The functions $\Phi(x,.)$ and E are analytical functions on the surface K with the indicated cuts. The function E has real values on the real axis, on the imaginary axis and on the edges of the cuts. Let us introduce the points $\sigma_0 = 0$, $\sigma_1 = k_1 - 0$, $\sigma_2 = k_1 + 0$, $\sigma_3 = k_2 - 0$, $\sigma_4 = k_2 + 0, \ldots$ on the surface K. The function E, restricted on the interval $[\sigma_{m-1},\sigma_m]$, is a strictly monotonically increasing function. The images $\delta_l = [E_{m-1},E_m]$ of the intervals $[\sigma_{m-1},\sigma_m]$, $m = 2l-1$, $l = 1,2,\ldots$, constitute the spectrum of the operator H_0. The spectral intervals are divided by the gaps $\underline{\delta}_l = (E_m,E_{m+1})$. The spectrum of the operator H_0 is continuous of the multiplicity two and the functions $\Phi(x,k)$, $\Phi^*(x,k)$, $k \in R_+$, can be choosen as the eigenfunctions.

The function E, restricted on the interval $[\sigma_{m-1},\sigma_m]$, can be analytically continued on the whole real axis R. After this continuation we obtain an even smooth periodic function $\mathcal{E}_1$ on the axis with the period $a^* = 2\pi/a$. The curve representing the dependence between $E = \mathcal{E}(k)$ and $\mathrm{Im}(k)$ on the loop, enveloping the edges of a cut c_1, is a smooth oval.

2.2. The asymptotic behavior of the solutions of Eq. (1.3)

Let us consider the so called isoenergy curve,

$$\mathcal{E}(k) + v(y) = E, \tag{2.3}$$

where $E \in R$ is a fixed parameter. The isoenergy curve is considered as a curve on the plane $\{(y,k): y \in R, k \in K\}$. An interval of the isoenergy curve

$$\gamma = \{(y,k): k = k(y), y \in \delta = (a,b)\}$$

is called a regular branch of the isoenergy curve if $k \in C^\infty(\delta)$. With any regular branch γ it is possible to associate a formal solution f of the Eq. (1.3), see [1]. The main term f_0 of this solution is given by the formula:

$$f_0 = |\mathcal{E}_k(k)|^{-1/2}\,\phi(x,k)\exp i\int_{\gamma(y)} \Omega\ , \quad \Omega = 1/\varepsilon\, k\,dy + \Omega_1,$$

$$\Omega_1 = \langle d\phi(.,k),\phi(.,-k)\rangle\,[\langle\phi(.,k),\phi(.,-k)\rangle]^{-1},$$

$$d\phi(.,k) = \phi_k(.,k)dk,\quad \langle f,g\rangle = \int_0^a f(x)g(x)dx,$$

$$k = k(y),\ \gamma(y) = \{(y',k): (y',k) \in \gamma,\ y' \in (a,y)\}.$$

In the following we will suppose for simplicity that the function v consists of a finite number of strictly monotonous branches divided by nongenerated critical points. Consider the graphs of the functions

$$E = E_m + v(y),\ m = 0,1,\ldots$$

Each strip lieing between two neighbouring graphs corresponds to the spectral interval or the gap. The y-coordinates of the intersection points of the line $\mathcal{E} = E$ with the indicated graphs are called turning points. They are determined by the equations

$$E_m + v(y) = E.$$

The turning points separate the projections of the real and complex branches of the isoenergy curve to y-axis, on the real branches $\mathrm{Im}(k) = 0$, on the complex branches $\mathrm{Im}(k) \neq 0$.

There are three general cases.

1) Let two turning points bound a real branch and correpond different neighbouring numbers m - 1 and m. In fact in this case there are two separate unclosed branches bounded by the same turning points. They can be combined in a simple closed real curve γ on a cylinder P by means of identification of the points $(y,k + na^*)$, $n \in Z$. The curve γ surrounds the cylinder P. It can be treated as a period of the periodic curve

$$E_1(k) + v(y) = E, \; m = 21 - 1.$$

2) Let two turning points bound a real branch and correspond a same number m. Again in this case there are two unclosed separate branches bounded by the turning points, which can be combined in a simple closed real curve γ on the cylinder P. This curve does not surround the cylinder.

3) Let two turning points bound a complex branch. There are two separate closed complex isoenergy curves bounded by the turning points, on the first of them $Re(k) = k_1$, on the second one $Re(k) = -k_1$. They can be mathched each other by means of identification of the points $(y,k+ na^*)$, $n \in Z$. In this case we will receive a simple closed curve γ on the direct product P*R of the points $\{(Y,k'): Y \in P, \; k' = Im(k)\}$, on the curve γ $Re(k) = const$.

All described curves γ are invariant with respect to the transformation $k \to -k$.

2.3. Quantization conditions

The solution f corresponding to the real branch has the same order in the different points y as $\varepsilon \to 0$. In the case of a complex branch this order differs in the different points y exponentially.

In some asymptotically small vicinity of the isolated turning point the formal solution f must be changed to a more complicated formal solution, see [1].

Consider some interval (α,β) bounded by the turning points and covered by the real branch of the isoenergy curve. Using the mentioned complicated formal solutions it is possible to construct linear combinations f_+ and f_- of the solutions f and f^* which after continuation through the vicinities of the points α and β respectively will become exponentially small as $\varepsilon \to 0$.

The solutions f_+ and f_- coincide up to a constant factor if certain conditions, which might be called quantization conditions, are fulfilled. These conditions are of the form

$$1/\varepsilon \int_\gamma k dy + \varepsilon \Omega_1 + \ldots = 2\pi n + ind(\gamma), \; n \in Z. \qquad (2.4)$$

Here γ denotes the closed branch of the isoenergy curve, described in the previous subsection, $ind(\gamma)$ denotes the Maslov index of the curve γ. It is not difficult to see that in the cases 1) and 2) of the previous subsection $ind(\gamma)$ equals 0 and 2 respectively.

The quantization conditions can be regarded as a restriction imposed to the spectral parameter E. The values of the spectral parameter obeying the quantization conditions can be understood in another way. They can provide the asymptotic description of the eigenvalues or the resonances of the operator H. For such E the corresponding formal solutions $f_+ = f_-$ give the asymptotic description either of the eigenfunctions or of the resonance states as $\varepsilon \to 0$. In a natural sense their supports lie essentially in the interval (α,β).

Turning again to the linear combination f_+ we continue this solution through the interval (β,α_1), covered by the complex branch γ_1 of the isoenergy curve, to the next interval (α_1,β_1), covered by the real branch. The relative order of the solution in the interval (α_1,β_1) is $\exp(-1/2\ \mu)$,

$$\mu = 1/\varepsilon \int_{\gamma_1} \mathrm{Im}(k)dy. \tag{2.5}$$

Such a type of the continuation of the solution can be called a tunnelling.

3. ON THE ASYMPTOTIC BEHAVIOR OF THE SPECTRUM OF H IN THE CASE A

3.1. Contribution of a separate branch

If $v(y) \to +\infty$ as $y \to \infty$ the spectrum of the operator H is simple and purely discrete. In this case the periodic term $p(x)$ plays role of a perturbation to the operator H_1. This perturbation has a substantial influence on the structure of the spectrum.

Let us suppose that $v(0) = 0$ and that $v(y)$ is strictly decreasing, when $y < 0$, and strictly increasing, when $y > 0$. In addition we suppose that $y = 0$ is a nondegenerated critical point.

We point out that the operator H_1 has the standard semiclassical form, i.e., the solutions of the equation

$$-z_{xx} + v(\varepsilon x)z = Ez \tag{3.1}$$

can be described asymptotically in terms of the standard semiclassical constructions. In other words, we have to introduce the standard isoenergy curve γ,

$$k^2 + v(y) = E. \tag{3.2}$$

It has a non-trivial real branch if $E > 0$, in this case the curve γ is an oval. We have to select the ovals γ obeying the quantization conditions

$$1/\varepsilon \int_{\gamma} kdy + O(\varepsilon) = 2\pi n + \pi. \tag{3.3}$$

These conditions determine asymptotically the eigenvalues of H_1. The supports of the corresponding eigenfunctions coincide asymptotically with the projections of the ovals to the y-axis.

Switching the perturbation p(x) in transforms the procedure.

In this case each spectral interval $\delta_1 = [E_{m-1}, E_m], m = 2l-1$, of the operator H_0 yields a branch of the discrete spectrum of the operator H. The spectral branch lies in the interval $[E_{m-1}, \infty)$. If $E \in \delta_1$ the supports of the corresponding eigenfunctions are some single intervals containing the point $y = 0$. If $E > E_m$ the supports are pairs of intervals which do not contain the point $y = 0$, one of them lies on the semi-axis R_-, other lies on the semi-axis R_+.

Let us introduce the turning points which are corresponded to a given number m, $E_m + v(y) = E$. If $E < E_m$ such points are absent, if $E = E_m$ the only turning point is 0, and there are two such points: $y^-_m < 0$ and $y^+_m > 0$ if $E > E_m$.

Now let us introduce the strip s_1 contained between two curves

$$E = E_{m-1} + v(y), \ E = E_m + v(y) \tag{3.4}$$

on the plane $\{(y,E)\}$. In the natural sense the strip s_1 is generated by the interval δ_1, see 2.2. Let us consider the real isoenergy curves γ connected with s_1. If $E < E_{m-1}$ such curves are absent. It means that the spectrum of H which can be connected with s_1 is absent if $E < E_{m-1}$. If $E = E_{m-1}$ the isoenergy curve is degenerating and given by the point $(y = 0, k = k_1)$. We can say that the point E_{m-1} is the lower bound of the part of the spectrum of the operator H connected with s_1. In some asymptotically vanishing vicinity of the point E_{m-1} the asymptotic behavior of the eigenvalues and the corresponding eigenfunctions can not be described in the terms of the above constructions. To give the asymptotic behavior in this small vicinity we need some additional notions which can be found in [3]. We have no possibilities to discuss these detailes in the present short report.

If $E_{m-1} < E < E_m$ there is a closed isoenergy curve γ for each E, it has the type 2), see 2.2, and its projection to the y-axis is bounded by the points y^-_{m-1} and y^+_{m-1}. The further reasons are almost evident. Following the example of Eq. (3.1) we have to select the curves γ which obey the quantization conditions (2.4). These conditions determine asymptotically eigenvalues E_n of H. The corresponding eigenfunctions z_n are given by the expressions f_+ ($= f_-$), their supports coincide asymptotically with the intervals $[y^-_{m-1}, y^+_{m-1}]$.

The case $E = E_m$ is transitional, in small vicinity of the point E_m we have to use again the constructions mentioned above in connection with the case $E = E_{m-1}$.

If $E > E_m$ the isoenergy curve consists of two closed curves γ_- and γ_+, surrounding the cylinder P, type 1) in 2.2. Their projections are bounded by the points y^-_{m-1}, y^-_m and y^+_m, y^+_{m-1} respectively. Each of these curves have to be considered independently. As a result we receive two sets E^-_n and E^+_n of eigenvalues of H which asymptotic behaviors are given by the quantization conditions

$$1/\varepsilon \int_{\gamma_-} k dy + \varepsilon\Omega_1 + \ldots = 2\pi n, \tag{3.5}$$

$$1/\varepsilon \int_{\gamma_+} k dy + \varepsilon\Omega_1 + \ldots = 2\pi n, \; n \in Z. \tag{3.6}$$

The asymptotical supports of the corresponding eigenfunctions z^-_n and z^+_n are the intervals $[y^-_{m-1}, y^-_m]$ and $[y^+_{m-1}, y^+_m]$ respectively.

So we have received the results described in the beginning of the subsection.

3.2. General picture of the spectrum

To receive the total structure of the asymptotic behavior of the spectrum we have to consider the natural union of the contributions generated by all spectral intervals δ_1.

In fact the last statement is a *theorem* but for the precise formulating we would have to add some essential comments.

For the general values of the spectral parameter E the total isoenergy curve,

$$\mathcal{E}(k) + v(y) = E, \tag{3.7}$$

treated as a curve on the cylinder P, consists of some finite number of the separate closed real curves γ. These curves have the separate projections on the y-axis divided by the gaps. Above each gap there lies the complex isoenergy curve γ of the type 3), see 2.2.

The quantization conditions for each real curve γ give the description of the asymptotic behavior of the corresponding eigenvalues which is correct in all power orders with respect to ε. If all these asymptotic series are different they give the exhaustive description of the asymptotic properties of the eigenvalues and, as result, they lead to the exhaustive description of the eigenfunctions. However, if some of these series are equal we have to take into account the tunnelling effects to describe their splitting. For example, such a situation arises if the potential v(y) is an even function. The effect of the tunnelling can lead to displacement of the eigenvalues along the real spectral axis on exponentially small values. To estimate this displacement we have to solve a system which takes into account all chains of the tunnelling through the gaps and the transitions through the projections of the real curves γ. It is not difficult to write down the system. Of course, the main correction is determined by the tunnelling through the single gaps: for the gap, covered by the compex curve γ, the exponential order of the correction is given by the factor $\exp(-\mu)$, see (2.5), for a chain containing some number of neighbouring gaps the exponential order of the correction is given by the product of the such type factors. Sometimes it is impossible to limit the consideration by the single gaps: to estimate the main order of the displacement of the eigenvalues E^-_n and E^+_n in the case of the even potential v(y) we have to take into account the whole direct chain between the corresponding real curves γ_- and γ_+.

The principal consequence of the tunnelling is the non-uniformity of the asymptotic formulas with respect to the parameter E. The fact is that the lengths of the spectral gaps $\underline{\delta}_l$ go to 0 as $l \to \infty$. As result the corresponding factors $\exp(-\mu_l)$ tend to 1. It means that the quantization conditions for the real isoenergy curves γ_l with the great numbers l can not be considered independently and the splitting of the eigenvalues in the different series, connected with the single curves γ_l, becomes invalid. However, we can investigate the case of the large E using the asymptotic information on the dispersion function $E(k)$ and the Bloch solutions $\Phi(x,k)$ as $k \to \infty$. For example, in the main order $E(k) = k^2 + \ldots$

So the eigenvalues E_n have to be described as functions of two parameters, n and ε, and the object of investigation in this situation is the asymptotic behavior of the function of two variables $E_n(\varepsilon)$ as $n \to \infty$ and $\varepsilon \to 0$. We can say that for sufficiently small ε and $n \to \infty$ the asymptotic behavior of this function coincides in the main order with the asymptotic behavior of the eigenvalues of the operator H_1. We will not discuss the situation in more detail here.

Concerning the asymptotic behavior of the eigenfunctions of the operator H we will give only the following result. Denote by $P_l(\delta)$ the spectral projector of the operator H with respect to the interval δ and to l-th branch of the eigenvalues, let f and g are smooth finite functions of x. Then

$$(P_l(\delta)f,g) \to (P_0(\delta \cap \delta_l)f,g) \tag{3.8}$$

as $\varepsilon \to 0$. Here P_0 is the spectral projector of H_0.

4. ON THE ASYMPTOTIC BEHAVIOR OF THE SPECTRUM OF H IN THE CASE B

4.1. Contribution of a separate branch, the first case

Let v(y) be a decreasing function as $y \to \infty$ satisfying the condition

$$|v^{(r)}(y)| \leq C\, |y|^{-c-r}\ , \ c > 0,\ r = 0,1,\ldots \tag{4.1}$$

It appears that the roles of the potentials p(x) and v(y) are reversed in a sense in comparison with the previous case: the operator H_0 has to be regarded now as the unperturbed one while the potential $v(\varepsilon x)$ must be treated as a perturbation. The operators H and H_0 have identical essential spectra which are continuous and twice degenerated in the intervals δ_1, δ_2, ... In addition the operator H can have eigenvalues in the gaps $\underline{\delta}_0 = (-\infty, E_0)$, $\underline{\delta}_1$, $\underline{\delta}_2, \ldots$

Let us specify the assumptions concerning v. We will assume that it has only one non-generating critical point at y = 0 such that $v(0) = -v_0 < 0$ and, moreover, that it is a strictly de(in)creasing function if y < 0 (> 0). Sometimes we will suppose additionally that the potentials p and v are analytical functions in strips surrounding the real axis.

Again each spectral interval $\delta_1 = [E_{m-1}, E_m]$, $m = 2l - 1$, of the operator H_0 yields a branch of the spectrum of H. In order to understand the structure of this branch we have to introduce the corresponding strip (3.4) on the plane $\{(y,E)\}$. The positions of the turning points can be determined as the y-coordinates of the points of the intersections of the line $E = E$ and the boundaries of the strip. So the considered spectral branch lies on the interval $[E_{m-1} - v_0, E_m]$. The structure of the isoenergy curves depends on the correlation of v_0 and $|\delta_1| = E_m - E_{m-1}$.

Let $v_0 < |\delta_1|$. If $E < E_{m-1} - v_0$ the real isoenergy curve is absent and , as a result, the corresponding part of the spectrum of H is absent too. If $E = E_{m-1} - v_0$ we have the transitional case. If $E_{m-1} - v_0 < E < E_{m-1}$ the real part of the isoenergy curve is an oval γ_1 on the cylinder P lieing above the interval $[y^-_{m-1}, y^+_{m-1}]$, see the case 2) in 2.2. The corresponding branch of the spectrum consists of discrete eigenvalues E_n asymptotic formulas of which can be obtained from the quantization conditions. The asymptotical support of the eigenfunction z_n is exactly $[y^-_{m-1}, y^+_{m-1}]$.

It is worth to note that the curve γ_1 expands to infinity as $E \to E_{m-1}$, i.e., $y^-_{m-1} \to -\infty$, $y^+_{m-1} \to +\infty$. The case $E = E_{m-1}$ is also transitional. Let $E_{m-1} < E < E_m$. In this situation the isoenergy curve γ_1 becomes unbounded. Infiniteness of the isoenergy curve means that the corresponding part of the spectrum of H is continuous.

The structure of the infinite isoenergy curve is quite different in two cases: $E < E_m - v_0$ and $E > E_m - v_0$, the case $E = E_m - v_0$ is transitional. If $E < E_m - v_0$ the isoenergy curve γ_1 on the cylinder P consists of two separate parts γ^+_1 and γ^-_1, both of which cover the whole y-axis, and they can be matched each other by the transformation $k \to -k$. The quantization conditions lose their force and for *each* E there are two eigenfunctions of the continuous spectrum which correspond two separated parts of γ. The supports of γ^-_1 and γ^+_1 are the whole axis R.

If $E_m - v_0 < E < E_m$ the isoenergy curve also consists of two parts γ^-_1, γ^+_1 but in this case they do not cover the y-axis and there exists the gap $[y^-_m, y^+_m]$ between them. Both of these parts are closed and invariant with respect to the transformation $k \to -k$. We can again introduce two eigenfunctions of the continuous spectrum of H for each E but now their supports are $(-\infty, y^-_m] \cup [y^+_m, +\infty)$. As $E \to E_m$ the isoenergy curve tends to infinity, the case $E = E_m$ is transitional and for $E > E_m$ the real isoenergy curve is absent such that the spectrum of H is absent asymptotically if $E > E_m$.

So the asymptotical structure of the spectrum is clear: as in the previous section we can describe the total spectrum of the operator H joining the contributions of the separate spectral intervals δ_1. But now the influence of the joining process on the structure of the total spectrum is not trivial. We will discuss this process later. Nevertheless let us note here that the continuous parts can be combined without any peculiarities.

With the infinite isoenergy curve and with the corresponding part of continuous spectrum of H we can connect the scattering picture. It can be described by the unitary scattering matrix $s(E)$, $E \in \delta_1$,

$$s(E) = \{ s_{ij}(E) \}_{i,j=1,2}. \tag{4.2}$$

The basis in the unproper eigenspace of H for the given $E \in \delta_1$ can be naturally choosen such that s_{11} and s_{22} will have the sense of the transition coefficients and s_{12}, s_{21} will have the sense of the reflection coefficients. It is not difficult to describe their asymptotic behaviors as $\varepsilon \to 0$.

If $E < E_m - v_0$ the reflection coefficients asymptotically are equal 0:

$$s_{12}, s_{21} = O(\varepsilon^\infty); \ |s_{11}|, |s_{22}| = 1 + O(\varepsilon^\infty). \tag{4.3}$$

On the contrary if $E > E_m - v_0$ the transition coefficients are equal 0:

$$s_{11}, s_{22} = O(\varepsilon^\infty); \ |s_{12}|, |s_{21}| = 1 + O(\varepsilon^\infty). \tag{4.4}$$

The terms $O(\varepsilon^\infty)$ in these formulas can be estimated more exactly: in both cases they are given by the tunnelling through the gaps separating two parts of the isoenergy curve γ_1. In the case (4.3) the corresponding complex isoenergy curve lies in the domain of complex values of the variable y, in the case (4.4) this curve lies in the domain of complex values of the variable k.

4.2. Contribution of a separate branch, the second case

The spectral picture in the case $v_0 > |\delta_1|$ is rather different. Neglecting the transitional points we have to consider three following situations: $E \in (E_{m-1} - v_0, E_m - v_0)$, $E \in (E_m - v_0, E_{m-1})$, $E \in (E_{m-1}, E_m)$. In the first the isoenergy curve γ_1 on the cylinder P is an oval, in the second situation there are two separate isoenergy curves surrounding the cylinder, in the third case there are two infinite curves divided by the real gap. The first two situations describe asymptotically the discrete points of the spectrum of H, the latter describes the continuous part of the spectrum.

On the continuous part of the spectrum, i.e., on the interval δ_1 we can again consider the scattering matrix $s(E)$, as before it is a matrix of the second order. Since there exists a real gap between the projections of two branches of the unbounded isoenergy curve on the y-axis their elements have the asymptotic property (4.4).

4.3. General picture of the spectrum

In this case the simple unification of the contributions, generated by all spectral intervals δ_1, can not give the total picture of the spectrum of H. The reason is almost evident. Indeed on the intervals δ_1 we can construct the asymptotic formulas for the total system of two independent eigenfunctions of the continuos spectrum and since the differential equation for the eigenfunctions has the second order the discrete spectrum can not lie in the intervals δ_1. But if the parameter v_0 is sufficiently large the discrete points generated by the interval δ_1 lie partially in the interval δ_{1-1} which is an interval of the continuous spectrum of H. So there is a paradox.

The fact is that the tunnelling interaction between the closed finite isoenergy curve γ_1, generating the set of discrete points, and the infinite isoenergy curve γ_{1-1}, generating the continuos spectrum, shifts the discrete points in the complex spectral plane and transforms them in the resonances. On the contrary the interaction between different finite isoenergy curves shifts the discrete points only along the real spectral axis.

The resonances of the operator H can be defined using the form

$$F(E) = (R(E)f,g), \; R(E) = (H - E)^{-1}. \qquad (4.5)$$

Let us suppose that f and g are finite functions then under our suppositions about the analytical properties of p and v the function F(E) admits the analytical continuations through the intervals δ_{1-1} from the upper half-plane in some strips under the intervals in the lower half-plane and inversely from the lower half-plane in some strips in the upper one. The resonances of H can be defined as the poles of these continuations, they do not depend on the functions f and g and the set of the resonances is symmetrical with respect to the real axis.

In general we can describe only a subset of resonances of H, their imaginary parts can be estimated asymptotically in terms of the tunnelling between the corresponding finite isoenergy curves and the infinite isoenergy curves through the natural direct chains of the transitions and the tunnellings. The asymptotical orders of the imaginary parts are exponentially small and given by the same tunnelling factors as the shifts of the real eigenvalues.

As in the previous case the other principal consequence of the tunnelling is the non-uniformity of the asymptotic formulas with respect to the parameter E. But we can overcome this difficulty by the same way using the asymptotic formulas for the dispersion function and for the Bloch solutions as $k \to \infty$.

4.4. Some remarks

Let us return to the question about the asymptotic behavior of the eigenvalues E_n of a given branch. We restrict our discussion to the case $v_0 < |\delta_1|$. If in (4.1) $c > 2$ then the whole area bounded by the infinite isoenergy curve for $E = E_{m-1}$ is finite and the total number of the eigenvalues E_n in the interval $[E_{m-1} - v_0, E_{m-1}]$ is also finite. In the case $c < 2$ this area is infinite and the number of eigenvalues E_n is infinite. The main term of their asymptotic behavior as $n \to \infty$ is the same as for the operator

$$E_{m-1} - 1/2m_{m-1}\, d_x^2 + v(\varepsilon x),$$

where m_{m-1} is the parameter from the representation

$$E(k) = E_{m-1} + 1/2m_{m-1}(k - \sigma_{m-1})^2 + \dots, \quad k \to \sigma_{m-1}.$$

Our last remark is concerned the asymptotic behavior of the eigenfunctions. The main result says that

$$(P_1(\delta)f,g) \to (P_0(\delta\cap[E_{m-1}-v_0,E_m-v_0]f,g), \tag{4.6}$$

as $\varepsilon \to 0$. The unproper spectral projector P_1 includes the contributions of the spectral points and the resonances of H which can be connected with the spectral interval δ_1.

5. ON THE ASYMPTOTIC BEHAVIOR OF THE SPECTRUM OF H IN THE CASE C

5.1. Previous notions

In this section the potential $v(y)$ is supposed to be of the special form: $v(y) = -y$.

The spectrum of the operator H coincides with the spectrum of the operator H_1, i.e., it is simple continuous on the whole axis. In the following we will suppose that

$$\int_0^a p(x)dx = 0, \tag{5.1}$$

it does not restrict generality of the consideration. Let us introduce the kernel $R(x,x';E)$ of the resolvent $R(E) = (H - E)^{-1}$ of the operator H. Above of the real spectral axis, $Im(E) > 0$, it admits the representation

$$R(x,x';E) = \begin{cases} M^{-1}(E)\, f(x,E)\, g(x',E), & x < x', \\ M^{-1}(E)\, g(x,E)\, f(x',E), & x' < x. \end{cases} \tag{5.2}$$

In the formula (5.2) the functions $f(x,E)$ and $g(x,E)$ are solutions of the differential equation

$$-z_{xx} + p(x)z + v(\varepsilon x)z = Ez, \tag{5.3}$$

which are from $L_2(R_-)$ and $L_2(R_+)$ respectively. The function M is given by the known formula

$$M(E) = -\{f(x,E)g_x(x,E) - f_x(x,E)g(x,E)\}. \tag{5.4}$$

It is possible to normalize the function f such way that it will be an analyical function of E in the whole plane. The function g can be constructed as an analytical function of E in the upper plane E. If the potential p is an analytical function in the strip $|Imp(y)| < q$, $q > 0$, then the solution g admits the analytical continuation in the half-plane $Im(E) > -\varepsilon q$. The following will be supported on this condition. The denominator $M(E)$ can be assumed to be an analytical function of E in the half-plane $Im(E) > -\varepsilon q$ and a periodic function with the period εa. We can suppose that it has no roots if $Im(E) \geq 0$ but it can have roots in the strip $-\varepsilon q < Im(E) < 0$. Each root E_n generates the whole periodic series of roots: $E_n \to E_n + n\varepsilon a$. They are the resonances of the operator H. In the half-plane $Im(E) > -\varepsilon q'$, $q' < q$, the function M can be represented as the product

$$M(E) = M_1(E) \prod_{n \in N} [\exp(2\pi/\varepsilon a\,(E - E_n) - 1], \qquad (5.5)$$

where the function M_1 has no roots and N is a set of roots in the fundamental domain: $-\varepsilon q' < Im(E) < 0$, $0 \leq Re(E) < \varepsilon a$.

The scattering matrix in this case is reduced to the reflection coefficient $r(E)$,

$$r(E) = M(E)^*/M(E). \qquad (5.6)$$

It is also a periodic function and has the property: $|r(E)| = 1$.

All above facts of this subsection do not depend on the value of ε if $\varepsilon > 0$.

5.2. Contribution of a separate branch

The initial objects of the asymptotical constructions are a spectral interval δ_1 and the corresponding strip bounded two lines:

$$E = E_{m-1} - y, \; E = E_m - y \qquad (5.7)$$

on the (y,E) -plane. It is clear that in this case for each E we have only the single closed isoenergy curves γ_1 of the type 1). The quantization conditions (2.4) can be written down in more explicit form:

$$E = a/2\pi \int_{\gamma_1} [E_1(k)dk - \varepsilon\, \Omega_1] + \varepsilon a n + \dots, \; n \in Z. \qquad (5.8)$$

It means that the quantization conditions lead to a periodic ladder in the E-axis. In the physics literature such ladders are called Stark-Wannier ladders. The omitted terms have the ε^2-order, in principle we can give the whole asymptotic series instead of them, it does not depend on n.

5.3. General picture of the spectrum

So for each spectral interval δ_1 of the operator H_0 we can construct the periodic Stark-Wannier ladder. What is the spectral sense of these ladders? It becomes clear only after the unification of the contributions of the different spectral intervals δ_1. Their interaction transforms each of these real ladders to the complex periodic ladder with the exponentially small imaginary part.

To describe the picture in more detail let us introduce an additional assumption: namely let us suppose that the operator H_0 has only a finite number of the non-trivial spectral intervals δ_1. A spectral interval will be called non-trivial if its length is not equal 0. It is well known that such set of potentials can be described by some explicit formulas but we will not use these formulas here. We have to indicate that for such potentials one of the spectral intervals δ_N becomes infinite: $\delta_N = [E_{2N-2},\infty)$.

For the potentials of this class we can construct only the finite number of the periodic ladders. For the infinite spectral interval δ_N the corresponding isoenergy curves are also infinite and asymptotically describe the continuous spectrum of the operator H and the scattering.

We can claim that the operator H has *exactly* N periodic chains of the resonances and their asymptotic behavior as $\varepsilon \to 0$ is described by the corresponding Stark-Wannier ladders up to exponentially small complex terms.

However it should be mentioned that this proposition is true not for arbitrary small ε, the parameter ε can tend to 0 only along special subsequences.

Concerning the imaginary parts of the resonances we have to say that for each ladder they are estimated by the tunnellings and the transitions through the direct chain of the complex and real isoenergy curves connecting the considered real finite curve γ_1 and the infinite real curve γ_N. So the orders of the imaginary parts are exponentially small and can be characterized as the products of the corresponding tunnelling factors $\exp(-\mu_{1'})$, $1 \leq 1' \leq N-1$.

The following asymptotical formula is valid also:

$$(P_1(\delta)f,g) \to (P_0(\delta\cap\delta_1)f,g) \qquad (5.9)$$

as $\varepsilon \to 0$. The unproper spectral projector P_1 includes only the contributions of the resonances from the lower half-plane belonged to the 1-th ladder.

6. REFERENCES

We give only a very short list of papers, the extensive ,but not complete, list of works can be recovered with a help of [3-6].

1. Buslaev V.S. (1984) 'Adiabatic perturbation of a periodic potential', Teor. Mat. Fiz. <u>58</u>, 233-243
2. Buslaev V.S., Dmitrieva L.A. (1987) 'Adiabatic perturbation of a periodic potential. II', Teor. Mat. Fiz. <u>73</u>, 430-442
3. Buslaev V.S. (1987) 'A semiclassical aproximation to the equation with periodic coefficients', Sov. Math. Uspekhi, <u>42</u>, 77-96
4. Guillot J.C., Ralston J., Trubovitz E. (1988) 'Semiclassical asymptotics in solid state physics', Comm. Math. Phys., <u>116</u>, 401- 415

5. Buslaev V.S., Dmitrieva L.A. (1989) 'Bloch electron in an external field', Algebra and analysis, 1, N2, 1-29
6. Buslaev V.S., Dmitrieva L.A. (1989) 'Bloch electrons in an external electric field' in P. Exner, P. Seba (eds.), Schrodinger operators, standard and nonstandard, World Scientific, Singapore, N. J., L., Hong Kong, pp. 101-129

QUANTUM TUNNELLING FOR BLOCH ELECTRONS IN SMALL ELECTRIC FIELDS

J.M. Combes (*)
and
P. Hislop (**)

Abstract : We consider the quantum Hamiltonian describing a one dimensional particle in a periodic potential plus constant electric field and show that it exhibits an infinite ladder of resonances in the semi-classical regime. The lifetime of these resonances is related to tunnelling phenomena. This provides in this particular case a positive answer to the question of the existence of Stark Ladders.

Mars 1990
CPT-90/P.2371

(*) and PHYMAT, Departement de Mathématiques, Université de Toulon et du Var, La Garde, FRANCE.

(**) Permanent address : Mathematics Department, University of Kentucky, Lexington, KY 40506, USA

A. Boutet de Monvel et al. (eds.), Recent Developments in Quantum Mechanics, 113–131.

I INTRODUCTION

Consider an electron moving in a periodic potential and submitted to an electric field Fx. One expects that after tunnelling through a finite number of potential barriers, the number of which might be quite large if F is small, the electron will escape to infinity under the action of the electric field. This is supported by mathematical results about the absolute continuity of the spectrum of the corresponding Hamiltonian and existence of suitable wave-operators. Due to the long time needed for tunnelling and in analogy with the usual Stark effect in atoms one expects in addition that resonances should occur. Then due to the periodicity of the potential it is easy to show that such resonances form infinite ladders of the form $\{E_0 + F\gamma, \gamma \in \Gamma\}$ where E_0 has non zero imaginary part and Γ is the lattice group. The problem of the existence of such ladders has been the source of controversy in the solid state physics literature for about two decades ([Z]) and rigourous arguments in favour of their existence have been given ([AF], [B], [BD], [N]). In this talk we present some new results ([CH]) which use two sets of techniques recently developped in the semi-classical analysis of Schrodinger Operators. One concerns rigorous quantitative estimates of tunnelling in multiple well potentials (see below) and the other concerns the theory of shape-resonances ([CDKS], [HS1], [HiSi], [Si1]) which has already been successfully applied to the Stark effect in atoms ([Si2]). In the case of Bloch electrons some intrinsic new difficulties appear due to the fact that for small electric fields the electron has to tunnel through a large number of barriers before the escape, some of them being small when the electron gets close to the escape, while some of the intermediate wells can be "resonant" in a sense described below, thus producing oscillations in electronic motion. In comparison Stark effect in a one center Coulomb potential only has one large barrier at small electric field. To cope with these difficulties we consider only the one dimensional case in the semi-classical regime fo $g^2 = h^2/2m$ small. It has to be mentionned that it is precisely in one dimensional structures, the superlattice devices, that the first experimental evidence for existence of Stark ladders has recently been obtained [BBV].

So the quantum Hamiltonian is

$$H(g, F) = - g^2 \frac{d^2}{dx^2} + \mathcal{V}(x) + Fx \tag{1}$$

acting on $L^2(\mathbb{R})$. Here $\mathcal{V}$ is a real periodic potential with period τ satisfying :

(A1) $\mathcal{V}$ is restriction to $\mathbb{R}$ of a function $\mathcal{V}(z)$ analytic in a strip $|\mathrm{Im}\, z| < b$ for some $b > 0$.

We also assume that F is not too large in the following sense

(A2) $0 < F < ||\mathcal{V}||_\infty$

This guarantees that the full potential

(2) $V(x) = \mathcal{V}(x) + Fx$

has local minima and maxima which will trap the particle. The following spectral properties of H hold :

- H is essentially self-adjoint on $C_0^\infty(\mathbb{R})$ by the Faris-Lavine theorem ([RS])
- The spectrum of H is absolutely continuous and fills the whole real axis ([RS])
- Assume $\int_0^\tau \mathcal{V}(x)\,dx = 0$; then the wave-operators $W_\pm\left(H, -g^2\frac{d^2}{dx^2} + Fx\right)$ exist and are unitary [J].
- For a dense set $\mathcal{A}$ of $L^2(\mathbb{R})$ the functions $< u, (H-z)^{-1} u >$, $u \in \mathcal{A}$, have meromorphic continuations from $\mathrm{Im}z > 0$ through $\mathbb{R}^+$ up to $\mathrm{Im}z > -b$ ([HH]). Furthermore the set $\mathcal{A}$ is translation invariant ; hence if E_0 is a pôle of a matrix element $< u, (H-z)^{-1} u >$ for some $u \in \mathcal{A}$, $\mathrm{Im}\,E_0 < 0$, it follows that $E_0+kF\tau$, $k \in \mathbb{Z}$, is a pôle of $< u_{k\tau}, (H-z)^{-1} u_{k\tau} >$ where $u_{k\tau}$ is translate of u by $k\tau$. Following the conventional interpretation of these pôles as resonances it follows that resonances form ladders :

$$L(E_0) = \{E_0 + kF\tau,\ k \in \mathbb{Z}\}$$

Then some important problems to be solved are :

- Do resonances really exist ?
- If they do where are they located ; in particular does the Oppenheimer behaviour of Stark effect hold ([Av]) :

$$\mathrm{Im}\,E_0 = O\left(e^{-a(gF)^{-1}}\right) \text{ as } F \to 0 \quad \text{for some } a > 0.$$

After reviewing the shape resonance theory (ch II) and providing the necessary semi-classical estimates (ch III) we will analyse these problems in ch IV.

II SHAPE RESONANCE THEORY

The basic idea of the one dimensional geometric theory of resonances is the following (which can be generalized to n dimensions). Consider a partition $\vartheta_{int} \cup \vartheta_{ext} = \mathbb{R}$ where ϑ_{int} and ϑ_{ext} are two open sets with intersection

(3) $\Omega = \vartheta_{int} \cup \vartheta_{ext}$

Let $H = -g^2\frac{d^2}{dx^2} + V$ and $H_\alpha = -g^2\Delta + V_\alpha$, α being int or ext, such that $V_\alpha = V$ on ϑ_α and V_α extends V on $\mathbb{R}$ such that H_α is self-adjoint. Assume that in some interval $I = [I^-, I^+]$ such that

(4) $\Omega \subset G(I^+) = \{x \in \mathbb{R},\ V(x) > I^+\}$

the spectrum $\sigma(H_{int})$ of H_{int} (resp. $\sigma(H_{ext})$) is discrete (resp. absolutely continuous). Then

$$H_d = H_{int} \oplus H_{ext}$$

has eigenvalues embedded in a continuum. Geometric perturbation theory allows us to show that in some sense to be specified below H is a perturbation of H_d by "tunnelling through Ω". Then one expects such eigenvalues to turn into resonances. To prove this using ordinary perturbation theory is quite easy with the help of **spectral deformation** techniques separating the continuous spectrum from the eigenvalues so that they become isolated.

Spectral deformations.

Consider a family of unitary operators [Hu] :

$$(U_t v)(x) = s'_t(x)^{1/2}\, v(s_t X),\ v \in L^2(\mathbb{R}^n)$$

where

$$s_t(x) = x + t\, f(x)$$

with f a smooth function. Later we will require also that :

(5)$_i$ $\quad$ $\operatorname{Supp} f \subset \mathbb{R} / \vartheta_{int}$

(5)$_{ii}$ $\quad$ $|t| < \|f'\|_\infty$

Then $s_t(\vartheta_{ext}/\Omega) = \vartheta_{ext}/\Omega$ whereas s_t is identity on ϑ_{int}. We assume now that the family of self-adjoint operators

$$H(t) = U_t\, H\, U_t^{-1}$$

has an analytic continuation in a ball $B_{t_0} = \{t \in \mathbb{C},\ |t| < t_0\}$ for some $t < \|f'\|_\infty$ (e.g. $\mathcal{D}(H(t)) = \mathcal{D}(H)$ and $H(t)u$ is analytic for all $u \in \mathcal{D}(H)$).

Notice that $\sigma(H(t)) = \sigma(H)$ for t real but this is no more true in general if Imt $\neq 0$. However one can show that $\sigma_p(H(t))$, the point spectrum of H(t), is locally invariant (see e.g. [AC], [Hu], [RS]) i.e. $\forall t \in B_{t_0} : \sigma_p(H(t+\varepsilon)) = \sigma_p(H(t))$ for $|\varepsilon| < \varepsilon_0(t)$. This is the key to the separation of point and continuous spectrum. Typical examples of transformations s_t are the translations $x \rightarrow x-t$ used by Herbst and Howland [HH] ; they show that in this case the continuous spectrum of H(t) is just $\mathbb{R}$ - (Imt)F. Invariance of point spectrum is also the crucial ingredient in the interpretations of complex eigenvalues of H(t), Imt > 0, as resonances ; with the

Definition 1

Spectral resonances of H are complex isolated eigenvalues of H(t), Imt $\neq 0$, connected to $i\infty$ in the resolvent set $\rho(H(t))$.

The main statement of spectral deformation theory is the property of spectral resonances to be complex pôles of functions $< u, (H-z)^{-1}v >$ where $u, v \in \mathcal{A}$, the dense set of analytic vectors with respect to $\{U_t, |t| < t_o\}$. It is worth mentionning at this point that two transformations functions f_1 and f_2 such that the sets $\mathcal{A}_1$ and $\mathcal{A}_2$ have dense intersection lead to the same spectral resonances. This gives a more intrinsic character to the above definition.

Non trapping

The main task now is to identify the resolvent set of H(t) for Imt ≠ 0. In the case of complex scaling [AC] or complex translations ([HH]) this can be done using global conditions of V and perturbation arguments since it is easy to control the case V = 0. For general f one needs a new approach ; in practice one is interested only into what happens in the neighbourhood of some given energy e_0. A usefull tool is then provided by the so-called non-trapping conditions ([BCD1,I], [HS1], [Kl], [Si1], [Na], [DeBHi]). They are mainly the statement that the corresponding classical system does not have trapped trajectories in a given region of phase space. One condition of this type is

(6) $$2f'(V-e) + fV' \leq -S$$

for some $S > 0$ and e near e_0. With some type of numerical range analysis one can show that it implies absence of spectrum for H(t) in a neighbourhood of e_0 when Imt ≠ 0. More precisely one has the following result (see [BCD1, III] for it's n dimensional version ; also [Si1] for a related method) :

Theorem 2

Assume there exists $f^* \in C^2(\mathbb{R})$ such that :

i) $-1 \leq f^*(x) \leq f'(x) \leq 1$

ii) $\exists\, S > 0$ such that if $t = i\tau$, $0 \leq \tau < \tau_0$ one has $2 f^* (V- e) + f V' \leq - S$.

Then with $\upsilon_t = \left\{ z \in \mathbb{C},\ \mathrm{Im}(1+t f^*)^2 (e_0 - z) < \frac{\tau S}{2} \right\}$ one has for g small enough :

(7) 1) $\upsilon_t \subset \rho(H(t))$

2) $\|(H(t)-z)^{-1}\| \leq 8/\tau S \qquad \forall\, z \in \upsilon_t$

Before sketching the proof let us make the following remarks :

1) If $f^* = f'$ condition (ii) is just (6) ; however arbitrariness of f^* allows to make (ii) also work in the classically forbidden region.

2) (6) implies $\{p^2+V, fp\} \geq S > 0$ on the energy shell $p^2+V = 0$ where $\{\ ,\ \}$ is Poisson bracket. So (6) says that fp increases along classical trajectories, indicating that they should be unbounded.

3) When $f = x$ (complex scaling) (6) is just the well-known virial condition. It says that x^2 is a convex function of time along a classical trajectory. Such a condition has already been used (see [RS]) to show absence of eigenvalues.

Sketch of the proof of Theorem 2 :
One has for all $u, v \in L^2(\mathbb{R})$

(8) $\qquad ||v||\ ||(H(t)-z)u|| \geq -\mathrm{Im} < v, (H(t)-z)\, u >$

Let $r_t = 1+t f^*$ and $v = \bar{r}_t\, r_t^{-1} u$; then :

$\mathrm{Im} < v, (H(t)-z)\, u > = \mathrm{Im} < w,\ r_t(H(t)-z)\, r_t\, w >$ with $w = r_t^{-1} u$; notice that for $|t|$ small enough one has $\frac{1}{2}\ ||u|| \leq ||v||,\ ||w|| \leq 2||u||$.

Now by a straight forward calculation :

$r_t\,(H(t)-z)\, r_t = -g^2\, r_t\, T_t\, r_t + r_t^2\, (V_t - z)$ with $T_t = U_t\, \dfrac{d^2}{dx^2}\, U_t^{-1}$ and $V_t = U_t\, V\, U_t^{-1}$ for $t \in \mathbb{R}$.

Direct calculation yields :

$$r_t\, T_t\, r_t = \frac{d}{dx}\mathrm{Re}(r_t^2\, s'^{-2}_t)\frac{d}{dx} + i\frac{d}{dx}\mathrm{Im}(r_t^2\, s'^{-2}_t)\frac{d}{dx} + r_t\,(T_t\, r_t)$$

The assumption on f^* and f' has been arranged so that $\mathrm{Im}\, r_t^2\, s'^{-2}_t \leq 0$ for $\mathrm{Im}t > 0$ small enough ; hence

$\mathrm{Im} < w,\ r_t(H(t) - z)\, r_t\, w >$

$$\geq - < w, [\mathrm{Im}\, r_t^2\, (V_t - z) - g^2\, \mathrm{Im}\, r_t\, (T_r\, r_t)]\, w >$$

$$\geq < w, [\tau S - \mathrm{Im}\, r_t^2 (e_0 - z) - g^2\, \mathrm{Im}\, r_t\, (T_r\, r_t)]\, w >$$

$$\geq \tau S/8\ ||u||^2$$

for g small enough and $z \in \upsilon_t$. Hence by (8) $||(H(t) - z)u|| \geq \tau S/8\ ||u||$.
Since $H^*(t) = H(\bar{t})$ the same proof yields $z \in \rho(H(t))$ if $z \in \upsilon_t$ and estimate (7).

This theorem will be used later to show that the spectral deformation $H_{ext}(t)$ of some "external" Hamiltonian H_{ext} has no spectrum in the vicinity of e_0 which is some isolated point in the spectrum of the "internal" Hamiltonian H_{int}. So we are in the situation described at the beginning of this chapter. We now have to show in which sense H(t) is a perturbation of $H_{int}(t) \oplus H_{ext}(t)$.

Geometric Perturbation Theory.

Roughly speaking the aim of geometric perturbation theory is to relate spectral properties of the Hamiltonian $H = -g^2 \frac{d^2}{dx^2} + V$ to local properties of the potential V or some "localized" Hamiltonians coinciding with H in some regions of configuration space (see e.g. [Si3]). The form presented here is adapted from [BCD2] where it was applied to the analysis of tunnelling phenomena in multiple well Schrödinger operators.

Let ϑ be an open subset of $\mathbb{R}$ and $J \in C^1(\mathbb{R})$ such that Supp $J \subset \vartheta$. A **local Hamiltonian** for ϑ is an operator $H_\vartheta = -g^2 \frac{d^2}{dx^2} + V_\vartheta$ on $L^2(\mathbb{R})$ such that $V_\vartheta = V$ in ϑ. One obtains easily for $z \in \rho(H) \cap \rho(H_\vartheta)$ the "geometric resolvent equation" (GRE) :

(9) $$R(z)\, J = J\, R_\vartheta(z) + R(z)\, W(J)\, R_\vartheta(z)$$

where $R(z) = (H-z)^{-1}$, $R_\vartheta(z) = (H_\vartheta-z)^{-1}$ and

$$W(J) = +\, g^2 \left[\frac{d^2}{dx^2}, J\right]$$

is a first order differential operator with the same support as J'. If one has a partition

$\mathbb{R} = \vartheta_{int} \cup \vartheta_{ext}$

$\Omega = \vartheta_{int} \cap \vartheta_{ext}$

and two functions J_{int} and J_{ext} with Supp $J_\alpha = \vartheta_\alpha$, α = int or ext, and $J_{int} + J_{ext} = 1$ (so that Supp $J'_\alpha \subset \Omega$) then applying (9) for both ϑ_{int} and ϑ_{ext} one obtains the G.R.E. :

(10) $$R(z)\, J = J\, R_d(z) + R(z)\, W_d\, R_d$$

where $J : H_d \equiv L^2(\mathbb{R}) \oplus L^2(\mathbb{R}) \to L^2(\mathbb{R})$ is given by

(11) $$J(\oplus u_\alpha) = \sum_\alpha J_\alpha u_\alpha$$

and

$$R_d(z) = (H_d - z)^{-1} \text{ with } H_d = \oplus H_\alpha$$

$$W_d(\oplus u_\alpha) = \sum_\alpha W(J_\alpha)\, u_\alpha .$$

Consider now a spectral deformation generated by a function f satisfying $(5)_i$ and $(5)_{ii}$. Then (10) still holds for the transformed Hamiltonians H(t), $H_{ext}(t)$ and $H_{int}(t) = H_{int}$. Notice that since s_t leaves Ω invariant and $W(J_\alpha)$ has support in Ω these operators are invariant under U_t. Considering now (10) in the neighbourhood of an energy e_0 such that e_0 in non-trapping for H_{ext} and H_{int} has isolated point spectrum we are faced with a standard problem in perturbation theory of point spectrum. Of course this requires that one has good control on the kernel of this equation. This will be shown to be a consequence of the existence of a potential barrier separating the internal and external regions ϑ_{int} and ϑ_{ext}. The relevant estimates will be given in Ch III together with more details on the way to solve (10).

Width of resonances

Assuming then that (10) has solutions for energies z near e_0, the isolated eigenvalues of H_{int} should become spectral resonances for H. To estimate their imaginary part one can use the following simple argument ; let $H(t)\psi = E\psi$, $\|\psi\| = 1$, then :

$$\begin{aligned} \operatorname{Im}E &= \operatorname{Im} < \psi, H(t)\psi > \\ &= <\psi, \operatorname{Im} H(t)\psi > \end{aligned}$$

Now by $(5)_i$ one has $\operatorname{Im} H(t) = \operatorname{Im} J_{ext} H(t) J_{ext}$ since f is zero on $\vartheta_{int} = \operatorname{Supp} J_{int}$. So

$$\operatorname{Im}E = < J_{ext}\psi, (\operatorname{Im} H(t))\, J_{ext}\psi >.$$

Now from perturbation theory one expects that eigenstates of H(t) should be close to those of H_{int}. Hence the lifetime of resonances depends crucially on the support and decay properties of eigenstates of the internal Hamiltonian H_{int}. A general result from shape resonance theory obtained in this way is :

$$|\operatorname{Im}E| \leq C\, e^{-2\rho(U_{int}, U_{ext})}$$

where $U_\alpha = \vartheta_\alpha \setminus G(e_0)$ and ρ is "Agmon's distance" which will be defined below. This estimate is good enough to give Oppenheimer's behaviour for Stark effect in atoms and molecules ([Si2]). However it is not strong enough to apply also to Stark ladders since in this case $\rho(U_{int}, U_{ext})$ is not $O((g\ F)^{-1})$ but only $O((g^{-1})$. So a more refined analysis of the localisation properties of eigenstates of H_{int} will be needed. The key fact here is that although the wave-functions of a system of identical potential wells are delocalized over all these wells, this is no more true if the equivalence of wells is destroyed and this is precisely what the electric field does. Then the wave-functions become localized only on resonant wells, that is in those wells having a local Hamiltonian with spectrum close to the reference energy e_0. We describe this fact in more mathematical terms now.

Spectral properties of multiple well Schrödinger operators.

It is worth recalling at this point some basic facts about the simplest model namely the symmetric double well with a non degenerate minimum at $x = \pm x_m$. Imagine that the two wells are separated by an infinite potential barrier (e.g. a Dirichlet boundary condition at x = 0). Then one has a doubly degenerate spectrum with eigenvalues

$$e_n(g) = (2n+1)\, g\omega + O(g^2)\ (g \to 0), \quad n = 0, 1, \ldots \tag{12}$$

with $\omega^2 = V''(x_m)$ (see e.g. [Sim]). Removing the infinite barrier leads to a splitting of the doubly degenerate e_n into two eigenvalues $e_n^{(s)}$ and $e_n^{(a)}$ such that

$$e_n^{(a)} - e_n^{(s)} = \delta_n(g) = O\left(\exp\left(-g^{-1}\int_{S^-}^{S^+}\sqrt{V-e_n}\right)\right)$$

with S^+ and S^- the turning points at energy e_n. The corresponding antisymmetric and symmetric eigenstates are equally localized in both wells. A remarkable fact, first noticed by Jona-Lasinio et al ([JMS]) is that this localization is very unstable againt small perturbations of the symmetry. Namely it is enough that the symmetry is broken by a perturbation W with $||W|| > C\delta_n(g)$ for some constant C in order that the perturbed wave-functions become localized in either one of the wells. The same occurs with periodic potentials when symmetry is broken by a small electric field and this will play a crucial rôle in the discussion of resonance widths. First let us explain the mathematical mechanism behind this instability of localization. For a given potential V and energy interval $I = [I^-, I^+]$ consider the partition into wells

$$\mathrm{Supp}\,(V-I^+)_- = \bigcup_i W_i$$

where $f_\pm = \pm\sup(\pm f, 0)$ and the wells W_i are assumed to be connected. Wells are separated by barriers and there is a natural metric associated to tunnelling accross these barriers given by

$$ds^2 = g^{-2}(V-I^+)_+\, dx^2 \tag{13}$$

Agmon's distance $\rho(x,y)$ is the geodesic distance associated to this metric. The semi-classical theory of multiple well Schrodinger operators relates the spectral properties of H to those of local Hamiltonians having only one well ([BCD2], [HS2]). More precisely let $\rho_i = \inf_{j\neq i} \rho(W_i, W_j)$ and choose for example :

$$\vartheta_i = \{x, \rho(x, W_i) \leq \frac{1}{2} \rho_i\}$$

Let $V_i = V$ in ϑ_i whereas V_i is growing outside ϑ_i and $H_i = -g^2 \frac{d^2}{dx^2} + V_i$; then one has the following [BCD2] :

Stability Theorem 3

There exists a constant C independent of g such that for $\Delta \geq \exp(-\inf \rho_j)$ and if $\mathrm{dist}(I^{\pm}, \sigma(H_i)) \geq \Delta \quad \forall i$ one has

i) $\dim P(I) = \sum_i \dim P_i(I)$

ii) $\forall \varphi_i \in P_i(I) L^2(\mathbb{R})$, $||\varphi_i|| = 1$, there exists $\tilde{\varphi}_i \in P(I) L^2(\mathbb{R})$ such that

$||\tilde{\varphi}_i - \varphi_i|| \leq \Delta^{-1/2} e^{-\rho_i/2}$.

Here P(.) (resp. $P_i(.)$) denotes the spectral projections for H(resp. H_i). This theorem says in particular that eigenstates of H corresponding to $\sigma(H) \cap I$ are localized into **resonant wells** for which $\sigma(H_i) \cap I \neq \emptyset$.

We will refer to this result later to show that the internal Hamiltonian H_{int} has eigenstates localized at a distance $O((gF)^{-1})$ from the external region ; this in turn is the basis of Oppenheimer's estimate for resonance width.

III SEMI - CLASSICAL ESTIMATES

We provide now the technical tools for the solution of the geometric resolvent equation (10). The main point is that the kernel of the resolvent $(H-z)^{-1}$ is exponentially small in the classically forbidden region. Details on the derivation of these estimates can be found in [BCD2] and [CH]. Agmon's distance was defined in (13) ; for an open set Ω let :

$\rho_\Omega(x) = \rho(x, \Omega)$

Then ([Ag]) ρ_Ω is a.e. differentiable and

(14) $$|\rho'_\Omega(x)| \leq g^{-1}\,(V(x) - I^+)_+^{1/2}$$

Consider the positive self-adjoint operator C_Ω such that C_Ω^2 is associated by the representation theorem to the positive quadratic form on $\mathcal{H}^1(\Omega) \oplus L^2(\mathbb{R}\backslash\Omega)$:

$$a_\Omega[u] = g^2 \left\| \frac{d}{dx} u \right\|_\Omega^2 + \langle u, (V - I^+)_+ u \rangle_\Omega$$

where $\langle u, v \rangle_\Omega = \int_\Omega \bar{u}(x)\, v(x)\, dx$.

Consider the potential $V^{(0)}$ obtained by "filling up the wells" up to energy I^+ :

$V^{(0)} = \sup(V, I^+)$

and $H^{(0)} = -g^2 \frac{d^2}{dx^2} + V^{(0)}$, $R^{(0)}(z) = (H^{(0)}-z)^{-1}$; one has :

Lemma 4

Let $z \in \mathbb{C}$, Re $z < I^+$; then

i) $\| C_\Omega R^{(0)}(z) C_\Omega \| \leq 1$

ii) If Ω_1, Ω_2 are disjoint open sets such that $\rho(\Omega_1, \Omega_2) > 0$ then

$$\| C_{\Omega_1} R^{(0)}(z) C_{\Omega_2} \| \leq e^{-\rho(\Omega_1, \Omega_2)}$$

The proof of i) is straight forward using positivity of $V^{(0)}-I^+ = (V, I^+)_+$. As to ii) it is a simple exercise using the elementary identity $\| \frac{d}{dx}(\xi v) \|^2 = \mathrm{Re} \langle \xi^2 v, -\frac{d^2}{dx^2} v \rangle + \| \, |\frac{d}{dx}\xi| v \|^2$

and applying it to $\xi = e^{-\rho_1}$ where $\rho_1(x) = \inf(\rho(x, \Omega_1), \rho(\Omega_1, \Omega_2))$ so that by (14) :

$$\left| \frac{d}{dx}(e^{-\rho_1}) \right|^2 \leq g^{-2}(V - I^+)_+ e^{-2\rho_1} \qquad \text{on } \mathbb{R}\backslash(\Omega_1 \cup \Omega_2)$$

$$= 0 \text{ on } \Omega_1 \cap \Omega_2$$

Then for all $v \in \mathcal{H}^1(\Omega_1) \oplus L^2(\mathbb{R}\backslash\Omega_1)$ one has

$$\| C_{\Omega_1} v \|^2 = \| C_{\Omega_1} (e^{-\rho_1} v) \|^2$$

$$\leq \mathrm{Re} \langle e^{-2\rho_1} v, (H^{(0)}-Z)\, v \rangle$$

Chosing $v = R^{(0)}(z) C_{\Omega_2} u$ where $u \in \mathcal{H}^1(\Omega_2) \oplus L^2(\mathbb{R}\backslash\Omega_2)$ gives :

$$|| C_{\Omega_1} R^{(0)}(z) C_{\Omega_2} u ||^2 \leq e^{-2\rho(\Omega_1, \Omega_2)} ||u|| \, ||C_{\Omega_2} R^{(0)}(z) C_{\Omega_2} u |$$
$$\leq e^{-2\rho(\Omega_1, \Omega_2)} ||u||^2$$

This extends to any u by a limiting procedure. From this Lemma one obtains the main estimate :

Theorem 5

Let $\mathcal{U} = \mathrm{Supp}(V\text{-}I^+)$. (union of wells) and Ω_i, i = 1, 2, be open sets such that $\rho(\Omega_i, \mathcal{U}) > \delta$ for some $\delta > 0$. Then if $\mathrm{Re}\, z \leq I^+$ one has :

$$|| C_{\Omega_1} R(z) C_{\Omega_2} || \leq e^{-\rho(\Omega_1, \Omega_2)} + C\, e^{-[\rho(\Omega_1,\mathcal{U})+(\Omega_2, \mathcal{U}]} ||R(z)||$$

for some constant C depending only on δ.

The proof uses the geometric resolvent equation (9) with $\vartheta = \{x, \rho(x,\mathcal{U}) > \delta/2\}$ so that one can take $H_\vartheta = H^{(0)}$; we refer to [CH] for the details. The interpretation of (15) is as follows in terms of path integrals : among all geodesics going from Ω_1 to Ω_2 some lie entirely in the classically forbidden region where the potential is $V^{(0)}$ and give the term $e^{-\rho(\Omega_1, \Omega_2)}$ as expected from Lemma 4. Other geodesics cross some wells but since quantum eigenstates are localized there the price to pay in the singular factor $||R(z)||$.

To conclude this chapter let us describe a factorisation trick for the local pertubation W(J) appearing in the geometric resovent equation (9) which clarifies our introductin of the operators C_Ω . This trick plays a usefull rôle in the proof of the stability theorem and will be used in the next chapter. Assume the open set ϑ is of the form :

(16)i $\quad \vartheta = \{x, \rho(x, A) < \delta\}$ for some $A \subset \mathbb{R}$, $\delta > 0$.

Define then $J(x) = j(\rho(x, A))$ where : i) j is piecewise linear ii) $f(0) = 1$ iii) $f(\rho) = 0$ if $\rho > \delta$. Then supp $J \subset \vartheta$ and supp $J' \subset \Omega = \{x, 0 < \rho(x, A) < \delta\}$ with

(16)ii $\quad |J'(x)| \leq (g\delta)^{-1} (V(x) - I^+)_+^{1/2}$

Now define the following maps from $\mathcal{H}^1(\Omega) \oplus L^2(\mathbb{R}\backslash\Omega)$ to $L^2(\mathbb{R}) \oplus L^2(\mathbb{R})$:

(17)i $\quad m_1(J)\, u = \delta^{1/2} g J' u \oplus \delta^{-1/2} \chi_\Omega g\, u'$

(17)ii $\quad m_2(J)\, v = \delta^{-1/2} g \chi_\Omega v' \oplus \delta^{1/2} g J' v$

where χ_Ω is the characteristic function of Ω. Then one has

(18) $\quad W(J) = m_2^*(J)\, m_1(J)$

(19) $\| m_i(J)\, u \| \leq \delta^{-1/2} \| C_\Omega\, u \|$, $i = 1, 2$.

In this last inequality one uses (13). The factorisation (18) together with estimates (19) and (15) are obviously very appropriate tools to solve the geometric resolvent equation .

IV EXISTENCE AND ESTIMATES FOR STARK LADDERS

We now consider the quantum Hamiltonian given by (1). The potential $V(x) = \mathcal{V}(x) + Fx$ satisfies (A1) and (A2) ; we fix the x coordinate such that $V(0) = \mathcal{V}(0) = \mathcal{V}_0$ where $\mathcal{V}_0 = \max_{X \in \mathbb{R}} \mathcal{V}(x)$. A unit cell for V is an interval $U_k \equiv [-\tau(k+1), -\tau k]$, $k \in \mathbb{Z}$. Associated with U_k is a single cell Hamiltonian $H_k = -g^2 \frac{d^2}{dx^2} + V_k$ where $V_k = V$ on U_k and is monotone increasing on $\mathbb{R}\backslash U_k$. We denote by $e_0 \leq e_1 \leq$ ....the eigenvalues of H_0 ; up to a constant linear in F the e_n's admit an expansion (12) ; the spectrum of H_k, $k \geq 0$, is $\sigma(H_0) - kF\tau$.

Let $S_T(e)$: inf $\{x, V(x) = e\}$ and $\tilde{S}_T(e) = \inf \{x > S_T(e), V(x) = e\}$, the left most turning points at energy e. We assume for simplicity that for the energy interval $I = [I^-, I^+]$ one has $\rho_0 = \rho(S_T(I^+), \tilde{S}_T(I^+)) > 0$; although I^+ might be g dependent (see Lemma 6 below) one can also assume that ρ_0 becomes large as g tends to zero (see (13)) so that for a fixed $\delta > 0$ one has $\rho_0 \geq 2\delta$ for g small enough. The internal region is taken as

$$\vartheta_{int} = \{x, \rho(x, S_T(I^+)) > \delta\}$$

A partition of unity is obtained by chosing J_{int} as in (16) and $J_{ext} = 1 - J_{int}$; we let

$$\vartheta_{ext} = \text{supp } J_{ext}.$$

Notice $\Omega = \vartheta_{int} \cap \vartheta_{ext}$ satisfies (4). To solve the GRE(10) we need some preliminary statements about H_{int} and H_{out} ; recall that $H_\alpha = -g^2 \frac{d^2}{dx^2} + V_\alpha$, α = int or out, with $V_\alpha = V$ in ϑ_α and V_α is increasing outside ϑ_α. First to get localisation properties for eigenstates of H_{int} through the stability theorem one needs :

Lemma 6

For any $N > 0$ and $\varepsilon > 0$ there exists in $J = [e_0 - g^N, e_0 + g^N]$ a subinterval $I = [I^-, I^+] \ni e_0$ such that dist $(I^{\pm}, \sigma(H_k)) \geq F g^{N+2+\varepsilon}$ for g small enough and all k such that $U_k \subset \vartheta_{int}$.

The proof uses Bargman's bound [RS] for the number N(e) of eigenvalues of $\oplus H_k$, $U_k \subset \vartheta_{int}$, below $e > e_0$; it shows $N(e) \leq C(g^2F)^{-1}$ for some constant C ; then the proof of the Lemma is by contradiction.

It follows from the stability theorem that H_{int} has at least one eigenvalue $\tilde{e_0} \in I$, close to e_0 to an arbitrary order in g for h small. One can't exclude of course that the full multiplicity of the spectrum of H in I is larger than one i.e. that there is more than one resonant well for H_{int}.

Next we refer to [CH] for the construction of a function f satisfying $(5)_i$ and such that theorem 2 can be applied to H_{out} for all $e \in I$ as constructed above. Then $H_{out}(t)$, for Imt > 0 and t satisfying $(5)_{ii}$, has no spectrum near I and it's resolvent satisfies an estimate (7) uniformly in a neighbourhood of I.

Consider then the G.R.E (10) with $R(z) = (H(t) - z)^{-1}$, $R_{out}(z) = (H_{out}(t) - z)^{-1}$ and $R_{int}(z) = (H_{int} - z)^{-1}$; to solve it we use a variant of the factorization trick as follows. Let $\tilde{J}_\alpha$ be the characteristic function of ϑ_α and $\tilde{J} : \mathcal{H}_d \to L^2(\mathbb{R})$ be defined as in (11) for J ; notice $J\tilde{J}^* = 1$. Then simple calculation gives :

$$\tilde{J}^* W_d = m^*_{d,2}\, m_{d,1} T$$

with

$$m_{d,i}(\oplus v_\alpha) = \oplus (m_i (J_{int}) v_\alpha) \quad , i = 1, 2$$

and $m_i(J)$ given by (17) whereas

$$T(u_{int} \oplus u_{out}) = (u_{int} - u_{out}) \oplus (u_{int} - u_{out}).$$

From (10) it follows that

$$R(z)\, J\, m^*_{d,2} = J\, R_d(z)\; m^*_{d,2} + R(z)\, J\, m^*_{d,2}\; K(z)$$

with

$$K(z) = m_{d,1} T\, R_d(z)\; m^*_{d,2}$$

Using estimates (19) one obtains

$$||K(z)|| \le 2\delta^{-1} \sup_\alpha ||C_\Omega R_\alpha(z) C_\Omega||$$

Applying theorems 2 and 5 one has then for g small enough and $t = i\tau$:

$$(20) \qquad ||K(z)|| \le \frac{C}{\delta}\, [1 + e^{-2\rho_0} \sup(8/\tau S, ||R_{int}(z)||]$$

for some constant C independent of z and g provided $z \in \upsilon_t$, Re z < I^+. In particular $||K(z)|| < \frac{1}{2}$ if h is small enough and $z \in \upsilon_t$ with dist $(z, \sigma(H_{int})) \ge e^{-2\rho_0}$; under such conditions the solution to (10) is given by :

(21) $R(z) = J R_d(z) \tilde{J}^* + J R_d(z) m_{d,2}^* (1-K(z))^{-1} m_{d,1} R_d(z) \tilde{J}^*$.

We refer to[CH] for the almost standard perturbation arguments leading with the help of (21) to

Theorem 7
Let $\Gamma \subset \mathbb{C}$ be a simple closed contour passing through $I^{\pm}$ and lying in υ_t where t satisfies $(5)_{ii}$, Im t > 0. Then for g small enough one has $\Gamma \subset \rho(H_t)$ and if

$$P_t = \frac{-1}{2i\pi} \int_\Gamma R(z)dz$$

one has :

$$\dim P_t = \dim P_{int}(I)$$

Furthermore if $E_0 \in \sigma(H_t P_t)$ then

$$|\mathrm{Im}\, E_0| < C e^{-2\rho_0}$$

for some constant C independent of g.

Since H_{int} has at least one eigenvalue $\tilde{e}_0 \in I$ and H has no point spectrum it follows that there is at least one spectral resonance of H inside Γ with imaginary part exponentially small in g. As mentionned before this shows that Stark ladder exist in the semi-classical regime but this does not show the expected Oppenheimer's behaviour for Stark resonances. Further arguments are necessary which we will sketch now. They are based on the remark that eigenfunctions of H(t) should be close to those of H_{int} which in turn are localized in the resonant cells U_k such that $\sigma(H_k) \cap I \neq \emptyset$ by the stability theorem. Let U_{r_0} (= U_0), U_{r_1}, ..., U_{r_N} be these resonant cells and φ_0, ..., φ_N be real normalized eigenfunctions :

$$H_{r_i} \varphi_i = e_i \varphi_i$$

with $e_i \in I$. From the stability theorem and perturbation theory it is easy to show that if η_i is the characteristic function of H_{r_i} and $\phi_i = P_t \eta_i \varphi_i$ then $\{\varphi_0, \ldots, \varphi_n\}$ is a basis of $P_t\mathcal{H}$. The ϕ_i's are approximate resonance wave-functions. Notice that $\phi_i^* \equiv P_t^* \eta_i \varphi_i = P_{\bar{t}} \eta_i \varphi_i = \bar{\phi}_i$ since $H(\bar{t})$ is obtained from H(t) by complex conjugation. Defining

$$S_0 = \begin{cases} \inf_j (\rho(U_0, U_{r_j})) & \text{if } N \geq 1 \\ \rho(U_0, \Omega) & \text{if } U_0 \text{ is the only resonant cell} \end{cases}$$

one has

Lemma 8

For g small enough there exists a constant $a > 0$ independent of g such that :

1) $|< \phi_k^*, H_t^\alpha \phi_0 >| \leq e^{-aS_0} \; \forall \, k = 1, ..., N$ and $\alpha = 0, 1$

2) $\mathrm{Im} < \phi_0^*, H_t \phi_0 > \leq e^{-aS_0}$

We refer to [CH] for the proof of this Lemma. What remains to do is to show that among all eigenvectors of H(t) with eigenvalues inside Γ one of them, say ψ_0, must have a component over ϕ_0 which is not too small. Using the equality $\mathrm{Im}\, E_0 = \mathrm{Im} (< \psi_0^*, H_t \phi_0 > / < \psi_0^*, \varphi_0 >$ one obtains ([CH]) :

Theorem 9

Assume that the range of P_t is spanned by eigenvectors of H(t). Then there exists at least one $E_0 \in \sigma(H(t)\, P_t)$ such that for g small enough and some constant $a > 0$ independent of g :

$$|\mathrm{Im}\, E_0| < e^{-aS_0}$$

We conclude with a discussion on the order of magnitude of S_0. If $\mathcal{U}_0$ is the only resonant cell, which is the generic case for g small and F fixed then $S_0 = 0((g\, \tau\, F)^{-1})$.

In case there are many resonant wells and F itself is polynomialy small in g then the closest resonant cell to U_0 should satisfy according to the harmonic approximation (12) $\rho(U_0, U_{r_1}) = S_0 = 0((\tau\, F)^{-1})$. Finaly if F is exponentialy small in g then one expects clusters $C_0 \ni U_0, C_1, ..., C_N$ with $\rho(C_0, C_1) = 0((\tau\, F)^{-1})$; in this case the conclusion of theorem 9 with $S_0 = \rho(C_0, C_1)$ still holds but it's proof requires a suitable adaptation of Lemma 8. On the other hand if one lets F vary with g in such way that F is O(g) then our construction of a function f satisfying the conditions of theorem 2 for $H = H_{out}$ does not apply anymore. This is linked to the fact that a particle with energy near E_0 will meet a large number of barrier tops before making it to the exterior region ϑ_{ext}. Although individual barrier tops are known to be non-trapping in a certain sense (see [BCD1,II]) it is still an open problem whether the same is true when there are many of them. Apart from this technical restriction it appears that in all regimes Oppenheimer's behaviour for the ground state Stark ladders associated to e_0 is due to the fact that the Agmon's distance between the resonant region where the particle is initialy localized and the next one is of the order of $(\tau\, F)^{-1}$.

[AC] Aguilar J., Combes J.M. : A class of analytic perturbations for one body Schrödinger Operators. Com. Math. Phys. **22**, 269-279 (1971)

[Ag] Agmon S. : Lectures on exponential decay of second order elliptic Equation. Princeton Math. Notes. **29** (1982)

[AF] Agler J., Froese R. : Existence of Stark ladder resonances. Com. Math. Phys. **100**, 161-171 (1985)

[Av] Avron J. : The lifetime of Wannier Ladder States. Ann. Phys. **143**, 33-53 (1982)

[B] Bentosela F. : This volume.

[BVD] Bleuse J., Bastard G., Voisin P. Phys. Rev. Letters **60**, 220 (1988)

[BCD1] Briet Ph., Combes J.M., Duclos P. : On the location of resonances for Schrödinger Operators.
I. Resonance free domains. J. Math. Anal. Appl. **126**, 90-99 (1987)
II. Barrier top resonances. Com. in P.D.E. **12**, 201-222 (1987)
III. Shape resonances. To appear

[BCD2] Briet Ph., Combes J.M., Duclos P. : Spectral Stability under tunnelling. Com. Math. Phys. **126**, 133-156 (1989)

[BD] Buslaev V.S. : Dimitrieva L.A. This volume

[CDKS] Combes J.M., Duclos P., Klein M., Seiler R. : The shape resonance. Com. Math. Phys. **110**, 215-236 (1987)

[CH] Combes J.M., Hislop P. : Stark ladder resonances for small electric fields. Preprint, Univ. Kentucky (1989)

[DeBHi] De Bièvre S., Hislop P. : Spectral resonances for the Laplace-Beltrami operator. Ann. Inst. H. Poincaré **48**, 105-145 (1988)

[HH] Herbst I.W., Howland J.S. : The Stark ladder and other one dimensional external field problems. Com. Math. Phys. **80**, 23-40 (1981)

[HS1] Hellfer B., Sjostrand J. : Resonances en limite semi-clasique. Supplem Bulletin SMF **114** (1986)

[HS2] Hellfer B., Sjostrand J. : Multiple wells in the semi-classical limit I. Com. in P.D.E. **9**, 337-408 (1984)

[HiSi] Hislop P., Sigal I.M. : Semi-classical theory of shape resonances in quantum mechanics, Memoirs. Am. Math. Soc. **399** (1989)

[Hu] Hunziker W. : Distortion analyticity and molecular resonance curves. Ann. Inst. M. Poincaré, **45**, 339-358 (1986)

[J] Jensen A. : Asymptotic completeness for a new class of Stark effect Hamiltonians. **107**, 21-28 (1986)

[NA] Nakamura S. : A note on the absence of resonances for Schrödinger Operators. Lett. Math. Phys. **16**, 217-223 (1988)

[N] Nenciu G. : This volume

[JMS] Jona-Lasinio, Martinelli F., Scoppola E. : New approach to the semi-classical limit of quantum mechanics I. Com. Math. Phys. **80**, 223 (1981)

[Kl] Klein M. : On the absence of resonances for Schrödinger Operators in the semi-classical limit. Com. Math. Phys. **106**, 485-494 (1985)

[RS] Reed M., Simon B. : Methods of Modern Mathematical Physics IV : Analysis of Operators (Academic, N.Y, 1978)

[Si1] Sigal I.M. : Sharp exponential bounds on resonance states and width of resonances. Adv. Appl. Math. **9**, 127-166 (1988)

[Si2] Sigal I.M. : Geometric Theory of Stark resonances in multielectron systems. Com. Math. Phys. **119**,287-314 (1988)

[Si3] Sigal I.M. : Geometric methods in the quantum many body problems. Com. Math. Phys. **85**, 309-324 (1982)

[Sim] Simon B. : Semi-classical analysis of low lying eigenvalues I. Ann. Inst. H. Poincaré **38**, 295-307 (1983)

[Z] Zak J. : Stark ladders in solids ? Phys. Rev. Lett. **20**, 1477-1481 (1968)

ASYMPTOTIC INVARIANT SUBSPACES, ADIABATIC THEOREMS AND BLOCK DIAGONALISATION

G. Nenciu
Institute of Atomic Physics
Theoretical Physics Department
Bucharest, P.O. Box MG6 - Romania

ABSTRACT. We consider the evolution $i\varepsilon \frac{d}{ds}U(s) = H(s)U(s)$, $U(0) = 1$ for $\varepsilon \to 0$. A recurrence procedure providing arbitrary order asymptotic invariant subspaces of U(s) corresponding to the isolated parts of the spectrum of H(s) is written down. The point of the construction is that at the kth step the invariant subspaces at the *"time"* s are constructed from H and its first k derivatives at the same time. As consequences we give a hierarchy of adiabatic theorems as well as a block diagonalisation scheme.

1. INTRODUCTION

It turns out that the adiabatic theorem of Quantum Mechanics [1-5], the theory of spectral concentration [4], as well as the theory of simplification and asymptotic diagonalisation of differential evolution equations [6-8] can be viewed as particular cases of the following *"invariant subspace problem"*. Consider the evolution equation

$$i\varepsilon\, \partial_s U_\varepsilon(s,s_o) = H(s)U_\varepsilon(s,s_o);\quad U_\varepsilon(s_o,s_o) = 1 \tag{1.1}$$

in the limit $\varepsilon \to 0$. The problem is to construct out of H(s) (without integrating (1.1)) families of subspaces which are invariant or at least almost invariant under $U_\varepsilon(s,s_o)$. This amount to construct families, $P_\varepsilon(s)$, of projection operators satisfying

$$P_\varepsilon(s) - U_\varepsilon(s,s_o)P_\varepsilon(s_o)U^{-1}(s,s_o) \text{ - as small as possible.} \tag{1.2}$$

If H(s) does not depend on s, H(s) = H, the solution of the above problem is well known: all the spectral projections of H are invariant under $U_\varepsilon(s,s_o)$.

If H(s) has for all s an isolated part of the spectrum, the corresponding family of spectral projections, Q(s), satisfies

A. Boutet de Monvel et al. (eds.), Recent Developments in Quantum Mechanics, 133–149.

$$Q(s) - U_\varepsilon(s,s_o)Q(s_o)U_\varepsilon^{-1}(s,s_o) \quad \text{- is of order } \varepsilon \quad . \tag{1.3}$$

This is the content of the adiabatic theorem of Quantum Mechanics in its simplest form [2].

The basic result of this lecture is (under appropriate technical conditions on H(s)) a recurrent construction of families of subspaceswhich are invariant up to errors of order ε^k, $k = 1,2,\ldots$ in the limit $\varepsilon \to 0$.

For definiteness, we shall work in a Hilbert space setting although extensions to Banach spaces or indefinite metric spaces (appearing naturally in linear Hamiltonian mechanics) are possible (see e.g. [9]). The existence of the invariant subspaces invariant up to errors of order ε^k, $k = 1,2,\ldots$ has been pointed out for the first time by Garrido [3]. The results and the proofs below are refinements of the theory developed in [10]. Actually this lecture is the final form of the result announced in the Remark 7 in [4]. The application to the Stark-Wannier ladder problem is discussed in [15], [16].

2. THE RESULTS

Let $s \in I = (a,b) \subset \mathbb{R}$, and H(s) be a family of semi-bounded self-adjoint operators in a Hilbert space $\mathcal{H}$. The conditions about H(s) we shall impose at various stages are

H_1. $(H(s) - i)^{-1}$ is indefinitely norm differentiable.

H_2. The spectrum, $\sigma(s)$, of H(s) consists for all $s \in I$ of n isolated pieces n-1 of them being uniformly (with respect to s) bounded sets. More exactly

$$\sigma(s) = \bigcup_{j=1}^{n} \sigma_j(s), \quad \sigma_j(s) \text{ uniformly bounded for } j = 1,2,\ldots,n-1.$$

$$\min_{j \neq k} \operatorname{dist}(\sigma_j(s), \sigma_k(s)) = d(s) > 0 \quad \text{all } s \in I.$$

H_3. The domain of H(s) is independent of s, $\mathcal{D}(h(s)) = \mathcal{D}$, and for all $f \in \mathcal{D}$, $f(s) = H(s) f$ is continuously norm differentiable.

H_4. $\inf_{s \in I} d(s) = d > 0$

$$\sup_{s \in I} \|\partial_s^\ell (H(s) - i)^{-1}\| < \infty \qquad \ell = 1,2,\ldots$$

H_5. If $I = \mathbb{R}$ then H_4 hold and in addition

$$\|\partial_s^\ell (H(s) - i)^{-1}\| \in L^1(\mathbb{R}), \qquad \ell = 1,2,\ \ldots$$

Remarks. 1. One can relax H_1, imposing $(H(s) - i)^{-1}$ to be only N times differentiable, $N < \infty$. In this case the recurrent procedure below stops at the step N-1.

2. By the resolvent formula

$$(H(s) - z)^{-1} = (H - z_o)^{-1} [1 - (z - z_o)H(s)]^{-1}$$

$(H(s) - z)^{-1}$ is indefinitely norm differentiable for all $z \in \rho(H(s))$.

3. One can extend the results below to non semi-bounded operators (e.g. Dirac operator) but in this case conditions on $\partial_s^{\ell}(H(s) - z)^{-1}$ as $z \to \infty$ are needed.

The recurrent construction of $P_{k,j}(s)$ below can be viewed as the "*spectral analysis*" of the family H(s).

Let $H_k(s,\varepsilon)$, $P_{k,j}(s,\varepsilon)$ $j = 1,\ldots,n$, $k = 0,1,\ldots$ as given by the following recurrence procedure:

$$H_o(s) = H(s) \tag{2.1}$$

Let $P_{0,j}(s)$ be the spectral projections of $H_0(s)$ corresponding to $\sigma_j(s)$. Due to the norm differentiability of $(H(s) - z)^{-1}$ and to the Riesz formula

$$P_{0,j}(s) = (2\pi i)^{-1} \oint_{C_j} (H_o(s) - z)^{-1}dz \ , \quad j = 1,\ldots, n-1 \tag{2.2}$$

where C_j are bounded contours surrounding $\sigma_j(s)$, $P_{0,j}$ are indefinitely norm differentiable. Since

$$P_{0,n} = 1 - \sum_{j=1}^{n-1} P_{0,j} \tag{2.3}$$

$P_{0,j}$ is also indefinitely differentiable.

Consider the self-adjoint operators

$$H_1(s;\varepsilon) = H_0(s) + B_0(s;\varepsilon)$$

$$B_0(s;\varepsilon) = i\varepsilon \sum_{j=1}^{n} P_{0,j}(s)\ \partial_s\ P_{0,j}(s) \quad . \tag{2.4}$$

Obviously $B_o(s;\varepsilon)$ is bounded and moreover $\lim_{\varepsilon\to 0}||B_o(s;\varepsilon)|| = 0$. Let $\varepsilon_1(s)$ be chosen such that $\sup_{\varepsilon\in(0,\varepsilon_1)} ||B_o(s;\varepsilon)|| \le d(s)/4$. Then for $0 \le \varepsilon \le \varepsilon_1$

by the regular perturbation theory $H_1(s;\varepsilon)$ satisfies H_2 and if $\sigma_{1,j}(s;\varepsilon)$ are the parts of the spectrum of $H(s;\varepsilon)$ which coincide with $\sigma(s)$ in limit $\varepsilon \to 0$ then

$$\min_{j\neq k} \operatorname{dist}(\sigma_{1,j}(s;\varepsilon), \sigma_{1,k}(s;\varepsilon) \geq d(s)/2$$

One can repeat the procedure by defining

$$H_2(s;\varepsilon) = H_1(s;\varepsilon) + B_1(s;\varepsilon)$$

$$B_1(s;\varepsilon) = \sum_{j=1}^{n} P_{1,j}(s;\varepsilon)\left\{i\varepsilon\, \partial_s P_{1,j}(s;\varepsilon) + [P_{1,j}(s;\varepsilon), H_o(s)]\right\}.$$

Noting that $[P_{1,j}, H_o] = [P_{1,j} - P_{o,j}, H_o]$ and that by perturbation theory $\lim_{\varepsilon\to 0}||P_{1,j} - P_{o,j}|| = 0$ one concludes that again for ε small enough $H_2(s;\varepsilon)$ satisfies H_2.

Obviously the procedure can be continued indefinitely: if $H_k(s;\varepsilon)$ is already constructed then

$$H_{k+1}(s;\varepsilon) = H_k(s;\varepsilon) + B_k(s;\varepsilon) \tag{2.5}$$

$$B_k(s;\varepsilon) = \sum_{j=1}^{n} P_{k,j}(s;\varepsilon)\left\{i\varepsilon\partial_s P_{k,j}(s;\varepsilon) + \left[P_{k,j}(s;\varepsilon), H_o(s)\right]\right\}. \tag{2.6}$$

In (2.6) $P_{k,j}(s;\varepsilon)$ are spectral projections of $H_k(s;\varepsilon)$ corresponding to the parts of the spectrum coinciding with $\sigma_j(s)$ in the limit $\varepsilon\to 0$. If $\varepsilon_{k+1}(s)$ is chosen such that

$$\sup_{\varepsilon\in[0,\varepsilon_k(s)]} ||\sum_{m=o}^{k} B_m(s;\varepsilon)|| \leq d(s)/4$$

then $H_{k+1}(s;\varepsilon)$ satisfies H_2 with $d(s)$ replaced by $d(s)/2$.

One can formulate now the basic estimation.

Theorem 1. Suppose H_1, H_2 hold true. Then for $s \in I$, $k = 0,1,\ldots$ there exist constants $0 < b_k(s) < \infty$ such that

i. If $\varepsilon_k(s)$ are defined by

$$\varepsilon_o(s) = \infty, \qquad \sum_{j=0}^{k-1} \varepsilon_k(s)^{j+1} b_j(s) = d(s)/4$$

then for $0 \le \varepsilon \le \varepsilon_k(s)$

$$||B_k(s;\varepsilon)|| \le \varepsilon^{k+1} b_k(s) \tag{2.7}$$

ii. If in addition H_4 holds true then

$$b_k = \sup_{s\in I} b(s) < \infty \tag{2.8}$$

and consequently

$$\varepsilon_k = \inf_{s\in I} \varepsilon_k(s) > 0 \tag{2.9}$$

iii. If in addition H_4, H_5 hold true then

$$\int_{-\infty}^{\infty} b_k(t)dt < \infty \tag{2.10}$$

iv. If for $s_o \in I$, $\partial_s^\ell (H(s) - i)^{-1} = 0$ for $\ell = 1,2,\dots,m$; then $B_\ell(s_o;\varepsilon) = 0$, $\ell = 0,1,\dots,m-1$ and consequently $b_\ell(s_o) = 0$, $\varepsilon_\ell(s_o) = \infty$ for $\ell = 0,1,\dots,m-1$.

From now on we shall suppose $H_1 - H_4$ hold true. The standard results about evolution equations [11] and H_3 assure the existence, unicity and strong continuity of the solution, $U_\varepsilon(s,s_o)$ of equation (1.1). Moreover, since B_k are bounded and norm differentiable the families $H(s) - B_k(s;\varepsilon)$, defined for $0 \le \varepsilon \le \varepsilon_k$ still satisfy H_3 and then the equations

$$i\varepsilon\partial_s U_k^A(s,s_o;\varepsilon) = \Big(H(s) - B_k(s;\varepsilon)\Big)U_k^A(s,s_o;\varepsilon) ;$$

$$U_k^A(s_o,s_o;\varepsilon) = 1 \tag{2.11}$$

have unique solutions.

The following is the main result of our lecture.

Theorem 2. For $s, s_o \in I$; $k = 0, 1,\dots$; $0 \le \varepsilon \le \varepsilon_k$

i. $$P_{k,j}(s;\varepsilon) = U_k^A(s,s_o;\varepsilon)P_{k,j}(s_o;\varepsilon)U_k^{A*}(s,s_o;\varepsilon) \quad . \tag{2.12}$$

ii. Let $\Omega_k(s,s_o;\varepsilon)$ be the *Möller operator*

$$\Omega_k(s,s_o;\varepsilon) = U_k^A(s,s_o;\varepsilon)^* U_\varepsilon(s,s_O) \quad . \tag{2.13}$$

Then

$$||U_k^A(s,s_o;\varepsilon) - U_\varepsilon(s,s_o)|| =$$

$$||1 - \Omega_k(s,s_o;\varepsilon)|| \le \varepsilon^k \int_{s_o}^{s} b_k(t)dt \quad . \tag{2.14}$$

iii. The inequality (2.14) can be improved

$$||\Omega_k(s,s_o;\varepsilon) - 1|| \le \varepsilon^{k+1} \; \omega_k(s,s_o) \tag{2.15}$$

with

$$\omega_k(s,s_o) = (n+1)\int_{s_o}^{s} b_{k+1}(t)dt + \Big(n(n-1)/2\Big)G\Big(b_k(s)+b_k(s_o)\Big), \quad G < \infty \; .$$

Let us rewrite some of the above results in more familiar forms. From (2.12) and (2.13)

$$||\Big(1 - P_{k,j}(s;\varepsilon)\Big)U_\varepsilon s,s_o)P_{k,j}(s_o;\varepsilon)|| =$$

$$= ||\Big(1 - P_{k,j}(s_o,\varepsilon)\Big) \; \Omega_k(s,s_o;\varepsilon) \; P_{k,j}(s_o;\varepsilon)|| =$$

$$= ||\Big(1 - P_{k,j}(s_o,\varepsilon)\Big)\Big(\Omega_k(s,s_o;\varepsilon) - 1\Big) \; P_{k,j}(s_o;\varepsilon)|| \quad . \tag{2.16}$$

From (2.16) and (2.15) one obtains

Corollary 1. For all s, $s_o \in I$, $k = 0, 1, \ldots$

$$||\Big(1 - P_{k,j}(s;\varepsilon)\Big)U_\varepsilon(s,s_o)P_{k,j}(s_O;\varepsilon)|| \le \varepsilon^{k+1}\omega_k(s,s_o). \tag{2.17}$$

In a similar way

$$||P_{k,j}(s;\varepsilon)U_\varepsilon(s,s_o)\Big(1 - P_{k,j}(s_o;\varepsilon)\Big)|| \le \varepsilon^{k+1}\omega_k(s,s_o). \tag{2.18}$$

Suppose now that H(s) is constant outside $J \subset I$ i.e. supp $\partial_s (H(s)-i)^{-1} \subset \subset J$. If $s_1, s_2 \notin J$ then from *Theorem 1* **iv.**, $P_{k,j}(s_i;\varepsilon) = P_{o,j}(s_i)$ all $k = 1,2, \ldots$ Then from (2.12) and (2.13) for k = 0

$$||\left(1 - P_{k,j}(s_2;\varepsilon)\right)U_\varepsilon(s_2,s_1)P_{k,j}(s_1;\varepsilon)|| =$$

$$= ||\left(1 - P_{o,j}(s_2)\right)U_\varepsilon(s_2,s_1)P_{o,j}(s_1)|| =$$

$$= ||\left(1 - P_{o,j}(s_1)\right)\Omega_o(s_2, s_1;\varepsilon)P_{o,j}(s_1)||$$

which together with (2.17) gives

Corollary 2. Suppose $s_1, s_2 \notin$ supp $\partial_s (H(s) - i)^{-1}$. Then for $k = 0,1,\ldots$

$$||\left(1 - P_{o,j}(s_1)\right)\Omega_o(s_2,s_1;\varepsilon)P_{o,j}(s_1)|| \le \varepsilon^{k+1}\omega_k(s_2,s_1). \quad (2.19)$$

$$||\left(1 - P_{o,j}(s_2)\right)U_\varepsilon(s_2,s_1)P_{o,j}(s_1)|| \le \varepsilon^{k+1}\omega_k(s_2,s_1) . \quad (2.20)$$

Similarly from (2.18)

$$||P_{o,j}(s_1)\Omega_o(s_2,s_1;\varepsilon)\left(1 - P_{o,j}(s_1)\right)|| \le \varepsilon^{k+1}\omega_k(s_2,s_1). \quad (2.21)$$

From (2.19) and (2.21)

$$||\left[P_{o,j}(s_1), \ \Omega_o(s_2,s_1;\varepsilon)\right]|| \le 2\varepsilon^{k+1}\omega_k(s_2,s_1) \quad . \quad (2.22)$$

Suppose now that $I = \mathbb{R}$ and $H_1 - H_5$ hold true. Then $\lim_{s\to\pm\infty} P_{k,j}(s;\varepsilon)$ exist and are equal to $\lim_{s\to\pm\infty} P_{o,j}(s)$. Then from *Theorem 1* **iii.** and (2.15)

Corollary 3. For k = 0, 1, 2,... there exist constants $\omega_k < \infty$ such that

$$||\left(1 - P_{o,j}(+\infty)\right)U_\varepsilon(\infty,-\infty;\varepsilon)P_{o,j}(-\infty)|| \le \varepsilon^{k+1}\omega_k \quad . \quad (2.23)$$

By perturbation theory there exist constants $P_k(s)$ such that

$$||P_{k,j}(s;\varepsilon) - P_{o,j}(s)|| \le \varepsilon P_k(s) \quad . \quad (2.24)$$

From (2.24) and (2.17)

Corollary 4. For all s, $s_o \in I$, k = 0,1, ...

$$\left\|\left(1-P_{o,j}(s)\right)U_{\varepsilon}(s,s_o)P_{o,j}(s_o)\right\| \le \varepsilon\left(P_k(s)+P_k(s_o)\right)+\varepsilon^{k+1}\omega_k(s,s_o). \tag{2.25}$$

The above results are various forms of the adiabatic theorem. Actually in physical literature the starting equation is

$$i\partial_t U_{\varepsilon}(t,t_o) = H(\varepsilon t)U_{\varepsilon}(t,t_o); \quad U_{\varepsilon}(t_o,t_o) = 1 .$$

The equation (1.1) is obtained by going to the *reduced time* $s = \varepsilon t$.

Corollary 1 is what we call *arbitrary order adiabatic theorem.* Let us stress that it is true for *all* $s \in I$, and is by far more powerfull than *Corollary 2* which is also called *adiabatic theorem to arbitrary order* [5]. The result in *Corollary 4* has an interesting consequence. Suppose $I = \mathbb{R}$ and $H_1 - H_4$ hold true. Then from (2.25)

$$\left\|\left(1-P_{o,j}(s)\right)U_{\varepsilon}(s,s_o)P_{o,j}(s_o)\right\| \le \varepsilon\left(P_k(s)+P_k(s_o)\right)+\mathrm{const}\,\varepsilon^{k+1}|s-s_o|$$

which implies that

$$\left\|\left(1-P_{o,j}(s)\right)U_{\varepsilon}(s,s_o)P_{o,j}\right\| \sim \varepsilon \quad , \tag{2.26}$$

for intervals of *reduced time* of order ε^{-k}, while the usual adiabatic theorem (i.e. (2.17) for k = 0) gives (2.26) only on intervals of order $\varepsilon^{o} = 1$. Let us remark also that $U_o^A(s,s_o;\varepsilon)$ coincides with the *adiabatic evolution* as defined and studied in [5] (see Eq. III.14 in R. Seiler's lecture). Yet another form of *Theorem 2* **ii.** is the following. Let $\psi \in \mathcal{H}$, $\|\psi\| = 1$, $\psi(s;\varepsilon) = U_{\varepsilon}(s,0)\psi$ and

$$I^{\psi}_{k,j}(s;\varepsilon) = \left(\psi(s;\varepsilon),\ P_{k,j}(s;\varepsilon)\psi(s;\varepsilon)\right) \quad . \tag{2.27}$$

Corollary 5.

$$\left|I^{\psi}_{k,j}(s;\varepsilon) - I^{\psi}_{k,j}(s_o;\varepsilon)\right| \le 2\varepsilon^{k+1}\omega_k(s,s_o) \quad . \tag{2.28}$$

In other words, up to errors of order ε^{k+1}, $P_{k,j}(s;\varepsilon)$ are constants of motion for the evolution given by (1.1).

We shall not discuss here the geometric content of the adiabatic theorem since this topic is nicely covered in the lectures of R. Seiler.

We shall only mention the fact the whole discussion can be repeated starting from $P_{k,j}(s;\varepsilon)$ instead of $P_{o,j}(s)$ (giving in particular the whole asymptotic expansion of the Berry's phase).

As emphasized by Wasow [7] the adiabatic invariance is related to the asymptotic diagonalisation for equation (1.1). Let $A_k(s,s_o;\varepsilon)$ as given by (see *Lemma 1* below)

$$i\partial_s A_k(s.s_o;\varepsilon) = \left(-i\sum_{j=1}^{n} P_{k,j}(s;\varepsilon)\partial_s P_{k,j}(s;\varepsilon)A_k(s,s_o;\varepsilon\right) \quad ,$$

$$A_k(s_o,s_o;\varepsilon) = 1 \tag{2.29}$$

The operators $A_k(s,s_o;\varepsilon)$ are unitary and

$$P_{k,j}(s;\varepsilon) = A_k(s,s_o;\varepsilon)P_{k,j}(s_o;\varepsilon)A_k^*(s,s_o;\varepsilon) \quad . \tag{2.30}$$

The following result shows that, up to errors of order ε^{k+1} the equation for $A_k^* U_\varepsilon$ is diagonal.

Theorem 3. Let $R(s,s_o;\varepsilon)$ be defined by

$$i\varepsilon\partial_s A_k^*(s,s_o;\varepsilon)U_\varepsilon(s,s_o) = R(s,s_o;\varepsilon)A_k^*(s,s_o;\varepsilon)U_\varepsilon(s,s_o) \tag{2.31}$$

Then

$$R(s,s_o;\varepsilon) = \sum_{j=1}^{n} P_{k,j}(s_o;\varepsilon)A_k^*(s,s_o;\varepsilon)H(s)A_k(s,s_o;\varepsilon)P_{k,j}(s_o;\varepsilon) + $$
$$+ R_1(s,s_O;\varepsilon) \tag{2.32}$$

where

$$||R_1(s,s_o;\varepsilon)|| \le \varepsilon^{k+1}b_k(s) \quad . \tag{2.33}$$

3. THE PROOFS

The first step in proving the basic estimate (2.7) is to derive a convenient form for $B_k(s;\varepsilon)$. To this end we make use of theory of transformation functions (which goes back to Daletsky and Krein, Sz.Nagy and Kato [6,12]). The basic construction goes as follows:

Consider a continuously differentiable family of orthogonal projections $Q_j(s)$, $j = 1,\dots,$ $j = 1,\dots,m$ with

$$Q_l(s)Q_m(s) = \delta_{l,m}Q_m(s), \quad \sum_{j=1}^{n} Q_j(s) = 1. \tag{3.1}$$

A transformation function [12,13] for the $Q_j(s)$ is a family $T(u,s)$ of unitary operators satisfying

$$Q_j(u) = T(u,s)Q_j(s)T^*(u,s), \qquad T(s,s) = \mathbb{1} \tag{3.2}$$

Lemma 1 [12,14]. If $K(s)$ is defined by

$$K(s) = -i\sum_{j=1}^{n} Q_j(s)\partial_s Q_j(s) \tag{3.3}$$

then

i. $K(s)$ is self-adjoint

ii. $Q_j(s)\, K(s)\, Q_j(s) = 0$ (3.4)

iii. The (unique) solution of

$$i\partial_u A(u,s) = K(u)A(u,s), \qquad A(s,s) = 1 \tag{3.5}$$

is a transformation function for $Q_j(s)$.

Consider now $K^\circ(u)$, $M^\circ(u,s;\varepsilon)$ as given by *Lemma 1* applied to $P^\circ_j(u) \equiv P_{o,j}(u)$. Define, for u in a small neighbourhood V of s

$$H^1(u,s;\varepsilon) = M^{\circ *}(u,s;\varepsilon)\left(H(u) - \varepsilon K^\circ(u)\right) M^\circ(u,s;\varepsilon) \tag{3.6}$$

If ε is sufficiently small, more exactly if

$$\varepsilon \sup_{u\in V} ||K^\circ(u)|| \le \inf_{u\in V} d(u)/4 \equiv \tilde{d}(s)/4 \tag{3.7}$$

by perturbation theory, the spectrum of H^1 still consists for all $u \in V$ of n disconnected pieces separated by intervals of length larger than $\tilde{d}/2$. Call the corresponding spectral projections $P^1_j(u,s;\varepsilon)$. Repeat the procedure starting from $P^1_j(u,s;\varepsilon)$. One can continue this process defining $H^2, H^3, \ldots$. The reason is that if $H^k(u,s;\varepsilon)$ is constructed then for

$$0 < \varepsilon < \tilde{\varepsilon}_k(s) = \tilde{d}(s)/4 \left(\sum_{\ell=0}^{k-1} \sup_{u\in V} ||K^\ell(u,s;\varepsilon)||\right)^{-1} \tag{3.8}$$

its spectrum is still well separated. So starting with k = 0 we have

$$K^k(u,s;\varepsilon) = -i\sum_{j=1}^{n} P_j^k(u,s;\varepsilon)\,\partial_u P_j^k(u,s;\varepsilon) \tag{3.9}$$

$$i\partial_u M^k(u,s;\varepsilon) = K^k(u,s;\varepsilon)M^k(u,s;\varepsilon); \qquad M^k(s,s;\varepsilon) = 1 \tag{3.10}$$

$$H^{k+1}(u,s;\varepsilon) = M^{k*}(u,s;\varepsilon)\left(H^k(u,s;\varepsilon)-\varepsilon K^k(u,s;\varepsilon)\right)M^k(u,s;\varepsilon). \tag{3.11}$$

Lemma 2.

$$B_k(s,\varepsilon) = -\,\varepsilon\, K^k(s,s;\varepsilon) \tag{3.12}$$

Proof. Let

$$Z^k = M^\circ M^1 \ldots M^k \tag{3.13}$$

By induction

$$i\varepsilon\partial_u Z^k = (H - Z^k H^{k+1} Z^{k*})Z^k \tag{3.14}$$

Indeed for k = 0 $\quad Z^\circ = M^\circ$ and from (3.11)

$$i\varepsilon\partial_u Z^\circ = \left[H - M^\circ M^{\circ *}(H - \varepsilon K^\circ)M^\circ M^{\circ *}\right]Z^\circ = \varepsilon K^\circ Z^\circ .$$

For k + 1, using (3.14), (3.10) and (3.11)

$$i\varepsilon\partial_u Z^{k+1} = i\varepsilon\partial_u Z^k M^{k-1} = \left(H - Z^k H^{k+1} Z^{k*} + \varepsilon Z^k K^{k+1} Z^{k*}\right)Z^{k+1} =$$

$$= \left(H - Z^{k+1} H^{k+2} Z^{k+1*}\right)Z^{k+1}$$

Again by induction

$$H_k(u;\varepsilon) = Z^{k-1}(u,s;\varepsilon)H^k(u,s;\varepsilon)Z^{k-1*}(u,s;\varepsilon) \tag{3.15}$$

$$P_{k,j}(u;\varepsilon) = Z^{k-1}(u,s;\varepsilon)P_j^k(u,s;\varepsilon)Z^{k-1*}(u,s;\varepsilon) \quad . \tag{3.16}$$

For k = 1 using (3.11), (3.13), (3.9) and the fact that $P_j^\circ = P_{o,j}$

$$Z^\circ H^1 Z^{\circ *} = M^\circ M^{\circ *}(H - \varepsilon K^\circ)M^\circ M^{\circ *} = H + i\,\varepsilon \sum_{j=1}^{n} P_{o,j}\,\partial_u P_{o,j} = H_1 .$$

For k + 1

$$Z^k H^{k+1} Z^{k*} = Z^k M^{k*}(H^k - \varepsilon K^k)M^k Z^{k*} = H_k - \varepsilon Z^{k-1} K^k Z^{k-1*} =$$

$$= H_k + i\varepsilon \sum_{j=1}^{n} P_{k,j} Z^{k-1}(\partial_u P^k_j) Z^{k-1*} =$$

$$= H_k + i\varepsilon \sum_{j=1}^{n} P_{k,j} \left\{ \partial_u P_{k,j} - (\partial_u Z^{k-1}) P^k_j Z^{k-1*} - Z^{k-1} P^k_j (\partial_u Z^{k-1*}) \right\} =$$

$$= H_k + \sum_{j=1}^{n} P_{k,j} \left\{ i\varepsilon\partial_u P_{k,j} - (H - H_k)P_{k,j} + P_{k,j}(H - H_k) \right\} = H_{k+1} .$$

In the calculation we have used (3.11), (3.13), (3.9) and (3.14).

Now by (2.5), (3.15) and (3.11)

$$B_k(s,\varepsilon) = H_{k+1}(s;\varepsilon) - H_k(s,\varepsilon) = H^{k+1}(s,s;\varepsilon) - H^k(s,s;\varepsilon)$$

$$= - \varepsilon K^k(s,s;\varepsilon)$$

and the proof of (3.12) is finished.

We shall estimate now $K^k(s,s;\varepsilon)$.

Lemma 3. For $j = 1,2,\ldots,n-1$ let C_j be contours (of finite lengths) surrounding $\sigma_j(s)$ such that $\mathrm{dist}(C_\ell, \sigma_j(s)) \ge d(s)/2$ for all ℓ, j. Then there exist:

$$0 \le b_{m,k}(s) < \infty , \qquad k = 0,1,\ldots; \quad m = 1,2, \ldots$$

$$0 \le d_{m,k}(s) < \infty , \qquad k,m = 0,1, \ldots$$

such that for $0 \le \varepsilon \le \tilde{\varepsilon}_k(s)$ as defined by (3.8) (with V an arbitrarily small neighbourhood of s)

$$\max_j \left|\left| \partial^m_u P^k_j(u,s;\varepsilon)\Big|_{u=s} \right|\right| \le b_{m,k}(s)\,\varepsilon^k \tag{3.17}$$

$$\sup \left|\left| \partial_u^m \left(H^k(u,s;\varepsilon) - z\right)^{-1} \right|_{u=s} \right|\right| \le d_{m,k}(s) \tag{3.18}$$

where the supremum is taken with respect to all $z \in \bigcup_{j=1}^{n-1} C_j$.

Proof. The proof is by induction over k. For $k = 0$, (3.18) is true by H_1, H_2 and the fact that C_j are compacts. For (3.17) and $j = 1,\ldots, n-1$ use the Riesz formula (2.2). For P_n^o use (2.3). Assume now that (3.17), (3.18) are true for $0,1,\ldots,k$. From the resolvent formula

$$(A + B - z)^{-1} = (A - z)^{-1} \left[1 + B(A - z)^{-1}\right]^{-1} \tag{3.19}$$

and (3.11) one gets:

$$(H^{k+1} - z)^{-1} = M^{k*}(H^k - z)\left[1 - \varepsilon K^k(H^k - z)^{-1}\right]^{-1} M^k \quad .$$

From this, taking the derivatives, using (3.9), (3.10) as well as the induction hypothesis (3.18) follows.

Now, using (3.19) and (3.11) we have

$$P_j^{k+1}(u,s;\varepsilon) - P_j^k(s,s;\varepsilon) = \frac{\varepsilon}{2\pi i} M^{k*}(u,s;\varepsilon) \int_{C_j} dz$$

$$\left(H^k(u,s;\varepsilon) - z\right)^{-1} \left[1 - \varepsilon K^k(u,s;\varepsilon)\left(H^k(u,s;\varepsilon) - z\right)^{-1}\right]^{-1}$$

$$K^k(u,s;\varepsilon)\left(H^k(u,s;\varepsilon) - z\right)^{-1} M^k(u,s;\varepsilon) \quad . \tag{3.20}$$

Differentiating (3.20) with respect to u one finds that all terms contain either K^k or its derivatives and therefore by (3.9) the derivatives of P_j^k. From this one obtains (3.17) by the induction hypothesis.

Proof of Theorem 1. Observe that

$$||K^\ell(s,s;\varepsilon)|| \le \sqrt{n-1} \max_j \left|\left| \partial_u P_j^\ell(u,s;\varepsilon)\right|_{u=s} \right|\right|$$

$$\le \sqrt{n-1}\ \varepsilon^\ell b_{1,\ell}(s) \equiv b_\ell(s)\ \varepsilon^\ell \quad ,$$

and that $\tilde{d}(s) \longrightarrow d(s)$ when V shrinks to s. The second point of *Theorem 1* is obvious by contruction.

Proof of Theorem 2. Consider for $f \in \mathcal{D}$

$$f(u) = U_k^{A*}(u,s;\varepsilon)\ P_{k,j}(u,s;\varepsilon)U_k^A(u,s;\varepsilon)f \quad .$$

One can easily see that f(u) is differentiable and

$$i\varepsilon\partial_u f(u) = U_k^{A*}\Big\{[B_k,\ P_{k,j}] + [H,\ P_{k,j}] + i\varepsilon\partial_u P_{k,j}\Big\} U_k^A\ f \quad . \tag{3.21}$$

Taking into account that $\sum_{j=1}^{n}\Big(P_{k,j}\partial_u P_{k,j} + (\partial_u P_{k,j})P_{k,j}\Big) = \partial_u \sum_{j=1}^{n} P_{k,j}^2 = 0$,

$P_{k,j}\partial_u P_{k,j} + (\partial_u P_{k,j})P_{k,j} = \partial_u P_{k,j}$ and using the definition (2.6) of B_k one can verify that the expression inside the curly bracket in (3.21) vanishes and then $\partial_u f(u) = 0$. It follows that $f(u) = f(s) = P_{k,j}(s;\varepsilon)f$ whereof the first part of *Theorem 2* follows.

By direct computation

$$i\varepsilon\partial_u\Omega_k(u,s;\varepsilon) = U_k^{A*}(u,s;\varepsilon)\ B_k(u;\varepsilon)\ U_k^A(u,s;\varepsilon)\Omega_k(u,s;\varepsilon) \tag{3.22}$$

whereof (2.14) follows taking into account (2.7).

Consider now (2.15)

$$||\Omega_k - 1|| \le ||\Omega_{k=1} - 1|| + ||\Omega_{k+1} - \Omega_k|| =$$

$$= ||\Omega_{k+1} - 1|| + ||\Omega_{k+1}\Omega_k^* - 1|| \quad . \tag{3.23}$$

We shall estimate now the last term in (3.23). Using (2.13)

$$||\Omega_{k+1}\Omega_k^* - 1|| \le \sum_{j,\ell} P_{k,j}(s_o;\varepsilon)\Big(U_{k+1}^A U_k^{A*} - 1\Big)P_{k,\ell}(s_o;\varepsilon)|| \tag{3.24}$$

Using (2.12) we get for the off diagonal terms $j \neq \ell$:

$$||P_{k,j}(s_o;\varepsilon)U^A_{k+1}(s,s_o;\varepsilon)U^{A*}_k(s,s_o;\varepsilon)P_{k,\ell}(s_o;\varepsilon)|| \le$$

$$\le ||P_{k,j}(s_o;\varepsilon)-P_{k+1,j}(s_o;\varepsilon)|| + ||P_{k+1,j}(s;\varepsilon)-P_{k,j}(s;\varepsilon)|| \tag{3.25}$$

From (2.7) and the regular perturbation theory there exists a constant G such that

$$\max_j ||P_{k+1,j}(s;\varepsilon) - P_{k,j}(s;\varepsilon)|| \le G\, b_k(s)\, \varepsilon^{k+1} \quad . \tag{3.26}$$

Consider now the diagonal part. From the differential equation for U^A_k one obtains

$$i\varepsilon\partial_s U^A_{k+1} U^{A*}_k = U^A_{k+1}(B_k - B_{k+1})U^{A*}_k \tag{3.27}$$

whereof using again (2.12)

$$||P_{k,j}(s_O;\varepsilon)\left(U^A_{k+1}(s,s_o;\varepsilon)U^{A*}_k(s,s_o;\varepsilon) - 1\right)P_{k,j}(s_o;\varepsilon)|| \le$$

$$\le \varepsilon^{-1}\int_{s_o}^{s} ||P_{k,j}(t;\varepsilon)\left(B_k(t;\varepsilon) - B_{k+1}(t;\varepsilon)\right)P_{k,j}(t;\varepsilon)||\, dt \quad . \tag{3.28}$$

From (3.12) and (3.4) applied to $P_{k,j}$

$$P_{k,j}(s;\varepsilon)\, B_k(s;\varepsilon)\, P_{k,j}(s;\varepsilon) = 0 \tag{3.29}$$

whereof the r.h.s. of (3.28) is bounded by $\varepsilon^{k+1}\int_{s_o}^{s} b_{k+1}(t)\, dt$.

Combining this with (3.23), (2.14), (3.24), (3.25) and (3.26) one obtains (2.15).

Proof of Theorem 3.

Consider

$$\Phi_k(s,s_o;\varepsilon) = A^*_k(s,s_o;\varepsilon)U^A_k(s,s_o;\varepsilon) \quad .$$

Due to (2.12) and (2.30)

$$[\Phi_k(s,s_o;\varepsilon),\ P_{k,j}(s_o;\varepsilon)] = 0 \quad . \tag{3.30}$$

The differential equation for Φ_k can be found from (2.11) and (2.29)

$$i\varepsilon\partial_s\Phi_k = A_k^*(H - B_k + i\varepsilon \sum_{j=1}^{n} P_{k,j}\partial_s P_{k,j})A_k\Phi_k \quad . \tag{3.31}$$

This can be simplified. Due to (3.30)

$$\Phi_k(s,s_o;\varepsilon) = \sum_{j=1}^{n} P_{k,j}(s_o;\varepsilon)\Phi_k(s,s_o;\varepsilon)P_{k,j}(s_o;\varepsilon) \quad . \tag{3.32}$$

Now using again (3.4), (3.29) and (2.30) it follows from (3.31) and (3.32)

$$i\varepsilon\partial_s\Phi_k(s,s_o;\varepsilon) = \left(\sum_{j=1}^{n} P_{k,j}(s_o;\varepsilon)A_k^*(s,s_o;\varepsilon)H(s)A_k(s,s_o;\varepsilon) \cdot \cdot P_{k,j}(s_o;\varepsilon)\right)\Phi_k(s,s_o;\varepsilon) \quad . \tag{3.33}$$

Then *Theorem 3* follows from the fact that

$$A_k^*U_\varepsilon = \Phi_k\Omega_k$$

(3.33), (3.22) and (2.7).

REFERENCES

1. Born, M. Fock, V., Beweis des Adiabatensatzes, Z.Phys. **5**, 165-180 (1928).
2. Kato, T., On the adiabatic theorem of quantum mechanics, J.Phys.Soc.Japan **5**, 435-439 (1950).
3. Garrido, L.M., Generalized adiabatic invariance, J.Math.Phys. **5**, 335-362 (1964).
4. Nenciu, G., Adiabatic theorem and spectral concentration, Commun.Math.Phys. **82**, 121-135 (1981).
5. Avron, J.E., Seiler, R., Yaffe, L.G., Adiabatic theorem and applications to the quantum Hall effect, Commun. Math.Phys. **110**, 33-49 (1987).

6. Krein, S.G., Linear differential equations in Banach spaces, A.M.S. Translations of Mathematical Monographs, vol. **29**, Providence, 1971.

7. Wasow, W., Topics in the theory of linear ordinary differential equations having singularities with respect to a parameter, Series de Mathématiques Pures et Appliquées, IRMA Strasbourg, 1978.

8. Jdanova, G.V., Fedoriuk, M.V., Asymptotic theory for the systems of second order differential equations and the scattering problem, Trudy Mosk.Mat.Ob. **34**, 213-242 (1977).

9. Boutet de Monvel, A., Nenciu, G., On the theory of adiabatic invariants for linear Hamiltonian systems, C.R.Acad.Sci. Paris (in press).

10. Nenciu, G., Rasche, G., Adiabatic theorem and Gell-Mann-Low formula, H.P.A. **62**, 372-388 (1989).

11. Tanabe, H., Equations of Evolution, Pitman, Berlin, 1966.

12. Kato, T. Perturbation Theory for Linear Operators, Springer, Heidelberg, 1976.

13. Reed, M., Simon, B., Methods of Modern Mathematical Physics, IV, Academic Press, New York, 1978.

14. Messiah, A., Quantum Mechanics, II, North Holland, Amsterdam, 1969.

15. Nenciu, A., Nenciu, G., Dynamics of Bloch Electrons in Electric Fields I, II, J.Phys. **A14**, 2817-2835 (1981); J.Phys. **A15**, 3313-3331 (1982).

16. Nenciu, G., Dynamics of band electrons in electric and magnetic fields: Rigorous justification of the effective Hamiltonians, Rev.Mod.Phys. to be published.

On the Quantum Hall-Effect

Ruedi Seiler
Technische Universität Berlin MA7-2
Straße des 17. Juni 136
D-1000 Berlin 12

I. Introduction

In these lectures I will try to give you an impression about the Quantum Hall-Effect. The theoretical point of view will mainly be the one developed by Y. Avron and myself [1,2,3]. The section on corrections to the Kubo formula is based on some work together with M. Klein [4]. During the three hours of these lectures it will be impossible to do justice to all authors who have contributed to this field.
It will be even impossible to give technical details of our own approach. I hope this will have the positive effect of exposing more clearly the structure and main points, serving by that as reading guide to the literature about this subject.

The discovery of the Quantum Hall-Effect by v. Klitzing in an experiment at the Grenoble laboratory is now nine years old [5]. In essence it showed integrality of the Hall conductivity and at same time vanishing of the direct resistivity in a two dimensional system in a large magnetic field, at low temperature and over a wide range of the magnetic field strength.

The models used to describe this effect are with a few exceptions all within the framework of nonrelativistic quantum mechanics. There are the general many-body type systems with arbitrary interactions. They will be the main subject of these lectures. Special many body systems have been discussed mostly to account for the Fractional Quantum Hall-Effect [6]. The most popular approach is in the framework of one particle theories. There interactions are taken into account by means of an effective potential. In many cases impurities are represented by random potentials. To the one particle category with effective potential belong some particularly important contributions: The first one is by Laughlin [7] — where the relevance of a topologically nontrivial configuration space became clear — and the second one by Thouless, Kohomoto, Nightingale and de Nijs [8], where for the first time integrality of Hall conductivity as given by the Kubo Formula became clear (for periodic potentials). The topological aspect of particles in magnetic fields was anticipated by Dubrovin and Novikov [9] and Novikov [10]. However this work passed unnoticed in the west to a large extent.

The main results of our work which parallels the one by Niu and

A. Boutet de Monvel et al. (eds.), Recent Developments in Quantum Mechanics, 151–173.

Thouless [11] is the following : If the many body Hamiltonian of the Quantum Hall System has a gap separating the groundstate from the rest of the spectrum, then Hall conductivity is quantized. This holds because Hall conductivity, given by Kubo's formula, is equal to the integral of the first Chern class over a torus divided by the degeneracy of the groundstate [1,11,12]. It was furthermore possible to prove that Kubo's formula represents Hall conductivity [2] and more recently, that the corrections to Ohm's law relating the Hall current to the voltage V are of infinite order in V in the limit $V \rightarrow 0$. This limit is essentially the same as the adiabatic limit as we will see later. Hence the adiabatic theorem, as it has been explained in the lectures by G. Nenciu, is crucial in this approach [16,2,4, see also the lecture notes by Nenciu].

Let me now tell you what is the main structure of the theory about the Quantum Hall-Effect: Dynamics of a general Quantum Hall System is naturally formulated in terms of a family of Schrödinger operators parametrized by the coordinates of a two dimensional torus. The parameters can be interpreted as two magnetic fluxes through two holes of configuration space or as the boundary conditions in a rectangular geometry of configuration space. Since the Hamiltonian is an operator valued function on this parameter space, the spectral projection onto the groundstate defines a line, respectively a vector bundle over the torus with a canonical connection. Miraculously its first Chern class turns out to be Kubo's formula for the Hall conductivity.

The fact that the structurally simple experimental result discovered by von Klitzing can be understood in terms of very elementary concepts of topology, geometry and quantum mechanics is very satisfactory. After all it is the strong desire of all of us here to rely in our interpretation of the world on fundamental laws. This ultimate goal, which seems to me the only way of coping with the complexities of this world and in particular the one just here, can fortunately be achieved in the context of the Quantum Hall-Effect.

II.Experiment

A typical Hall sample as it is presently used is the Gallium - Arsenide Aluminum –Gallium –Arsenide heterostructure (Fig. 1). The Aluminum content is about half of the Gallium.

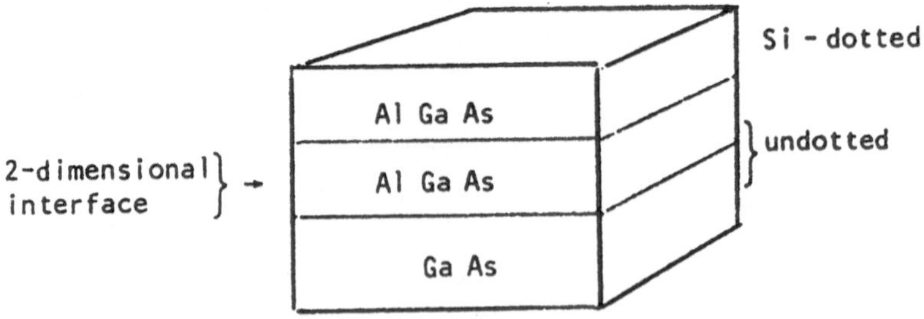

Fig. 1:

The Hall geometry is schematically the following (Fig. 2).

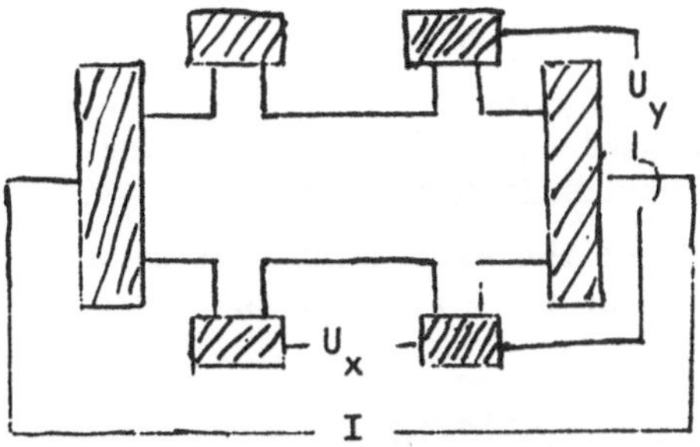

Fig. 2:

Using the most popular, phenomenological one particle band picture one would draw the following figure 3 :

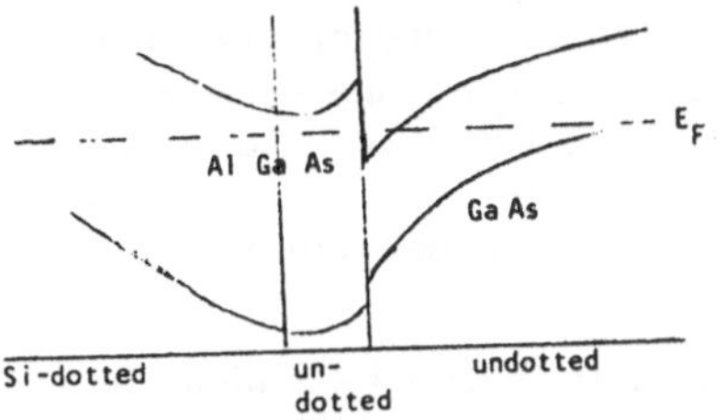

Fig. 3:

Experiments are typically done with the following parameter values :

Magnetic field B	:	1 — 12 Tesla
Temperature T	:	0.5 — 4 Kelvin
Impurities	:	$10^4/cm^3$ (distance between impurities ~ 1000 Å)
Hall current	:	~ 1 μA
Hall voltage	:	~ 10 mV
Effective dielectric constant in GaAs	:	~ 10
Distance between Gallium Atoms	:	~ 2 Å

The main experimental result is summarized in the following schematic plot (Figure 4) :

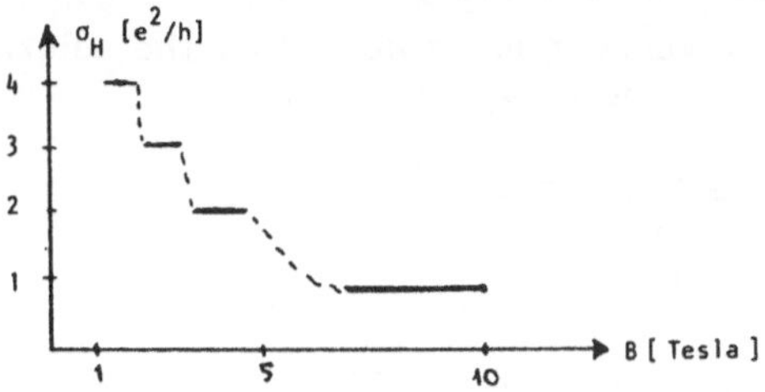

Fig. 4:

It shows the following remarkable facts : The Hall conductivity in units e^2/h is an integer over large ranges of the magnetic field strengt B. This holds with the precision of about $1:10^8$. Hall conductivity takes small integer values and is a decreasing function of B. For very low temperatures and high purity of the sample, plateaus at fractional values develop, approaching something like a devils staircase in the limit of no impurities (translational invariance; see remark 1 at the end of these lectures).

AlGaAs—GaAs is not the only material known to show the Quantum Hall Effect. In fact the material used in v. Klitzing's original experiments was different. He used a Silicon-Siliconoxide transistor.

For more information about the experimental situation and some of the theory, I refer you to the book edited by Prange [14].

III. General Quantum Hall-Systems

1. The model

Now I should like to describe the model used here to describe the Quantum Hall-Effect. It is formulated in terms of non relativistic many body Quantum theory. In our original article we used a two hole geometry for configuation space Λ in $\mathbb{R}^2$ or $\mathbb{R}^3$. Here I will use the conventional rectangular geometry $\Lambda = [0,L_1] \times [0,L_2]$ in $\mathbb{R}^2$ as in [4]. The two hole geometry is a way more fundamental. However its interpretation requires more time than I can spend just for this point.

The magnetic field perpendicular to the 1—2 plane in $\mathbb{R}^3$ is described by minimal coupling $p \to p - eA$, where p denotes the momentum operator, e the electric charge and A the vector potential for the magnetic field $B = \operatorname{rot} A$. The electric field E is again introduced by minimal coupling to a time dependent vector potential $\phi_1(t)a_1$,

$$E(t,x) = -\frac{d}{dt}\,\phi_1(t)\,a_1(x)\ .$$

ϕ_1 is a time dependent function and $a_1 = e_1/L_1$ a multiple of the unit vector e_1 in the 1—direction. By fiat the electric field is switched on and off by means of a smooth switching function $f \in C^\infty(\mathbb{R},\mathbb{R})$. It has the following graph (Fig. 5) :

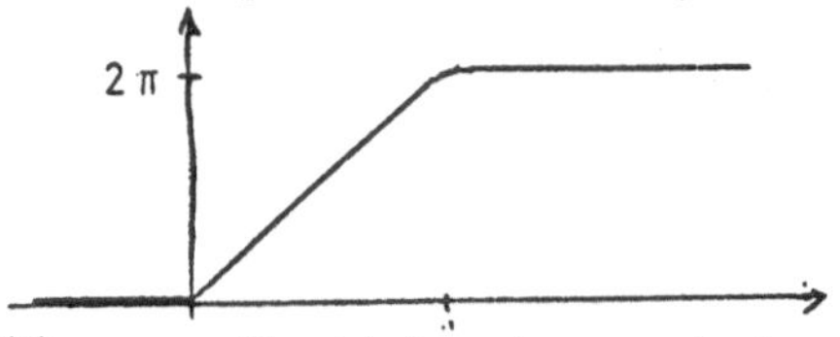

Fig. 5:

We set $\phi_1(t) := -f(s=t/\tau)$, where τ is by definition the adiabatic parameter used later to define dynamics. Doing so we get for E :

$$\begin{aligned} E(t,x) &= \frac{1}{\tau}\, f'(s=t/\tau)\, a_1 \\ &= \frac{1}{\tau}\, f'(s)\, \frac{e_1}{L_1} \quad . \end{aligned}$$

Hence the voltage is $(1/\tau)f'(s)$. Since in the interval of scaled time [0,1] this function is mostly $2\pi/\tau$, we introduce the notation $V := 2\pi/\tau$. V denotes the voltage which is present across the sample in the 1—direction between switching on and switching off of the electric field E.

Dynamics of a general Quantum Hall System is formulated in terms of the following Hamiltonian :

(III.1) $$\tilde{H}(\phi) = \sum_{i=1}^{N} \tfrac{1}{2} \left[(p_i - A(x_i)) - \phi_1 a_1 - \phi_2 a_2\right]^2 + W(x_1, \ldots, x_N)$$

$$, \phi \in \mathbb{R}^2 .$$

Mathematically it is defined as the form perturbation of the Laplacian with periodic boundary conditions on $\Lambda^N \subset \mathbb{R}^{2N}$. Charge and mass of the electron as well as the speed of light are one in the units used here. The vector potentials are those introduced earlier. W stands for the interaction of the electrons with the nuclei and with impurities as well as the electrons among themself. It is a pure multiplication operator and typically of the form

$$W(x_1, \ldots, x_N) = \sum_{i=1}^{N} V(x_i) + \sum_{i<j} \frac{1}{|x_i - x_j|} ,$$

where V is assumed to be form bounded with respect to the 2-dimensional Laplacien with relative bound zero. The proper definition of $\tilde{H}$ requires some technical machinery for which I should like to refer you to [4]. It turns out that under physically reasonable conditions $\tilde{H}(\phi)$ is a family of selfadjoint operators, analytic in $\phi \in \mathbb{R}^2$ and of type A in Kato's terminology.

The Schrödinger equation for the interacting many body system in terms of the scaled time $s = t/\tau$ is given by

(III.2) $$\begin{aligned} i\, \partial_s U_\tau(s,\varphi) &= \tau\, H(s)\, U_\tau(s,\varphi) \quad , \quad (s,\varphi) \in \mathbb{R}^2 \\ U_\tau(0,\varphi) &= 1 \end{aligned}$$

where we used the notation

(III.3) $$H(s,\varphi) = \tilde{H}(\phi_1 = f(s), \phi_2 = \varphi) .$$

The existence of a unitary propagator which is sufficiently smooth in s and φ is a consequence of a well known theorem by Kato and Yoshida supplemented by an additional argument explained in [2]. (Kato's theorem theorem was discussed in detail by K. Yajima in his lectures).
The operator U_τ acts on the Hilbert $\mathcal{H} := L^2(\Lambda)^{\otimes N}$. The technical properties just mentioned are important in the following. Smoothness in the scaled time s and type A analyticity are needed to apply the adiabatic theorem. Smoothness in the parameter φ is crucial to define and analyze the Hall current (see below).

The following property of $\tilde{H}(\phi)$ is instrumental in the following : $\tilde{H}(\phi)$ is up to unitary equivalence doubly periodic in ϕ. More precisely there exists a family of unitary operators $G(\phi)$, $\phi \in \mathbb{R}^2$, such that

$$\tilde{\tilde{H}}(\phi) := G(\phi)\, \tilde{H}(\phi)\, G^{-1}(\phi)$$

is doubly periodic in ϕ with period $(2\pi, 2\pi)$. G is a multiplication operator and explicitly given by the formula

$$G(\phi) := \exp i(\phi_1 F_1 + \phi_2 F_2)$$

$$F_i(x) := \sum_{l=1}^{N} (a_i, x_l) \quad , \quad x = (x_1, \ldots, x_N) \in \Lambda^N, \quad i=1,2 \ .$$

This claim is not difficult to prove. Notice that $\hat{H}(\phi)$ has a ϕ –independent symbol. The domain of $H(\hat{\phi})$ is given by

$$D(\hat{H}(\phi)) = G(\phi)\, D(\hat{H}(\phi)) = G(\phi)\, D(H(0)) \ .$$

Hence the elements of $D(\hat{H}(\phi))$ are elements of the first Sobolev space over Λ^N, which satisfy the following boundary conditions :

$$\psi(x_1, \ldots, x_k + L_r e_r, \ldots, x_N) = e^{i(a_r, L_r e_r)\phi_r}\, \psi(x_1, \ldots, x_N)$$

$$= e^{i\phi_r}\, \psi(x_1, \ldots, x_N) \ ; \qquad k = 1 \ldots N \ ; \ r = 1,2 \ .$$

Hence the domain of $\hat{H}(\phi)$ is periodic in ϕ.

To end this section let me explain in more detail, what is meant if we talk about the gap condition as it was mentioned in the introduction. This is most easily done by means of a drawing (Fig. 6) :

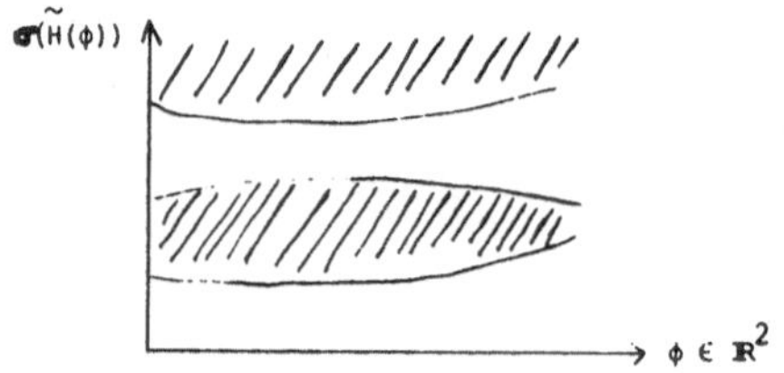

Fig. 6:

The shaded areas possibly contain spectrum of $\tilde{H}(\phi)$ or $\hat{H}(\phi)$. We assume that there is a ground state or a "ground band" of $\tilde{H}(\phi)$ of finite dimension separated from the rest of the spectrum by a gap. This allows to define the by assumption finite dimensional spectral projectors

$$\tilde{P}(\phi) := -\frac{1}{2\pi i} \int_{\Gamma(\phi)} (\tilde{H}(\phi) - z)^{-1}\, dz$$

$$\hat{P}(\phi) := -\frac{1}{2\pi i} \int_{\Gamma(\phi)} (\hat{H}(\phi) - z)^{-1}\, dz \quad ,$$

where $\Gamma(\phi)$ is a circle in the complex plane circling the "ground band" in the positive sense. Now we are ready to define what we shall call the "physical state" : It is the following nonnormalized, zero temperature, density operator :

$$P_T(s,\varphi) := U_T(s,\varphi)\, P_\varphi\, U_T^*(s,\varphi) \quad .$$

Here P_φ stands for $\tilde{P}(0,\varphi)$.

2. The Current

Now let me introduce the current operator. Its analysis, in particular its limiting behavior for $V \to 0$, will be crucial for formulation and proof of the main result : The validity of Ohm's law for Hall current and Hall voltage up to very small corrections (infinit order) and integrality of Hall conductivity in the case $\dim P_\varphi = 1$.

Before I will go into the analysis of the current operator in the framework of the above model, let me give you an informal and heuristic argument about why one might expect Ohm's law to hold in such a simple quantum mechanical model. This is by no means obvious. In fact Quantum-Hall-Systems are the only quantum mechanical models, where the validity of Ohm's law can be proved.

Consider a particle in a constant magnetic field B in the 3–direction and constant electric in the 1–direction. The Hamiltonian for this system is

$$H := \tfrac{1}{2} v^2 - E x_1 \tag{III.4}$$

where $v = p - A$ denotes the velocity and $p = i\nabla$ the momentum operator. A is the vector potential and $B = \operatorname{rot} A$.

If $E = 0$ it is easy to see that the vector operator

$$c := x + v \wedge b \,, \qquad b := \frac{B}{|B^2|}$$

is an integral of motion. It is the observable related to the Landau center of motion and generates magnetic translations [see also remark 1].

Notice that all commutators between the vector components of v and c are zero except

$$[v_1, v_2] = -i|B|, \qquad [c_1, c_2] = -i/|B| \tag{III.5}$$

If $E \neq 0$, c is not anymore an integral of motion. It satisfies the Heisenberg equations

$$\dot{c} = D := E \wedge b \,.$$

D is called the drift vector. The above equation can be solved and leads to the following result :

$$v(t) := R(t)\,(v_0 - D) + D$$
$$x(t) := Dt + R(t)\,(v_0 - D) \wedge b + (c_0 - D \wedge b)$$

R(t) is a time dependent rotation. The above equation tells us, that the particle moves on a circle around a center and this center has a constant drift velocity D in the direction perpendicular to B and E. The radius of the circle is proportional to $1/|B|$.

If we now consider the case of many independent particles of the above type the Hall current is

$$j := \rho D$$

where ρ denotes the density of particles. It can be "computed" as follows : v_1 and v_2 obey the same commutation relations as the position and momentum operator of a harmonic oscillator

$$[v_2,v_1]: = i|B| \sim [p,q] = i\hbar .$$

Under the substitution $v_2 \to p$, $v_1 \to q$ and $B \to \hbar$ the Hamiltonian (III.4) becomes the Hamiltonian of a simple harmonic oscillator

$$\tfrac{1}{2}(p^2+x^2)$$

with spectrum

$$\sigma: = \{ (n+\tfrac{1}{2})\hbar \mid n=0,1,\dots \} .$$

Hence the Hamiltonian (III.4) for a particle in a constant magnetic field has the same spectrum. The Hamiltonian (III.4) for $E=0$ commutes with the vector operator c. Hence every spectral point is infinitly degenerate. This comes about since the center of gyration of a particle can be translated to every point in the 1—2 plane. The generators of such translations are the operators c_1 and c_2. We know their commutation relations. If we substitute $-1/|B|$ by $\hbar$ the commutation relation (III.5) turn into Heisenberg's commutation relation.

If we now adopt the rule "One particle of each Landau level into every cell of phase space or 1-2 plane with volume h or $2\pi/B$ respectively", we get the following density of particles for a state with the first n Landau levels occupied :

$$\rho: = \frac{n}{h} = \frac{n}{2\pi/B} = \frac{|B|}{2\pi} n .$$

Hence the Hall current in absolut value turns out to be

$$\begin{aligned} j: &= \frac{|B|}{2\pi} n|D| \\ &= \frac{|B|}{2\pi} n \frac{E\cdot B}{|B|^2} \\ &= \frac{n}{2\pi} |E|. \end{aligned}$$

Conductivity is therefore

$$\sigma: = \frac{n}{2\pi} .$$

After this heuristic argument for quantization of the Hall conductivity let me come back to the proper definition of the Hall current in the framework of our model.

The observable for the Hall current is by definition the multiparticle Hamiltonian (III.3) differentiated with respect to its "dual variable" φ. The Hall current is the expectation value with respect to the normalized physical state:

$$I(s,\varphi): = \frac{1}{q} \text{Trace } P_7(s,\varphi)\, \partial_\varphi H(s,\varphi),$$

where q denotes the dimension of $P(s,\varphi)$.

It is useful for the following to note that the current can be written as a time derivate :

(III.6) $$I(s,\varphi) := \frac{i}{q\tau}\, \partial_s \,\mathrm{Trace}\, P_\varphi\, U_\tau^*(s,\varphi)\, \partial_\varphi\, U_\tau(s,\varphi)\ .$$

To prove this equation one has to make use of the Schrödinger equation (III.2) and regularity of U_τ in s and φ. (III.6) can be interpreted in the following manner: Electric current is equal to time derivative of the transported charge. The factor $1/\tau$ comes about since we used the derivative with respect to scaled time s and not physical time t.

Substituting $V = 2\pi/\tau$ for τ in the above expression for the current leads to

$$I(s,\varphi) : = V \frac{i}{2\pi q}\, \partial_s \,\mathrm{Trace}\, P_\varphi\, U_\tau^*(s,\varphi)\, \partial_\varphi\, U_\tau(s,\varphi)\ .$$

It is therefore natural to define conductivity by

(III.7) $$\sigma(s,\varphi;V) : = \frac{i}{2\pi q}\, \partial_s \,\mathrm{Trace}\, P_\varphi\, U_\tau^*(s,\varphi)\, \partial_\varphi\, U_\tau(s,\varphi)\ .$$

It will be the purpose of the following section to analyse this expression in the limit $V \to 0$.

Conductivity is related to transported charge

$$\begin{aligned} Q(\varphi;V) : &= \tau \int_0^1 ds\ I(s,\varphi) \\ &= \frac{i}{q}\ \mathrm{Trace}\, P_\varphi\, U_\tau^*(s,\varphi)\, \partial_\varphi\, U_\tau(s,\varphi)\, \Big|_{s=0}^{1} \end{aligned}$$

as follows :

(III.8) $$\int_0^1 \sigma(s,\varphi;V)\ ds = \frac{1}{2\pi}\, Q(\varphi)\ .$$

It tells us that the scaled time average of the Hall conductivity equals the transported electric charge.

Let me end this section by giving you a nice but useless alternative formula for the current. It displays an interesting geometrical and dynamical aspect. The formula is given in terms of an identity between skew symmetric two forms on $\mathbb{R}^2$:

$$I(s,\varphi)\ ds\ d\varphi = (\partial_\varphi\, E(s,\varphi))\ ds\ d\varphi + \frac{i}{q\tau}\ \mathrm{Trace}\, P_\tau\, dP_\tau\, dP_\tau\ .$$

The last term is explicitly given by

$$\mathrm{Trace}\, P_\tau\, dP_\tau\, dP_\tau = \mathrm{Trace}\, P_\tau(s,\varphi) \left(\frac{\partial P_\tau(s,\varphi)}{\partial \varphi} \frac{\partial P_\tau(s,\varphi)}{\partial s} - \frac{\partial P_\tau(s,\varphi)}{\partial \varphi} \frac{\partial P_\tau(s,\varphi)}{\partial s} \right) ds\ d\varphi$$

$E(s,\varphi)$ denotes the instantaneous energy,

$$E(s,\varphi) := \frac{1}{q}\,\mathrm{Trace}\,P_\tau(s,\varphi)\,H(s,\varphi)\ .$$

The first term is reminiscent of the expression for the group velocity of a wave package in a lattice. It is of dynamical nature. The second term is geometrical. This will become clear in the next section. In fact the expression is already very close to a first Chern class.

3. Main results

Now we are ready to formulate the main results about the Hall conductivity for general Quantum Hall Systems. Recall that $\tilde{H}(\phi)$ denotes a family of many body Hamiltonians and that $\tilde{H}(\phi)$ is unitary equivalent to another family of Schrödinger operators

$$\hat{H}(\phi) = G(\phi)\tilde{H}(\phi)G(\phi)^*$$

which is periodic in both arguments ϕ_1 and ϕ_2, i.e. $\hat{H}(\phi)$ is defined on the 2-torus. The same is of course true for the spectral projection $\hat{P}(\phi)$ of the "ground band" of $\hat{H}(\phi)$, which is defined provided the gap condition holds.

In the following we will not make a statement about conductivity $\sigma(s,\varphi;V)$ itself but about its time and flux average

$$\langle\sigma\rangle := \int_0^1 ds \int_0^{2\pi} \frac{d\varphi}{2\pi}\,\sigma(s,\varphi;V)\ .$$

Main Result:
If the gap condition and some technical assumptions hold then

(III.9) $$\langle\sigma\rangle := \frac{1}{2\pi q}\cdot\frac{i}{2\pi}\int_T \mathrm{Trace}\,\hat{P}d\hat{P}d\hat{P} + O(V^\infty)\ .$$

The integral runs over the torus. q denotes the dimensionality of range $\hat{P}$ as before.

Since

(III.10) $$\frac{i}{2\pi}\int_T \mathrm{Trace}\,\hat{P}d\hat{P}d\hat{P}$$

is the integral of the first Chern class of a vector bundle over T it is an integer (see remark 3 at the end of these lectures).

We will sketch the proof of both statements, the asymptotic expression (III.9) for $\langle\sigma\rangle$ and the integrality of (III.10). The latter is of course only of pedagogical value since it is a well known fact. In remark 3 at the end of these lectures I will give you the connection between the traditional manner of writing vector bundles, connections etc. and the one used here in terms of projection valued functions.

Since conductivity is related to charge transport, the above result (III.9) can be reformulated as follows :

$$\text{(III.11)} \qquad <Q> := \int_0^{2\pi} \frac{d\varphi}{2\pi}\, Q(\varphi)$$

$$:= \frac{i}{2\pi q} \int_T \text{Trace } \hat{P} d\hat{P} d\hat{P} \;+\; O(1/\tau^{\infty}) \,.$$

Recall that the adiabatic parameter τ (time scale) is related to the electric potential V by $\tau = 2\pi/V$. Hence equation (III.11) is a consequence of (III.9).

There are lots of questions related to these results. On a technical level they do not say anything about the actual size of the correction terms; i.e. they just tell us that for every integer n there is a constant c(n), such that

$$\left| <\sigma> - \frac{1}{2\pi q} \cdot \frac{i}{2\pi} \int \text{Trace } \hat{P} d\hat{P} d\hat{P} \right| \leq c(n)\, V^n \quad , \quad (V \to 0) \,.$$

Although it is in principle possible to give a rough estimate of c(n) in terms of the gap energy, we have not done so.

We have nothing to say at present about the actual value of the integer. In particular we can not understand why experimentally it is a small number and why it is decreasing in B ?

What about the gap condition. Is it physically reasonable or can it even be proved for the present model ? In the context of general Quantum Hall System this is out of reach. The gap condition can be motivated in the framework of a non interacting particle model with effective potential. In fact in this context the gap condition can even be relaxed considerably. It can be replaced by the assumption that there is a mobility edge in the bands and localized states. This makes the results better suited for an explanation of the plateaus.

4. Sketch of proof

The main ingredients of proof are the adiabatic theorem explained in the lectures of G. Nenciu and a simple version of the Chern-Simons formula (Step2).

A. We start by the following chain of arguments leading to the statement :

$$\text{Ch} := \frac{1}{2\pi i} \int_T \text{Trace } \hat{P} d\hat{P} d\hat{P} \qquad \text{is an integer.}$$

I shall divide the proof of this statement into 5 steps.

Step 1: There exists a family of unitary operators $U(\phi)$, $\phi \in \mathbb{R}^2$ such that

$$\text{(III.12)} \qquad \begin{aligned} &\text{i)} \;\; U(0,\phi_2) = 1 \;,\; \phi_2 \in \mathbb{R} \\ &\text{ii)} \;\; U(\phi)\hat{P}(0,\phi_2)U^{-1}(\phi) = \hat{P}(\phi) \;\;,\;\; \phi \in \mathbb{R}^2 \,. \end{aligned}$$

This result is due to Kato. It is proved easily by analyzing the following initial value problem :

$$i\partial_1 U(\phi_1,\phi_2) = i\,[\,\partial_1 \hat{P}(\phi_1,\phi_2)\,,\,\hat{P}(\phi_1,\phi_2)\,]\,U(\phi_1,\phi_2)$$

$$U(0,\phi_2) = 1 \;,\; \phi_2 \in \mathbb{R}\,.$$

Due to the regularity of $\hat{P}(\phi)$ it has a unique solution (Theorem of Kato and Yoshida, see also lectures of K.Yajima). It satisfies equations (III.12). Notice that $U(\phi)$ is not necessarily periodic in ϕ_1 and ϕ_2 even though $\hat{P}(\phi)$ is. The periodicity of $\hat{P}$ implies however that,

$$U(\phi_1,0) = U(\phi_1,2\pi) \;,\; \phi_1 \in \mathbb{R}\,.$$

U is therefore defined on the cylindre $\mathbb{R} \times S^1$.

Step 2: $\quad \mathrm{Trace}\,\hat{P}d\hat{P}d\hat{P} = d(\mathrm{Trace}\,\hat{P}(\phi_1=0,\cdot)U^{-1}dU)$:

This statement is a consequence of the following

Lemma: Let $U(\phi)$ and $P(\phi)$ be smooth families of unitary operators and projections acting on a Hilbert space. Then the following identity between 2 forms on $\mathbb{R}^2$ holds :

$$\mathrm{Trace}\,P_u dP_u dP_u = \mathrm{Trace}\,PdPdP + d(\mathrm{Trace}\,PU^{-1}dU)$$

where P_u is defined by

$$P_u := UPU^{-1}\,.$$

Substituting $\hat{P}(\phi_1=0,\phi_2)$ for P and $\hat{P}$ for P_u, the above lemma says just what we want to show in this second step. Notice that

$$\mathrm{Trace}\,\hat{P}(\phi_1=0,\cdot)d\hat{P}(\phi_1=0,\cdot)d\hat{P}(\phi_1=0,\cdot) = 0$$

and that $\mathrm{Trace}\,\hat{P}d\hat{P}d\hat{P}$ is a two form on the torus (periodic in ϕ), whereas $\mathrm{Trace}\,\hat{P}(\phi_1=0,\cdot)U^{-1}dU$ is a two form on $\mathbb{R}^2$.

Step 3:

$$\int_Q d(\mathrm{Trace}\,\hat{P}(\phi_1=0,\cdot)U^{-1}dU) = \int_0^{2\pi} d\phi_2\;\mathrm{Trace}\,\hat{P}(\phi_1=0,\phi_2)\;U^{-1}(\phi_1=2\pi,\phi_2)\;\partial_2 U(\phi_1=2\pi,\phi_2)\;,$$

Q denotes the square of length 2π:

This is an immediate consequence of the remark made in step 1: U is defined on the cylindre with value 1 on the circle $\phi_1=0$.

Step 4:

$$\text{Trace}\,\hat{P}(\phi_1=0,\phi_2)\; U^{-1}(\phi_1=2\pi,\phi_2)\; \partial_2 U(\phi_1=2\pi,\phi_2) \;=\; \text{Trace}\,\hat{P}(0)\; W^{-1}(\phi_2)\; \partial_2 W(\phi_2) \;,$$

where W is defined by

$$W(\phi_2) \;:=\; \hat{P}(0)\; U(\phi_1,\phi_2)\; V(\phi_2)\; \hat{P}(0) \,\Big|_{\phi_1=0}^{2\pi} .$$

V is a solution of the initial value problem

$$i\partial_2 V(\phi) \;=\; i\;[\; \partial_2\hat{P}(0,\phi)\;,\; \hat{P}(0,\phi)\;]\; V(\phi)$$

$$V(\phi=0) \;=\; \mathbf{1}$$

Again by Kato's theorem (see step 1) V has the property

$$\hat{P}(0,\phi_2) \;=\; V(\phi_2)\; \hat{P}(0)\; V(\phi_2)^{-1} \;.$$

Notice that W^{-1} denotes the inverse in the subspace range $\hat{P}(0)$ and $\text{Trace}\,\hat{P}(0)$ the trace in the same subspace. The proof of the above statement is purely algebraic and straight forward.

Putting all 4 steps together one gets the following result :

$$\text{Ch} \;=\; \frac{1}{2\pi i}\int_0^{2\pi} d\phi_2\; \partial_2 \text{Trace}\,\hat{P}(0)\;\log\; W(\phi_2)$$

$$\;=\; \frac{1}{2\pi i}\int_0^{2\pi} d\phi_2\; \partial_2 \log\;\det\;\hat{P}(0)\; W(\phi_2)$$

where W is defined by

$$W(\phi_2) \;=\; \hat{P}(0)\; U(\phi_1,\phi_2)\; V(\phi_2)\; \hat{P}(0) \,\Big|_{\phi_1=0}^{2\pi} .$$

Now we are ready to do the 5-th and last step :

Step 5: The function $\det\,\hat{P}(0)\;W(\phi_2)$ maps the circle S^1 - parametrized by ϕ_2 - on the unit circle $\{\, z \mid |z| = 1\}$ in $\mathbb{C}$. The winding number of this map is the Chern number Ch defined previously:

To prove this statement it is enough to show that $\det\hat{P}(0)\,W(\phi_2)$ is 2π-periodic. This is a straight forward computation. I will not present it here.

B. Next we look at the proof of the statement :

(III.13) $$\langle\sigma\rangle \;=\; \frac{\text{Ch}}{2\pi q} \;+\; O(V^{\infty}) \;.$$

The main instrument is the adiabatic theorem. It is used here in the form: Let

(III.14) $$H_A(s,\varphi) := H(s,\varphi) + \frac{i}{\tau}\,[\,\partial_s P(s,\varphi)\,,\,P(s,\varphi)\,]$$

be the adiabatic Schrödinger operator and $U_A(s,\varphi)$ the corresponding propagator. The wave operator, which compares the physical and the adiabatic time evolution is defined by the equation

(III.15) $$\Omega(s,\varphi) := U_A^*(s,\varphi)\,U_\tau(s,\varphi) \quad .$$

In the limit $V \to 0$ U_A approximates the physical evolution in the following sense :

$$\Omega(s,\varphi) = 1 + O(V) \,.$$

Under the assumption that the switching function f(s) and therefore also the Hamiltonian $H(s,\varphi)$ is constant in the vicinity of $s=0$ and $s=1$ for all φ the adiabatic theorem implies

(III.16) $$\Omega(1,\varphi)\,P_\varphi = P_\varphi\,\Omega(1,\varphi) + O(V^\infty)$$

uniformly in φ. In other words $\Omega(1,\varphi)$ and P_φ commute modulo a term of arbitrary high power in V respectively $1/\tau$.

The proof of (III.13) is again divided into four steps :

Step 1:

(III.17) $$<\sigma> = \frac{i}{2\pi q}\int_0^{2\pi}\frac{d\varphi}{2\pi}\,\mathrm{Trace}\,P_\varphi U_A^*\partial_\varphi U_A\,\Big|_{s=0}^{1} + \\ + \frac{i}{2\pi q}\int_0^{2\pi}\frac{d\varphi}{2\pi}\,\mathrm{Trace}\,P_\varphi \Omega^*\partial_\varphi \Omega\,\Big|_{s=0}^{1}$$

To proof of this statement is just a matter of plugging the expression $U_\tau = U_A\Omega$,
which is a consequence of the defining equation (III.15), into equation (III.7).

Step 2: The first term of the rhs of (III.17) is $\frac{1}{2\pi q}\int_T \mathrm{Trace}\,\hat{P}d\hat{P}d\hat{P}$:

Reading the Chern-Simons formula backwards and using the intertwining property of the adiabatic propagator which reads

$$P(s,\varphi) = U_A(s,\varphi)\,P_\varphi\,U_A^{-1}(s,\varphi) \quad , \quad (s,\varphi)\in\mathbb{R}^2$$

one gets the equation

$$d\,(\,\mathrm{Trace}\,P_\varphi U_A^* dU_A\,) = \mathrm{Trace}\,P\,dP\,dP \;.$$

Since $U_A(s,\varphi=0) = U_A(s,\varphi=2\pi)$ for all s and $U_A(s=0,\varphi) = 1$ for all φ one gets

$$\int_{Q=[0,1]\times[0,2\pi]} d(\operatorname{Trace} P_\varphi U_A^* dU_A) = \int_0^{2\pi} d\varphi \operatorname{Trace} P_\varphi U_A^* \partial_\varphi U_A$$

$$= \int_Q \operatorname{Trace} PdPdP \ .$$

Hence it remains to show that

$$\int_Q \operatorname{Trace} PdPdP = \int_T \operatorname{Trace} \hat{P}d\hat{P}d\hat{P} \ .$$

I will not prove this in detail because the argument is technical and no new ideas are needed. Let me just mention that again the Chern-Simons formula is very useful.

Step 3: The second term of (III.17) equals

$$\frac{i}{2\pi q}\int_0^{2\pi} \frac{d\varphi}{2\pi} \operatorname{Trace} Z^{-1}\partial_\varphi Z$$

where Z is defined by the equation $Z := P_\varphi \Omega P_\varphi$ and Z^{-1} is the inverse in range P_φ.

Here the adiabatic theorem in the variant (III.16) will be crucial. Consider the equations mod V^∞

$$\operatorname{Trace} P_\varphi \Omega^* \partial_\varphi \Omega = \operatorname{Trace} P_\varphi \Omega^* P_\varphi (\partial_\varphi \Omega) P_\varphi$$

$$= \operatorname{Trace} Z^* \partial_\varphi (P_\varphi \Omega P_\varphi) - \operatorname{Trace} Z^* (\partial_\varphi P_\varphi) \Omega P_\varphi -$$

$$- \operatorname{Trace} Z^* P_\varphi \Omega \partial_\varphi P_\varphi \ .$$

The two last terms are both of order V^∞ because - again mod V^∞-

$$\operatorname{Trace} Z^* (\partial_\varphi P_\varphi) \Omega P_\varphi = \operatorname{Trace} P_\varphi \Omega P_\varphi (\partial_\varphi P_\varphi) P_\varphi \Omega P_\varphi$$

$$\operatorname{Trace} Z^* P_\varphi \Omega \partial_\varphi P_\varphi = \operatorname{Trace} \Omega P_\varphi \Omega P_\varphi (\partial_\varphi P_\varphi) P_\varphi$$

and

$$P_\varphi (\partial_\varphi P_\varphi) P_\varphi = 0 \ .$$

The last equation is a trivial but important consequence of $P_\varphi^2 = P_\varphi$.

Step 4: The second term of (III.17) is of order V^∞ :

Similarly to the argument in step 4 of the proof of integrality (part A), one can derive the formula

$$\text{(III.18)} \quad \int_0^{2\pi} \frac{d\varphi}{2\pi} \, \mathrm{Trace}\, P_\varphi \Omega^* \partial_\varphi \Omega \Big|_{s=0}^{1} = \int_0^{2\pi} \frac{d\varphi}{2\pi} \, \partial_\varphi \log \det P_\varphi Z(\varphi) + O(V^\infty)$$

Due to the adiabatic theorem we know that

$$\det P_\varphi Z(\varphi) = 1 + O(V) \ .$$

Furthermore $Z(0) = Z(2\pi)$. So the image of the circle S^1 under the map

$$\begin{aligned} S^1 &\to \mathbb{C} \\ \varphi &\mapsto \det P_\varphi Z(\varphi) \end{aligned}$$

is a closed curve in the complex plane in the vicinity of 1. Hence the logarithm of $\det P_\varphi Z(\varphi)$ is uniquely defined. The integral on the right hand side of (III.18) is therefore zero.

IV. Remarks

In this chapter I will present some complementary remarks to the results about general Quantum Hall Systems.

1. Degeneracy of ground state :

As special case of the result about the averaged Hall conductivity we have the following implication :

Ground state non degenerate $\Rightarrow$ $<\sigma> = \frac{1}{2\pi}$ times an integer.

The question arises, under what conditions is the ground state of a general Quantum Hall System non degenerate.

In answering this question one can adopt two point of views. On the one hand the ground state of a two parametric family of selfadjoint Hamiltonians is generally non degenerate because the codimension of selfadjoint matrices with degenerate spectrum among selfadjoint matrices is three. This is the von Neumann-Wigner theorem. On the other hand the typical splitting of energy eigenvalues for a multiparticle system is much too small to be of any significance since it is much smal ler than the thermal energy.

Whether the groundstate of a Quantum Hall System is degenerate or not can only be discussed in the frame work of a one particle theory with effective potential. There is one interesting exception to this statement. It concerns a general Quantum Hall System in rectangular configuration space with twisted boundary conditions (see below) and translationally invariant interactions. There it can be shown that every discrete eigenvalue has to be degenerate. The degree of degeneracy depends upon the so called filling factor ν. The result is due to Halperin and independently to Avron and Zak.

Consider the Hamiltonian

$$H = \frac{1}{2}\sum_{i=1}^{N_e} v_i^2 + \sum_{i<k} V(x_i - x_k) \quad , \quad v_i = p_i - A(x_i) \ ,$$

for N_e electrons in the magnetic field $B = \mathrm{rot}\,A$, with the translational invariant potential V. Let us denote the magnetic translations by t(a), $a \in \mathbb{R}^2$. They obey the commutation relations

$$t(a)\ t(b) = t(b)\ t(a)\ \exp[i\det(B,a,b)] \ ,$$

for all a, b in the 1-2 plane of $\mathbb{R}^3$ and B in the 3-direction. We choose the magnetic field so, that $\det(B,e_1,e_2) = 2\pi N_f$ where N_f is an integer called the number of fluxes through the rectangle spanned by the unit vectors e_1 and e_2 in the 1-2 plane. Obviously $t(e_1)$ and $t(e_2)$ commute.

The magnetic translations on the multiparticle space are defined in terms of the one particle operator by:

$$T(a) = \underbrace{t(a) \otimes \ldots\ldots \otimes t(a)}_{N_e}$$

By construction one gets the commutation relation

$$T(e_1)\ T(e_2) = T(e_2)\ T(e_1)\ \exp[2\pi i N_f N_e] \ .$$

Since H commutes with the magnetic translations T(a) we can decompose the Hilbert space of states according to a direct integral. Every fibre is characterized by the boundary condition

$$T(e_1)\ \psi = e^{i\phi_1}\psi \ ,$$
$$T(e_2)\ \psi = e^{i\phi_2}\psi \ .$$

The corresponding family of Schrödinger operator is denoted by $H(\phi_1,\phi_2)$. Its about the eigenvalues of this two parametric family of selfadjoint operators which I am going to make a statement on the degeneracy.

Let us denote by t_1 and t_2 the following operators

$$t_1 := T(\tfrac{1}{N_f} e_1) \ ,$$
$$t_2 := T(\tfrac{1}{N_f} e_2) \ .$$

The obey the commutation relation

$$t_1 t_2 t_1^{-1} t_2^{-1} = e^{2\pi i\nu} \quad , \quad \nu := \frac{N_e}{N_f} \ .$$

Furthermore they commute with H, $T(e_1)$ and $T(e_2)$. Hence they are mapping every fibre of the direct integral decomposition into itself. Denoting by $t_1(\lambda,\phi)$ the restriction to the eigenspace of $H(\phi)$ with eigenvalue λ inside the fibre labeled by ϕ we get the result :

$$\det t_1(\lambda,\phi)\, t_2(\lambda,\phi)\, t_1^{-1}(\lambda,\phi)\, t_2^{-1}(\lambda,\phi) = e^{2\pi i \nu d(\lambda,\phi)} = 1$$

where d denotes the dimensionality of the eigenspace (λ,ϕ). Hence we get the result :

$$\nu = \frac{p}{q} \;, \quad p \wedge q = 1 \qquad \Rightarrow \qquad q \text{ divides } d(\lambda,\phi) \;.$$

This very strange result shows among other things that the family of selfadjoint operators $H(\phi)$ is non generic.

2. One particle picture

In this remark let me describe the general picture about the integer quantum Hall effect in the effective one particle model. The spectrum of the one particle Hamiltonian with a (random) potential modelling impurities gives the following spectral density

Fig. 7:

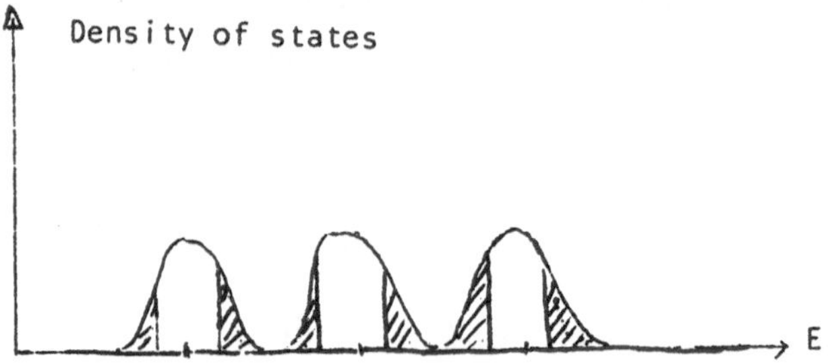

The location of the bands are given by the Landau energies which are multiples of $\hbar\omega_c$. The impurities are expected to broaden the peaks of the purely magnetic system. The one particle states in the shaded areas are localized and do not contribute to the Hall current. Only the states in the center of the bands are responsible for charge transport. If the Fermi energy is put between the bands the multiparticle Hamiltonian has a non degenerate ground state. The Hall conductivity is an integer divided by 2π. In the one particle picture integrality can also be shown if the Fermi energy is in the shaded area of localized states. In this picture the size of the effective adiabatic parameter can be estimated very crudely [4]. It is in this setting only that one can discuss the infinite volume limit. This has been done by Bellissard [15]. He used some powerful techniques developed by Connes in the field of noncommutative geometry.

3. Geometry of Adiabatic Dynamics

It is a remarkable fact that time dependent quantum dynamics aquires in the adiabatic limit surprising geometric properties. They are at the root of the results about the Quantum Hall Effect and intimately connected with what is called Berry's phase [13, 17, 18, 2].

Consider the parameter space $\mathbf{X}$. It is typically a manifold, say the torus or an n-dimensional real space $\mathbb{R}^n$. Let $\tilde{H}(x)$, $x \in \mathbf{X}$, be an operator valued function acting on elements in the Hilbert space $\mathcal{H}$, such that

i) $\tilde{H}(x) = \tilde{H}(x)^*$, $x \in \mathbf{X}$

ii) $\tilde{H}$ satisfies the gap condition (Fig. 6) and $\tilde{P}(x)$ is the affiliated spectral projection.

iii) $(\tilde{H}(x)+i)^{-1}$ is smooth in $x \in X$

iv) $D(\tilde{H}(x))$ is an x-independent domain in $\mathcal{H}$.

Let ω be a smooth path in X and $H = \tilde{H} \circ \omega$ the trace of $\tilde{H}$ on ω. Then we can define the initial value problem

$$\text{(IV.1)} \qquad i\partial_s U(s) = (f(H) + i\,[\dot{P},P])\, U(s)$$
$$U(0) = 1 .$$

where $P := \tilde{P} \circ \omega$ and f belongs to an appropriate function space. The above initial value problem has a unique solution U(s) (see the lectures by K. Yajima). It satisfies the intertwining property :

$$\text{(IV.2)} \qquad P(s) = U(s)\, P(0)\, U^{-1}(s).$$

This is easy to prove by differentiating both sides of the equation.

Notice that we have already encounterd a special case of the situation just mentioned. In the case of general Quantum-Hall systems the manifold X is $\mathbb{R}^2$. The operator valued function $H(s,\varphi)$ is defined by equation (III.3) and the function f is the identity. Equation (IV.1) specializes to the initial value problem for the adiabatic time evolution $U_A(s,\varphi)$ with the adiabatic Hamiltonian $H_A(s,\varphi)$ defined previously (III.15). Let me now argue that U is the parallel transport along the curve ω with respect to a canonical connection ∇ on the trivial vector bundle $\mathfrak{B} = X \times \mathcal{H}$.

On $\mathfrak{B}$ we consider two connections. Recall first that by definition a connection on a vector bundle E with base manifold X is a complex linear map from the smooth sections $\Gamma(E)$ to the sections $\Gamma(T^*M \otimes E)$ satisfying the Leibniz rule:

$$\text{(IV.3)} \qquad \nabla(f\psi) = (df)\psi + f\,\nabla\psi , \qquad f \in C^\infty(X) ,\ \psi \in \Gamma(E) .$$

The first and trivial connection on $\mathfrak{B}$ is just the differentation d on X. (IV.3) is obviously true. Next we define the canonical connection ∇ :

$$\text{(IV.4)} \qquad \nabla := \tilde{P}\, d\tilde{P} + (1-\tilde{P})\, d(1-\tilde{P}) .$$

A straightforward computation shows that ∇ can also be written as follows

$$\nabla = d - [(d\tilde{P}) , \tilde{P}] .$$

The covariant differentiation along ω is therefore given by

$$\text{(IV.5)} \qquad (\dot{\omega}\nabla) = (\dot{\omega}d) - [\dot{\omega}(d\tilde{P}) , \tilde{P}]$$
$$= \partial_s - [\dot{P} , P] .$$

Hence the section $\tilde{\psi} \in \Gamma(E)$ is horizontal along ω or equivalently the vectors $\psi(s) = \tilde{\psi} \circ \omega(s)$, $s \in [0,1]$, are parallel if

$$\partial_s \psi(s) = [\dot{P}(s) \, , \, P(s)] \, \psi(s) \, .$$

Hence if we set $f = 0$ in the differential equation (IV.1) we get the result

$$\psi(s) = U(s) \, \psi(0) \, .$$

The case of a general function f can be treated analogously.

Now I want to explain a most remarkable property of ∇ which is related to the intertwining property (IV.2) of U. To formulate this consider the subbundle $\tilde{\mathfrak{B}}$ of the trivial bundle $\mathfrak{B}$ defined by

$$\tilde{\mathfrak{B}} := \{ \langle x,v \rangle \, | \, x \in X, \; v = P(x)v \in \mathcal{H} \} \subset \mathfrak{B}$$

On $\tilde{\mathfrak{B}}$ there is again a canonical connection

$$\tilde{\nabla} := \tilde{P}d \, . \tag{IV.6}$$

It is the restriction of ∇ in the following sense

$$\tilde{\nabla}\tilde{\psi} = \nabla\tilde{\psi} \, , \; \tilde{\psi} \in \Gamma(\tilde{\mathfrak{B}}) \tag{IV.6}$$

The above equation is an immediate consequence of the definitions (IV.4) and (IV.6) and can be interpreted as the infinitesimal version of the intertwining property (IV.2). With the following picture I should like to visualize the situation :

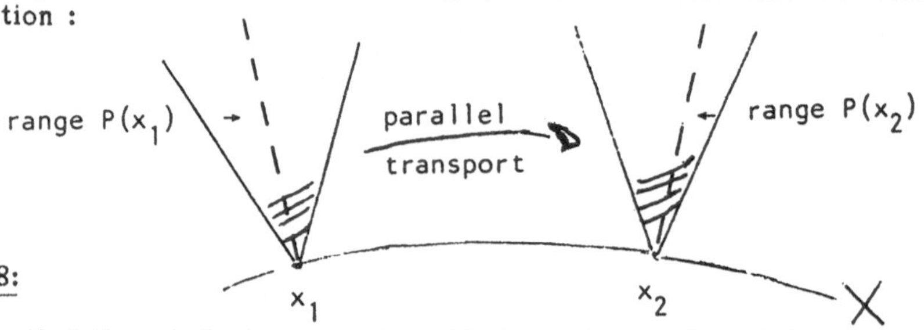

Fig. 8:

$\tilde{\nabla}$ is called the adiabatic connection. It gives rise to the adiabatic curvature

$$\begin{aligned} \tilde{\Omega} &:= \tilde{\nabla}^2 \\ &= \tilde{P} \, (d\tilde{P}) \, (d\tilde{P}) \, \tilde{P} \, . \end{aligned} \tag{IV.7}$$

From $\tilde{\Omega}$ the Chern classes and the Chern character are derived by taking traces of the elementary symmetric polynomials in $\tilde{\Omega}$. The first Chern class is by definition

$$c_1(\Omega) = - \frac{1}{2\pi i} \, \text{Trace} \, \tilde{\Omega} \, .$$

To end this section let me indicate the connection between the formalism used here and the conventional one of differential geometry. Let X be a manifold, $\mathcal{H}$ a Hilbert space and P a smooth function of $x \in X$ with values in the finite dimensional projections of $\mathcal{H}$. We shall now construct a vector bundle E with base manifold X, projection

$$\pi : \quad E \to X \,,$$

covering $\{U_\alpha\}$, $\alpha \in I$, coordinate functions

$$\phi_\alpha : \quad \pi^{-1}(U_\alpha) \to \mathbb{R}^n$$

and transition functions

$$g_{\alpha\beta} := \phi_\alpha \phi_\beta^{-1} \,.$$

Let $\{U_\alpha\}$, $\alpha \in I$ be a covering of X and $\{s_\alpha^1(x), \ldots, s_\alpha^n(x)\}$, $\alpha \in I$, be an orthonormal basis of range P(x), $x \in U_\alpha$, smooth in X. (The existence of such a covering and such a basis is guaranteed under very weak conditions). Then we construct the coordinate map
as follows :

$$\phi_\alpha : \quad \pi^{-1}(U_\alpha) \to \mathbb{R}^n$$

$$\psi \quad \to \{(s_\alpha^i, \psi)\}_{i=1}^n \,.$$

This construction allows the translation of the formalism used here into the standard one.

4. The Fractional Quantum Hall Effect

In the sense of Mathematical Physics there is no satisfactory theory about the Fractional Quantum Hall Effect. Many interesting elements exist but the picture is much less complete as the one for the Integer Quantum Hall Effect. There are very exciting connections in particular with two-dimensional quantum field theories which are still far from being understood. The main problem however is to prove stability of the spectrum of the many body Hamiltonian for rational filling numbers [3]. This might be the subject of some lectures in the future.

References

[1] Avron, J.E., Seiler, R.:
Quantisation of the Hall conductance for general multiparticle Schrödinger Hamiltonians.
Phys. Rev. Lett. 54, 259-262 (1985).

[2] Avron, J.E., Seiler, R., Yaffe, L.G.:
Adiabatic Theorems and Applications to the Quantum Hall Effect.
Commun. Math. Phys. 110, 33-49 (1987).

[3] Avron, Y., Seiler, R., Shapiro, B.:
General Properties of Quantum Hall Hamiltonians for Finite Systems.
Nucl. Phys. B 265, 364-374 (1986).

[4] Klein, M., Seiler, R.:
Power-law Corrections to the Kubo Formula Vanish in Quantum Hall Systems.
Technische Universität Berlin, Fachbereich Mathematik MA 7-2, D-1000 Berlin 12, Reprint 214 (1989).

[5] von-Klitzing, K., Dorda, G., Pepper, M.:
New method for high accuracy determination of the fine structure constant based on the quantized Hall effect.
Phys. Rev. Lett. 45, 494-497 (1980).

[6] Laughlin, R.B.:
Phys. Rev. Lett. 50, 1395-1398 (1983).
Phys. Rev. B 27, 3383-3389 (1983).

[7] Laughlin, R.B.:
Quantized Hall Conductivity in Two Dimensions.
Phys. Rev. B 23, 5652-5654 (1981).

[8] Thouless, D.J., Kohmoto, M., Nightingale, M., den-Nys, M.:
Quantized Hall Conductance in a two dimensional periodic potential.
Phys. Rev. Lett. 49, 405-408 (1982).

[9] Dubrovin B.A., Novikov S.P.:
Ground states of two-dimensional electron in a periodic magnetic field.
Sov.Phys.JETP 52, 511-516 (1980)

[10] Novikov S.P.:
Magnetic Bloch functions and vector bundles.
Sov. Math. Dokl. 23, 298-303 (1981)

[11] Niu, Q., Thouless, D.J.:
Quantized adiabatic charge transport in the presence of substrate disorder

and many body interactions.
J. Phys. A 9, 30-49 (1984).

[12] Avron, J.E., Seiler, R., Simon, B.:
Homotopy and Quantization in Condensed Matter Physics.
Phys. Rev. Lett. 51, 51-53 (1983).

[13] Nenciu, G.:
Adiabatic Theorem and Spectral Concentration.
Commun. Math. Phys. 82, 121-135 (1981).

[14] Prange, R.E., Girvin, S.M. (Editors):
The Quantum Hall Effect.
Graduate Texts in Contemporary Physics, Springer-Verlag (1987).

[15] Bellissard, J.:
C^*-Algebras in Solid State Physics.
2D Electrons in a uniform magnetic field.
Operator Algebras and Applications, Vol. 2.
D.E. Evans and M. Takesaki Eds.,
London Math. Soc. Lecture Notes 136, Cambridge (1988).

[16] Kato, T.:
On the Adiabatic Theorem of Quantum Mechanics.
J. Phys. Soc. Jap. 5, 435-439 (1950).

[17] Berry, M.:
Quantal Phase Factors Accompagnying Adiabatic Changes.
Proc. Roy. Soc. London A 392, 45-57 (1984).

[18] Simon, B.:
Holonomy, the Quantum Adiabatic Theorem and Berry's Phase.
Phys. Rev. Lett. 51, 2167-290 (1983).

Magnetic Schrödinger operators and effective Hamiltonians

Johannes Sjöstrand
Dépt. de Mathématiques, Bât. 425,
Université de Paris Sud, F-91405 Orsay, France
and URA 760, CNRS

0. Introduction

In this survey we discuss some recent progress in the semiclassical analysis for magnetic Schrödinger operators with periodic electric potentials and a constant magnetic field. Most of the material is based on joint works with B. Helffer and this part has been covered by the lecture notes [S1]. Some more recent work with C. Gérard and A. Martinez [GMS] is here surveyed for the first time.

We shall discuss the relation between Schrödinger operators, magnetic matrices and effective Hamiltonians, and in particular explain the so called Peierls substitution [P], used by physisists for a long time. In the three dimensional case we give a result about the density of states which is related to the de Haas-van Alphen effect in solid state physics, explained by Onsager [O]. In the two dimensional case, we obtain in many cases effective Hamiltonians which are small perturbations of the so called Harper's operator, and one can obtain results about the structure of the spectrum of such operators (see [HS2-5]],[Bel1,2],[S1]). Finally we describe a more direct approach to effective Hamiltonians, which was inspired by works of Guillot-Ralston-Trubowitz [GuRT] and by Buslaev [B] and we explain how a small perturbation of Harper's operator may appear from the case of a quasiperiodic operator in one dimension. Of independent interest are the relations between the various types of operators, roughly indicated by the diagram:

A. Boutet de Monvel et al. (eds.), Recent Developments in Quantum Mechanics, 175–193.

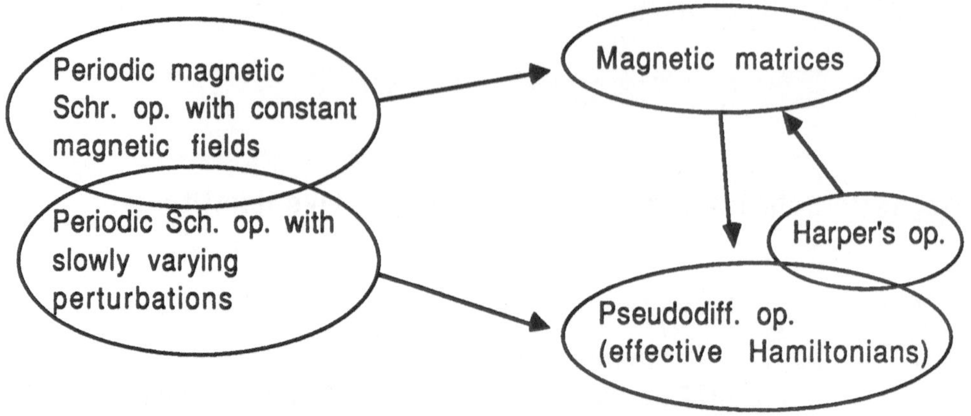

The point of view of C^*-algebras is not developed here, and we refer to the surveys of Bellissard [Bel 1,2]. The reduction from periodic magnetic Schrödinger operators can be done in many asymptotic regimes (see [HS2,5]), but we shall here concentrate on the case of weak magnetic fields (see Nenciu [Ne2], Bellissard [Bel1,2], Helffer-Sjöstrand [HS1,2], [S1]).

1. Reduction to magnetic matrices

Let us first make a quick review of standard Floquet (-Bloch) theory in the case of a vanishing magnetic field. (Fore more details see Reed-Simon [ReSi], Sjöstrand [S1].) Let $e_1,..,e_n$ be a basis in $\mathbb{R}^n$ and consider the lattice $\Gamma=\oplus_1^n\mathbb{Z}e_j$. The dual lattice is then defined by: $\Gamma^*=\{\gamma^*\in\mathbb{R}^{n*};\ \gamma\cdot\gamma^*\in 2\pi\mathbb{Z}\}=\ \oplus_1^n\mathbb{Z}2\pi e_j^*$, where $\{e_j^*\}$ is the standard dual basis of $\{e_j\}$. The operator U defined by:

(1.1) $Uu(x,\theta)=\Sigma_{\gamma\in\Gamma}\, e^{i\gamma\cdot\theta}u(x-\gamma)$

is unitary from $L^2(\mathbb{R}^n)$ to $L^2(\mathbb{T}^*;\mathcal{H}_\theta)=_{def.}\{u(x,\theta)\in L^2_{loc}(\mathbb{R}^n\times\mathbb{T}^*);\ u(x+\gamma,\theta)=e^{i\gamma\cdot\theta}u(x,\theta),\ \gamma\in\Gamma\}$, where $\mathbb{T}^*=\mathbb{R}^{n*}/\Gamma^*$ is the dual torus. We also introduce $\mathbb{T}=\mathbb{R}^n/\Gamma$. The inverse of U is given by:

(1.2) $U^{-1}v(x)=\int_{\mathbb{T}^*}v(x,\theta)\,d\theta/\mathrm{Vol}(\mathbb{T}^*)$.

Let $V\in C^\infty(\mathbb{R}^n;\mathbb{R})$ satisfy: $V(x+\gamma)=V(x)$ for all $x\in\mathbb{R}^n$, $\gamma\in\Gamma$. Put $P=P_{0,V}=\ -\Delta+V$ (which is an essentially selfadjoint operator from

C_0^∞ whose domain is the standard Sobolev space $H^2(\mathbb{R}^n)$. In the following we shall mostly neglect domain questions.) Then UPU^{-1} can be viewed as a direct integral over $\mathbb{T}^*$ of the operators P_θ, $\theta\in\mathbb{T}^*$, where P_θ is formally (in the sense of distributions) the same operator as $P_{0,V}$ but now acting in the space of θ-Floquet periodic functions:

$$\mathcal{H}_\theta =_{def.} \{u(x)\in L^2_{loc}(\mathbb{R}^n);\ u(x+\gamma)=e^{i\gamma\cdot\theta}u(x),\ \gamma\in\Gamma\}.$$

(The domain is $\mathcal{H}^2_\theta = H^2_{loc}(\mathbb{R}^n)\cap\mathcal{H}_\theta$.) P_θ has discrete spectrum: $\lambda_1(\theta)\le\lambda_2(\theta)\le\ldots$, where the λ_k depend continuously on θ and tend to $+\infty$ when $k\to+\infty$, uniformly with respect to θ. The spectrum of $P_{0,V}$ is then equal to $\cup_1^\infty\Lambda_j$, where $\Lambda_j=\lambda_j(\mathbb{T}^*)$.

We now add a constant small magnetic field:

(1.3) $B=\frac{1}{2}\Sigma\Sigma b_{j,k}dx_j\wedge dx_k$, where $b_{j,k}=-b_{k,j}(=\text{const.})$, and put

(1.4) $A_k(x)=\Sigma_j\frac{1}{2}b_{j,k}x_j$,

so that $d(\Sigma A_k(x)dx_k)=B$. Introduce the magnetic Schrödinger operator:

(1.5) $P_{B,V}=\Sigma(D_{x_j}+A_j(x))^2+V(x)$.

(This is again an essentially selfadjoint operator whose domain is a magnetic Sobolev space H^2_B, defined in [HS2].) $P_{B,V}$ will in general not commute with ordinary translations, but it does commute with the magnetic translations used by Zak [Z] and many others:

(1.6) $T^B_\alpha u(x)=e^{(i/2)\langle B,x\wedge\alpha\rangle}\,u(x-\alpha),\quad \alpha\in\Gamma.$

These unitary operators do not commute in general:

(1.7) $T^B_\alpha\, T^B_\beta=e^{-i\langle B,\alpha\wedge\beta\rangle}\,T^B_\beta\, T^B_\alpha.$

Avron-Simon [AvSi] and Nenciu [Ne1] showed gap stability: If $z_0\notin\sigma(P_{B_0,V})$ then $z\notin\sigma(P_{B,V})$ for all (z,B) in some neighborhood of (z_0,B_0) (Here $\sigma(\cdot)=$"spectrum of".). This is not quite as trivial as it may look, since the domain of the operator changes when we vary B. Our problem is to study spectral properties of $P_{B,V}$ near the energy level z_0 where z_0 is fixed and in $\sigma(P_{0,V})$.

We start by discussing the case when $B=0$.

Proposition 1.1 ([HS1]). With z_0 given and fixed there exist $N\in\mathbb{N}$ and functions $\varphi_j:\mathbb{T}^*\to\mathcal{H}_\theta\cap C^\infty(\mathbb{R}^n)$ which are analytic for the C^∞ topology in a complex neighborhood of $\mathbb{T}^*$, such that for all $(\theta,z)\in\mathbb{T}^*\times\mathrm{neigh}(z_0)$, the operator

$$\mathcal{P}(\theta,z)=\begin{pmatrix} P_\theta-z & R_\theta^- \\ R_\theta^+ & 0 \end{pmatrix} : \mathcal{H}_\theta^2\times\mathbb{C}^N\to\mathcal{H}_\theta\times\mathbb{C}^N$$

is bijective. Here $R_\theta^-=(R_\theta^+)^*$, $R_\theta^+u(j)=(u|\varphi_j(\cdot,\theta))_{\mathcal{H}_\theta}$.

We notice here that if

$$\mathcal{E}=\mathcal{P}^{-1}=\begin{pmatrix} E & E_+ \\ E_- & E_{-+} \end{pmatrix},$$

then we have the equivalence:

$$z\in\sigma(P_\theta) \Longleftrightarrow 0\in\sigma(E_{-+}(\theta,z)).$$

This follows from the identities:

$$(P_\theta-z)^{-1}=E-E_+E_{-+}^{-1}E_-$$

$$E_{-+}^{-1}=-R_+(P_\theta-z)^{-1}R_-,$$

which are easy consequences of the fact that $\mathcal{E}$ is the inverse of $\mathcal{P}$. In the context of degenerate elliptic operators, this type of general observation was used very much by Grushin about 20 years ago, and this is why we call $\mathcal{P}$ a Grushin operator associated to P.

Suppose that z_0 belongs to a single band: $z_0\in\Lambda_{k_0}$, $\Lambda_{k_0}\cap\Lambda_k=\emptyset$ for all $k\neq k_0$. Then it was showed in [HS2] and by Nenciu [Ne2] that we can take $N=1$ in Proposition 1.1, and more precisely so that $\|\varphi_1(\cdot,\theta)\|_{\mathcal{H}_\theta}=1$, $(P_\theta-\lambda_{k_0}(\theta))\varphi_1=0$, $\bar{\varphi}_1(x,\theta)=\varphi_1(x,-\theta)$. Then E_{-+} is a scalar: $E_{-+}(\theta,z)=z-\lambda_{k_0}(\theta)$. Write $\varphi=\varphi_1$ and define for $\gamma\in\Gamma$: $\varphi_\gamma(x)=U^{-1}(e^{-i\gamma\cdot\theta}\varphi(x,\theta))=$

$\varphi_0(x-\gamma)$. These are the so called Wannier functions and they form an orthonormal basis for the spectral subspace associated to $P_{0,V}$ and the interval Λ_{k_0}. Moreover the analyticity in θ implies that there exists $C>0$ such that $|\varphi_0(x)| \leq C e^{-|x|/C}$. See also Bellissard [Bell,2].

We return to the general case, and put $\varphi_{j,\gamma}(x)= U^{-1}(e^{-i\gamma\cdot\theta}\varphi_j(x,\theta))=\varphi_{j,0}(x-\gamma)$, for $\gamma\in\Gamma$. From Proposition 1.1, we get:

Proposition 1.2 ([HS1]). For $z\in \mathrm{neigh}(z_0)$,

$$\mathcal{P}_0(z)=\begin{pmatrix} P_{0,V}-z & R_-^0 \\ R_+^0 & 0 \end{pmatrix} : H^2(\mathbb{R}^n)\times \ell^2(\Gamma;\mathbb{C}^N)\longrightarrow L^2(\mathbb{R}^n)\times \ell^2(\Gamma;\mathbb{C}^N)$$

is bijective. Here by definition: $R_-^0=(R_+^0)^*$, $R_+^0 u(\gamma,j)=(u|\varphi_{j,\gamma})_{L^2}$.

When $B\neq 0$, we introduce the magnetic Wannier functions $\varphi^B_{j,\gamma}=T^B_\gamma\varphi_{j,0}$, and the new auxiliary operators R_+^B : $L^2\longrightarrow \ell^2(\Gamma;\mathbb{C}^N)$, $R_+^B u(\gamma,j)=(u|\ \varphi^B_{j,\gamma})$, and $R_-^B=(R_+^B)^*$. We also introduce the Grushin operator:

$$\mathcal{P}_B(z)=\begin{pmatrix} P_{B,V}-z & R_-^B \\ R_+^B & 0 \end{pmatrix} : H_B^2\times \ell^2(\Gamma;\mathbb{C}^N)\longrightarrow L^2\times \ell^2$$

where H_B^2 is the domain of $P_{B,V}$ and can be described as a magnetic Sobolev space. We then have:

Theorem 1.3(HS1]). For (z,B) in a neighborhood of $(z_0,0)$ in $\mathbb{C}\times\mathbb{R}_B^{n(n-1)/2}$, we have:

1) $\mathcal{P}_B(z)$ is bijective with inverse:
$$\mathcal{E}(B,z)=\begin{pmatrix} E & E_+ \\ E_- & E_{-+} \end{pmatrix}.$$

2) We have the equivalence: $z\in\sigma(P_{B,V}) \Leftrightarrow 0\in\sigma(E_{-+}(B,z))$.

3) The (block-)matrix of E_{-+} is given by

$$E_{-+}(B,z;\alpha,\beta)=e^{(i/2)\langle B,\alpha\wedge\beta\rangle} f(B,z;\alpha-\beta)$$

where f is C^∞ in B holomorphic in z and of exponential decrease with all its derivatives:

$$|\partial_B{}^{\gamma} f(B,z;\alpha)|\leq C_\gamma e^{-|\alpha|/C_0}.$$

4) $f(0,z;\alpha)=\hat{E}_{-+}(\cdot,z)(-\alpha)$, where the E_{-+} to the right is the quantity appearing in Proposition 1.1.

For the single band case of this result we refer to Nenciu [Ne2], Bellissard [Bel1,2], Helffer-Sjöstrand [HS2].

2. Magnetic matrices and associated pseudodifferential operators

For $f\in \ell^1(\Gamma;\mathrm{Mat}(N))$, where Mat(N) denotes the space of complex $N\times N$ matrices equipped with the standard operator norm, we put: $\mathfrak{M}_B(f)(\alpha,\beta)=e^{(i/2)\langle B,\alpha\wedge\beta\rangle} f(\alpha-\beta)$, which is the matrix of a bounded operator $\mathfrak{M}_B(f)$: $\ell^2(\Gamma;\mathbb{C}^N)\to \ell^2(\Gamma;\mathbb{C}^N)$. For a fixed N, these matrices form a C^*-algebra:
$\mathfrak{M}_B(f)^*=\mathfrak{M}_B(f^{\otimes})$ where $f^{\otimes}(\alpha)=f(-\alpha)^*$, $\mathfrak{M}_B(f)\mathfrak{M}_B(g)=\mathfrak{M}_B(f\#_B g)$ where $f\#_B g(\alpha)=\Sigma_{\alpha'+\alpha''=\alpha}\exp((i/2)\langle B,\alpha'\wedge\alpha''\rangle)f(\alpha')g(\alpha'')$, so that $\|f\#_B g\|_{\ell^1}\leq \|f\|_{\ell^1}\|g\|_{\ell^1}$. We denote this algebra by $\mathcal{Q}_B(N)$ and consider it sometimes as $\ell^1(\Gamma;\mathrm{Mat}(N))$ equipped with the "multiplication" $\#_B$. If we define the magnetic translations on $\ell^2(\Gamma;\mathrm{Mat}(N))$ as on L^2: $\tau_\gamma^B u(\alpha)=e^{(i/2)\langle B,\alpha\wedge\gamma\rangle}u(\alpha-\gamma)$, then $\tau_\gamma^{-B}=\mathfrak{M}_B(\delta_\gamma)$, where δ_γ is the function on Γ which is equal to 1 at the point γ and 0 elsewhere. (When $N\geq 2$, we identify $\delta_\gamma(\alpha)$ with $\delta_\gamma(\alpha)\otimes \mathrm{id}_{\mathbb{C}^n}$.)

We want to map $\mathcal{Q}_B(N)$ onto an algebra of pseudodifferential operators on some space $\mathbb{R}^m$. Let us recall the definition of the space of symbols $S^0_{0,0}(\mathbb{R}^{2m})$.

Definition 2.1. $S^0(\mathbb{R}^{2m})=\{a\in C^\infty(\mathbb{R}^{2m});\ \partial_x^\alpha\partial_\xi^\beta a\in L^\infty$ for all $\alpha,\beta\in\mathbb{N}^n\}$.

In this definition, we can also incorporate the case when a takes its values in Mat(N). For $a\in S^0(\mathbb{R}^{2m})$, we define Op(a): $L^2(\mathbb{R}^m)\to L^2(\mathbb{R}^m)$ by the oscillatory integral:

$$\text{(2.1)}\quad \mathrm{Op}(a)u(x)=(2\pi)^{-m}\iint e^{i(x-y)\cdot\theta}a(\tfrac{1}{2}(x+y),\theta)u(y)\,dyd\theta.$$

This is the Weyl quantization of the symbol a, and the L^2-boundedness of this operator is assured by the classical Calderon-Vaillancourt theorem. (See Hörmander [Hö] for more general symbol classes.) These operators also form a C^*-algebra. We have
$\mathrm{Op}(a)^*=\mathrm{Op}(a^*)$, $\mathrm{Op}(a)\circ\mathrm{Op}(b)=\mathrm{Op}(c)$, where $c\in S^0$ is given by:

$$\text{(2.2)}\quad c(x,\xi)= =\exp((i/2)\sigma(D_x,D_\xi;D_y,D_\eta))(a(x,\xi)b(y,\eta))|_{(y,\eta)=(x,\xi)}.$$

Here $\sigma(x,\xi;y,\eta)=\xi\cdot y-x\cdot\eta$ and we shall write $c=a\# b$.

If p and q are real valued linear forms on $\mathbb{R}^{2m}$, then (2.2) simplifies to

$$e^{ip}\# e^{iq}=e^{(i/2)\{p,q\}}\,e^{i(q+p)},$$

which implies the commutation relation

$$\text{(2.3)}\quad e^{ip}\# e^{iq}=e^{i\{p,q\}}\,e^{iq}\# e^{ip}.$$

Here the Poisson bracket $\{p,q\}=\Sigma(\partial_{\xi_j}p)(\partial_{x_j}q)-(\partial_{x_j}p)(\partial_{\xi_j}q)$ is constant.

We compare (2.3) with the commutation relations

$$\text{(2.4)}\quad \tau_\alpha^{-B}\tau_\beta^{-B}=e^{i\langle B,\alpha\wedge\beta\rangle}\,\tau_\beta^{-B}\tau_\alpha^{-B}=e^{i\frac{1}{2}\langle B,\alpha\wedge\beta\rangle}\,\tau_{\alpha+\beta}^{-B}$$

The example 1 below shows that it is always possible to find m and a surjective linear map $\ell:\mathbb{R}^{2m}\to\mathbb{R}^{n*}$, such that

$$\text{(2.5)}\quad \{\alpha\cdot\ell,\beta\cdot\ell\}=\langle B,\alpha\wedge\beta\rangle \text{ for all } \alpha,\beta\in\mathbb{R}^n.$$

Here $(\alpha\cdot\ell)(x,\xi)=\alpha\cdot\ell(x,\xi)$.

Example 1. $m=n, \ell(x,\xi)=\xi+A(x)$, where $A(x)=(A_1(x),..,A_n(x))\in \mathbb{R}^{n*}$. This is related to the Peierls substitution.

Example 2. $n=2$, $B=h\, dy_1\wedge dy_2$, $h\neq 0$. Here y_1,y_2 are some linear coordinates on $\mathbb{R}^2$. Then we can take $m=1$, and using dual coordinates to y_1, y_2, we can take $\ell(x,\xi)=(\ell_1(x,\xi),\ell_2(x,\xi))=(h\xi,x)$.

Example 3. $n=3$, $B\neq 0$. Choose linear coordintaes y_1,y_2,y_3 such that $B=h\, dy_2\wedge dy_1$. Using the correponding dual coordinates on $\mathbb{R}^{3*}$, we can take $m=2$ and put: $\ell(x,\xi)=(x_1,h\xi_1,x_2)$, for $(x,\xi)=(x_1,x_2;\xi_1,\xi_2)\in\mathbb{R}^4$.

Introduce the map: $\tau_{\alpha}^{-B}\mapsto \mathrm{Op}(e^{i\alpha\cdot\ell})$, $\alpha\in\Gamma$, and notice that the τ_{α}^{-B} and the $\mathrm{Op}(e^{i\ell})$ verify the same commutation relations. This implies that we can extend the map to linear combinations in such a way that it commutes with compositions of operators: We map:

$$\mathcal{a}_B\ni \mathcal{M}_B(f)=\Sigma_{\alpha\in\Gamma} f(\alpha)\tau_{\alpha}^{-B}\mapsto \Sigma_{\alpha\in\Gamma} f(\alpha)\mathrm{Op}(e^{i\alpha\cdot\ell})=$$
$$=\mathrm{Op}(g\circ\ell)=_{def.}\mathcal{R}(f),$$

where g is the function on $\mathbb{T}^*$ whose Fourier coefficients are given by $\hat{g}(\alpha)=f(\alpha)$. We then have: $\mathcal{R}_\ell(f_1\#_B f_2)=\mathcal{R}_\ell(f_1)\mathcal{R}_\ell(f_2)$, $\mathcal{R}_\ell(f)^*=\mathcal{R}_\ell(f^{\otimes})$, $\mathcal{R}(\delta_0)=\mathrm{Op}(1)=\mathrm{id}$.

Theorem 2.2. If $f\in\ell^1(\Gamma;\mathrm{Mat}(N))$, then $\sigma(\mathcal{M}_B(f))=\sigma(\mathcal{R}_\ell(f))$, where $\mathcal{M}_B(f)$ and $\mathcal{R}_\ell(f)$ are considered as bounded operators in ℓ^2 and in L^2 respectively.

In the case when f is of exponential decrease, this result was proved in [HS2], and the easy extension to general f in ℓ^1 was given in [S1]. An important ingredient of the proof is a characterization of pseudodifferential operators of R.Beals [Be]. $\mathcal{M}_B(f)$ and $\mathcal{R}_\ell(f)$ have no reason to be unitarily equivalent, but it is quite possible that one can establish a more explicit link between them which also leads to Theorem 2.1. (This is indicated by the results in [GMS].)

Applying Theorem 2.2 to the magnetic matrix E_{-+} appearing after Proposition 1.1, we get:

Theorem 2.3 ([HS1,2]). Same assumptions as in Theorem 1.3. Define $g(B,z;\theta)=\Sigma f(B,z;\alpha)e^{i\alpha\cdot\theta}$ and choose ℓ as above. Then:

1) g is C^∞ in B and holomorphic in (z,θ) in a neighborhood of $\{0\}\times\{z_0\}\times\mathbb{T}^*$ in $\mathbb{R}^{n(n-1)/2}\times\mathbb{C}\times(\mathbb{T}^*+i\mathbb{R}^{n*})$. Moreover, $g(0,z;-\theta)=E_{-+}(\theta,z)$ is the matrix defined after Proposition 1.1.
2) For z in a neighborhood of z_0, we have $z\in\sigma(P_{B,V})$ iff $0\in\sigma(Op(g(B,z,\cdot)\circ\ell))$.

Here, it is justified to call $Op(g(B,z,\cdot)\circ\ell)))$ an effective Hamiltonian.

Remarks.

a) Replacing ℓ by $-\ell$, we get rid of the minus sign in 1) .

b) In the single band case, we can arrange to have $g(0,z;\theta)=z-\lambda_{k_0}(\theta)$ and if we choose $\ell=\xi+A(x)$, we get a scalar effective Hamiltonian $\approx\lambda_{k_0}(D_x+A(x))-z$. This justifies the so called Peierls substitution.

c) Fix B and let ℓ be a corresponding choice. To the field hB, we can then associate $\ell_h=\ell(x,h\xi)$, and we get a "semiclassical" effective Hamiltonian, $Op(g(hB,z;\ell(x,h\xi)))$.

3. Density of states and the de Haas-van Alphen effect

We first recall how to define the density of states measure in our context. (See Shubin [Sh] for more general cases and Helffer-Sjöstrand [HS1,2], for more details in the present case.) If $f\in C_0(\mathbb{R})$, then $f(P_{B,V})$ has a C^∞ distribution kernel $K_{f,B}(x,y)$ which satisfies:

$$K_{f,B}(x+\gamma,y+\gamma)=e^{(i/2)\langle B,(x-y)\wedge\gamma\rangle}K_{f,B}(x,y),\ \gamma\in\Gamma,$$

and in particular:

$$K_{f,B}(x+\gamma,x+\gamma)=K_{f,B}(x,x),\ \ \gamma\in\Gamma.$$

We define $\tilde{tr} f(P_{B,V})$ as the mean value of $K_{f,B}(x,x)$:

$$\tilde{tr} f(P_{B,V})=(\mathrm{Vol}\,\mathbb{T})^{-1}\int_{\mathbb{T}} K_{f,B}(x,x)\,dx.$$

It is known (see [HS2], [S1] and further references given there) that we have the following properties:

- $f\geq 0 \Rightarrow \tilde{tr} f(P_{B,V})\geq 0$ so there exists a unique Borel measure $\rho_{B,V}$ on $\mathbb{R}$ (the density of states measure) such that $\tilde{tr} f(P_{B,V})=\int f(t)\rho_{B,V}(dt)$.
- The function $B\mapsto \int f(t)\rho_{B,V}(dt)$ is C^∞ for every $f\in C_0^\infty(\mathbb{R})$.
- $\mathrm{Supp}\,\rho_{B,V}=\sigma(P_{B,V})$.
- $\rho_{B,V}(]-\infty,t])$ can only take very special values when $t\in\mathbb{R}\setminus\sigma(P_{B,V})$. ("Gap labbeling".)

Theorem 3.1 ([HS1,2], see also [S1]). Same assumptions as in Theorem 2.3. Let $f\in C_0^\infty(\mathbb{R})$ have its support in a sufficiently small neighborhood of z_0. Then:

$$\int f(t)\rho_{B,V}(dt)=-(\pi\,\mathrm{Vol}(\mathbb{T}))^{-1}\int(\partial_{\bar z}\tilde f(z))\,tr(Q^{-1}\partial_z Q)L(dz),$$

where $\tilde f\in C_0^\infty(\mathbb{C})$ is an extension of f with small support and with the property that $\partial_{\bar z}\tilde f=0$ on $\mathbb{R}$. Moreover $Q=Op(g(B,z;\cdot)\circ\ell)$ and $tr(Op(a))$ is by definition the meanvalue of the trace of the symbol of a. $L(dz)$ is the standard Lebesgue measure.

Some ideas of the proof. We have

$$(z-P_{B,V})^{-1}=-E(B,z)+E_+(B,z)E_{-+}(B,z)^{-1}E_-(B,z),$$

and

$$f(P_{B,V})=-(\pi)^{-1}\int\partial_{\bar z}\tilde f(z)(z-P_{B,V})^{-1}L(dz).$$

Since $E(B,z)$ is holomorphic in z, this quantity gives no contribution to the integral, and we get

$$f(P_{B,V})=-(\pi)^{-1}\int\partial_{\bar z}\tilde f\ E_+E_{-+}^{-1}E_-\,L(dz)$$

Define $\hat{tr}\,\mathfrak{M}_B(f)=(\mathrm{Vol}\,\mathbb{T})^{-1}\,tr\,f(0)$. One can then show that for

$\mathrm{Im}\, z \neq 0$, $\tilde{tr}(E_+E_{-+}^{-1}E_-) = \hat{tr}(E_{-+}^{-1}E_-E_+)$. On the other hand $E_-E_+ = \partial_z E_{-+}$, so

$$\tilde{tr} f(P_{B,V}) = -(\pi)^{-1} \int (\partial_{\bar{z}} \tilde{f})\, \hat{tr}(E_{-+}^{-1} \partial_z E_{-+}) L(dz).$$

It then only remains to notice that

$$\hat{tr}(\mathfrak{M}_B(f)) = tr(\mathcal{R}(f))/\mathrm{Vol}(\mathbb{T}),$$

so that

$$\hat{tr}(E_{-+}^{-1} \partial_z E_{-+}) = \mathrm{Vol}(\mathbb{T})^{-1}\, tr(Q^{-1}\partial_z Q). \qquad \square$$

This result can be used in the study of certain oscillations in the density of states measure in the three-dimensional case, related to the de Haas-van Alphen effect, explained heuristically by Onsager [O]. Assume in the remainder of this section that $n=3$ and fix $z_0 \in \mathbb{R}$. For z real and close to z_0 we introduce the "Fermi surface":

(3.1) $\qquad \mathcal{F}(z) = \{\theta \in \mathbb{T}^*;\ z \in \sigma(P_\theta)\}$.

Here P_θ was introduced in section 1. We need a certain number of assumptions:

(H1) $\qquad \dim(\mathrm{Ker}(P_\theta - z_0)) = 1$ on $\mathcal{F}(z_0)$.

We have then a Floquet eigenvalue $\lambda(\theta)$ depending analytically on θ in a neighborhood of $\mathcal{F}(z_0)$ with $\lambda|_{\mathcal{F}(z_0)} = z_0$. We also assume:

(H2) $\qquad d\lambda \neq 0$ on $\mathcal{F}(z_0)$.

For simplicity assume that $\mathcal{F}(z_0)$ has precisely one component per (dual) lattice point.

Fix a magnetic field $B_0 \neq 0$. We are interested in the asymptotic behaviour of $\rho_{hB_0,V}$ when $h \to 0$. Let $\mathcal{H} = \mathrm{Ker}\, B_0^{\perp}$ which is also given by $\eta_3 = 0$, if η_1, η_2, η_3 are the dual coordinates to the coordinates y_1, y_2, y_3 introduced in the example 3 of section 2 (corresponding also to the plane $x_2 = 0$ in the coordinates x_1, ξ_1, x_2 of the very same example). We then get an effective Hamiltonian: $\mathrm{Op}(\tilde{g}(x_1, h\xi_1, x_2, z; h))$ which we shall view as a one-dimensional pseudodifferential operator depending on x_2 as an additional parameter.

(H3) Let $\mathcal{H}_t$ denote the translate: $\eta_3=t$. If $\eta_0\in\mathcal{F}(z_0)\cap\mathcal{H}_t$, then either $d(\lambda_{|\mathcal{H}_t})(\eta_0)\neq 0$, or the Hessian $\nabla^2(\lambda_{|\mathcal{H}_t})(\eta_0)$ is definite.

(After this conference the author managed to relax this condition, and details will appear in the proceedings of the international conference on partial differential equations and mathematical physics at the University of Alabama in Birmingham, March 1990.) Let $K_0(z_0)$ be a component of $\mathcal{F}(z_0)$ and let $K_0(z)$ be the neighboring component of $\mathcal{F}(z)$. For x_2 in the projection onto the η_3-axis of $K_0(z)$, we let $a_0(z,x_2)$ be the area of the compact set in $\eta_3=x_2$ which is bounded by the intersection of this plane with $K_0(z)$:

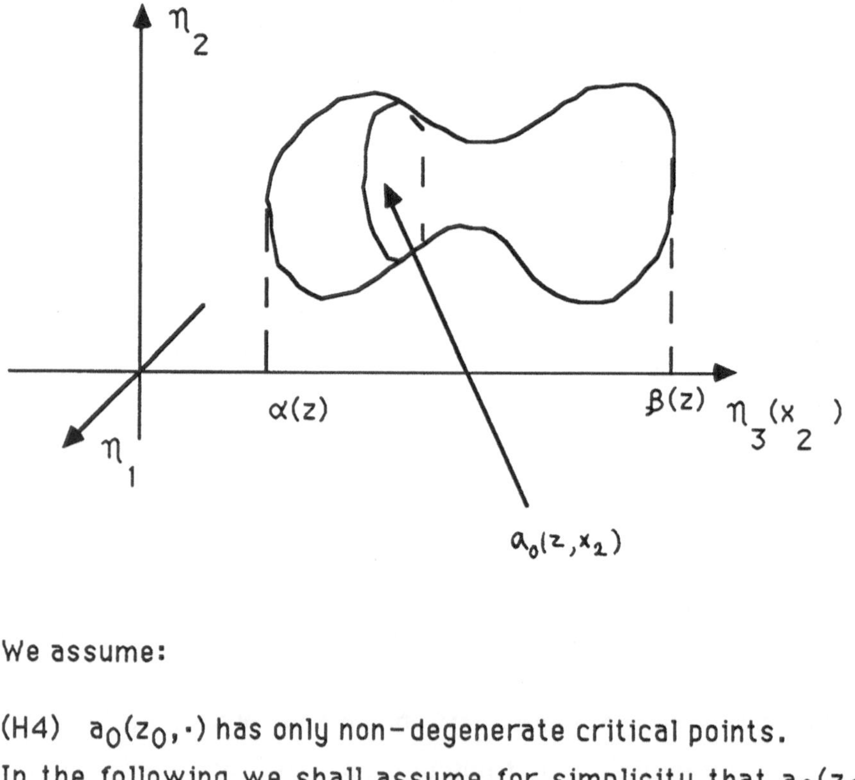

We assume:

(H4) $a_0(z_0,\cdot)$ has only non-degenerate critical points.

In the following we shall assume for simplicity that $a_0(z_0,\cdot)$ has only one critical point (necessarily the maximum). We then have :

Theorem 3.2 ([HS1]). There exists a realvalued classical symbol $a(z,x_2;h)\sim a_0(z,x_2)+a_1(z,x_2)h+\dots$, such that if we put $\tilde{a}(z;h)=$

$\sup_{x_2} a(z,x_2;h)$, and define $\zeta(kh;h)$ for kh in a neighborhood of $\tilde{a}(z_0;0)$ by the Bohr-Sommerfeld condition: $\tilde{a}(\zeta(kh;h);h)=kh$, $(k\in\mathbb{N})$ then for every $f=f(z;h)$ with z-support in some small fixed neighborhood of z_0 and $\|f(\cdot;h)\|_{C^\alpha}\le C_0 h^{-N_0}$ for some $\alpha>0$, $C_0>0$, $N_0>0$, we have:

$$\int f(z;h)\,\rho_{hB_0,V}(dz)=\int f(z;h)b(z;h)dz+\mathcal{O}(h^\infty),$$

where b is independent of f and is of the form: $b=b_{reg}+b_{sing}$, where $\partial_z{}^k b_{reg}=\mathcal{O}(1)$, $h\to 0$,

$$b_{sing}(z)=\sum_{\{k\in\mathbb{N};\ |kh-\tilde{a}(z_0;0)|\le\delta_0\}} hm(z,kh;h)|z-\zeta(kh;h)|^{-\frac{1}{2}}H(\pm(z-\zeta(kh;h))).$$

Here $\delta_0>0$ is a constant and $m>0$ is a classical elliptic symbol of order 0. The $\pm$ sign depends on the sign of $\partial_z\tilde{a}$.

4. Effective Hamiltonians in more general situations

In Guillot-Ralston-Trubowitz [GuRT] and Buslaev [B] an implicit use of effective Hamiltonians is made in the asymptotic construction of approximate eigenfunctions. However in those papers no definition of the effective Hamiltonian is given. The results below are taken from a joint paper with C.Gérard and A.Martinez [GMS] and partly our approach follows the one of [B].

Consider,

(4.1) $\quad \mathrm{Op}(P(hy,y,\eta+A(hy)))$,

where $P\in C^\infty(\mathbb{R}^{3n};\mathbb{R})$, $A\in C^\infty(\mathbb{R}^n;\mathbb{R}^{n*})$ satisfy:

(4.2) $\quad P(x,y,\eta)$ is Γ-periodic in y,

(4.3) $\quad P=\sum_{|\alpha|\le m} a_\alpha(x,y)\eta^\alpha$ is a polynomial of degree m in η

and if we put $p=\sum_{|\alpha|=m} a_\alpha(x,y)\eta^\alpha$, then

$p(x,y,\eta)\ge|\eta|^m/C_0$, for some $C_0>0$.

(4.4) $\quad |\partial^\beta_{x,y}a_\alpha(x,y)|\le C_\beta$ for all $x,y\in\mathbb{R}^n$, $\beta\in\mathbb{N}^{2n}$,

(4.5) $|\partial_x^\beta A(x)|\le C_\beta$ for all $x\in\mathbb{R}^n$, and all $\beta\in\mathbb{N}^n$ with $|\beta|\ge 1$.

Here we use standard multiindex notation: $\eta^\beta=\eta_1^{\beta_1}\cdot\ldots\cdot\eta_n^{\beta_n}$, $|\beta|=|\beta_1|+..+|\beta_n|$.

Example 1. If V is Γ-periodic, then $\Sigma_1^n(D_{y_j}+A_j(hy))^2+V(y)$ corresponds to $P(x,y,\eta)=\eta^2+V(y)$, and if A_j are linear functions, then we are in the preceding situation, up to a slight change of notation.

Example 2. If $V(x,y)$ is $\tilde\Gamma\times\Gamma$ periodic, where $\tilde\Gamma$ is another lattice, then the quasiperiodic operator $-\Delta_y+V(hy,y)$ corresponds to $P=\eta^2+V(x,y)$, $A=0$.

Following Buslaev, we consider on $\mathbb{R}^{2n}_{x,y}$ the operator

(4.6) $\qquad \mathrm{Op}(P(x,y;h\xi+A(x)+\eta)).$

We notice that the operator (4.1) can be identified with the operator (4.6) acting on the space of $u(x)\delta(x-hy)$ for $u\in L^2(\mathbb{R}^n)$. (We sometimes neglect domain questions and we refer to [GMS] for more details.) It is also of interest to consider (4.6) acting in $L^2(\mathbb{R}^{2n})$, and when doing so we first make a Floquet reduction in y. Put :

$$Uu(x,y,\theta)=\Sigma_\Gamma e^{i\gamma\theta}u(x,y-\gamma).$$

Then $e^{i((x/h)-y)\cdot\theta}U$ is unitary: $L^2(\mathbb{R}^{2n})\to\tilde{\mathcal{H}}_0=_{\mathrm{def.}}\{v\in L^2_{loc}(\mathbb{R}^{3n})$; $v(x,y+\gamma,\theta)=v(x,y,\theta)$, $v(x,y,\theta+\gamma^*)=\exp(i((x/h)-y)\cdot\gamma^*)\,v(x,y,\theta)$, $\gamma\in\Gamma$, $\gamma^*\in\Gamma^*$, $\displaystyle\iiint_{\mathbb{R}^n\times\mathbb{T}\times\mathbb{T}^*}|v(x,y,\theta)|^2\,dxdyd\theta<\infty\}$, and the conjugate of (4.6) is formally the same operator, now acting in $\tilde{\mathcal{H}}_0$. Notice that our operator commutes with the multiplication operators $T_{\gamma^*}=\exp(i((x/h)-y)\cdot\gamma^*)$, $\gamma^*\in\Gamma^*$. We conclude that (4.6) acting in $L^2(\mathbb{R}^{2n})$ and (4.6) acting in $\mathcal{H}_0=_{\mathrm{def.}}$

$L^2(\mathbb{R}^n \times \mathbb{T})$ have the same spectrum.

We now apply vectorvalued pseudodifferential operators. Put $P(x,\xi)=Op_{(y,\eta)}(P(x,y,\xi+\eta)): \mathcal{K}_{m,\xi} \to \mathcal{K}_{0,\xi}$, where $\mathcal{K}_{0,\xi}=\mathcal{K}_0$ and $\mathcal{K}_{m,\xi} \subset \mathcal{K}_{0,\xi}$ is a natural domain of definition. Then the operator (4.6) acting on $\mathcal{K}_0$ can be identified with the vector valued pseudodifferential operator $Op_h(P(x,\xi+A(x)))=Op(P(x,h\xi+A(x)))$. Note that

$$e^{iy\cdot\xi}P(x,\xi)e^{-iy\cdot\xi}=Op_{(y,\eta)}(P(x,y,\eta))): \mathcal{H}_\xi^m \to \mathcal{H}_\xi \,.$$

We fix $z_0 \in \mathbb{R}$. (The following results are valid also with obvious modifications in the case when z_0 is replaced by a fixed compact interval.) First we have a simple adaptation of Proposition 1.1.

Proposition 4.1. There exist $\varphi_1(x,\xi;y),..,\varphi_N(x,\xi;y) \in C^\infty \cap \mathcal{H}_\xi$, Γ^*-periodic in ξ with $\|\partial^\alpha_{x,\xi}\varphi_j\|_{\mathcal{H}_\xi(E)} \le C_\alpha$, where E is some bounded Γ-fundamental domain, such that

$$\begin{pmatrix} Op_{y,\eta}(P(x,y,\eta))-z & R^0_-(x,\xi) \\ R^0_+(x,\xi) & 0 \end{pmatrix} : \mathcal{H}_\xi^m \times \mathbb{C}^N \to \mathcal{H}_\xi \times \mathbb{C}^N$$

is bijective for x,ξ real and z in some complex neighborhood of z_0. Here $R^0_+(x,\xi)u(j)=(u|\varphi_j(x,\xi;\cdot))$ and $R^0_-=R^{0*}_+$.

Put $\psi_j(x,\xi;y)=e^{-iy\cdot\xi}\varphi_j(x,\xi;y) \in \mathcal{K}_{0,\xi}$. Then

$$\mathcal{P}(x,\xi,z)= \begin{pmatrix} P(x,\xi)-z & R_-(x,\xi) \\ R_+(x,\xi) & 0 \end{pmatrix} : \mathcal{K}_{m,\xi} \times \mathbb{C}^N \to \mathcal{K}_{0,\xi} \times \mathbb{C}^N$$

is bijective, where we define $R_+(x,\xi)u(j)=(u|\psi_j(x,\xi,\cdot))$, $R_-=R_+^*$. Applying the theory of vectorvalued pseudodifferential operators, we get:

Theorem 4.2 ([GMS]). For z in some complex neighborhood of z_0 and for h>0 small enough, the operator $Op_h(\mathcal{P}(x,\xi+A(x),z))$: $\mathcal{K}_m\times\mathbb{C}^N\to\mathcal{K}_0\times\mathbb{C}^N$ has a uniformly bounded inverse $Op_h(\mathcal{E}(x,\xi+A(x),z;h))$, where $\mathcal{E}\in S^0(\mathbb{R}^{2n};\mathcal{L}(\mathcal{K}_{0,\xi}\times\mathbb{C}^N,\mathcal{K}_{m,\xi}\times\mathbb{C}^N))$, $\mathcal{E}\sim\sum_0^\infty\mathcal{E}_j(x,\xi,z)h^j$, with $\mathcal{E}_0(x,\xi,z)=\mathcal{P}(x,\xi,z)^{-1}$. (For $j\geq 2$, $\mathcal{E}_j$ may depend on A.)

Corollary 4.3 ([GMS]). z belongs to the spectrum of the operator (4.6) acting in $L^2(\mathbb{R}^{2n})$ iff $0\in\sigma(Op_h(E_{-+}(x,\xi+A(x),z;h)))$, where $Op_h(E_{-+}(x,\xi+A(x),z;h))$ acts in $L^2(\mathbb{R}^n;\mathbb{C}^N)$. Here $E_{-+}(x,\xi,z;h)$ denotes the lower right entry in the matrix $\mathcal{E}(x,\xi,z;h)$.

It is also easy to show that $E_{-+}(x,\xi+\gamma^*,z;h)=E_{-+}(x,\xi,z;h)$ for all $\gamma^*\in\Gamma^*$.

We now return to the operator (4.1), which can be identified with the operator (4.6), acting on $\{u(x)\delta(x-hy);\ u\in L^2(\mathbb{R}^n)\}$. One can then go through the same procedure as above, modify the spaces, check continuity and finally get:

Theorem 4.4 ([GMS]). For z in a complex neighborhood of z_0 and for h>0 sufficiently small, we have $z\in\sigma(Op(P(hy,y,\eta+A(hy)))$ iff $0\in\sigma(Op_h(E_{-+}(x,\xi+A(x),z;h)))$, where E_{-+} is the same symbol as in Theorem 4.2 and $Op_h(E_{-+}(..))$ now acts as a bounded operator in λ_0^N. Here $\lambda_0 =_{def.}\{\sum_\Gamma u_\gamma\delta(x-h\gamma);\ (u_\gamma)\in\ell^2(\Gamma)\}$. If all the operators $Op(P(z+hy,y,\eta+A(z+hy))$, $z\in\mathbb{R}^n$ are isospectral, then we can replace λ_0 by $L^2(\mathbb{R}^n)$.

This result can be viewed as a generalization of Theorem 2.3. Let us examine what it gives in the quasiperiodic case under some more assumptions. Take n=1, $V_\lambda(x,y)=U_\lambda(x)+\lambda^2W(y)$, where $\lambda>>1$. We assume that W is 1-periodic, $W(y)\geq 0$ with equality precisely on $\mathbb{Z}$ and that $W''(0)>0$. $U_\lambda(x)$ is a 2π-periodic function to be chosen later. Consider $D_y^2+\lambda^2W(y)$. Applying the results of

Harrell [Ha], Outassourt [Ou], and Simon [Si], we know that the first band in the spectrum of this Hill's operator is of the form $[\lambda E(\lambda)-a(\lambda)e^{-S_0\lambda}, \lambda E(\lambda)+a(\lambda)e^{-S_0\lambda}]$, where $E(\lambda)\sim E_0+E_{-1}\lambda^{-1}+\ldots$, $E_0>0$, $a(\lambda)\sim\lambda^{5/2}(a_0+a_{-1}\lambda^{-1}+\ldots)$, $a_0>0$, $S_0>0$. Moreover this band is generated by the Floquet eigenvalue, $\mu_\lambda(\xi)=\lambda E(\lambda)-a(\lambda)e^{-S_0\lambda}(\cos\xi+r(\lambda,\xi))$, where $|\partial_\lambda{}^k\partial_\xi{}^\ell r(\lambda,\xi)|\leq C_{k,\ell}e^{-\frac{1}{2}S_0\lambda}$.

Take $U_\lambda(x)=\mu_\lambda(x)-2\lambda E(\lambda)$. If $z_0=0$, then for λ sufficiently large, we can take N=1 in Proposition 4.1, and we get $E_{-+0}(x,\xi,z)=z+a(\lambda)e^{-S_0\lambda}(\cos\xi+\cos x+r(\lambda,\xi)+r(\lambda,x))$. This means that we have a symbol very close to that of Harper's operator.

For a survey of the semiclassical analysis of Harper's operator and further references, we refer to [S1].

References

[AvSi] J.Avron, B.Simon, 'Stability of gaps for periodic potentials under variation of a magnetic field', J.Phys. A : Math.Gen.18(2)(1985), 2199-2205.

[Be] R.Beals, 'Characterization of P.D.O. and applications', Duke Math. J. 44(1977),45-57.

[Bel1] J.Bellissard, 'C*-Algebras in solid state physics', 2D-Electrons in a uniform magnetic field. Proceedings of the Warwick conference on operator algebras (July 1987).

[Bel2] J.Bellissard, 'Almost periodicity in solid state physics and C* algebras', Proceedings of the H.Bohr centennary conf. On almost periodic functions (April 1987).

[B] V.Buslaev, 'Semiclassical approximation for equations with periodic coefficients', Uspeki Mat. Nauk 42(6)(1987),77-98, Russ. Math. Surv. 42(6)(1987),97-125.

[GMS] C.Gérard, A.Martinez, J.Sjöstrand, 'A mathematical approach to the effective Hamiltonian in perturbed periodic problems', Prépublication Univ. de Paris Sud 1990.

[GuRT] J.C.Guillot, J.Ralston, E.Trubowitz, 'Semi-classical methods in solid state physics', Comm. in Math. Phys. 116(1988),401-415.

[Ha] E.M.Harrell,'The band structure of a one-dimensional periodic system in a scaling limit', Ann. Physics 119(1979),351-369.

[HS1] B.Helffer,J.Sjöstrand, 'On diamagnetism and de Haas-van Alphen effect', Ann. de l'IHP, physique théorique, 52(3) (1990).

[HS2] B.Helffer,J.Sjöstrand, 'Equation de Schrödinger avec champ magnétique et équation de Harper', Springer LN in phys. 345(1989).

[HS3] B.Helffer,J.Sjöstrand, 'Analyse semiclassique pour l'équation de Harper', Bull. de la S.M.F.116(4)(1988), Mémoire n° 34.

[HS4] B.Helffer,J.Sjöstrand, 'Analyse semiclassique pour l'équation de Harper II. Comportement semi-classique près d'un rationnel', Bull. de la S.M.F., Mémoire , to appear.

[HS5] B.Helffer,J.Sjöstrand, 'Semi classical analysis for Harper's equation III. Cantor structure of the spectrum', Bull. de la S.M.F., Mémoire, to appear.

[Hö] L.Hörmander, 'The analysis of linear partial differential operators I-IV', Grundlehren, Springer 1983-85.

[Ne1] G.Nenciu, 'Stability of energy gaps under variation of the magnetic field', Letters in Math. Ph. 11(1986),127-132.

[Ne2] G.Nenciu, 'Bloch electrons in a magnetic field: Rigourous justification of the Peierls-Onsager effective Hamiltonian', Letters in Math. Ph. 17(1989), 247-252.

[O] L.Onsager, 'Interpretation of the de Haas-van Alphen effect', Phil.Mag., 43(1952),1006-1008.

[Ou] A.Outassourt, 'Comportement semi-classique pour l'opérateur de Schrödinger à potentiel périodique', J.Funct. Anal. 72(1987),65-93.

[P] R.Peierls, 'Zur theorie des diamagnetismus von Leitungselectronen', Z. für Physik 80(1933),763-791.

[ReSi] M.Reed, B.Simon, 'Methods of modern mathematical physics, I-IV', Academic Press, 1974-1979.

[Sh] M.Shubin, 'The spectral theory and the index of elliptic operators with almost periodic coefficients', Usp.Mat.Nauk 34(2)(1979), 95-135, Russian Math. Surv., 34(2)(1979), 109-157.

[Si] B.Simon, 'Semiclassical analysis of low lying eigenvalues III.Width of the ground state band in strongly coupled solids', Ann. Physics 158(1984), 415-420.

[Sl] J.Sjöstrand, 'Microlocal analysis for the periodic magnetic Schrödinger equation and related questions', Lecture notes CIME, Montecatini July 1989, Springer LN Math. to appear.

[Z] J.Zak, 'Magnetic translation group', Physical Review, 134(6A),June 1964.

PERTURBATIONS OF SUPERSYMMETRIC SYSTEMS IN QUANTUM MECHANICS

B. BAUMGARTNER
Institut für Theoretische Physik
Universität Wien
Boltzmanngasse 5, A-1090 Vienna, Austria

ABSTRACT. The methods of supersymmetry are extended to the factorization method. The degeneracy of levels in factorizable systems is broken under perturbations. With the methods of supersymmetry it is possible to state laws on the order of these perturbed energy levels. One proof of a confirmation of these laws has a more algebraic touch and works for first order perturbation theory. The proof of the laws beyond perturbation theory is hard, more of an analytic spirit and exploits convexity properties of the potentials. The convexity properties serve also for an intuitive argument. An important application is the law on the ordering of energy levels in atoms.

1 Introduction

The main themes of this lecture can be traced back almost to the origin of quantum mechanics. We will not use any sophisticated new mathematics, so that all of the following could have been done, in principle, about 40 years ago, when Infeld and Hull revised Schrödinger's factorization method, [1,2,3], and recast it into a form which is now reborn with supersymmetric quantum mechanics [4]. Why then are the following results new? Probably because half a century ago everybody was interested in finding exact solutions or approximations. Now we are interested in finding inequalities, yielding certain laws, for example, for the ordering of those atomic energy levels which are degenerate in the hydrogen atom. (The law which governs this level ordering holds also for the bound states of electrons at donors in semiconductors [5].)

The old intuitive argument to explain this level ordering is the remark that the classical orbits with lower angular momentum dive deeper into the charge cloud of the core electrons. There they feel a stronger attraction to the nucleus so that the effective potential is lower than the Coulomb potential generated by the shielded nucleus, as it is experienced by the outermost electron when it has high angular momentum. For the outermost electron this is the right argument, but it is not sufficient for an explanation of the ordering of Roentgen levels of the core electrons. So we complete this intuitive argument [6]: As the angular momentum decreases, the classically allowed region (also the region, where the quantum probability for the location of the electron is large) increases in all directions, both closer to

A. Boutet de Monvel et al. (eds.), Recent Developments in Quantum Mechanics, 195–208.

the nucleus and to farther distances. Now the effective potential at intermediate distances acts like a comparison potential $V_c(r) = constant - Z_{eff}/r$. But everywhere in the newly acquired regions, the effective potential lies below this comparison potential. So the energies, which would be equal in $V_c(r)$, drop with decreasing angular momentum. This is the result of a "concavity property" of the effective potential, stated as $V''(r) + (2/r)V'(r) \leq 0$ or $(\Delta V)(r) \leq 0$, or $(V'(r)/r^2)' \leq 0$.

This concavity property gives not only an intuitive argument, it is also an essential ingredient in the proof of strict inequalities. This proof uses the tools of supersymmetric quantum mechanics, but it proceeds working with functional analysis.

In first order perturbation theory one can prove this level ordering in a more algebraic spirit. There one needs not only supersymmetry, but its extension to the factorization method.

By a certain transformation of coordinates one can exchange the rôles of coupling constants and eigenvalues. This trick transforms a supersymmetric pair of Hamiltonians into a new pair of isospectral operators. It is applicable to some factorizable systems, yielding a new class of factorization. Applied to the hydrogen atom, it established the relationship between Coulomb potentials and harmonic oscillators. Not only the equations for the supersymmetric and factorizable systems, but also the inequalities for the perturbed systems and the strict inequalities beyond perturbation theory can be transformed in this way.

Also these class-II-inequalities can be applied to a fundamental physical problem: The explanation of the level ordering of nucleons in atomic nuclei. Here one compares the Woods-Saxon potential with the potential of a harmonic oscillator [6].

2 Factorizable Systems

2.1 The Necessary Tools from Supersymmetric Quantum Mechanics

We work on $\mathcal{L}^2(a,b)$, where a may be finite or $-\infty$, also b may be finite or $+\infty$. One can reconstruct a Hamiltonian out of $\phi(x)$, the positive ground state wave function, or another positive, not necessarily square integrable solution of the differential equation

$$(-\frac{d^2}{dx^2} + V(x) - \eta)\phi(x) = 0. \tag{2.1}$$

From the theorem on nodes (the Sturm oscillation theorem) we know that $\phi(x)$ can be chosen positive, only if η is not above the ground state energy.

The negative logarithmic derivative of ϕ

$$G(x) := -\phi'(x)/\phi(x), \tag{2.2}$$

(by some authors called the "superpotential") serves to define the operator

$$A = \frac{d}{dx} + G(x), \tag{2.3}$$

(first in $C_0^1(a,b)$, then taking its closure) and its adjoint

$$A^* = -\frac{d}{dx} + G(x). \tag{2.4}$$

Taking the Friedrichs extension of A^*A, we get the Hamiltonian

$$H = A^*A + \eta = -\frac{d^2}{dx^2} + G^2(x) - G'(x) + \eta \tag{2.5}$$

(with Dirichlet boundary conditions in case of finite a or b).

With the relations

$$A\phi = 0, \qquad (H - \eta)\phi = 0, \tag{2.6}$$

one observes that the potential is reconstructed by

$$V(x) = G^2(x) - G'(x) + \eta, \tag{2.7}$$

which is the Riccati equation.

Now one constructs a supersymmetric partner

$$H_1 = AA^* + \eta, \tag{2.8}$$

with

$$V_1(x) = G^2(x) + G'(x) + \eta. \tag{2.9}$$

By use of intertwining relations

$$AH = H_1 A, \qquad HA^* = A^* H_1, \tag{2.10}$$

it is easily confirmed that A maps bound states and scattering states φ_E of H to bound states and scattering states $\varphi_{1,E}$ of H_1.

$$\begin{aligned} \varphi_{1,E}(x) &= (E-\eta)^{-1/2}(A\varphi_E)(x), \\ \varphi_E(x) &= (E-\eta)^{-1/2}(A^*\varphi_{1,E})(x), \end{aligned} \tag{2.11}$$

$$H\varphi_E = E\varphi_E, \qquad H_1\varphi_{1,E} = E\varphi_{1,E}, \tag{2.12}$$

$$\|\varphi_{1,E}\| = \|\varphi_E\|. \tag{2.13}$$

So H and H_1 are isospectral, or, if $\phi \in L^2(a,b)$, essentially isospectral [7], since ϕ is annihilated by A, so η is not an element of the spectrum of H_1.

2.2 Factorizable Systems, Class I

The construction of a supersymmetric partner may be iterated. This gives a family $G_\ell(x)$, η_ℓ, such that for $\ell = 1, 2, 3\ldots$ with

$$A_\ell = \frac{d}{dx} + G_\ell(x), \tag{2.14}$$

the Hamiltonian

$$H_\ell = A_{\ell-1}A_{\ell-1}^* + \eta_{\ell-1} \tag{2.15}$$

is also represented as

$$H_\ell = A_\ell^* A_\ell + \eta_\ell. \tag{2.16}$$

The function, number and operators G, η, A, H we started with are now denoted as G_0, η_0, A_0, H_0.

The eigenfunctions $\{\varphi_{\ell,E}\}$ to $\{H_\ell\}$ with energies E are related by

$$\begin{aligned} \varphi_{\ell+1,E} &= (E-\eta_\ell)^{-1/2} A_\ell \varphi_{\ell,E}, \\ \varphi_{\ell,E} &= (E-\eta_\ell)^{-1/2} A_\ell^* \varphi_{\ell+1,E}. \end{aligned} \tag{2.17}$$

As an abstract construction, this procedure is always possible. If the functions $G_\ell(x)$ can be written down explicitly, one speaks of a "factorizable system". Infeld and Hull worked out a classification into types A to F, which is represented here in a table. (Since we use only real parameters, we split their types A and E into three subtypes.)

Factorizable Systems of Class I

Type		Domain (a,b)	$-G_\ell(x)$ $(\mu=\ell+c)$	η_ℓ	convertible into class II	$F(x)$	Applications
A	α	$(0,\pi)$	$\mu\cot x+\dfrac{\kappa}{\sin x}$	$4\kappa\mu+\mu^2$	no	$\cot x$	Harmonic oscillator in
	β	$\mathbf{R}_+$	$\mu\coth x+\dfrac{\kappa}{\sinh x}$	$4\kappa\mu-\mu^2$	no	$\coth x$	curved space
	γ	$\mathbf{R}$	$\mu\tanh x+\dfrac{\kappa}{\cosh x}$	$-\mu^2$	no	$\tan x$	
B		$\mathbf{R}$	$\mu+e^{-x}$	$-\mu^2$	yes	1	
C		$\mathbf{R}_+$	$\mu/x-\kappa x$	$4\kappa\mu$	no	$1/x$	Harmonic oscillator in more than 1 dimension
D		$\mathbf{R}$	$-x$	2μ	no	0	1-dimensional harmonic oscillator
E	α	$(0,\pi)$	$\mu\cot x-q/\mu$	μ^2-q^2/μ^2	no	$2\cot x$	Hydrogen atom
	β	$\mathbf{R}_+$	$\mu\coth x-q/\mu$	$-\mu^2-q^2/\mu^2$	yes	$2\coth x$	in curved space
	γ	$\mathbf{R}$	$\mu\tanh x-q/\mu$	$-\mu^2-q^2/\mu^2$	yes	$2\tanh x$	
F		$\mathbf{R}_+$	$\mu/x-q/\mu$	$-q^2/\mu^2$	yes	$2/x$	Hydrogen atom

In the most important applications η_ℓ is chosen as the ground state energy of H_ℓ. Because of the isospectral property η_ℓ gives then the energy of the first excited state of $H_{\ell-1}$, of the second excited state of $H_{\ell-2}$ ... and finally of the ℓ-th excited state of H_0. So the energies of all the H_ℓ can be labeled by a main quantum number N.

In the case of the hydrogen atom we have $a = 0$, $b = \infty$, $x \in \mathbf{R}_+$,

$$G_\ell(x) = -\frac{\ell+1}{x} + \frac{Z}{2(\ell+1)}, \tag{2.18}$$

$$\eta_\ell = -\frac{Z^2}{4(\ell+1)^2}, \tag{2.19}$$

$$H_\ell = -\frac{d^2}{dx^2} + \frac{\ell(\ell+1)}{x^2} - \frac{Z}{x}. \tag{2.20}$$

The energy of states with the main quantum number N is equal to the ground state energy η_ℓ with $\ell = N - 1$:

$$E_N = -\frac{Z^2}{4N^2}. \tag{2.21}$$

I remark that, in order to comply with the usual convention, one could change the appearance of indices, writing $\varphi_{N,\ell}$ instead of $\varphi_{\ell,E}$ for $E = E_N$. Also the sign of every second wave functions could be changed. Then (2.17) would read as

$$\begin{aligned} \varphi_{N,\ell+1} &= -(E_N - \eta_\ell)^{-1/2} A_\ell \varphi_{N,\ell}, \\ \varphi_{N,\ell} &= -(E_N - \eta_\ell)^{-1/2} A_\ell^* \varphi_{N,\ell+1}. \end{aligned} \tag{2.22}$$

This would of course have no effect on the considerations of the energies.

2.3 Exchange of Coupling Constants and Eigenvalues

Any pair (H_0, H_1) of supersymmetric essentially isospectral Hamiltonians can be considered as an element of a family of supersymmetric partners $(H_0(\xi), H_1(\xi))$, $\xi \in \mathbf{R}$, each pair connected by similar intertwining operators

$$A(\xi) = \frac{d}{dx} + G(x) + \xi/2, \tag{2.22}$$

so that, with $\eta = -\xi^2/4$,

$$\begin{aligned} H_0(\xi) &= -\frac{d^2}{dx^2} + G^2(x) - G'(x) + \xi G(x), \\ H_1(\xi) &= -\frac{d^2}{dx^2} + G^2(x) + G'(x) + \xi G(x). \end{aligned}$$

With

$$\begin{aligned} A(\xi)\varphi_{0,\xi,\varepsilon} &= (\varepsilon + \xi^2/4)^{1/2}\varphi_{1,\xi,\varepsilon}, \\ A^*(\xi)\varphi_{1,\xi,\varepsilon} &= (\varepsilon + \xi^2/4)^{1/2}\varphi_{0,\xi,\varepsilon}, \end{aligned} \tag{2.23}$$

the equations

$$(H_0(\xi) - \varepsilon)\varphi_{0,\xi,\varepsilon} \tag{2.24a}$$

are related to

$$(H_1(\xi) - \varepsilon)\varphi_{1,\xi,\varepsilon}. \tag{2.24b}$$

The parameter ξ enters as a coupling constant for the superpotential $G(x)$. If $G(x)$ does not change sign and is negative, the eigenvalues $\varepsilon(\xi)$ are decreasing functions of the coupling constant ξ. These functions may be inverted to $\xi(\varepsilon)$. So we pose the problem to find a family of operators $L_0(\varepsilon)$, $L_1(\varepsilon)$, such that the equations (2.24) may be transformed into

$$(L_i(\varepsilon) - \xi)\psi_{i,\varepsilon,\xi} = 0. \tag{2.25}$$

The solution to this problem is found in two steps. The first step is to define

$$\hat{L}_i(\varepsilon) - \xi := |G(x)|^{-1/2}(H_i(\xi) - \varepsilon)|G(x)|^{-1/2}, \tag{2.26}$$

$$\hat{\psi}_{i,\varepsilon,\xi}(x) := |G(x)|^{1/2}\varphi_{i,\xi,\varepsilon}(x). \tag{2.27}$$

So

$$(\hat{L}_i(\varepsilon) - \xi)\hat{\psi}_{i,\varepsilon,\xi} = 0 \tag{2.28}$$

holds, $\hat{L}_i(\varepsilon)$ are symmetric operators and do not depend on ξ.

In the second step one makes a coordinate transformation $x \to y$, connected with a unitary transformation $T : \mathcal{L}^2(a,b), dx) \to \mathcal{L}^2(c,d), dy)$, such that $-|G|^{-1/2}\frac{d^2}{dx^2}|G|^{-1/2}$ is mapped to $-\frac{d^2}{dy^2} + U(y)$, and

$$T : \hat{L}_i(\varepsilon) \to L_i(\varepsilon) = -\frac{d^2}{dy^2} + U(y) + (G^2 - G' - \varepsilon)/|G|. \tag{2.29}$$

The appropriate transformation is

$$\frac{dy}{dx} = |G(x)|^{1/2}, \qquad y(x) = c + \int_a^x |G(x')|^{1/2}dx', \tag{2.30}$$

$$T : \hat{\psi}(x) \to \psi(y) = |G(x)|^{-1/4}\hat{\psi}(x) = |G(x)|^{1/4}\varphi(x), \tag{2.31}$$

$$U(y) = -G''/4G^2 + 5G'^2/16G^3 = \ddot{G}/4G - (3/16)(\dot{G}/G)^2, \tag{2.32}$$

where we denote $\frac{dG}{dx} = G'$, $\frac{dG}{dy} = \dot{G}$.

We remark that the transformation $\hat{\psi}(x) \to \psi(y)$ is unitary, but $\varphi(x) \to \hat{\psi}(x)$ is not. So the $\hat{\psi}$ defined in (2.31) is not normalized, when φ is.

The action of the intertwining operators $A(\xi)$, $A^*(\xi)$ in (2.23) is transformed to

$$\begin{aligned} B^-(\xi)\psi_{0,\varepsilon,\xi} &= (\xi^2/4 + \varepsilon)^{1/2}\psi_{1,\varepsilon,\xi}, \\ B^+(\xi)\psi_{1,\varepsilon,\xi} &= (\xi^2/4 + \varepsilon)^{1/2}\psi_{0,\varepsilon,\xi}, \end{aligned} \tag{2.33}$$

with

$$B^-(\xi) = |G|^{1/2}\frac{d}{dy} - G'/4G + G + \xi/2, \tag{2.34}$$

$$B^+(\xi) = -|G|^{1/2}\frac{d}{dy} + G'/4G + G + \xi/2, \tag{2.35}$$

$$\|\psi_{1,\varepsilon,\xi}\| = \|\psi_{0,\varepsilon,\xi}\|. \tag{2.36}$$

2.4 Factorizable Systems, Class II

This class of factorizations emerges by applying the change of coupling constants and eigenvalues to each supersymmetric pair of a class I factorizable system. Because of the conditions that all $G_\ell(x) + \xi$ should appear in the facorization, $G_\ell(x)$ purely negative, not all types of class I are amenable to such a change, only the types B, E_β, E_γ and F. Type B is transformed to Schrödinger's factorization of the harmonic oscillator.

The most important example involves again the hydrogen atom, type F, with $G_\ell(x) = -(\ell+1)/x, \xi = Z/(\ell+1)$. With $y = 2((\ell+1)x)^{1/2}$ the corresponding Schrödinger equation

$$(H_\ell(\xi) - \varepsilon)\varphi(x) = (-\frac{d^2}{dx^2} + \frac{\ell(\ell+1)}{x^2} - \frac{Z}{x} - \varepsilon)\varphi(x) = 0 \tag{2.37}$$

is transformed into the Schrödinger equation for the harmonic oscillator

$$(L_\ell(\varepsilon) - \xi)\psi(y) = \frac{1}{(\ell+1)^2}(-\frac{d^2}{dy^2} + \frac{(2\ell+1/2)(2\ell+3/2)}{y^2} + \kappa^2 y^2 - \xi)\psi(y) = 0 \tag{2.38}$$

where

$$\kappa^2 = -\varepsilon/4 \tag{2.39}$$

is positive, since the disjoint eigenvalues ε of the hydrogen atom are negative.

The type F factorization for the hydrogen atom works also for non-integer values of ℓ. We have to choose

$$\ell = \ell_{HA} = -\frac{1}{4}, \frac{1}{4}, \frac{3}{4}, \frac{5}{4}, \ldots \tag{2.40}$$

to get for the oscillator

$$\ell_{HO} = 2\ell_{HA} + \frac{1}{2} = 0, 1, 2, 3, \ldots. \tag{2.41}$$

The operators A_ℓ, A_ℓ^*, which shift ℓ_{HA} by one unit, are transformed to operators $B^-(\xi)$, $B^+(\xi)$, which shift ℓ_{HO} by two units:

$$B_\ell^-(\xi) = (\ell + \frac{3}{2})(\frac{1}{y}\frac{d}{dy} - \frac{\ell+1}{y^2} - \frac{\xi}{2\ell+3}), \tag{2.42}$$

$$B_\ell^+(\xi) = (\ell + \frac{3}{2})(-\frac{1}{y}\frac{d}{dy} - \frac{\ell+2}{y^2} - \frac{\xi}{2\ell+3}), \tag{2.43}$$

here $\ell = \ell_{HO}$.

In subsection 3.3 we will extend this relation between the unperturbed Hydrogen atom and the unperturbed harmonic oscillator to the perturbed systems.

3 First Order Perturbation Theory for Factorizable Systems

3.1 Level Splittings of a Perturbed Supersymmetric Pair

We consider a pair of Hamiltonians $(H_\ell, H_{\ell+1})$, each one perturbed by λV. We want to compare the energies in the perturbed system, in first order perturbation theory:

$$E(N, \ell, \lambda) = E_N + \lambda\langle\varphi_{\ell,E}|V|\varphi_{\ell,E}\rangle. \tag{3.1}$$

The $\varphi_{\ell,E}$ are the wave functions of the unperturbed system, with energies $E = E_N$.

We multiply the level splittings

$$\delta_{N,\ell} = E(N,\ell,\lambda) - E(N,\ell+1,\lambda) \tag{3.2}$$

by a weight factor $E_N - \eta_\ell$ and use the algebra of supersymmetry to obtain

$$\begin{aligned}(E_N - \eta_\ell)\delta_{N,\ell} &= \frac{\lambda}{2}\langle\varphi_{\ell,E}|(A_\ell^* A_\ell V + V A_\ell^* A_\ell - 2A_\ell^* V A_\ell)|\varphi_{\ell,E}\rangle = \\ &= \frac{\lambda}{2}\langle\varphi_{\ell,E}|(A_\ell^*[A_\ell, V] + [V, A_\ell^*]A_\ell)|\varphi_{\ell,E}\rangle = \\ &= \lambda\langle\varphi_{\ell,E}|(-V''/2 + G_\ell V')|\varphi_{\ell,E}\rangle\end{aligned} \tag{3.3}$$

and similarly

$$\begin{aligned}(E_N - \eta_\ell)\delta_{N,\ell} &= \frac{\lambda}{2}\langle\varphi_{\ell+1,E}|(2A_\ell V A_\ell^* - A_\ell A_\ell^* V - V A_\ell A_\ell^*)|\varphi_{\ell+1,E}\rangle = \\ &= \lambda\langle\varphi_{\ell+1,E}|(V''/2 + G_\ell V')|\varphi_{\ell+1,E}\rangle.\end{aligned} \tag{3.4}$$

These two simple tricks are the decisive clue to theorems on the level ordering in first order perturbation theory.

3.2 Level Splittings of the Perturbed Hydrogen Atom

Combining the formula (3.3), expressing $\delta_{N,\ell+1}$, and (3.4) as it stands, we can get rid of the ℓ-dependence of the different operators acting on V. It depends on the detailed form of G_ℓ, hence on the type of factorization, how to combine them. In the case of the hydrogen atom we have to multiply the equations containing G_ℓ by $\ell + 1$:

$$(\ell+1)(E_N - \eta_\ell)\delta_{N,\ell} - (\ell+2)(E_N - \eta_{\ell+1})\delta_{N,\ell+1} = \lambda(\ell+1/2)\langle\varphi_{\ell+1,E}|(V'' + (2/x)V')|\varphi_{\ell+1,E}\rangle. \tag{3.5}$$

This recurrence relation for $\delta_{N,\ell}$ can be completed by

$$(E_N - \eta_{N-1})\delta_{N,N-1} = 0, \tag{3.6}$$

so we get

$$(\ell+1)(E_N - \eta_\ell)\delta_{N,\ell} = \lambda \sum_{\nu=1}^{N-\ell-1} (\ell+\nu-1/2)\langle\varphi_{\ell+\nu,E}|(V'' + (2/x)V')|\varphi_{\ell+\nu,E}\rangle, \tag{3.7}$$

and conclude that we can state

Theorem 1: For a hydrogen atom, perturbed by a spherically symmetric potential λV, there appears an ordering of the levels

$$E(N,\ell,\lambda) > E(N,\ell+1,\lambda) \quad \text{if } \Delta V = V'' + (2/x)V' > 0,$$

$$E(N,\ell,\lambda) < E(N,\ell+1,\lambda) \quad \text{if } \Delta V = V'' + (2/x)V' < 0.$$

For the other factorizable systems there hold similar theorems. The factor in front of V' depends on the type of factorization. It is listed as $F(x)$ in the table of factorizations.

3.3 Level Splittings of the Perturbed Three-Dimensional Harmonic Oscillator

It is an elementary but space-consuming procedure to perform similar calculations as above, with the B-operators replacing the A-operators. The result is:

$$E(N,\ell,\lambda) - E(N,\ell+2,\lambda) = \tag{3.8}$$

$$= \lambda\left(\frac{E_N^2}{2\ell+3} - \kappa^2(2\ell+3)\right)^{-1} \sum_{\nu=1}^{(E/\kappa-3-2\ell)/4} (2\ell+4\nu+1)\langle\psi_{\ell+2\nu,E}|(V'' - (1/y)V')|\psi_{\ell+2\nu,E}\rangle,$$

so we have

Theorem 2: For a three-dimensional harmonic oscillator, perturbed by a spherically symmetric potential λV, there emerges an ordering of the energy levels:

$$E(N,\ell,\lambda) > E(N,\ell+2,\lambda) \quad \text{if } V'' - (1/y)V' > 0,$$
$$E(N,\ell,\lambda) < E(N,\ell+2,\lambda) \quad \text{if } V'' - (1/y)V' < 0.$$

3.4 Harmonic Oscillator in One Dimension: Level Spacings

This is the simplest example of a factorizable system. It allows moreover an extension of our methods to find inequalities of higher order. Apart from a factor, the intertwinors are the usual annihilation and creation operators:

$$A_\ell = \frac{d}{dx} - x, \qquad A_\ell^* = -\frac{d}{dx} - x, \qquad \eta_\ell = 2\ell + 1, \tag{3.9}$$

$$H_\ell = -\frac{d^2}{dx^2} + x^2 + 2\ell. \tag{3.10}$$

So the sequence of supersymmetric partners consists just of the harmonic oscillator shifted in energy by the amount of level spacings: energies and states depend on only one parameter n, the number of nodes of the wave function, instead of N, ℓ:

$$n = N - \ell - 1, \qquad \varphi_n := \varphi_{\ell,E} \quad \text{for } E = 2N, \tag{3.11}$$

$$E(n,\lambda) = E(N,\ell,\lambda) = 2n + 1 + \lambda\langle\varphi_n|V|\varphi_n\rangle. \tag{3.12}$$

The changes of the level spacings

$$\delta_n = \delta_{N,\ell} = E(n,\lambda) - E(n-1,\lambda) - 2 \tag{3.13}$$

obey

$$2n\delta_n = \lambda\langle\varphi_n|(-V''/2 - xV')|\varphi_n\rangle, \tag{3.14}$$

and

$$2n\delta_n = \lambda\langle\varphi_{n-1}|(V''/2 - xV')|\varphi_{n-1}\rangle, \tag{3.15}$$

the rewritten form of (3.3) and (3.4).

By subtracting (3.14) from (3.15) for $n+1$ we get

$$(n+1)\delta_{n+1} - n\delta_n = (\lambda/2)\langle\varphi_n|V''|\varphi_n\rangle, \tag{3.16}$$

and

$$n\delta_n = (\lambda/2)\sum_{\nu=0}^{n-1}\langle\varphi_\nu|V''|\varphi_\nu\rangle, \tag{3.17}$$

a very simple formula, implying that all the level spacings of a harmonic oscillator increase under the perturbation of a potential with positive second derivative, or decrease if $V'' \leq 0$.

Starting with (3.16) we can iterate the procedure and do with $\langle\varphi_\nu|V''|\varphi_\nu\rangle$ what we have done with $\langle\varphi_\nu|V|\varphi_\nu\rangle$, arriving at

$$2n[((n+1)\delta_{n+1} - n\delta_n) - (n\delta_n - (n-1)\delta_{n-1})] = (\lambda/2)\sum_{\nu=0}^{n-1}\langle\varphi_\nu|V''''|\varphi_\nu\rangle, \tag{3.18}$$

which gives the recursion relation

$$n(n+1)(\delta_{n+1} - \delta_n) = (n-1)n(\delta_n - \delta_{n-1}) + (\lambda/4)\sum_{\nu=0}^{n-1}\langle\varphi_\nu|V''''|\varphi_\nu\rangle, \tag{3.19}$$

and finally

$$n(n+1)(\delta_{n+1} - \delta_n) = (\lambda/4)\sum_{\nu=0}^{n-1}(n-\nu)\langle\varphi_\nu|V''''|\varphi_\nu\rangle. \tag{3.20}$$

So the sign of the fourth derivative of the perturbing potential tells us about the increase or decrease of the level spacings with increasing n. This procedure can be iterated again and again, yielding formulas for k-th order differences of the energies related to the $2k$-th differential of the potential.

4 Exact Inequalities Beyond Perturbation Theory

4.1 The Main Theorem

When we intend to compare the eigenvalues of two Hamiltonians H, H_1 defined on the same set in $\mathbf{R}$, with the same kinetic energy operator, but with different potentials, we use some supersymmetry relations and also a set of comparison potentials. So the intuitive argument of comparing the effective atomic potentials with Coulomb comparison potentials is turned into a strict proof.

In order to find comparison potentials for the general case, we have to define a function $G(x)$, such that the difference of the potentials, which is the difference of the Hamiltonians,

$$V_1 - V = H_1 - H = 2G'(x) \tag{4.1}$$

holds. For technical reasons we need a strict inequality

$$G'(x) > 0. \tag{4.2}$$

Then we can compare the energies $E_n(V)$ with $E_{n-1}(V+2G')$, where n counts the eigenvalues, it denotes the number of nodes of the corresponding wave functions.

Now we know of a family of comparison potentials V_c, where

$$E_n(V_c) = E_{n-1}(V_c + 2G') \tag{4.3}$$

because of supersymmetry:

$$V_c = G_c^2 - G_c' + \eta, \tag{4.4}$$

$$G_c(x) = G(x) + c/2, \qquad \eta \text{ arbitrary}. \tag{4.5}$$

If, for example, we wish to compare $E_n(V + \frac{\ell(\ell+1)}{x^2})$ to $E_{n-1}(V + \frac{(\ell+1)(\ell+2)}{x^2})$, we have $2G' = 2(\ell+1)/x^2$, $G_c(x) = -(\ell+1)/r + c/2$, so we have the Coulomb potentials with $Z = (\ell+1)c$ as comparison potentials.

As in the intuitive argument we need the convexity or concavity property of V, that V lies above or below the comparison potential V_c. V_c is that comparison potential which has the properties of touching V at x_0, with the same slope:

$$V_c(x_0) = V(x_0), \qquad V_c'(x_0) = V'(x_0) \tag{4.6}$$

at a fixed point x_0. For the intuitive argument we had to choose for x_0 some mean value. For the strict proof of inequalities we need comparison for each point x_0.

The convexity or concavity property which is needed, is that $V(x)$ lies above or below all the comparison potentials tangent to V. If the comparison potentials were linear, this would be the usual convexity or concavity property. Here we may use the fact that $V_c - V_0 = cG + c^2/4 + \eta_c - \eta_0$ is linear in G. Then we may consider $V - V_0$ as a function of G and observe that we have to look for the convexity or concavity of $(V - V_0)\circ G^{-1}$. When stated in derivatives in x:

$$\left(\frac{d^2}{dx^2} - \frac{G''(x)}{G'(x)}\frac{d}{dx}\right)(V - V_0)\left\{ \begin{matrix} \geq \\ \leq \end{matrix} \right\} 0 \text{ iff } (V - V_0)\circ G^{-1} \text{ is } \left\{ \begin{matrix} \text{convex} \\ \text{concave} \end{matrix} \right. \tag{4.7}$$

Then we may also construct a supersymmetric partner to $H = -\frac{d^2}{dx^2} + V(x)$. For this we need

$$g(x) = -u'(x)/u(x), \tag{4.8}$$

with $u(x)$ a positive solution of

$$\left(-\frac{d^2}{dx^2} + V(x)\right) u = Eu.$$

The hard work to be done is the proof of the following

Proposition: If u is the positive square integrable wave function of the ground state of $-\frac{d^2}{dx^2}+V(x)$, and if $(V-V_0)\circ G^{-1}$ is $\left\{\begin{array}{c}\text{convex}\\ \text{concave}\end{array}\right\}$ and not affine, then

$$g'(x)\left\{\begin{array}{c}>\\ <\end{array}\right\}G'(x). \tag{4.9}$$

For the inequalities between g' and G' to hold strictly everywhere, it sufficies that the convexity inequalities (4.7) hold strictly at least at one single point x.

With the use of this proposition one can prove

Theorem 3: If $(\frac{d^2}{dx^2}-\frac{G''}{G'}\frac{d}{dx})(V-V_0)\left\{\begin{array}{c}\geq\\ \leq\end{array}\right\}0$, not everywhere $=0$, then

$$E_n(V)\left\{\begin{array}{c}>\\ <\end{array}\right\}E_{n-1}(V+2G'). \tag{4.10}$$

Proof:

$$E_n(V)\overset{\text{supersymmetry}}{=}E_{n-1}(V+2g')\quad\overset{\text{Proposition+Mini-Max-Principle}}{\left\{\begin{array}{c}>\\ <\end{array}\right\}}\quad E_{n-1}(V+2G').$$

4.2 Exchange of Coupling Constants and Eigenvalues

Now also in this case can we make an exchange of coupling constants and eigenvalues if G has a definite sign, $G(x)<0$. If $(V-V_0)\circ G^{\text{inv}}$ is convex (concave), then for each c also $(V+cG-V_0)\circ G^{\text{inv}}$ is convex (concave). So the discrete eigenvalues $\tilde{b}_n(c)$ of $-d^2/dx^2+V+cG+2G'$ and $b_n(c)$ of $-d^2/dx^2+V+cG$ obey the inequalities

$$b_n(c)\left\{\begin{array}{c}>\\ <\end{array}\right\}\tilde{b}_{n-1}(c). \tag{4.11}$$

These eigenvalues are decreasing functions of c which can be inverted to $c_n(b)$ and $\tilde{c}_n(b)$ with the property

$$c_n(b)\left\{\begin{array}{c}>\\ <\end{array}\right\}\tilde{c}_{n-1}(b). \tag{4.12}$$

Transforming the Hamiltonians as we did for factorizable systems, we transform to the coordinate y with

$$\frac{dy}{dx}=|G|^{1/2}. \tag{4.13}$$

We define

$$W_0(y)=(1/4)\ddot{G}/G-(3/16)(\dot{G}/G)^2-G. \tag{4.14}$$

Then the convexity of $(V - V_0) \circ G^{\text{inv}}$ transforms to the convexity of $(W - W_0) \circ (1/G)^{\text{inv}}$, also to be expressed by

$$\left(\frac{d^2}{dy^2} + (2\dot{G}/G - \ddot{G}/\dot{G})\frac{d}{dy}\right)(W - W_0) \geq 0. \tag{4.15}$$

So we arrive at

Theorem 4: If $(W - W_0) \circ (1/G)^{\text{inv}}$ is $\left\{ \begin{array}{c} \text{convex} \\ \text{concave} \end{array} \right\}$ and not affine, then

$$E_n(W) \left\{ \begin{array}{c} > \\ < \end{array} \right\} E_{n-1}(W + 2\dot{G}/|G|^{1/2}). \tag{4.16}$$

This theorem applies to the broken degeneracies of the levels of the harmonic oscillator. Again it gives the same inequalities and under the same conditions on the perturbing potential, as theorem 2, but here beyond perturbation theory.

4.3 An Outline of the Proof of the Proposition

We consider the case of convexity of $(V - V_0) \circ G^{\text{inv}}$ and indicate the procedure of the proof:

a) For fixed x_0, choose c such that $V_c'(x_0) = V'(x_0)$.

b) Assume w.l.o.g. $G(x_0) + c \geq 0$ (otherwise change $x \to -x$).
If there exists no γ, such that

$$(G(x_0) + \gamma)^2 - G'(x_0) = V(x_0) - E, \tag{4.17}$$

continue with f).
If there exist such γ, choose the larger one.

c) Deduce from the convexity of $V - V_0$ and the Mini-Max-Principle, applied to the ground state, that $\gamma < c$.

d) $\forall x > x_0$ you have

$$V_\gamma(x) < V_c(x) - V_c(x_0) + V(x_0) - E \leq V(x) - E \tag{4.18}$$

since $V_\gamma'(x) < V_c'(x) \leq V'(x)$.

e) Let ϕ_γ be the ground state wave function of V_γ, so

$$G + \gamma = -\phi_\gamma'/\phi_\gamma. \tag{4.19}$$

Due to (4.18) you have the Wronskian relation

$$u'(x_0)\phi_\gamma(x_0) - u(x_0)\phi_\gamma'(x_0) < 0 \tag{4.20}$$

which implies

$$g(x_0) = -u'/u > -\phi_\gamma'/\phi_\gamma = G(x_0) + \gamma > 0 \tag{4.21}$$

and

$$g(x_0)^2 > (G(x_0) + \gamma)^2. \tag{4.22}$$

f) The Schrödinger equation for u is equivalent to the Riccati equation

$$g'(x_0) = g^2(x_0) - V(x_0) + E. \tag{4.23}$$

There exists no γ solving (4.17) iff $G'(x_0) < E - V(x_0)$. Inserting this inequality in (4.23) gives immediately $g'(x_0) > G'(x_0)$. If such a γ exists, insert (4.22) into (4.23), use (4.17) and you have also finished the proof:

$$g'(x_0) > (G(x_0) + \gamma)^2 - V_0(x_0) + E = G'(x_0).$$

References

[1] E. Schrödinger, A method of determining quantum-mechanical eigenvalues and eigenfunctions, Proc. Roy. Irish Acad. *A46*, (1940) 9–16.

[2] E. Schrödinger, Further studies on solving eigenvalue problems by factorization, Proc. Roy. Irish Acad. *A46*, (1941) 183.

[3] L. Infeld, T.E. Hull, The factorization medhod, Rev. Mod. Phys. *23*, (1951), 21-68.

[4] E. Witten, Dynamical breaking of supersymmetry, Nucl. Phys. *B188*, (1981), 513-554.

[5] A.K. Ramdas, S. Rodriguez, Spectroscopy of the solid-state anologues of the hydrogen atom: donors and acceptors in semiconductors, Rep. Progr. Phys. *44*, (1981), 1297.

[6] B. Baumgartner, A. Pflug, A new approach to the understanding of level ordering in atoms and nuclei, preprint UWThPh-1989-11 to appear in Amer. J. Phys.

[7] P.A. Deift, Applications of a commutation formula, Duke Math. J. *45*, (1978), 267.

[8] B. Baumgartner, Level Comparison Theorems, Ann. Phys. *168*, (1986), 484-526.

ON THE EIGENVALUES OF A PERTURBED HARMONIC OSCILLATOR

Anne Boutet de Monvel-Berthier

Laboratoire de Physique Mathématique et Géométrie
Université Paris VII
2, place Jussieu, 75251 Paris Cedex 05

ABSTRACT

We study Schrödinger operators on $\mathbf{R}^n$ of the form

$$H = H_V = -\Delta + q(x) + V(x)$$

where q(x) is a positive definite quadratic form and V(x) a potential which may be considered as a perturbation. We show that the discrete spectrum of H has similar properties as that of the unperturbed $H_0 = -\Delta + q$. In particular the singular points of the distribution on **R**: $t \to \mathrm{Tr}\, e^{itH}$ are the lengths of the periodic orbits of the hamiltonian flow associated to H_0.

This work was done in collaboration with G.Lebeau and L.Boutet de Monvel.

1. INTRODUCTION AND RESULTS

The study of the relations between quantum and classical mechanics goes back to the very origin of quantum physics, i.e. to the years when the "old quantum theory" of Bohr, Einstein, Sommerfeld was developped and was to lead to what is now understood as quantum mechanics (cf. the works of Heisenberg, Schrödinger).

For history and extensive references to other works on topics connected with the present paper see [ABmL]. There is also litterature by the means of probability theory, e.g. Feynman integrals [ABHk] which make it possible to compute the way in which the solutions of quantum equations of motion are connected to the classical ones, in the precise sense of asymptotic expansions in powers of Planck's constant.

A. Boutet de Monvel et al. (eds.), Recent Developments in Quantum Mechanics, 209–222.

In this paper we study the singular support (or wave front set) of the distribution $t \longmapsto \mathrm{Tr}\, e^{itH}$. This somehow requires more refined techniques than what is needed to study many asymptotic properties of the spectrum, e.g. to prove the existence and compute the coefficients of the asymptotic expansions as mentionned above, and for this we will need information on a variant of the singular spectrum of the Schwartz distribution-kernel of e^{itH}, which we can study by the methods of microlocal analysis (i.e. the several variables analogue for P.D.E. theory of the W.K.B. method).

A first proof by one of us used an idea of J.Sjöstrand from the theory of semi-classical approximations; it consisted in introducing a "small" scaling parameter h, by replacing the operator H by its homothetic in the ratio h and studying the asymptotic propagation of singularities in that context [S], [BmBmL].

The method used here consists in the microlocal study of the operator $P_t = e^{itH}\, e^{-itH_0}$.

The problem for classical asymptotics is to understand what happens when the parameter $h \to 0$. Here we are rather interested in what happens at the high frequencies; there is no Planck constant in our problem and its role is played by the inverse of the energy, i.e. of the eigenvalues of the problem. To understand correctly the propagation of singularities we have to think of H or H_0 as living on $\mathbf{R}^n_{pol}$ and use the "polynomial" pseudodifferential operators and the related singular spectrum described in the lecture of L.Boutet de Monvel (in this context x and D_x are both of degree 1/2 and H or H_0 is of degree 1). All recent analysis concerning singularities of distributions go through the study of the singular support, and the techniques of "microlocal analysis" involved in this use methods closely related to those of constructions of asymptotic expansions by the W.K.B. methods. However P_t does not belong to a "good" class of pseudodifferential operators and here all these methods seem to fail (in fact the whole point is to prove that it is a pseudodifferential operator, i.e. that it preserves microsupports). We therefore need a trick to handle pseudodifferential operators whose symbols are too irregular to permit symbolic calculus, but which still allow functional calculus (for such operators see also Beals [B2]). In this problem this gives slightly more information than the more systematic semi-classical methods (the two are closely related anyway).

Notations and problem.

Let H_0 be the unperturbed operator:

$$H_0 = \frac{1}{2}(-\Delta + q(x)) \tag{1.1}$$

where q(x) is a positive quadratic form. We choose an orthonormal basis in which q is diagonal: $q(x) = \sum \mu_j^2 x_j^2$ with $\mu_j > 0$.

Let $H = H_V$ be the perturbed operator:

$$H = H_V = H_0 + V = \frac{1}{2}(-\Delta + q(x)) + V(x) \tag{1.2}$$

where V is a potential (in the complete problem V would also depend on the Planck constant and other parameters).

To really have a perturbation problem it is necessary to suppose that V is not too big: $V(x) = o(q(x))$ for $x \to \infty$ (V less than the quadratic form). Typically V could be a smooth periodic function such as $V(x) = \lambda \sin(a.x+b)$.

The spectrum of the hamiltonian H_0 is well known. The eigenvalues are the numbers indexed by multiindices $\alpha = (\alpha_1,\ldots,\alpha_n)$

$$\lambda_\alpha = \sum_{j=1}^{n} (\alpha_j + 1/2)\,\mu_j \tag{1.3}$$

(The eigenfunctions of H_0 are also well known: they are Hermite functions indexed by multiintegers, of the form $(\frac{\partial}{\partial x} - \mu.x)^\alpha . e^{-1/2\sum \mu_j x_j^2}$).

The order of magnitude of the eigenvalues of H_0 is the same as for an elliptic operator of order 1 (not 2) in n variables: the k-th eigenvalue has the size of $k^{1/n}$. It follows that the trace distribution $\mathrm{Tr}\, e^{itH_0}$ given by

$$\mathrm{Tr}\, e^{itH_0} = \sum_\alpha e^{it\lambda_\alpha} = e^{it/2\sum\mu_j} \prod_{j=1}^{n} (1 - e^{it\mu_j})^{-1} \tag{1.4}$$

is well defined as a distribution of the real variable t : the series converges as the Fourier series of a temperate distribution. The last term in formula (1.4) should be understood as the boundary value distribution of the product, which is well defined and holomorphic in the upper half plane Im t > 0.

Thus the trace distribution exists, and the formula above shows that it has polar singularities when t is the length of one of the closed integral curves of the hamiltonian vector field of H_0:

$$\mathfrak{X}_{H_0} = \sum \frac{\partial h}{\partial \xi_j}\frac{\partial}{\partial x_j} - \frac{\partial h}{\partial x_j}\frac{\partial}{\partial \xi_j} = \sum \xi_j \frac{\partial}{\partial x_j} - \mu_j^2 x_j \frac{\partial}{\partial \xi_j}. \tag{1.5}$$

Thus the set of singularities of the trace distribution e^{itH_0} is equal to the set of periods of the hamiltonian flow of the classical dynamical system associated with H_0 (whose hamiltonian is the symbol $1/2(\xi^2+q(x))$ of H_0): the set $\bigcup \mu_j \mathbf{Z}$ of all integral multiples of the fundamental lengths μ_j.

Now we want to know what remains true when we replace H_0 by the perturbed operator $H = H_0+V$. For instance if $V = o(\|x\|^2)$, H^{-1} is compact; and one checks easily that $H_0^{-1}(H-H_0)$ is a compact operator on L^2 so the eigenvalues of H and H_0 have the same order of magnitude. This is true in particular if $V = O(\|x\|^{2-\varepsilon})$, $\varepsilon>0$.

Problem: Can one still say something precise about the asymptotic behaviour of the eigenvalues of H; in particular can we still locate the singularities of the distribution e^{itH} ?

We expect that H and H_0 have analogous properties, at least in the case where V is smooth and periodic. In fact we will have a better result than that, and show that the distributions e^{itH} and e^{itH_0} have the same singularities, with rather mild growth conditions on V and its derivatives (we are however still far from what might seem optimal from the remarks above: $V = O(\|x\|^{2-\varepsilon})$, $\varepsilon>0$ and no condition on the derivatives): we will prove the following theorem:

Theorem 1.– If V and its derivatives satisfy suitable growth conditions (see theorem 3 and §3 below), then the distributions Tr e^{itH} and Tr e^{itH_0} have the same singular set.

We will try to reproduce for H_0 and H the method of Fourier integral operators used for partial differential operators (see Hörmander [H2], Colin de Verdière [Co], Duistermaat-Guillemin [DG], Chazarain [Ch])

Let X be a compact Riemannian manifold, A a first order elliptic positive pseudodifferential operator, e.g. $\sqrt{-\Delta}$, or $\sqrt{-\Delta+V}$ or $\sqrt{-\Delta}+V$. We want to look at the distributions: Schwartz kernel of e^{itA} and Tr e^{itA}. Hörmander defined Fourier integral operators and proved:

Theorem 2.– The operator e^{itA} is a Fourier integral operator.

This means that it has locally integral representations:

$$e^{itA}f(x) = \int e^{i\phi(t,x,y,\theta)}\, b(t,x,y,\theta)\, f(y)\, dy\, d\theta \tag{1.6}$$

so that the Schwartz kernel of e^{itA} is the distribution

$$K(t,x,y) = \int e^{i\phi(t,x,y,\theta)}\, b(t,x,y,\theta)\, d\theta. \tag{1.7}$$

The singular support SS(K) of K is the associated characteristic Lagrangian manifold Λ: This is the set of points

$$\Lambda = \{(t,x,y,\tau,\xi,\eta) \mid \exists\theta \text{ such that } d_\theta\phi=0,\ \tau=d_t\phi,\ \xi=d_x\phi,\ \eta=d_y\phi\}. \tag{1.8}$$

Here Λ is geometrically characterized as the flow out of the diagonal by the hamiltonian field of A i.e.

$$\Lambda = \{(t,x,y,\tau,\xi,\eta) \mid \tau=a(\xi)=a(\eta),\ (x,\xi) = \Phi_t(y,-\eta)\} \tag{1.8)bis}$$

where a is the symbol of A and Φ_t the hamiltonian field of a.

In "classical" cases the symbol b (as a) has an asymptotic expansion

$$b \sim \sum b_{m-k} \tag{1.9}$$

where the functions b_{m-k} are smooth, homogeneous of degree m–k with respect to θ (k varies among positive integers, m is a fixed number, which depends on the degree of the distribution under study and on the number of variables in the integral representation).

From (1.5) one gets for the distribution Tr e^{itA}

$$\text{Tr}\, e^{itA} = \int e^{i\phi(\tau,x,y,\theta)}\, b(t,x,x,\theta)\, dx\, d\theta. \tag{1.10}$$

This is again a Fourier integral distribution. The only points that contribute to its singular spectrum are those for which $d_\theta\phi = 0$ and $d_x\phi + d_y\phi = 0$, with $x=y$. They correspond to points on the Lagrangian Λ such that $x=y$, $\xi+\eta=0$. Since $(x,\xi) = \Phi_t(y,-\eta) = \Phi_t(x,\xi)$ this means exactly that (x,ξ) belongs to a closed integral curve of period t.

Now we want to imitate this for the operator H_0 and the perturbed operator H. There are several things to adapt:

i) $\mathbf{R}^n$ is not compact, and $H_0 = -\Delta + V$ is an operator of order two, even though it behaves as a first order elliptic operator as regards the order of magnitude of its eigenvalues. To compensate this we will work in the frame of "polynomial pseudodifferential operators" described in the talk of L. Boutet de Monvel, and count the degrees and adapt the definition of the singular spectrum accordingly.

ii) Comparison of H_0 and H: the idea is to study the operator

$$P_t = e^{-itH_0}\, e^{itH} \tag{1.11}$$

and prove that it is microlocal (i.e. preserves singular supports) even if it is not a pseudodifferential operator in the usual sense.

$P=P_t$ satisfies the differential equation

$$\frac{dP}{dt} = i\, V_t\, P \qquad \text{with } V_t = e^{-itH_0}\, V\, e^{itH_0}. \tag{1.12}$$

Let us recall schematically from Hörmander [H] and from the lecture of L.Boutet de Monvel the symbol spaces we are led to use: S^m_δ is the space of Hörmander symbols of degree m and type $(1-\delta,\delta)$ (we will say "of type δ" for short). Here since both x and ξ are of degree 1/2 so $\partial_x\, \partial_\xi$ are of degree $-1/2$, a smooth function a belongs to S^m_δ if and only if for all (x,ξ) multi-indices α one has

$$|\partial^\alpha_{x\xi}\, a| = O\left(<x\xi>^{m+(\delta-1/2)|\alpha|}\right) \qquad \text{with } <x\xi>=(1+x^2+\xi^2). \tag{1.13}$$

If the operator V was a good pseudodifferential operator of order ≤ 0 as in the pseudodifferential examples above, it would be easy to use the symbolic calculus of pseudodifferential operators to solve (1.12) for an asymptotic series and show that this gives an asymptotic of P. The precise condition is $V \in S^0_\delta$, with $\delta<1/2$; the case $\delta=1/2$ appears as usual as an uncomputable limit case, and the case $\delta>1/2$ seems out of reach by this method. Here unfortunately the total symbol of V does not depend on ξ, so in the best case (unless it is a polynomial of degree 0 or 1) it behaves as a symbol of type 1/2 for which the symbolic calculus does not work because the degrees of the successive terms one computes go up rather than down.

However we will show the following result, which is almost as good even though it does not compute explicitly the symbol of P by closed differential formulas:

Theorem 3.- If Im V is bounded and $V\in S^{\gamma}_{\delta}$ with $\nu=\gamma+\delta<1$, then the operator P, defined by (1.11) or (1.12), has the following properties:

i) it is of order 0 i.e. continuous $\mathcal{H}^s\to\mathcal{H}^s$

ii) if $Q_j\in OPS^1_0$, $j=1,\dots,n$, then $[Q_1,[\dots[Q_N,P]..]]$ is of order $\leq N\nu$.

From condition ii) it follows quite easily that P is microlocal, so that e^{itH}, e^{itH_0} have the same singular spectrum, so as Tr e^{itH} and Tr e^{itH_0} (see §3).

2. REVIEW OF SYMBOLIC CALCULUS

If $a(x,\xi)$ is a function of moderate growth on $\mathbf{R}^{2n}$ (in fact more generally a temperate distribution) we define the operator $A=a(x,D)$ by the formula

$$Af(x) = \int e^{ix.\xi}\, a(x,\xi)\, \hat{f}(\xi)d\xi \qquad \text{(with } \hat{f} \text{ the Fourier transform of f)} \tag{2.1}$$

or equivalently

$$\langle Af,g\rangle = \langle e^{ix.\xi}\, a\, ,\, \hat{f}(\xi)\, g(x)\rangle$$

(this being the definition when a is only a temperate distribution).

The operator A is thus well defined as a continuous operator from $\mathcal{S}$ to $\mathcal{S}'$, and from $\mathcal{S}$ to $\mathcal{S}$ if a is a "symbol" in a very mild sense, i.e. a and all its derivatives are of polynomial growth at infinity. If a is smooth and of rapid decrease ($a\in\mathcal{S}$) then A is continuous $\mathcal{S}'\to\mathcal{S}$ (regularizing — or of degree $-\infty$).

As indicated above, let S^m_δ be the space of smooth functions on $\mathbf{R}^{2n}$ such that for all (x,ξ) multi-indices α

$$|\partial^\alpha_{x\xi} a| \leq O\left(<x\xi>^{m+(\delta-1/2)|\alpha|}\right) \qquad \text{(with } <x\xi>=(1+x^2+\xi^2)) \tag{2.2}$$

(this corresponds to $S^m_{1-\delta,\delta}$ with the notation of Hörmander [H1], if we remember that x and ξ are both of degree 1/2).

Then S^m_0 corresponds to the space of "classical" symbols, and the set OPS^m_0 of corresponding operators a(x,D) to the set of "classical" operators. For these one has the usual formulas giving asymptotically the symbol of the composition, or the commutator of two operators, i.e.

$$a \circ b \sim \sum \frac{i^{-\alpha}}{\alpha!} \left(\frac{\partial}{\partial x}\right)^\alpha a \left(\frac{\partial}{\partial \xi}\right)^\alpha b \tag{2.3}$$

$\sigma([a,b]) = -i\,\{a,b\}$ (the Poisson bracket).

This formula is in fact valid whenever it makes sense, i.e. when the degrees of successive terms $\to -\infty$: this is the case if $a \in S^m_\delta$, $b \in S^m_{\delta'}$ if $\delta+\delta'<1$, in particular if δ or $\delta' = 0$ (we always assume δ and $\delta' <1$).

One has a notion of ellipticity (i.e. the symbol is inversible and the inverse belongs to S^{-m}_δ near ∞). There is also a notion of microlocalization and of singular support (but in this context this describes a combination of smoothness and growth at ∞). If $\delta \leq 1/2$ the operators a(x,D), $a \in S^0_\delta$ are still L^2 continuous (the Calderon-Vaillancourt proof still applies). This is again false if $\delta>1/2$, although for $\delta<1$ the corresponding operators a(x,D) still act continuously on S^∞ (the space of smooth rapidly decreasing functions), extend continuously to $S^{-\infty}$ (the space of temperate distributions), and diminish the singular support i.e. are "microlocal".

There is also a scale of Sobolev spaces associated to this situation: for s>0 $\mathcal{H}^s$ is the domain of A^s where A is the harmonic oscillator

$$A = 1/2(-\Delta + |x|^2)$$

and OPS^m_δ for $\delta \leq 1/2$ acts continuously $\mathcal{H}^s \to \mathcal{H}^{s-m}$ (for all s).

The operators in OPS^m_δ with $\delta>1/2$ do not usually operate continuously on the scale of Sobolev spaces $\mathcal{H}^s$ (with loss of m degrees), this being linked to the fact that the symbolic calculus does not work. In questions where such symbols seem to appear naturally it is therefore important to find something else. Following the idea of R.Beals we therefore replace the set OPS^m_δ by the set $OP\Sigma^m_\delta$ defined in the following manner (note that here the operators are simply defined, but the "total symbols" no longer are).

Definition 4.- We will say that an operator A is of degree m ($A\in \mathcal{L}^m = OP\Sigma^m_1$) if A is continuous $\mathcal{H}^s \to \mathcal{H}^{s-m}$ for all s. Then $OP\Sigma^m_\delta$ is defined by :

$A\in OP\Sigma^m_\delta$ if for any sequence of "classical" operators of degree 1, $Q_1,\dots,Q_N\in OPS^1_0$, we have :

$$[Q_1,[Q_2,[\dots[Q_N,A]\dots]]] \text{ is of degree} \leq m+N\delta. \tag{2.4}$$

Proposition 5.- If $\delta\leq 1/2$ then $OP\Sigma^m_\delta = OPS^m_\delta$.

This result is essentially due to Beals, although easier to prove in our context: if $A = a(x,D)\in OP\Sigma^0_{1/2}$ the iterated commutators $(Ad(x,D))^\alpha A$ all belong to $OP\Sigma^0_{1/2}$ so their symbols belong to a Banach space of distributions (the space of symbols of L^2 continuous operators) whose norm is translation invariant; it follows instantly that all derivatives of the symbol a are bounded functions i.e. $a\in OPS^0_{1/2}$.

The converse assertion follows from the Calderon-Vaillancourt theorem. The case $\delta<1/2$ then follows from this and the usual symbolic calculus which shows that the operator with symbol $a^{(\alpha)}$ is of type 1/2 and degree $\leq m-(1/2-\delta)|\alpha|$.

This equality $OP\Sigma^m_\delta = OPS^m_\delta$ is not true if $\delta>1/2$. However we have the obvious following partial result:

Proposition 6.- If $V = V(x)$ is independent of ξ and $V\in OPS^m_\delta$ with $m\geq 0$ and $\delta\geq 1/2$, then $V\in OP\Sigma^m_\delta$.

3. PERTURBATION OF A HARMONIC OSCILLATOR

We now return to the problem in the introduction, so H_0 is a harmonic oscillator:

$$H_0 = 1/2\,(-\Delta + q(x))$$

with q a positive definite quadratic form on $\mathbf{R}^n$, and

$$H = H_0 + V(x)$$

where V is a perturbing potential. We make the following hypotheses on V:

$$\partial_x^\alpha V = O\left(|x|^{\beta+\theta|\alpha|}\right) \tag{3.1}$$

i.e. in other terms $V \in OPS_\delta^\gamma$ with $\gamma=\beta/2$, $\delta=(1+\theta)/2$. We consider H and H_0 as "polynomial" pseudodifferential operators on $\mathbf{R}^n$ so the principal symbol is

$$\sigma(H) = \sigma(H_0) = 1/2\,(\xi^2 + q(x)). \tag{3.2}$$

Both H and H_0 are elliptic operators of order 1 in this context, therefore H has a parametrix $H^{-1} \in OP\Sigma_\delta^{-2}$.

Now we wish to study the Schwartz kernels $K_0(t,x,y)$ and $K(t,x,y)$ of e^{itH_0} and e^{itH} and in particular their microsupports as operators on $\mathbf{R}_{usual} \times \mathbf{R}^n_{pol}$. We have

$$(\partial_t - iH_x)K = (\partial_t - iH_y)K = 0. \tag{3.3}$$

We further wish to describe the singular spectrum (asymptotic singularity) of K as an operator on $\mathbf{R}_{usual} \times \mathbf{R}^n_{pol}$. From (3.3) it follows that SS(K) is contained in the characteristic set of (3.3):

$$SS(K) \subset \left\{(txy,\tau\xi\eta) \mid \tau=\sigma_H(x,\xi)=\sigma_H(y,\eta)\right\} \tag{3.4}$$

since $\partial_t - H_x$ or $\partial_t - H_y$ is elliptic outside of this set.

We then study the operator

$$P_t = e^{itH_0} - e^{itH} \tag{3.5}$$

As we already mentioned this satisfies the differential equation

$$\frac{dP}{dt} = i\, V_t\, P \qquad \text{with } V_t = e^{-itH_0}\, V\, e^{itH_0} \tag{3.6}$$

and the initial condition $P_0 = 1$.

From the property (3.1) of V and the fact that e^{itH_0} behave as Fourier integral operators (of degree and type 0), it follows that the V_t form a C^∞ family of pseudo-differential operators: $V_t \in OP\Sigma_\delta^\gamma$.

We now proceed to prove theorem 3 i.e. that if $V \in OP\Sigma_\delta^\gamma$ with $\gamma+\delta<1$, and the imaginary part of V belongs to $OP\Sigma_\delta^0$ (in particular is bounded) then P belongs to $OP\Sigma_{\gamma+\delta}^0$.

In fact P satisfies $\partial_t P = iV_tP$, $P(0)=1$, and since Im V is a bounded operator, e^{itH} and P are bounded operators on L^2 with bound

$\| P_t \| \leq e^{a|t|}$.

In the same manner the operator $\mathrm{Im}(H_0^s\, V_t\, H_0^{-s})$ is bounded in L^2 because V is bounded in $\mathcal{H}^s$, and it follows that $H_0^s\, P_t\, H_0^{-s}$ is bounded in L^2 for all s, so P_t is of degree 0.

To finish the proof we must check the successive commutators

$P_k = [Q_k,[\ldots[Q_1,P]..]]$.

These also satisfy differential equations: recursively

$$\partial_t P_k = i\, V_t P_k + i\, [Q_k, V_t]\, P_{k-1} \tag{3.7}$$

since the commutator $[Q_k,V_t]$ is of degree $\leq \nu=\delta+\gamma$ if $V \in OP\Sigma_\delta^\gamma$, it follows recursively that the k-th commutators P_k are of degree $k\nu$ i.e. $P \in OP\Sigma_\nu^0$ $(\nu=\gamma+\delta)$. (Note that contrarily to this functional analysis argument which is easy, the pseudodifferential symbolic or asymptotic calculus gives nothing in this situation).

We may now end the proof: our final result, $P_t \in OP\Sigma_\nu^0$ with $\nu<1$, implies easily that P_t is microlocal (preserves singular supports) in other words that AP_tB is of degree $-\infty$ if A and B are "classical" pseudodifferential operators with disjoint supports (proof: if A and B are of degree 0 and disjoint supports we may write, for any N and mod. operators of degree $-\infty$,

$A=A_N\ldots A_1$ where the A_j are also of degree 0 and their support is disjoint of that of B. If we commute and take on account $A_jB \sim 0$ we get

$$APB \sim [A_N,[\ldots[A_1,P]..]]\, B$$

where the left hand side is of degree $-(\delta-1)N$ i.e. arbitrarily small, i.e. APB is of degree $-\infty$.)

It follows that K and K_0 have the same singular support. Then we have the integrals of their restrictions

$$\mathrm{Tr}\, e^{itH} = \int K(t,x,x)dx \quad \text{and} \quad \mathrm{Tr}\, e^{-itH_0} = \int K_0(t,x,x)dx$$

(this always works for the singular spectrum — and not for the cruder singular support — in our context as well as in the classical context.) This ends the proof of theorem 1. Notice that the proof gives the precise location of the singularities, but cannot give a much more precise description of these in the form of an asymptotic expansion or otherwise — this would require some form of symbolic calculus.

REFERENCES

[ABmL] S.Albeverio, A.Boutet de Monvel-Berthier and G.Lebeau, "Green's function and the trace formula for the Schrödinger operator", in preparation.

[ABHk] S.Albeverio, Ph.Blanchard and R.Hoegh-Krohn, "Feynman paths integral and the trace formula for the Schrödinger operators", Comm. Math. Phys. 83, 49-76 (1982).

[BB] R.Balian and C.Bloch, "Solutions of the Schrödinger equations in terms of classical paths", Annals of Physics 85, 514-545 (1974).

[BF] R.Beals and C.Fefferman, "Spatially inhomogeneous pseudodifferential operators I", Comm. Pure. Appl. Math 27, 1-24 (1974).

[B1] R.Beals, "A general calculus of pseudodifferential operators", Duke. Math. J. 42, 1-42 (1975).

[B2] R.Beals, "Characterization of pseudodifferential operators and apllications, Duke Math.J. 44, 47-57, (1977), and correction 46, 215 (1979).

[BM] M.V. Berry and K.E.Mount, "Semi-classical approximations in wave mechanics", Rep. Prog. Phys. 35, 315-397 (1972).

[BmBmL] A.Boutet de Monvel-Berthier, L.Boutet de Monvel and G.Lebeau, "On the eigenvalues of the perturbed harmonic oscillator", in preparation.

[Bm1] L.Boutet de Monvel, "Revue sur la théorie des $\mathscr{D}$-modules d'opérateurs pseudodifférentiels", Ecole internationale d'été Franco-Roumaine, Brasov (1989), this volume.

[Bm2] L.Boutet de Monvel, "Hypoelliptic operators with double characteristics and related pseudodifferential operators", Comm. Pure. Appl. Math 27, 585-639 (1974).

[C1] J.Chazarain, "Formule de Poisson pour les variétés riemanniennes", Invent. Math. 24, 65-82 (1974).

[C2] J.Chazarain, "Spectre d'un hamiltonien quantique et période des trajectoires classiques", C.R. Acad. Sci. Paris 288, 725-728 (1979).

[C3] J.Chazarain, "Comportement du spectre d'un hamiltonien quantique", C.R. Acad. Sci. Paris 288, 895-897 (1979).

[C4] J.Chazarain, "Spectre d'un hamiltonien quantique et mécanique classique", Comm. in P.D.E. 5 (6), 595-644 (1980).

[Cv1] Y.Colin de Verdière, "Quasi-modes sur les variétés riemanniennes compactes", Invent. Math. 43, 15-52 (1977).

[Cv2] Y.Colin de Verdière, "Sur le spectre des opérateurs elliptiques à bicaractéristiques périodiques", C.R. Acad. Sci. Paris 288, 1195-1197 (1978).

[DG] J.J. Duistermaat and V.Guillemin, "The spectrum of positive elliptic operators and periodic geodesics", Proc. A.M.S. Summer Inst. Diff. Geom., Stanford (1973).

[GS] V.Guillemin and S.Stenberg, "Geometric asymptotics", A.M.S. (1977).

[H1] L.Hörmander, "Pseudodifferential operators and hypoelliptic equations", Amer. Math. Soc. Symp. on Singular Integrals, 138-183 (1966).

[H2] L.Hörmander, "Fourier integral operators I", Acta. Math. 127, 79-183 (1971).

[H3] L.Hörmander, "The analysis of linear pseudodifferential operators IV", Springer-Verlag (1985).

[L] J.Leray, "Analyse lagrangienne et mécanique quantique", Collège de France (1976-1977).

[M] V.P.Maslov, "Théorie des Perturbations et méthodes asymptotiques", Dunod (1972).

[S] J.Sjöstrand, Private communication.

[V1] A.Voros, "Développements semi-classiques", Thèse, Orsay (1972).

[V2] A.Voros, "Semi-classical approximation", Ann. Inst. H. Poincaré 24, 31-50 (1976).

[W] A.Weinstein, "Asymptotic of the eigenvalues cluster of the Laplacian plus a potential", Duke Math. J. 44, 883-892 (1972).

ON TOPICS IN SPECTRAL AND STOCHASTIC ANALYSIS FOR SCHRÖDINGER OPERATORS

Michael Demuth
Institute of Mathematics
Academy of Sciences of GDR
Mohrenstr. 39, 1086 Berlin, GDR

ABSTRACT. It is given a certain overview on results in spectral theory for Schrödinger and generalized Schrödinger operators obtained in the last years by means of stochastic analysis, in particular by the use of the Feynman-Kac formulae.

1. INTRODUCTION

The objective of this lecture is to give a collection of spectral theoretical results for Schrödinger operators since Simon has published his book on "Functional Integration and Quantum Physics" [28]. All these results are obtained by stochastic methods, e.g. in particular by the use of the Feynman-Kac formulae. Hence the present article is thought as an advertisement for the powerful tool of stochastic analysis in spectral theory of Schrödinger operators.

The general theory in the sections 2 and 3 is based on the articles by Aizenman, Simon [1], Demuth, van Casteren [15], and Simon [29]. Section 4 contains a series of applications obtained by several authors. The proofs are restricted to certain hints in order to emphasize the points where the stochastic analysis is used.

2. STOCHASTIC REPRESENTATION OF SEMIGROUPS

The main link between stochastic analysis and spectral theory is the transition probability function of a stochastic process. In order to avoid the abstract definition of a stochastic process we start with the definition of a Markov transition density function (see Friedman [17] p. 18). The reader can always have in mind the simplest example of the Wiener transition density function.

$$p_w(t,x,y) = (2\pi t)^{-n/2}\exp(-|x-y|^2/2t) . \tag{2.1}$$

A. Boutet de Monvel et al. (eds.), Recent Developments in Quantum Mechanics, 223–242.

Definition 2.1.: 1. Markov transition density function

Let p be a function mapping $(0,\infty) \times \mathbb{R}^n \times \mathbb{R}^n$ into $\mathbb{R}_+$. Let $0 < t < \infty$, $x,y \in \mathbb{R}^n$. It is assumed that $\int_\Delta p(t,x,y)dy$ is a Borel measurable function in x if t and Δ are fixed (Δ any Borel set on $\mathbb{R}^n$) and is a probability meausre in Δ for fixed t,x. p has to satisfy the Chapman-Komogorov - equation,

$$p(s+t,x,y) = \int_{\mathbb{R}^n} p(s,x,u)\, p(t,u,y)\, du, \quad 0 < s \le t < \infty \quad . \tag{2.2}$$

2. p is said to be a Feller density function if additionally for any bounded continuous function f the map $x \longrightarrow \int_{\mathbb{R}^n} p(t,x,y)f(y)\, dy$ is continuous for any $t > 0$.

3. p is said to be a symmetric (and bounded) Feller density function if $p(t,x,y) = p(t,y,x)$ and if $\sup_{x,y\in\mathbb{R}^n} p(t,x,y) < \infty$, $t > 0$, respectively. ■

Having a Feller transition function one can use the Kolmogorov construction (see [17] or Ginibre [18]) to obtain the Feller process $([0,\infty);\ \Omega_x, \mathcal{B}_x,\ P_F;\ w(\cdot))$. Here Ω_x is the set of all continuous functions $w(\cdot)\colon [0,\infty) \to \mathbb{R}^n$ with $w(0) = x$. $\mathcal{B}_x$ is a certain σ-field. P_F is called the Feller measure on $\mathcal{B}_x$. Then the transition probability is given by

$$P_F\{w\colon\ w(0) = x,\ w(t) \in \Delta\} = \int_\Delta p(t,x,y)\, dy \qquad \text{for any Borel set } \Delta \text{ of } \mathbb{R}^n.$$

Let G be a subset of Ω_x which is measurable with respect to $\{w(s)\colon 0 \le s \le \tau\}$, $\tau \le t$. Then one can introduce the conditional measure $P_F^{y,t}\{\cdot\}$ by

$$P_F^{y,t}\{G\} = \int_{\Omega_x} \chi\{w\colon\ w \in G\}\, p(t-\tau,\ w(\tau),\ y)\, P_F(dw), \qquad \tau \le t,$$

(see [15] Def.3.2.). $P_F^{y,t}$ is concentrated on $\Omega_x^{y,t} = \{w\colon\ w\in\Omega_x,\ w(t) = y\}$.

The Wiener transition density function p_W in (2.1) satisfies all the assumption in Definition 2.1. In this case we get the Wiener process $([0,\infty);\ \Omega_x, \mathcal{B}_x, P_W; w(\cdot))$ with the Wiener measure $P_W\{\cdot\}$ and the conditional Wiener measure $P_W^{y,t}\{\cdot\}$.

Having the measures one can introduce Feller or Wiener integrals with respect to certain functionals $F(\cdot)$ from $\Omega_x \to \mathbb{R}$. Here we consider only

$$F(w) = [\exp(- \int_{\Omega_x}^{t} V(w(s))\, ds)] \cdot f(w(t)) \tag{2.3}$$

where V are functions from $\mathbb{R}^n$ into $\mathbb{R}$ and $f \in L^p(\mathbb{R}^n)$. That includes the special cases $V \equiv 0$. i.e. $F(w) = f(w(t))$, and $V(x) = \infty$ for $x \in \Gamma \subset \mathbb{R}^n$ where F is determined by the indicator function of all trajectories avoiding Γ (see equation (2.10)).

For $V \equiv 0$ the integral

$$\int_{\Omega_x} f(w(t))\, P_F(dw) =: (P_0(t)f)(x) \tag{2.4}$$

is defined for $f \in C_0(\mathbb{R}^n)$ (continuous function with compact support). It can be extended to a strongly continuous contractive semigroup on L^p, $1 \le p < \infty$. Its generator will be denoted with K_0. For a symmetric Feller semigroup (Def. 2.1.3.) in L^2 K_0 is selfadjoint and positive. For Wiener semigroups in L^2 K_0 is given by the selfadjoint realization H_0 of $-\frac{\Delta}{2}$.

For nonvanishing potentials V one has to study the existence of

$$\int_{\Omega_x} [\exp(- \int_0^t V(w(s))ds] \cdot f(w(t))\, P_F(dw) =: (P_V(t)f)(x)\ . \tag{2.5}$$

That leads to the definition of Kato's class potentials.

Definition 2.2.: Kato's classes: K_F, $K_{F,loc}$; K, K_{loc}

We restrict us here to $n \ge 3$. Let $V: \mathbb{R}^n \to \mathbb{R}^1$. Then

$$V \in K_F(\mathbb{R}^n) \text{ iff } \lim_{\alpha\to 0} \sup_x \int_0^\alpha ds \int_{\mathbb{R}^n} p(s,x,y)|V(y)|dy = 0 \ ,$$

$$V \in K_{F,loc}(\mathbb{R}^n) \text{ iff } \chi_{|x|\le R} \cdot V \in K_F(\mathbb{R}^n), \qquad R > 0\ .$$

In particular for the Wiener case

$$V \in K(\mathbb{R}^n) \text{ iff } \lim_{\alpha\to 0} [\sup_x \int_{|x-y|\le\alpha} |V(y)|\ |x-y|^{-m+2}dy] = 0. \ \blacksquare$$

Kato's classes are sufficient in the following sense.

Theorem 2.3.: Let $V = V_+ - V_-$, $V_+ \in K_{F,loc}(\mathbb{R}^n)$, $V_- \in K_F(\mathbb{R}^n)$. Than $P_V(t)$ is a strongly continuous semigroup acting in $L^p(\mathbb{R}^n)$, $1 \le p < \infty$. ∎

A proof for Feller processes is given by van Casteren [34] or in [15] Theorem 3.5. In the Wiener case Hempel, Voight [19] proved that for some more general V (see Proposition 4.2.), see also Carmona [5].

The generator of $P_V(t)$ in L^p can be denoted with K_p. If we do not distinguish the different L^p -spaces we write $P_V(t) = \exp(-tK)$. If we consider L^2-spaces we always identify K_2 with K.

Theorem 2.4 a) Assume a Feller process with a bounded transition density function, such that K_0 is the generator of $P_0(t)$ in $L^p(\mathbb{R}^n)$, $1 \le p < \infty$. Let $V_+ \in K_{F,loc}$, $V_- \in K_F$. Then K_p is an extension of $K_0 + V$ (see [33] p.22). If we have a symmetric Feller process (Def. 2.1.3.) and consider p = 2, then K is selfadjoint (see [15] Theorem 3.3.).

For the Wiener process let $V_+ \in K_{loc}(\mathbb{R}^n)$, $V_- \in K(\mathbb{R}^n)$. Consider $L^2(\mathbb{R}^n)$. Then V is H_0 -form bounded with relative bound zero. Hence the KLMN-theoremensures that $K = H_0 \dot{+} V$ is a selfadjoint Schrödinger operator. Its corresponding semigroup satisfies the Feynman-Kac formula in (2.5) (see [29] p.459, Cycon et al. [9] p.13, or [15] Theorem 3.9.).∎

Moreover, Kato's class potentials are natural for considering exp(-tK) as operator from L^p to L^q, $1 \le p \le q \le \infty$.

Theorem 2.5.: Assume a symmetric Feller process (Def. 2.1.3.). Let $V \in K_{F,loc}$, $V_- \in K_F$. Then $P_V(t)$ (see (2.5)) is a bounded operator on L^p to L^q for t > 0 and $1 \le p \le q\ \infty$. It holds

$$\sup_{x\in\mathbb{R}^n} E_x^F \{\exp (\int_0^t V_-(w(s))ds)\} \le B\, e^{At} \tag{2.6}$$

with some A,B > 0. That implies for t > 0

$$\|e^{-tK}\|_{p,q} \le B^{1+\frac{1}{p}-\frac{1}{q}} \cdot e^{At} \cdot \sup_{x,y} p(t/2,x,y)^{\frac{1}{p}-\frac{1}{q}}, \tag{2.7}$$

where we wrote again formally e^{-tK} for $P_V(t)$. ∎

The proof is given in [15], Theorem 3.7. or for Schrödinger operators in [29] p.460. The main ingredient of the proofs is a lemma which was discovered several times (Berthier, Gaveau [4]; Khasmin'skii [20]; Portenko [26]): If we assume

$$\sup_x E_x^F \left\{ \int_0^t V_-(w(s))ds \right\} = :\alpha < 1 \quad ,$$

then

$$\sup_x E_x^F \left\{ \exp\left(\int_0^t V_-(w(s))ds \right) \right\} \le \frac{1}{1-\alpha} \quad . \tag{2.8}$$

Kato's class potentials are optimal for the estimate in (2.6). This estimate allows as to include obstacles.

Define a singularity region Γ as a certain closed region in $\mathbb{R}^n$ with $|\Gamma| > 0$ and piecewise $\mathcal{C}_1$ boundary. Over Γ we assume an additional potential or height $M > 0$, i.e. set

$$(Pf)(x): = \chi_\Gamma(x)f(x) \ , \qquad f \in L^2 \quad , \tag{2.9}$$

$$V_M := V + MP \quad .$$

Then we define

$$(P_M(t)f)(x) :=$$

$$= E_x^F\left\{\left[\exp(- \int_0^t V(w(s))ds)\right]\cdot\left[\exp(-MT_{t,\Gamma}(w))\right]\cdot f(w(t))\right\} \quad ,$$

where $T_{t,\Gamma}(w)$ is the spending time of the trajectory w in Γ. The semigroup $P_M(t)$ has a strong limit $U(t)$ as $M \to \infty$. $U(t) \mid_{L^2(\Sigma)}$ is a strongly continuous semigroup in $L^2(\Sigma)$. Its generator is denoted with $(K_0 + V)_\Sigma$. Very mild conditions on K_0, V, and Γ provide (see e.g. Baumgärtel, Demuth [3]) that $(K_0 + V)_\Sigma$ is the Friedrichs extention of $(K_0 + V) \mid [\mathrm{dom}(K_0 + V) \cap L^2(\Sigma)]$.

Theorem 2.6.: Assume a symmetric Feller process (Def. 2.1.3.) with the semigroup $\exp(-tK_0)$. Let $V_- \in K_F(\mathbb{R}^n)$, $V_+ \in K_{F,loc}(\mathbb{R}^n)$. Let Γ be a singularity region described above. Then the semigroup $\exp(-t(K_0 + V)_\Sigma)$ in $L^2(\Sigma)$ can be represented by

$$\left\{\left[\exp(-t(K_0 + V)_\Sigma)\right]\cdot f\right\}(x) \tag{2.10}$$

$$= E_x^F \left\{ \left[\exp(-\int_0^t V(w(s))ds\right] \cdot \chi\left\{w:\ w(s) \notin \Gamma,\ \forall s,\ s \in [0,t]\right\} \cdot f(w(t)) \right\},$$

$$f \in L^2(\Sigma), \quad x \in \Sigma .$$

This is the most general Feynman-Kac formula here including regular and singular potentials. By means of the conditional measure $P_F^{y,t}$ the integral kernel of the semigroup in (2.10) can be given explicitely:

$$\left[\exp(-t(K_0 + V)_\Sigma)\right](x,y) \tag{2.11}$$

$$= \int_{\Omega_x^{y,t}} \left[\exp(-\int_0^t V(w(s))ds)\right] \cdot \chi\left\{w:\ w(s) \notin \Gamma,\ \forall s,\ s \in [0,t]\right\} P_F^{y,t}(dw). \quad \blacksquare$$

3. SCHRÖDINGER SEMIGROUPS, KATO'S CLASS POTENTIALS

We restrict us now to Schrödinger semigroups although many arguments can be used also for Feller semigroups. One can find a series of reasons why it is useful to consider Schrödinger (or Feller) semigroups. Some of them are:

a) On account of the Feynman-Kac representation for exp $(-tH)$ one can use the powerful tool of the Markov process theory.

b) The integral kernels $[\exp(-t(H_0 + V)_\Sigma)](x,y)$ are given explicitely, such that the theory of integral operators is applicable.

c) The whole theory is not restricted to $H_0 = -\Delta/2$. It holds for more general generators(see also the final remarks in section 5).

d) Singular potentials can be included.

e) General properties of eigenfunctions can be investigated. Let f be an eigenfunction of H, $Hf = Ef$. Then $[\exp(-H)]f = [\exp(-E)]f$. By Theorem 2.5. follows for instance that $\exp(-H)$ maps L^2 into L^∞. Therefore $f \in L^\infty(\mathbb{R}^n)$.

f) The large time behaviour of $\exp(-tH)$ defines the ground state energy. Let $E_0 = \inf \sigma(H)$. Then (see Proposition 4.1)

$$E_0 = -\lim_{t\to\infty} [(1/t)\cdot \ln\|\exp(-tH)\|_{2,2}] \quad . \tag{3.1}$$

Also the ground state can be given by the large time behaviour of this semigroup.

g) Because of the Laplace transform

$$(H-z)^{-r} = \frac{1}{\Gamma(r)} \int_0^\infty \lambda^{r-1} e^{\lambda z} e^{-\lambda H} d\lambda \quad , \tag{3.2}$$

$r > 0$, $z \in \text{res } H$, $\text{Re}z < \inf(\text{Re}\sigma(H))$, one gets stochastic representations for the resolvents. That implies many spectral theoretical applications. Let me mention the Birman - Kuroda - Theorem for the existence and completeness of wave operators, the Weyl-Theorem for the stability of the essential spectra, or the theory of relatively compact perturbations. Moreover, one can investigate Sobolev estimates for $(H-z)^{-r}$.

Further advantages in investigating Schrödinger semigroups can be found in [29] p. 448.

In considering Schrödinger semigroups it is important to have the most general potentials. Kato's class potentials, given in Def. 2.2., are natural and up to some extend optimal. They are e.g. optimal because

$$V \in K \quad \text{iff} \quad \lim_{\lambda\to\infty} \|(H_0 + \lambda)^{-1} |V| \|_{\infty,\infty} = 0 \quad .$$

(This limit determines also the H_0-form-bound of V). $V \in K$ is also optimal for the property that $\exp(-tH)$ is a bounded operator on L^∞ to L^∞. On the other hand, Kato's class potentials are natural in the sense that they admit $\exp(-tH)$ to be a bounded operator from L^p to L^q (see Theorem 2.5.). Because of the inequality in (2.6) they are also natural for including obstacles.

One can find many necessary and sufficient conditions for certain potentials to belong to Kato's class (see [1] or [29]). Here we restrict us to mention the following properties:

Property 3.1.: Take $n \geq 3$. The $V \in K(\mathbb{R}^n)$ iff

$$\lim_{\alpha\to 0} \sup_x \int_0^\alpha ds\, s^{-n+1} \int_{|y|\leq s} |V(x+y)| dy = 0$$

(see [15] Proposition 3.14.). ■

That implies immediately two necessary conditions.

Proposition 3.2.: Let $V \in K(\mathbb{R}^n)$. Then

$$\int_{|y|\le\alpha} |V(y)|dy = o(\alpha^{n-2}), \qquad \alpha \to 0 \quad ,$$

and

$$\sup_x \int_{|x-y|\le 1} |V(y)|dy < \infty \qquad .$$

For spherically symmetric potentials $V(|x|) = |x|^\beta$ one gets $\beta > -2$. This is the same borderline as for the KLMN-Theorem. ■

Property 3.3.: It holds

$$L^q(\mathbb{R}^n) \subset K(\mathbb{R}^n) \quad \text{and} \quad L^q_{loc}(\mathbb{R}^n) \subset K_{loc}(\mathbb{R}^n) \qquad .$$

for $n/2 < q \le \infty$. Hence Coulomb potentials in $L^2(\mathbb{R}^3)$ are contained in $K(\mathbb{R}^3)$. □

Property 3.4.: Let $V_+ \in K_{loc}(\mathbb{R}^n)$, $V_- \in K(\mathbb{R}^n)$. Then for every $f \in L^p$, $1 \le p \le\infty$, the function $x \to [(\exp(-tH))f](x)$ is continuous (see [29] p. 466 or [15] Theorem 3.5. for $f \in C_\infty(\mathbb{R}^n)$). ■

Property 3.5.: Let $V_+ \in K_{loc}(\mathbb{R}^n)$, $V_- \in K(\mathbb{R}^n)$. Then the kernels of $\exp(-tH)$ are symmetric, i.e.

$$[\exp(-tH)](x,y) = [\exp(-tH)](y,x) \qquad .$$

For any $\varepsilon < 0$ there is a $C_\varepsilon > 0$ such that

$$[\exp(-tH)]x,y) \le C_\varepsilon t^{-n/2} e^{At} e^{-|x-y|^2/2(1+\varepsilon)t} \tag{3.3}$$

where $A > 0$ is taken from equation (2.6) (see [29] p. 474, [35], or [15] Theorem 3.7.). ■

For many applications the inequality in (3.3) is very useful. One corollary is contained in the final property.

Property 3.6.: Let $V_+ \in K_{loc}(\mathbb{R}^n)$, $V_- \in K(\mathbb{R}^n)$. Let $r \in \mathbb{N}$. Then the kernel of the resolvent can be estimated by

$$|[(H-z)^{-r}](x,y)|$$
$$\le \frac{c}{(r-1)!} \int_0^\infty d\lambda \cdot \lambda^{r-1-n/2} e^{\mathrm{Re}z\cdot\lambda} e^{A\lambda} e^{-|x-y|^2/2\rho\lambda} ,$$

with $\rho > 1$ and $\mathrm{Re}z < -A$. The properties of these integral kernels vary with r and n. ■

4. SPECTRAL CONSEQUENCES

As mentioned in Section 3 there are a lot of possibilities to get information about the spectra of Schrödinger operators by considering stochastically the semigroups or resolvents. From that variety some examples will be explained in the following.

The spectral theoretical results are always formulated without regarding to stochastic analysis. In order to recognize this aspect it is necessary to indicate some crucial parts of the proofs.

4.1. Spectra of Schrödinger Operators in L^p*-spaces*

Let $V_+ \in K_{loc}$, $V_- \in K$. Then the Feynman-Kac expression

$$\int_{\Omega_x} [\exp(-\int_0^t V(w(s))ds)]\ f(w)t))\ P_w(dw)$$

defines a strongly continuous semigroup $\exp(-tH_p)$ in $L^p(\mathbb{R}^n$. $1 \le p < \infty$ (see Theorem 2.3.) and is a bounded operator also for $p = \infty$ (see Theorem 2.5.). The spectrum of H_p, as infinitesimal generator in the Banach space L^p, can be given by

$$\sigma(H_p) := \{-\ln\mu,\ \mu \in \sigma(e^{-H_p})\} \qquad .$$

H_p are bounded below operators. Let $-a = \inf(\mathrm{Re}\ \sigma(H_p))$. Then the spectral radius of $\exp(-H_p)$ is given by exp a. On the other hand this spectral radius equals $\exp[\lim_{t\to\infty} (1/t)\cdot\ln\|\exp(-tH_p)\|_{p,p}]$. That implies

$$\inf(\mathrm{Re}\ \sigma(H_p)) = -\lim_{t\to\infty} [(1/t)\cdot\ln\|e^{-tH_p}\|_{p,p}] \qquad ,$$

$1 \le p < \infty$.

Proposition 4.1.: We set $\|\exp(-tH)\|_{p,p} := \|\exp(-tH)_p\|_{p,p}$ for $1 \le p < \infty$ and $\|\exp(-tH)\|_{\infty,\infty} := \|P_V(t)\|_{\infty,\infty}$(see (2.5) or Theorem (2.5.) For all $1 \le p \le \infty$ we define

$$\alpha_p := \lim_{t\to\infty} [(1/t)\ \ln\|\exp(-tH)\|_{p,p}]\ . \tag{4.1}$$

Then α_p is independent of p. That implies for instance

$$\inf \sigma(H_2) = -\lim_{t\to\infty} [(1/t)\cdot\ln\|\exp(-tH)\|_{\infty,\infty}] \tag{4.2}$$

Here H_2 denotes the usual Schrödinger operator in $L^2(\mathbb{R}^n)$. ∎

The proof is given by Simon [30] or [31]. By duality and interpolation one has

$$\|\exp(-tH)\|_{2,2} \le \|\exp(-tH)\|_{p,p} \le \|\exp(-tH)\|_{\infty,\infty} .$$

Then it is sufficient to show

$$\|\exp(-tH)\|_{\infty,\infty} \le c\ (1+t)^{n/2}\ \|\exp(-tH)\|_{2,2} . \tag{4.3}$$

It holds

$$\|\exp(-tH)\|_{\infty,\infty} = \sup_x E_x\left\{\exp(-\int_0^t V(w(s))ds)\right\} .$$

Here one introduces $\chi\{w: |x-w(t)| \le at\} + \chi\{w: |x-w(t)| > at\}$. Because of Theorem 2.5 it is

$$\|\exp(-tH_-\cdot\chi_{\le at}\|_\infty \le c\cdot t^{n/2}e^{\alpha_2 t} .$$

For the other term one can use Property 3.5. (equation (3.3)) and gets

$$\|\exp(-tH)\cdot\chi_{>at}\| \le c\ e^{At}\ e^{-a^2 t} .$$

Choosing the constant a large enough one obtains (4.3).

In 1982 it was an open problem whether the spectrum of H_p is independent of p (see [29] p.471). This question was solved by Hempel, Voigt [19]. They assumed slightly more general conditions on V.

Proposition 4.2.: Let $n \ge 3$, $V_+\in L^1_{loc}(\mathbb{R}^n\setminus\Lambda)$, ($\Lambda$ set of measure zero). Let $V_- \in L^1_{loc}(\mathbb{R}^n)$ and

$$\lim_{\alpha\to 0}\sup_x \int_{|x-y|\le\alpha} V_-(y)|x-y|^{-n+2}dy < 1 .$$

Then the Feynman-Kac formula (see (2.5) defines a strongly continuous semigroup on L^p, $1 \le p < \infty$. Its generator is denoted with H_p. Let $H_\infty := H_1^*$.

Then for all $1 \le p \le \infty$

$$\sigma(H_p) = \sigma(H_2) .$$

Moreover, if λ is an isolated eigenvalue of finite algebraic multiplicity for one H_p this is true for all $p \in [1,\infty]$. ∎

4.2. Eigenvalues and Eigenfunctions

In the following we will consider Schrödinger operators H in $L^2(\mathbb{R}^n)$. From Theorem 2.5 and Proposition 3.4 one has immediately the following consequences for the eigenfunctions of H.

Proposition 4.3: Let $V_+ \in K$. Let $f \in L^2(\mathbb{R}^n)$ be an eigenfunction of H. Then $x \to f(x)$ is a continuous function and $f \in L^p(\mathbb{R}^n)$, $1 \le p \le \infty$, in particular $f \in L^\infty(\mathbb{R}^n)$. ∎

One advantage of the Feynman-Kac-formulation is that one has an integral representation of the eigenfunctions. Let $Hf = Ef$, then

$$f(x) = \int_{\mathbb{R}^n} dy\, f(y) \int_{\Omega_x^{y,t}} e^{-\int_0^t [V(w(s))-E]ds}\, P_W^{y,t}(dw) \quad . \tag{4.4}$$

Similarly, one obtains stochastic representations for solutions of $(H_0+V)f = Ef$ in distributional sense on a bounded open set $\Omega \subset \mathbb{R}^n$, i.e. for $f \in L^1_{loc}$, $Vf \in L^1_{loc}$ and

$$(-\Delta\varphi, f) + ((V-E)\varphi, f) = 0$$

for all $\varphi \in C_0^\infty(\Omega)$.

Proposition 4.4.: Let $V \in K_{loc}$ and $Hf = 0$ in distributional sense in $\Omega \subset \mathbb{R}^n$. Then for any $x_0 \in \Omega$ there exists a ball $B := \{x: |x-x_0| \le R\} \subset \Omega$, such that for any $x \in B$ holds

$$f(x) = E_x\left\{[\exp(-\int_0^T V(w(s))ds]\cdot f(w(t))\right\} \tag{4.5}$$

where $T = T(w)$ is the first exit time of $w(\cdot)$ from B (see [1] p.218), i.e. $f(x)$ is given by its values on δB because $f(w(T)) \in \delta B$.

(4.5) implies (see [29] p.493)

$$|f(x)| \le c\cdot \int_{|x-y|\le R} |f(y)|dy \quad . \tag{4.6}$$

Here the constant c does not depend on f. But c can depend on x, V_-, and R. ∎

A nice consequence of the last proposition is.

Proposition 4.5.: Let $V_+ \in K_{loc}$, $V_- \in K$. Let $Hf = Ef$ and $f \in L^2(\mathbb{R}^n)$. Then $f(x)$ tends to zero as $|x| \to \infty$. ∎

That follows from (4.6) in the following way. Take $\varphi \in C_0^\infty$ such that $\|f-\varphi\|_{L^2}$ is arbitrarily small. Then

$$\|\chi\{|x-y| \le 1\} \cdot f\|_{L^1} \le \|f-\varphi\|_{L^2} + \|\chi\{|x-y| \le 1\}\cdot\varphi\|_{L^2} .$$

Further consequences of the representation in (4.5) are Harnack inequalities or the exponential decay of eigenfunctions for isolated eigenvalues (see e.g. [9]).

A recent application of (4.4) is given by Kirsch, Simon [22] in estimating the gap between the lowest eigenvalues of Schrödinger operators.

Proposition 4.6.: Let V be a bounded potential and assume that $H = H_0 + V$ has at least two eigenvalues E_0, E_1 below $\inf \sigma_{ess}(H)$. Let f_0 and f_1 be the corresponding eigenvectors. Fix $\delta > 0$ and denote $C_\delta = \{x: V(x) < E_1 + \delta^2\}$. Let B_R be the smallest ball of radius R containing C_δ. Set $\lambda \ge \sup_x \sup_{E\in[E_0,E_1]} |V(x)-E|^{1/2}$.

Then the gap can be estimated by

$$E_1 - E_0 \ge p(R)\, e^{-8\sqrt{2}\, n\, R\cdot\lambda} , \tag{4.7}$$

where $p(R)$ is polynomially bounded in R. ∎

Using a variational principle one has

$$E_1 - E_0 \ge \int_{B_R} |\nabla(f_1/f_0)|dx \cdot \inf_{x\in B_R} f_0(x)\cdot\|f_0\|^{-2}\|f_1\|^{-2} .$$

In order to obtain (4.7) one has to find upper and lower bounds for f_i. For instance, a lower bound for f_0 is given by

$$f_0(x) \ge e^{-t\lambda^2} E_x\{f_0(w(t))\} \ge e^{-t\lambda^2} E_x\left\{\chi\{w: |w(t)| \le a\}\cdot f_0(w(t))\right\}$$

$$\ge \frac{1}{2} e^{-t\lambda^2} E_x\left\{\chi\{w: |w(t)| \le a\}\right\}$$

$$= \frac{1}{2} e^{-t\lambda^2} \cdot \frac{1}{(2\pi t)^{n/2}} \int_{|y|\le a} \exp(-|x-y|^2/2t)dy ,$$

where one used the fact that, if we set $f_0(0) = 1$, $f_0(x) \geq 1/2$ as long as $|x| \leq a$ for some $a > 0$.

The stochastic theory is also applicable for the estimation of the number of eigenvalues $N(V)$ for the Schrödinger operator $H_0 + V$. Let $V \in K$ and $(1/p) + (1/q) = 1$; $p, q > 1$. Take $\lambda > 0$. Then as corollary from Theorem 2.5.

$$f \longrightarrow |V|^{1/p}(H_0+\lambda)^{-1}|V|^{1/q} f$$

is a continuous linear map from L^p into L^q. And it holds

$$\| \, |V|^{1/p}(H_0+\lambda)^{-1}|V|^{1/q} \|_{p,q} = \| (H_0+\lambda)^{-1}|V| \, \|_{\infty} \quad .$$

For $p = q = 2$ this is an estimation for the Birman-Schwinger-kernel.

Moreover the proof of the Cwickel-Lieb-Rosenblum bound (see Simon [28], p.95) is also based on functional integration methods. Let $n \geq 3$ and $V \in L^{n/2}(\mathbb{R}^n)$, then

$$N(V) \leq a_n \cdot \int_{R^n} |V_-(x)|^{n/2} dx \quad .$$

For every n one can find upper and lower bounds of a_n. This result can be used to prove the stability of matter in certain models.

4.3. Dirichlet Laplacian

As mentioned in Theorem 2.6. also Dirichlet operators can be represented stochastically. At the moment we set $V \equiv 0$. For the Dirichlet Laplacian $(H_0)_\Sigma$ one knows that $\exp(-t(H_0)_\Sigma)$ is a trace class operator if $|\Sigma| < \infty$. But this is not necessary. Davies [10] found a necessary and sufficient condition. He assumed certain regular $\delta\Sigma$. Denoting with $d(x)$ the distance between $x \in \Sigma$ and Γ he showed

$$\mathrm{trace}(\exp(-t(H_0)_\Sigma)) < \infty \quad \text{iff} \quad \int_\Sigma \exp(-ct/d^2(x))dx < \infty \quad . \tag{4.8}$$

The right hand side allows unbounded Σ with infinite volume.

For special regions Σ one can estimate the trace more explicitely. In [36] van den Berg (see also [11]) considered horn-shaped regions Σ. These are peaks in $\mathbb{R}^n$ with a decreasing cross section. Assume a peak in the direction of the x_1-axis. Let $\Sigma(x)_1$ be the cross section of Σ with plane through x_1 orthogonal to the x_1-axis. The main condition for horn-shaped regions is that $\Sigma(x_1) \subseteq \Sigma(x'_1)$ if $x_1 \geq x'_1 \geq 0$ or $x_1 \leq x'_1 \leq 0$. $\Sigma(x_1)$ is a bounded region in $\mathbb{R}^{n-1}$. Therefore the trace of the Dirichlet Laplacian $(H_0)_{\Sigma(x_1)}$ in $L^2(\mathbb{R}^{n-1})$ exists. Then one can estimate the difference between trace $\exp(-t(H_0)_\Sigma)$ and $\int_{-\infty}^{\infty} dx_1$ trace $\exp(-t(H_0)_{\Sigma(x_1)})$.

His main assumption is that $|\Sigma(x_1)| \in L^1_{loc}(\mathbb{R})$. In the proof he used mainly the decomposition of the Feynman-Kac formula. Set $\mathbb{R}^n = \mathbb{R}^1 \times \mathbb{R}^{n-1}$ and $x = (x_1, \vec{x}_{n-1})$. Let $P_w(dw_1)$, $P_w(d\vec{w}_{n-1})$ be the Wiener measures in $\mathbb{R}^1$ and $\mathbb{R}^{n-1}$, respectively. Then

$$\left\{[\exp(-t(H_0)_\Sigma)]f\right\}(x_1, \vec{x}_{n-1})$$

$$= \int_{\Omega_{x_1}} P_w(dw_1) \int_{\Omega_{\vec{x}_{n-1}}} P_w(d\vec{w}_{n-1})\, f(w_1(t), \vec{w}_{n-1}(t)) \cdot$$

$$\cdot\, \chi\left\{(w_1, \vec{w}_{n-1}) : \vec{w}_{n-1}(s) \in \Sigma(w_1(s)), \quad \forall s \ , \quad s \in [0,t]\right\} .$$

Also the asymptotics of trace $[\exp(-t(H_0)_\Sigma)]$ in t can be studied via the Feynman-Kac formula. The old result by Ray [27] that for $|\Sigma| < \infty$ trace $\exp(-t(H_0)_\Sigma) \leq (2\pi t)^{-n}|\Sigma|$ would follow obviously from (2.11). Further results for different Σ are given by Angelescu, Nenciu [2], Park [25] or van den Berg [37]. In [38] van den Berg gives a more detailed estimation for Σ with R-smooth boundaries.

$\delta\Sigma$ is called R-smooth if for any $x \in \delta\Sigma$ there are two open balls B_1 and B_2 with the radii R such that $B_1 \subset \Sigma$, $B_2 \subset \Gamma$, and $\delta B_1 \cap \delta B_2 = \{x\}$.

Proposition 4.7.: Let Σ be an open, bounded connected region in $\mathbb{R}^n$ with an R-smooth boundary $\delta\Sigma$. Then the asymptotics of trace $\exp(-t(H_0)_\Sigma)$ can be estimated by

$$\left|\text{trace } e^{-t(H_0)_\Sigma} - |\Sigma|/(2\pi t)^{n/2} - |\delta\Sigma|/4\cdot(2\pi t)^{(n-1)/2}\right|$$

$$\leq (n^4/\pi^{n/2}) \cdot |\Sigma|/(R^2 \cdot t^{(n-2)/2}) \qquad . \qquad \blacksquare$$

The proof uses the Kac's principle, i.e.

$$P_w\{w:\ w(0) = w(t) = x \in \Sigma,\quad w(s) \in \delta\Sigma \text{ for some } s \in [0,t]\}$$
$$\le 2n \exp(-d^2(x)/nt),$$

where d(x) is again the distance between x and Γ.

So far we have considered large singularity regions Γ. If Γ becomes smaller it is meaningless to consider the trace of $\exp(-t(H_0)_\Sigma)$ (see for instance (4.8)). But it becomes then interesting to compare it with the free semigroup $\exp(-tH_0)$ defined in $L^2(\mathbb{R}^n)$. Because this is a two-space situation we define an identification operator $Jf := f_\Sigma$. Then one wants to study

$$[\exp(-t(H_0+V)_\Sigma)] \cdot J - J \cdot \exp(-t(H_0+V) =: D_V(t) \quad .$$

Let us consider the Hilbert-Schmidt property of $D_V(t)$. If we have Kato's class potentials if follows from the inequality in (3.3) that it is sufficient to consider

$$[\exp(-t(H_0)_\Sigma)] \cdot J - J \cdot \exp(-tH_0) =: D(t) \quad .$$

Proposition 4.8.: Set n = 3. Let Γ be contained in a ball of radius R. Then D(t) and also $D_V(t)$ $(V_+ \in K_{loc},\ V_- \in K)$ are Hilbert-Schmidt operators. The Hilbert-Schmidt norm of D(t) can be estimated by

$$\|[\exp(-tH_0)_\Sigma)]\cdot J - J\cdot\exp(-tH_0)\|^2_{HS} \tag{4.9}$$

$$\le (2\pi t)^{-3/2}(2\pi Rt + 4R^2(2\pi t)^{1/2} + \frac{8\pi}{3}R^3) \quad . \quad ■$$

The proof is given by Demuth [14]. It follows because

$$\|D(t)\|^2_{HS}$$

$$\le (2\pi t)^{-3/2}[\frac{4\pi}{3}R^3 \quad +$$

$$\int_{\Omega_x} P_w(dw) \int_{\mathbb{R}^3} dx\chi\{x:\ |x-w(s)| \le R \text{ for some } s \le t\}].$$

This leads to the Wiener sausage problem. If one includes Kato's class potentials one obtains on the r.h.s of (4.9) an additional factor exp At. It is also possible to give the asymptotics for the trace norm of

$D(t)$. More general singularity regions Γ for the Wiener sausage problem were considered by Spitzer [32] or Le Gall [23].

Trace class conditions are useful in scattering theory to probe the stability of the absolutely continuous spectra.

Proposition 4.9.: Let V be a Kato's class potential and Γ be a bounded singularity region. Then

$$\sigma_{ac}(H_0+V) = \sigma_{ac}((H_0+V)_\Sigma) . \qquad \blacksquare$$

Because $P_V(t)$ is a trace class operator one has the existence of the two-space wave operators.

$$\underset{t\to\pm\infty}{s\cdot\lim}\, [\exp(it(H_0+V)_\Sigma)]\, J[\exp(-it(H_0+V))]P_{ac}(H_0+V) \quad ,$$

$$\underset{t\to\pm\infty}{s\cdot\lim}\, [\exp(it(H_0+V))]\, J^*[\exp(-it(H_0+V)_\Sigma)]\, P_{ac}((H_0+V)_\Sigma) \quad .$$

(P_{ac} - projection operator onto the absolutely continuous subspace). These wave operators are complete, because $JJ^* = 1_{L^2(\Sigma)}$ and because

$$\lim_{t\to\infty} \|P\, e^{-it(H_0+V)}\, P_{ac}(H_0+V)f\| = 0 \tag{4.10}$$

P was defined in (2.9). (4.10) follows because $P\, \exp(-s(H_0+V))$ is a Hilbert-Schmidt operator for any $s > 0$ and for $|\Gamma| < \infty$.

However one can include more general Γ (see also Demuth [14]).

Proposition 4.10.: Set $V \equiv 0$. Assume an unbounded singularity region Γ of the form

$$\int_\Gamma dx\, \chi\{x:\ at \le |x_i| \le bt\ ;\quad i = 1,2,\dots,n\} = 0 \tag{4.11}$$

for any $0 < a < b$ and for $t > \frac{c}{a}$ with an arbitrary constant c. Then

$$\sigma_{ac}(H_0) = \sigma_{ac}((H_0)_\Sigma) = [0,\infty) \quad . \qquad \blacksquare \tag{4.12}$$

Regular potentials of Kato's class could be included if they satisfy

$$\int_1^\infty dt\ t^{-n/2} \left[\int_{at\le|x_i|\le bt} |V(x)|^2 dx \right]^{1/2} < \infty \quad .$$

Similar results as in (4.11) are obtained by means of different methods also by Combes, Weder [7], Constantin [8] or for σ_{ess} by Leinfelder

[24]. The proof of (4.12) uses the Cook criterion. It can be shown that the corresponding wave operator exist, i.e. $\sigma_{ac}(H_0) \leq \sigma_{ac}((H_0)_\Sigma)$. On the other hand $(H_0)_\Sigma$ generates a contractive semigroup, such that (4.12) follows. The condition in (4.11) is sufficient for the wave operator existence because it implies

$$E_x\left\{\chi\{w:\ \exists s,\ s \leq t,\ w(s) \in \Gamma,\ at \leq |w_i(s)| \leq bt\}\right\} = 0$$

$$(i=1,2,\ldots,n)$$

The same argument is used to show the H_0-semicompleteness of the wave operator

$$\underset{t\to\infty}{s\text{-}\lim}\ [\exp(it(H_0)_\Sigma)]\ J\ [\exp(-itH_0)] \quad .$$

5. FINAL REMARKS

Because the number of pages of this article is limited the overview is stopped here although it is not at all finished. The stochastic analysis can be used for further aspects. The Feynman-Kac formulation admits a transformation of the coordinates (see [13]). That implies for instance that the singularity region in Proposition 4.10 can be translated or rotated arbitrarily. Also Laplacians with different masses are allowed, or N-body problems can be considered by transformation to clustered Jacobi coordinates. Probably, further applications can be found from Davies [12] although I have not seen this book till now.

Here we have considered only Schrödinger semigroups. But some of the spectral consequences remain true if we consider Feller semigroups (see e.g. von Casteren, Demuth [15], [16]). One obtains for instance the stability of the essential and absolutely continuous spectra of K_0, K_0+V, and $(K_0+V)_\Sigma$.

The stochastic analysis provides a unification of the formulation. As well as $-\Delta/2$ one can also consider $1/2\ (-i\nabla-\mathcal{A})^2$, i.e. problems with magnetic fields (see Simon [28] p.159). Also relativistic Hamiltonian can be investigated (see Carmona, Masters, Simon [6]). The whole theory is not restricted to $\mathbb{R}^n$. One can also take locally compact second countable Hausdorff spaces or Riemannian manifolds.

REFERENCES

[1] Aizenman, H.; Simon, B: Brownian motion and Harnack inequality for Schrödinger operators. Comm.Pure Appl.Math., Vol. XXXV, 209-273 (1982).

[2] Angelescu, N.; Nenciu, G.: On the independence of the thermodynamic limit on the boundary conditions in quantum statistical mechanics Comm.Math.Phys. **29**, 15-30 (1973).

[3] Baumgärtel, H; Demuth, M.: Decoupling by a projection. Rep.Math.Phys **15**, 173-186 (1979).

[4] Berthier, A.M.; Gaveau, B.: Critère de convergence des fonctionelle de Kac et application en mécanique quantique et en géométrie. J.Funct.Anal. **29**, 416-424 (1978).

[5] Carmona, R.: Regularity properties of Schrödinger and Dirichlet semigroups. J.Funct.Anal. **33**, 259-296 (1979).

[6] Carmona, R.; Masters, W.Ch.; Simon, B.: Relativistic Schrödinger operators: Asymptotic behaviour of the eigenfunctions. Preprint Cal.Inst. of Techn. (1989), to be published in J.Funct.Anal.

[7] Combes, J.M.; Weder, R.: New criterion for the existence and completeness of wave operators and applications to scattering of unbounded obstacles. Comm.Part.Equat. **6**, 1179-1223 (1981).

[8] Constantin, P.: Scattering for Schrödinger operators in a class of domains with non-compact boundaries. J.Funct.Anal. **44**, 87-119 (1981).

[9] Cycon, H.L.; Froese, R.G.; Kirsch, W.; Simon, B.: Schrödinger operators with applications to quantum mechanics and global geometry. Textbooks in Math.Phys., Springer-Verlag, 1986.

[10] Davies, E.B.: Trace properties of the Dirichlet Laplacian. Math.Z. **188**, 245-251 (1985).

[11] Davies, E.B.; van den Berg, M.: Heat flow out of regions in $\mathbb{R}^n$, Preprint 1988.

[12] Davies, E.B.: Heat kernels and spectral theory.Cambridge Univ. Press, 1988 (to appear).

[13] Demuth, M.: On transformations in the Feynman-Kac-formula and quantum mechanical N-body systems. Math.Nachr. **122**, 109-118 (1985).

[14] Demuth, M.: On spectral properties of semigroups with Dirichlet generators. In: Proceedings of Symp. "Part.Diff.Equat.", Holzhau 1988,.Teubner-Texte zur Mathematik, Vol. **112**, 52-62 (1988).

[15] Demuth, M.; van Casteren, J.: On spectral theory for Feller generators. Prep.Univ.Instelling Antwerpen, 88-18 (1988).

[16] Demuth, M.; van Casteren, J.: On differences of heat semigroups. Prep.Univ.Instelling Antwerpen, 88-13 (1988).

[17] Fridman, A.: Stochastic differential equations and applications, Vol.1, Academic Press, 1975.

[18] Ginibre, J.: Some applications of functional integration in statistical mechanics and quantum field theory. In: Statistical mechanics and quantum field theory, Les Houches 1970. Ed.C.DeWitt, R.Stora. Gordon and Breach, 327-427 (1971).

[19] Hempel, R.; Voigt, J.: The spectrum of a Schrödinger operator in $L^p(\mathbb{R}^\nu)$ is p-independent. Comm.Math.Phys. **104**, 243-250 (1986).

[20] Khas'minskii, R.S.: On positive solutions of the equation $\Delta u + Vu=0$ (Russian). Theor.Verojatnost.i Primenen. **4**, 332-341 (1959).

[21] Kirsch, W.; Simon, B.: Universal lower bounds on eigenvalues splitting for one-dimensional Schrödinger operators. Comm.Math.Phys. **97**, 453-460 (1985).

[22] Kirsch, W.; Simon, B.: Comparison theorems for a gap of Schrödinger operators. J.Funct.Anal. **75**, 396-410 (1987).

[23] Le-Gall, J.-F.: Sur une conjecture de M.Kac. Prob.Th.Rel.Fields **78**, 389-402 (1988).

[24] Leinfelder, H.: Gauge invariance of Schrödinger operators and related spectral properties. J.Operator Theory **9**, 163-179 (1983).

[25] Park, Y.M.: Bounds on exponentials of local number operators in quantum statistical mechanics. Comm.Math.Phys. **94**, 1-33 (1984).

[26] Portenko,N.I.: Diffusion processes with unbounded drift coefficient (Russian). Teor.Verojatnost i Primenen. **20**, 29-39 (1975).

[27] Ray, D.B.: On spectra of second-order differential operators. Trans.Amer.Math.Soc. **77**, 299-321 (1954).

[28] Simon, B.: Functional integration and quantum physics. Academic Press 1979.

[29] Simon, B.: Schrödinger semigroups. Bull.Amer.Math.Soc. **7**, 447-526 (1982).

[30] Simon, B.: Brownian motion, L^p-properties of Schrödinger operators and the localization of binding. J.Funct.Anal.**35**, 215-229 (1980).

[31] Simon, B.: Large time behaviour of the L^p-norm of Schrödinger semigroups. J.Funct.Anal. **40**, 66-83 (1981).

[32] Spitzer, F.: Electrostatic capacity, heat flow,and Brownian motion. Z.Wahrscheinlichkeitsth.u.verw. Gebiete **3**, 110-121 (1964).

[33] van Casteren, J.: Generators of strongly continuous semigroups. Pitman, 1985.

[34] van Casteren, J.: On generalized Schrödinger semigroups. Proc.of ISAM 88, Markovsche Processe und Steuerungstheorie, Gaussig GDR, 11.-15.Jan.1988.

[35] van Casteren, J.: Pointwise inequalities for Schrödinger semigroups. To appear in Lect.Notes Pure Appl.Math.; Preprint Univ.Instelling Antwerpen 87-27 (1987).

[36] van den Berg, M.: On the spectrum of the Dirichlet Laplacian for hornshaped regions in $\mathbb{R}^n$ with infinite volume. J.Funct.Anal. **58**, 150-156 (1984).

[37] van den Berg, M.: A uniform bound on trace $e^{t\Delta}$ for convex regions in $\mathbb{R}^n$ with smooth boundaries. Comm.Math.Phys. **92**, 525-530 (1984).

[38] van den Berg, M.: On the asymptotics of the heat equation and bounds on traces associate with Dirichlet Laplacian. J.Funct.Anal. **71**, 279-293 (1987).

ASYMPTOTIC OBSERVABLES IN THE N-BODY QUANTUM LONG RANGE SCATTERING

Jan Dereziński
Department of Mathematical Methods in Physics
Warsaw University
Hoza 74, PL-00-682 Warszawa - Poland

ABSTRACT. Certain observables converge in the Heisenberg picture to a limit as time goes to $\pm\infty$. These limits, which may be called asymptotic observables, are especially interesting in the context of long range N-body Schrödinger operators. By studying certain natural classes of asymptotic observables one can get a lot of insight in the quantum N-body scattering.

1. 2-BODY LONG RANGE SCATTERING

In the first part of our lecture we would like to describe the main concepts of the quantum scattering theory in the 2-body case. Let us fix the notation. We suppose that $H = -\frac{1}{2}\Delta + V(x)$ and $H_o = -\frac{1}{2}\Delta$ are self-adjoint operators on the Hilbert space $L^2(\mathbb{R}^n)$ with the domain $\mathcal{D}(H) = \mathcal{D}(H_o)$. We will assume that $V = V^s + V^L$ such that $(1+|x|)^{1+\mu_s} V^s (1-\Delta)^{-1}$ is compact for some $\mu_s > 0$ and $|\nabla^\alpha V^L| \le c_\alpha (1+|x|)^{-|\alpha|-\mu_L}$ for some $1 \ge \mu_L > 0$ and any multiindex α. The momentum operator is defined by $D = \frac{1}{i}\nabla$. The unit sphere will be denoted by S^{n-1}. The projection onto the continous spectrum of H will be denoted $E^c(H)$. $\sigma(B)$ will denote the spectrum of B.

If $V^L = 0$ then we say that the potential is short range. The scattering theory is much simpler in this case. Its basic objects are the so called wave operators defined by

$$\Omega^{\pm} := s - \lim_{t \to \pm\infty} e^{itH} e^{-itH_o} . \tag{1}$$

Unfortunately, if $V^L \neq 0$ then the limits (1) in general do not exist. This is why in the long range scattering theory we need a different formalism than that based on (1).

A. Boutet de Monvel et al. (eds.), Recent Developments in Quantum Mechanics, 243–255.

Actually, there exist two distinct formalisms of the long range scattering theory. The traditional one is based on the notion of modified wave operators. In order to define them we need to choose a certain function

$$[0,\infty] \times \mathbb{R}^n \ni (t,k) \to S_t(k) \in \mathbb{R}$$

such that the following limits exist:

$$s - \lim_{t \to \pm\infty} e^{itH} e^{-iS_t(D)} . \tag{2}$$

We shall not go into the details concerning the construction of the function S_t. It is enough to note that under our hypotheses such a function always exists but is nonunique. In the short range case there is one distinguished choice of S_t, namely $S_t(k) = \frac{t}{2}k^2$. In the general long range case no such a distinguished choice seems to exist.
We will denote the operators defined in (2) also by $\Omega^{\pm}$ and we will call them modified wave operators. The scattering operator is defined by:

$$S: = \Omega^{-*}\Omega^{+} \quad .$$

It is possible to show that in the momentum representation the kernel of S has the following form:

$$S(k_1,k_2) = e^{i\phi(k_1)}\delta(k_1 - k_2) + 2\Pi i\delta\left(\frac{1}{2}(k_1)^2 - \frac{1}{2}(k_2)^2\right)T(k_1,k_2), \tag{3}$$

where ϕ is a certain real function on $\mathbb{R}^n$. $T(k_1,k_2)$ for $|k_1| = |k_2|$ is called the scattering amplitude and $|T(k_1,k_2|^2$ - the scattering cross section.
It is clear that $\Omega^{\pm}$ and S are nonuniquely defined. By choosing different $S_t^{(1)}$ and $S_t^{(2)}$ we may obtain different modified wave operators $\Omega^{\pm}_{(1)}$ and $\Omega^{\pm}_{(2)}$. Note however that there always exists a function δ such that $\Omega^{\pm}_{(1)} = \Omega^{\pm}_{(2)}e^{i\delta(D)}$. Hence it is easy to see that the only nonuniqueness of $T(k_1,k_2)$ is contained in its phase and the scattering cross sections are defined uniquely. In fact, unlike scattering amplitudes, scattering cross sections have a physically measurable interpretation.

The formalism that we described above has the following drawback. We arrive at a uniquely defined object - the scattering cross section -only after a long chain of constructions which are nonunique. It turns out however that there exists an alternative formalism where the definition of the scattering cross sections is obtained in a more natural way. This is the formalism of asymptotic observables which we are going to describe below.

Let $C_\infty(\mathbb{R}^n)$ denote the space of all continuous functions on $\mathbb{R}^n$ that vanish at infinity. It can be shown that if $g \in C_\infty(\mathbb{R}^n)$ then the following limits exist:

$$\operatorname*{s-lim}_{t\to\pm\infty} e^{itH} g(D) e^{-itH}\, E^c(H) \qquad . \tag{4}$$

We denote (4) by $\gamma^\pm(g)$. It is easy to see that the maps

$$C_\infty(\mathbb{R}^n) \ni g \to \gamma^\pm(g) \in B\left(L^2(\mathbb{R}^n)\right)$$

are $*$-homomorphisms of C^*-algebras. They generate in an obvious way two projection valued Borel measures

$$\mathbb{R}^n \supset \Theta \to \Gamma^\pm(\Theta) \in \mathrm{Proj}\left(L^2(\mathbb{R}^n)\right).$$

Unlike modified wave operators, $\gamma^\pm$ and $\Gamma^\pm$ are uniquely defined. It also turns out that if we know $\Gamma^\pm$ then in principle we can compute the scattering cross sections.
In order to do this we need some additional definitions. Let $\mathcal{B}$ denote the W^*-algebra generated by all the spectral projections of H onto parts of its continuous spectrum. Clearly, $\mathcal{B} \simeq L^\infty([0,\infty))$. Let $\mathcal{M}$ denote the commutant of $\mathcal{B}$ reduced by the projection $E^c(H)$. Obviously, $\mathcal{M}$ is a type I von Neumann algebra and $\mathcal{B}$ is its center. Hence every $B \in \mathcal{M}$ possesses the following natural representation:

$$B = \int_{[0,\infty)}^{\oplus} d\lambda\, B_\lambda \qquad .$$

If $B \in \mathcal{M}_+$ then we can define:

$$[0,\infty) \in \lambda \to \mathrm{Tr}(B)(\lambda) := \mathrm{Tr}B_\lambda \qquad ,$$

which is a function on $[0,\infty)$ with values in $[0,\infty)$ defined everywhere up to a set of the Lebesgue measure zero.

Note that it is easy to show that $\Gamma^\pm(\Theta) \in \mathcal{M}_+$.

Now the relationship between the two formalisms described above can be expressed by the following identities.

Let χ_Θ denote the characteristic function of a set $\Theta \subset \mathbb{R}^n$. Then:

$$\Gamma^{\pm}(\Theta) = \Omega^{\pm}\chi_{\Theta}(D)\Omega^{\pm *} \quad . \tag{5}$$

Moreover if Θ_1 and Θ_2 are disjoint Borel subsets of $\mathbb{R}^n$ then for almost all $\lambda \in [0,\infty)$

$$\mathrm{Tr}\left(\Gamma^{-}(\Theta_1)\Gamma^{+}(\Theta_2)\right)(\lambda)$$

$$= (2\Pi)^2(2\lambda)^{n-2}\int\int_{S^{n-1}\times S^{n-1}} |T(\sqrt{2\lambda}s_1,\sqrt{2\lambda}s_2)|^2\chi_{\Theta_1}(\sqrt{2\lambda}s_1)\chi_{\Theta_2}(\sqrt{2\lambda}s_2)ds_1ds_2 \tag{6}$$

Let us stress that the second formalism is not designed to replace the first formalism alltogether and to make the notion of modified wave operators obsolete. Firstly, even if we know that the *-homomorphism $\gamma^{\pm}$ exist it does not mean by itself that they can be implemented by isometries. If we want to show that the formula (5) is actually true we need to construct modified wave operators.

Moreover, the knowledge of the functions S_t that appears in (2) tells us how to prepare *a distorted free state*, which for $t \to \pm\infty$ resembles a state that evolves under the full dynamics.

Finally, the expression on the lhs of (6) does not seem well suited for practical computations. If we want to investigate more closely the properties of the scattering cross sections then it seems to be more fruitful first to define modified wave operators with the help of the so called time independent modifiers and then to use formula (3), as in [IsoKi].

Let us remark that modified wave operators have been introduced in [Do]. The limits (4) have been first studied in [La, AMM]. For further references the reader may consult (RS, Pe, De2].

It is interesting to note that g(D) in (4) can be replaced with more general observables. We describe this in the following theorem, which will give us some hints on how to generalize the construction of $\gamma^{\pm}$ to the N-body case.

Theorem 1.1. Let $Q \in C(S^{n-1})$, $g \in C_\infty(\mathbb{R}^n)$ and $h \in C_\infty(\mathbb{R})$. Then

$$\operatorname*{s-lim}_{t\to\pm\infty} e^{itH}\, Q\left(\frac{x}{|x|}\right) g(D)h(H)\, e^{-itH}E^c(H) = \gamma^{\pm}(\psi_{\pm})$$

where $\psi_{\pm}(k) = Q\left(\pm\frac{k}{|k|}\right) g(k)\, h(\frac{1}{2}k^2)$.

2. N-BODY LONG RANGE SCATTERING

Now we are going to describe certain natural classes of asymptotic observables in the N-body case. Many of the results that we will talk about below seem to be new; they will be described in more detail and proved in our forthcoming paper [De2].

N-body Schrödinger operators are usually defined to be the operators of the form:

$$H = \sum_{1\le i\le N} -\frac{1}{2m_i}\Delta_i + \sum_{1\le i<j\le N} V_{ij}(x_i - x_j) \tag{0}$$

(see e.g. [RS]). From the mathematical point of view it is convenient to consider another class of operators, namely the class of the so called generalized N-body Schrödinger operators, which essentially includes all the operators of the form (0). Let us introduce this class and let us state the basic assumptions on the potentials that we will need for our purposes.

Let X be a certain Euclidean space. We will assume that $\{X_a : a \in \mathcal{A}\}$ is a fixed family of subspaces of X closed wrt taking the intersection, containing {0} and X. We will write $X_{a_{min}} = X$ and $X_{a_{max}} = \{0\}$. By $b \subset a$ we will mean $X_b \supset X_a$. We define $X_a^\perp =: X^a$. It is convenient to denote the dual space of X by K and X_a by K_a. The projection of K onto K_a will be denoted by π_a. We set #a equal to the maximum number of distinct $a_1,\dots,a_n \in \mathcal{A}$ such that $a_{max} = a_1 \subset a_2 \subset \dots \subset a_n$. We set $\mathcal{A}_n := \{a\in\mathcal{A}: \#a=n\}$ and $N := \max\{\#a: a\in\mathcal{A}\}$. The unit sphere in X will be denoted by S.

We assume that for every $a \in \mathcal{A}$ we choose a real function v_a on X^a such that $v_a = v_a^s + v_a^L$, $(1+|x^a|)^{1+\mu_s} v_a^s(x^a)(1-\Delta^a)^{-1}$ is compact on $L^2(X^a)$ for some $\mu_s > 0$ and $|\nabla^\alpha v_a^1(x^a)| \le c_\alpha(1+|x^a|)^{-|\alpha|-\mu_L}$ for some $1 \ge \mu_L > 0$ and any α.

The main object of our study will be the following self adjoint operator on $L^2(X)$ with the domain $\mathcal{D}(-\Delta)$:

$$H = -\frac{1}{2}\Delta + \sum_{a\in\mathcal{A}} v_a(x^a) \quad .$$

The operators of this form will be called generalized N-body Schrödinger operators (see [A, ABG]).

If $a \in \mathcal{A}$ then we can define the so called cluster Hamiltonian corresponding to the subsystem $a \in \mathcal{A}$:

$$H_a = -\frac{1}{2}\Delta + \sum_{b \subset a} v_b(x^b) \quad .$$

If we identify $L^2(X)$ with $L^2(X_a) \otimes L^2(X^a)$ then we have:

$$H_a = -\frac{1}{2}\Delta_a \otimes 1 + 1 \otimes H^a$$

where H^a is a certain self adjoint operator on $L^2(X^a)$. Note that H^a describes the inner degrees of freedom of the subsystem $a \in \mathcal{A}$.

$\mathcal{B}_a$ will denote the pure point spectrum of H^a and $\mathcal{B}_a^{disc}$ -the discrete spectrum of H^a. We define also $\mathcal{T}_a = \bigcup_{b \subsetneq a} \mathcal{B}_b$ - the set of thresholds of H^a.

It will be also useful to define the following sets:

$$AS = \bigcup_{a \neq a_{max}} (K_a \backslash \{0\}) \times \mathcal{B}_a$$

and if $\tau \in \mathcal{B}_a \cup \mathcal{T}_a$ then

$$AS_{a\tau} = \left(K_a \setminus \bigcup_{b \not\subset a} K_b\right) \times \{\tau\} \quad .$$

(The letters AS stand for the Asymptotic Set). We treat AS as a subset of $K \times \mathbb{R}$. Clearly $AS_{a\tau} \subset AS$. Note moreover that both AS and $AS_{a\tau}$ are endowed in an obvious way with a locally compact topology. Hence it makes sense to speak about the C^*- algebras of continuous functions on AS and $AS_{a\tau}$ that vanish at infinity.

The following two theorems will describe certain natural algebras of asymptotic observables associated with a generalized N-body Schrödinger operator. The first one seems to be new, the second one can be traced back to [AGM].

Theorem 2.1. There exist unique *-homomorphism

$$\gamma^{\pm}: C_\infty(AS) \to B(L^2(X))$$

that satisfy the following condition. Let $a \neq a_{max}$, $Q \in C(S)$, $g \in C_\infty(K_a)$ and $h \in C_\infty(\mathbb{R})$ such that $Q = 0$ on $\bigcup_{b \not\subset a} X_b \cap S$. We define $\psi_\pm \in C_\infty(AS)$ such that $\psi_\pm(k,\tau) = Q\left(\frac{\pm k}{|k|}\right) g(\pi_a k) h(\frac{1}{2}k^2 + \tau)$. Then

$$\gamma^{\pm}(\psi_\pm) = \text{s-}\lim_{t \to \pm\infty} e^{itH} Q\left(\frac{x}{|x|}\right) g(D_a) h(H) E^c(H) e^{-itH} \quad .$$

Now let us assume that $\tau \in \mathcal{B}_a$ for some $a \neq a_{max}$. We will assume additionally that for some $\nu > 1 - \mu_L$ the operator $(1 + |x^a|)^\nu E_{\{\tau\}}(H^a)$ is bounded ($E_{\{\tau\}}(H^a)$ denotes the spectral projection of H^a onto $\{\tau\}$).

Theorem 2.2. There exist unique *-homomorphisms

$$\gamma^{\pm}_{a\tau}: C_\infty(AS_{a\tau}) \to B(L^2(X))$$

that satisfy the following condition. Let $Q \in C(S)$, $g \in C_\infty(K_a)$ and $h \in C_\infty(\mathbb{R})$ such that $Q = 0$ on $\bigcup_{b \not\subset a} X_b \cap S$. We define $\psi_\pm \in C_\infty(AS_{a\tau})$ such that

$$\psi_\pm(k_a, \tau) = Q\left(\frac{\pm k_a}{|k_a|}\right)g(k_a)h(\frac{1}{2}k_a^2 + \tau).$$

Then

$$\gamma^{\pm}_{a\tau}(\psi_\pm) = \operatorname*{s-lim}_{t\to\pm\infty} e^{itH} Q\left(\frac{x}{|x|}\right)g(D_a)E_{\{\tau\}}(H^a)h(H)E^c(H)e^{-itH} .$$

Moreover if $(a_1, \tau_1) \neq (a_2, \tau_2)$ and $\psi^i_\pm \in C_\infty(AS_{a_i\tau_i})$ then

$$\gamma^{\pm}_{a_1\tau_1}(\psi^1_\pm)\gamma^{\pm}_{a_2\tau_2}(\psi^2_\pm) = 0 .$$

Now let $AS \supset \Theta \to \Gamma^{\pm}(\Theta)$ be the projection valued Borel measure generated by $\gamma^{\pm}$. Likewise, $AS_{a\tau} \supset \Theta \to \Gamma^{\pm}_{a\tau}(\Theta)$ will denote the projection valued Borel measure generated by $\gamma^{\pm}_{a\tau}$. Note that $\Gamma^{\pm}(AS) = E^c(H)$. Moreover $\Gamma^{\pm}_{a\tau}(AS_{a\tau})$ are mutually orthogonal projections less than $E^c(H)$.

We can use those projections to define the notion of the asymptotic completeness. Namely, we say that the system is asymptotically complete iff

$$E^C(H) = \sum_{a\neq a_{max}} \sum_{\tau\in\mathcal{B}_a} \Gamma^{\pm}_{a\tau}(AS_{a\tau}) . \tag{1}$$

In the more traditional approach the starting point of the long range N-body scattering theory is the construction of the modified channel wave operators:

$$\Omega^{\pm}_{a\tau} = \operatorname*{s-lim}_{t\to\infty} e^{itH} e^{iS_{at}(D_a)} E_{\{\tau\}}(H^a) ,$$

where S_{at} are suitably chosen functions on $[0,\infty) \times K$. The following identity explains the relationship between the asymptotic observables and modified wave operators. Let $\psi \in C_\infty(AS)_{a\tau})$. Then:

$$\gamma^{\pm}_{a\tau}(\psi) = \Omega^{\pm}_{a\tau}\ \psi(D_a)\Omega^{\pm *}_{a\tau} \quad .$$

Hence $\Gamma^{\pm}_{a\tau}(AS_{a\tau})$ is the projection onto Ran $\Omega^{\pm}_{a\tau}$. This means that (1) is equivalent to

$$\text{Ran}E^{c}(H) = \sum^{\oplus}_{a\neq a_{max}} \ \sum^{\oplus}_{\tau\in\mathcal{B}_a} \text{Ran}\Omega^{\pm}_{a\tau} \quad . \tag{2}$$

Note that traditionally (2) is used as the definition of the asymptotic completeness.

The asymptotic completeness of N-body systems is a very natural conjecture. It has been shown for short range systems in [SigSof1]. Unfortunately, in the long range case rigorous proofs of the asymptotic completeness contained in the literature are limited to 2-body and certain 3-body systems [E2]. Therefore it seems interesting that the formalism of asymptotic observables provides us with a possibility to ask certain questions about the N-body scattering which are more modest than the question about the validity of the asymptotic completeness.

An example of a concept that can be used to pose such questions is the so called local asymptotic completeness. To make its definition more clear first note that it is easy to show the following fact: if Θ is a Borel subset of AS then

$$\Gamma^{\pm}(\Theta) \ \geq \sum_{a\neq a_{max}} \ \sum_{\tau\in\mathcal{B}_a} \Gamma^{\pm}_{a\tau}(\Theta \cap AS_{a\tau}) \quad .$$

We will say that the system is (locally)asymptotically complete in Θ iff

$$\Gamma^{\pm}(\Theta) = \sum_{a\neq a_{max}} \ \sum_{\tau\in\mathcal{B}_a} \Gamma^{\pm}_{a\tau}(\Theta \cap AS_{a\tau}) \quad .$$

Note that if Θ_1 and Θ_2 are Borel subsets of AS such that $\Theta_1 \subset \Theta_2$ then the local asymptotic completeness in Θ_2 implies the local asymptotic completeness in Θ_1. Moreover, the asymptotic completeness is equivalent to the local asymptotic completeness in AS.

Now let us review what is known about the asymptotic completeness and local asymptotic completeness of various N-body systems.

1) $\mu_L > 0$.

Any system satisfying the assumptions specified at the beginning of this section is asymptotically complete in

$$\bigcup_{a \neq a_{max}} \bigcup_{\tau \in \mathcal{B}_a^{disc}} AS_{a\tau} \quad . \tag{3}$$

A particular case of this statement is the asymptotic completeness of 2-body systems (for which (3) is equal to AS).

The asymptotic completeness in (3) is relatively easy to prove. It seems that in a disguised form this fact was dicovered by V. Enss [E2].

2) $\mu_1 > \sqrt{3} - 1$.

In this case an N-body system is locally asymptotically complete in

$$\bigcup_{a \neq a_{max}} \bigcup_{\tau \in \mathcal{B}_a^{disc}} AS_{a\tau} \cup \bigcup_{\# a = N-1} AS_{a0} \quad . \tag{4}$$

A particular case of this statement is the asymptotic completeness of 3-body systems for $\mu_L > \sqrt{3}-1$, which was shown by V. Enss [E2]. (In this case (4) is equal to AS). The asymptotic completeness of N-body systems in (4) can be shown by a slight modification of the arguments contained in [E2].

3) $\mu_L > 1$ (or equivalently $v_a^L=0$ for all $a \in \mathcal{A}$ - this is the so called short range case).

In this case the asymptotic completeness has been shown by Sigal and Soffer [SigSof1].

One can define yet another interesting property of N-body systems related to the asymptotic completeness which merits our attention. We say that a system is asymptotically absolutely continuous iff the following condition is true. If Θ is any Borel subset of AS such that for any $a \neq a_{max}$ and $\tau \in \mathcal{T}_a \cup \mathcal{B}_a$ the Lebesgue measure of $\Theta \cap AS_{a\tau}$ is zero then $\Gamma^{\pm}(\Theta) = 0$.

It is easy to see that if a system is asymptotically complete then it is asymptotically absolutely continuous. In fact, in such a case

$$\Gamma^{\pm}(\Theta) = \sum_{a \neq a_{max}} \sum_{\tau \in \mathcal{B}_a} \Gamma^{\pm}_{a\tau} (\Theta \cap AS_{a\tau})$$

$$= \sum_{a \neq a_{max}} \sum_{\tau \in \mathcal{B}_a} \Omega^{\pm}_{a\tau} \chi_{\Theta \cap AS_{a\tau}} (D_a, \tau) \Omega^{\pm *}_{a\tau} = 0 \quad .$$

It can be shown that N-body systems with $\mu_1 = 1$ (which includes the Coulomb potentials) are asymptotically absolutely continuous. The proof of this fact is based on Sigal's and Soffer's result called the asymptotic clustering of N-body systems [SigSof3].

Now let us say a few words about the proofs of theorems 2.1.and 2.2. The main techniques used in these proofs can be grouped into three categories:

1) The study of observables in the Calkin algebra;
2) Velocity estimate;
3) Propagation theorem.

The latter two techniques are due to Sigal and Soffer [SigSof1,2,3, De1,2]. They are based on various ideas that originated in papers by Mourre and Enss [M1,2, E1,2]. The key ingredient of the proofs of velocity estimates and the propagation theorem is the Mourre estimate [M1,2, PSS, FH, CFKS, De1,2].

The use of the Calkin algebra in the scattering theory seems to be a new idea (although it is actually a francy reformulation of a well known technique which goes under the name of the geometric method [Sim, Sig, CFKS]). Let us describe this idea in more detail.

Let $B(\mathcal{H})$ denote the C^*-algebra of bounded operators on a Hilbert space $\mathcal{H}$ and $CB(\mathcal{H})$ - the closed ideal of compact operators. Then $B(\mathcal{H})/CB(\mathcal{H}) = \mathrm{Cal}(\mathcal{H})$ is called the Calkin algebra. The canonical homomorphism will be denoted by

$$B(\mathcal{H}) \ni B \to [B]_{Cal} \in \mathrm{Cal}(\mathcal{H}) \quad . \tag{5}$$

It turns out that certain classes of operators simplify their algebraic properties significantly if we look at their images in the Calkin algebra. We can illustrate this phenomenon by the following pedagogical example.

Let us consider two *-homomorphisms:

$$C(S) \ni Q \to \left[Q\left(\frac{x}{|x|}\right)\right]_{Cal} \in \mathrm{Cal}(L^2(X)) \quad , \tag{6}$$

and

$$C_\infty(K) \ni g \to \left[g(D)\right]_{Cal} \in \mathrm{Cal}(L^2(X)) \quad . \tag{7}$$

It is easy to show that $\left[Q\left(\frac{x}{|x|}\right),\ g(D)\right]$ is compact. Therefore the images of (6) and (7) commute. Hence by the well known theorem on the tensor product of two C^*-algebras (see e.g. [Ta]) we can define the tensor product of the *-homomorphisms (6) and (7). Thus we obtain the *-homomorphism

$$C_\infty(S \times K) \ni \psi \to \psi\left(\frac{x}{|x|}, D\right) \in \mathrm{Cal}(L^2(X))$$

such that if $Q \in C(S)$, $g \in C_\infty(K)$ and $\psi = Q \otimes g$ then

$$\psi\left(\frac{x}{|x|}, D\right) = \left[Q\left(\frac{x}{|x|}\right)g(D)\right]_{Cal} .$$

Now recall that the main object of our study are the limits of the form

$$\operatorname*{s-lim}_{t\to\pm\infty} e^{itH} B e^{-itH} E^c(H) . \qquad (8)$$

By the RAGE theorem [RS] (8) is zero if B is compact. Hence (8) does not depend on B as long as $[B]_{Cal}$ is the same. Therefore, if we want to prove theorem 2.1 it will be useful to study the images of operators

$$Q\left(\frac{x}{|x|}\right) g(D_a)h(H) \qquad (9)$$

inside the Calkin algebra. Let us describe how it can be done.

Let us define

$$es_a := \left\{(k_a,\lambda) \in K_a \times \mathbb{R}\colon \lambda - \frac{1}{2}k_a^2 \in \sigma(H^a)\right\}$$

and

$$ES := \bigcup_{a\neq a_{max}} \left(S \cap X_a \setminus \bigcup_{b\not\subset a} X_b\right) \times es_a$$

(The letters ES stand for the Energy Shell). We endow ES with the following topology

$$(x^n,k^n,\lambda^n) \to (x,k_a,\lambda) \in \left(S \cap X_a \setminus \bigcup_{b\not\subset a} X_b\right) \times es_a$$

iff $x^n \to x$, $\pi_a k^n \to k_a$ and $\lambda^n \to \lambda$. Note that in this way ES becomes a locally compact space. Now the following theorem provides a fairly detailed description of the images of operators of the form (9) inside the Calkin algebra and is a convenient starting point for the proof of theorem 2.1.

Theorem 2.3. There exists a unique *-homomorphism

$$\rho\colon C_\infty(ES) \to \mathrm{Cal}(L^2(X))$$

that satisfies the following condition. Let $a \neq a_{max}$, $Q \in C(S)$, $g \in C_\infty(K_a)$ and $h \in C_\infty(\mathbb{R})$ such that $Q = 0$ on $\bigcup_{b \not\subset a} X_b \cap S$. We define $\psi \in C_\infty(ES)$ such that if $(x_b, k_b, \lambda) \in \left(S \cap X_b \setminus \bigcup_{c \not\subset b} X_c \right) \times es_b$ then $\psi(x_b, k_b, \lambda) = Q(x_b) g(\pi_a k_b) h(\lambda)$. Then

$$\rho(\psi) = \left[Q\left(\frac{x}{|x|}\right) g(D_a) h(H) \right]_{Cal} .$$

REFERENCES

[A] Agmon, S.: Lectures on the exponential decay of solutions of second order elliptic equations, Princeton University Press, 1982.

[ABG] Amrein, W.O., Boutet de Monvel-Berthier, A.M. and Georgescu, V.: Notes on the N-body problem, preprint, Génève, 1989.

[AGM] Amrein, E.O., Georgescu, V. and Martin, Ph.A.: Approche algébrique de la théorie non-relativiste de la diffusion aux canaux multiples, in: *Physical Reality and Mathematical Description*, Enz/Mehra (eds), 255-276 (1974).

[AMM] Amrein, W.O., Martin, Ph.A. and Misra, B.: On the asymptotic condition of scattering theory, Helv.Phys.Acta **43**, 313-344 (1970).

[CFKS] Cycon, H.L., Froese, R., Kirsch, W. and Simon, B.: *Schrödinger Operators with Applications to Quantum Mechanics and Global Geometry*, Springer, Berlin, Heidelberg, New York, 1987.

[De1] Dereziński, J.: A new proof of the propagation theorem for N-body quantum systems, Comm.Math.Phys. **122**, 203-231 (1989).

[De2] Dereziński, J.: Algebraic Approach to the N-body Quantum Long Range Scattering, Preprint 1990.

[Do] Dollard, J.: Asymptotic convergence and Coulomb interaction, Journ.Math.Phys. **5**, 729-738 (1964).

[E1] Enss, V.: Asymptotic observables on scattering states, Comm. Math.Phys. **89**, 245-268 (1983).

[E2] Enss, V.: *Quantum scattering theory of two- and three-body systems with potentials of short and long range*, in: Schrödinger Operators, ed. by S.Graffi, Lecture Notes in Mathematics, vol.1159, (Springer, Berlin, Heidelberg, New York 1985).

[IsoKi] Isozaki, H. and Kitada, H.: Scattering matrices for two-body Schrödinger operators, Scientific Papers of the College of Arts and Sciences, Tokyo Univ. **35**, 81-107 (1985).

[La] Lavine, R.: Scattering theory for long range potentials, Journ.Func.Anal. **5**, 368-382, (1970).

[M1] Mourre, E.: Absence of singular continuous spectrum for certain self adjoint operators, Comm.Math.Phys. **78**, 391-408 (1981).

[M2] Mourre, E.: Opérateurs conjugués et propriétés de propagations, Comm.Math.Phys. **91**, 279-300 (1983).

[Pe] Perry, P.: *Scattering Theory by the Enss Method* (Harwood Academic London 1983).

[RS] Reed, M. and Simon B.: *Methods of Modern Mathematical Physics, III: Scattering Theory* (Academic Press, New York, 1979).

[Sig] Sigal, I.M.: Geometric methods in the quantum many-body problem. Nonexistence of very negative ions., Comm.Math.Phys. **85**, 309-324 (1982).

[SigSof1] Sigal, I.M. and Soffer, A.: The N-particle scattering problem: asymptotic completeness for short range systems, Anal.Math. **125**, 35-108, (1987).

[SigSof2] Sigal, I.M. and Soffer, A.: Local decay and velocity bounds, preprint, Princeton (1988).

[SigSof3] Sigal, I.M. and Soffer, A.: Long range many body scattering, Asymptotic clustering for Coulomb type potentials, preprint, Toronto (1988).

[Sim] Simon, B.: Geometric methods in multiparticle quantum systems, Comm.Math.Phys. **55**, 259-274, (1977).

[Ta] Takesaki, M.: *Theory of Operator Algebras I.* Springer Berlin, Heidelberg, New York 1979.

SPECTRAL PROPERTIES OF BENT QUANTUM WIRES

P. EXNER
Laboratory of Theoretical Physics
Joint Institute for Nuclear Research
141980 Dubna
U S S R

ABSTRACT. Spectral properties for Hamiltonians describing pure-semiconductor quantum wires are discusssed. The curvature-induced bound states that exist in thin infinitely long wires are shown to turn into resonances when a finite-length wire is joined to a pair of macroscopic electrodes.

1. EXISTENCE OF CURVATURE-INDUCED BOUND STATES

Quantum wires are stripes of a highly pure semiconductor material whose fabrication on a substrate is possible due to rapid progress in solid-state technology - cf., e.g., Temkin et al. (1987), Timp et al. (1988). If we assume that the semiconductor material is absolutely pure and take its crystallic structure into account, we can describe an electron within the wire as a free quantum particle with some effective mass subjected to Dirichlet boundary conditions.

The continuous spectrum starts in this case at some positive value. It has been observed recently (Exner and Seba, 1989) that if the wire is curved and thin enough, isolated eigenvalues below this threshold may exist.

Let us illustrate this assertion in the simplest case of a curved planar strip Ω of a width d whose shape is (up to Euclidean transformations of the plane) determined fully by the signed curvature γ of one of its boundaries which we choose as a reference curve; we denote it as Γ. The points of Ω can be expressed by means of the natural curvilinear coordinates s (the arc length of Γ) and u. We assume

(i) γ is infinitely smooth (in fact, $\gamma \in C^2$ is sufficient),

(ii) $\gamma(s) = O(|s|^{-3/2})$ as $|s| \to \infty$ (this assumption too may be weakened - cf. Exner and Seba (1989)).

The surprising result is now that a bent region on a quantum wire can bind electrons; if we exclude the trivial case $\gamma \equiv 0$, then the following assertion holds:

A. Boutet de Monvel et al. (eds.), Recent Developments in Quantum Mechanics, 257–264.

THEOREM 1: $\sigma_{ess}(H)_\Omega = [\lambda_1,\infty)$ holds for the Dirichlet Laplacian $H_\Omega = -\Delta_D^\Omega$, where $\lambda_1 = (\pi/d)^2$. There is a positive d_0 such that for all $d < d_0$, H_Ω has at least one eigenvalue $\varepsilon < \lambda_1$.

To prove the theorem, one has to pass to the curvilinear coordinates in which the operator under consideration turns into

$$H = -\frac{\partial}{\partial s}(1+u\gamma)^{-2}\frac{\partial}{\partial s} - \frac{\partial^2}{\partial u^2} + V(s,u) \quad , \tag{1a}$$

$$V(s,u) = -\frac{\gamma^2}{4(1+u\gamma)^2} + \frac{u\gamma''}{2(1+u\gamma)^3} - \frac{5}{4}\frac{u^2\gamma'^2}{(1+u\gamma)^4} \quad , \tag{1b}$$

and to perform suitable minimax estimates.

Let us remark that the smoothness of γ which is needed to pass from H_Ω to H by unitary equivalence, plays a rather technical role. It can be illustrated on the solvable example of the L-shapes strip of a width d. In this case, there is just one bound state at the energy

$$\varepsilon = 0.93\ (\pi/d)^2 \tag{2}$$

and the corresponding wavefunction can be calculated (Exner et al., 1989).

The minimax estimates mentioned above can be made more sophisticated to get other properties of the bound states. Let us mention two examples. Using Birman-Schwinger principle combined with the transversal-mode decomposition, one can derive (Exner and Seba, 1989) the following lower bound to d_0:

PROPOSITION 2: For a simply bent strip, $\gamma(s) \ge 0$, the critical width obeys

$$d_0 \ge d_+ := \frac{1}{2\|\gamma\|_\infty} = \left\{\left[1 + 4\,\frac{\|\gamma\|_\infty\ \gamma\|_2}{\|\gamma'\|_2}\right]^{-1/2} - 1\right\} .$$

On the other hand, combining the minimax estimates with the known results about the number of bound states of a one-dimensional Schrödinger operator (Klaus, 1977; Newton, 1983), we get

PROPOSITION 3: In addition to **(i)**, **(ii)**, assume that

$$d < \frac{1}{1 + c_\gamma\|\gamma_-\|_\infty} \quad ,$$

where $\gamma_\pm$ is the positive (negative) part, $\gamma = \gamma_+ - \gamma_-$, and

$$c_\gamma := \min\left\{ \frac{2\pi}{\|\gamma\|_\infty} , \left(\frac{2\pi^2}{\|\gamma''\|_\infty}\right)^{1/3} , \left(\frac{2\pi}{\sqrt{5}\|\gamma'\|_\infty}\right)^{1/2} \right\} .$$

Then the number of bound states (including multiplicity) fulfills

$$N(H) \le 1 + \frac{1}{8}\left(\frac{1+d\|\gamma\|_\infty}{1-d\|\gamma\|_\infty}\right)^2 \frac{\int_{\mathbb{R}^2} \gamma(s)^2 \; |s-t| \; \gamma(t)^2 \; ds\, dt}{\int_{\mathbb{R}^2} \gamma(s)^2 \; ds} .$$

Similar results can be derived also for curved three-dimensional tubes - for more details see Duclos and Exner (1990).

2. RESONANCES IN FINITE-LENGTH QUANTUM WIRES

Thinking about the ways in which the bound states described above could be observed, one has to realize that in fact every quantum wire has a finite length. It is typically attached to a pair of *"macroscopic"* wires. Since the continuous spectrum threshold for the latter is much lower, the bound state appears to be embedded into the continuum and one naturally expects it to turn into a resonance.

It is easy, however, to prove this assertion. We are going to do that in a model setting. First of all, we shall regards the system again as a two-dimensional one. In addition, we asssume that the quantum-wire width d may be neglected with respect to the width of the macroscopic electrodes, i.e., we shall model the latter by two halfplanes connected by a line segment of a length 2D.

Inspecting the formulae (1), we see that for small enough d the operator H is in a sense close to

$$H_\infty = -\frac{d^2}{ds^2} - \frac{1}{4}\gamma(s)^2 + \lambda_1 \qquad (3)$$

(this assertion can be made more precise - cf. Duclos and Exner (1990)), where $\lambda_1 = (\pi/d)^2$. The last named quantity blows to infinity, of course, as $d \to 0$, but for model purposes we shall keep it as a fixed positive number. In addition to **(i)**, **(ii)** we shall assume now that

(iii) γ is a non-zero function with a compact support such that

$$|\gamma(s)| \le 2\lambda_1^{1/2} .$$

The first question is how to connect the segment to the halfplanes. It can be done in a close analogy to the problem concerning a plane and a halfline (Exner and Seba, 1987). In the present case only the even partial waves are present in the decomposition with respect to the connection point (but it makes no difference because only the s-wave can be coupled non-trivially to the segment) and also the normalization is different. The coupling is described by boundary conditions which we shall write down below.

We consider therefore the nontrivial part of the problem concerning the s-wave component of the wavefunction in the halfplanes and write the electron wavefunction as $\psi = (u_1, f_2, u_3)$ with $u_j \in L^2(\mathbb{R}_+, r\,dr)$ and $f_2 \in L^2(-D,D)$. The Hamiltonian H_D of the model is then of the form

$$H_D \begin{pmatrix} u_1 \\ f_2 \\ u_3 \end{pmatrix} = \begin{pmatrix} -u_1'' - ru_1' - (4r^2)^{-1}u_1 \\ -f_2'' + [\lambda_1 - \gamma(s)^2/4]f_2 \\ -u_3'' - ru_3' - (4r^2)^{-1}u_3 \end{pmatrix} \tag{4}$$

with the boundary conditions

$$\begin{aligned} L_0(u_j) &= af_2(\mp D) \pm bf_2'(\mp D) , \\ L_1(u_j) &= cf_2(\mp D) \pm df_2'(\mp D) , \end{aligned} \tag{5}$$

where the quantities on the *lhs* are the boundary values regularized with respect to the logarithmic singularity - cf. Bulla and Gesztesy (1985). The requirement of probability current conservation at the junctions selects a four-parameter family of the boundary conditions (5), namely

$$\begin{aligned} a &= \frac{i}{\pi} 2^{5/4} e^{i(\delta-\alpha)} \frac{\mathcal{J}}{\sin\beta} , \\ b &= \frac{i}{\pi} 2^{5/4} e^{i(\delta-\alpha)} \frac{\mathcal{K}}{\sin\beta} , \\ c &= \frac{i}{\pi} 2^{5/4} e^{i(\alpha-\delta)} \frac{(\gamma - \ln 2) - \pi\mathcal{L}/4}{\sin\beta} , \\ d &= \frac{i}{\pi} 2^{5/4} e^{i(\alpha-\delta)} \frac{(\gamma - \ln 2) - \pi\mathcal{M}/4}{\sin\beta} . \end{aligned} \tag{6a}$$

for real α, β, ξ and $\beta \neq 0$, where $\gamma = 0.577\ldots$ is the Euler's constant and

$$\begin{aligned} \mathcal{J} &= \sin(\alpha + \delta + \tfrac{\pi}{4}) \cos\beta - \sin(\xi + \tfrac{\pi}{4}) , \\ \mathcal{K} &= \sin(\alpha+\delta) \cos\beta - \sin\xi , \\ \mathcal{L} &= \cos(\alpha + \delta + \tfrac{\pi}{4}) \cos\beta + \cos(\xi + \tfrac{\pi}{4}) , \\ \mathcal{M} &= \cos(\alpha+\delta) \cos\beta + \cos\xi . \end{aligned} \tag{6b}$$

We have adopted in (5) the natural assumption that both junctions are the same. In fact, one should choose some particular values of a, b, c, d, motivating the choice by a low-energy limit of a more realistic solution to the problem of injection of electronns into the quantum wire. This task is not easy and we avoid it showing existence of the resonances for any boundary conditions of the described type; the scattering on the junctions plays then role of a background.

The infinite wire is in our model described by the Hamiltonian (3) on $L^2(\mathbb{R})$. Under the assumptions **(i)-(iii)**, it has a finite number of simple eigenvalues,

$$0 < \varepsilon_1 < \varepsilon_2 < \ldots < \varepsilon_n < \lambda_1 \qquad , \tag{7}$$

where $n \geq 1$. In order to find the resonances of H_D and their relations to the eigenvalues (7), we use the factorization technique - cf., e.g., Kato (1966), Howland (1971), Baumgaertel and Demuth (1976). We write the Hamiltonian as

$$H_D = T_D - \frac{1}{2}|\gamma_D| \left(\frac{1}{2}|\gamma_D|\right) \qquad , \tag{8}$$

where T_D is the "*kinetic*" part and $\gamma_D := \gamma \upharpoonright [-D,D]$. The resonances are then the poles of the function

$$Q_1(z,D): = -\frac{1}{2}|\gamma_D|\ (H_D-z)^{-1} \left(\frac{1}{2}|\gamma_D|\right) \tag{9}$$

continued analytically to the lower complex halfplane, which can be expressed by means of the perturbation and the *free* resolvent as

$$Q_1(z,D) = I - [I + Q(z,D)]^{-1} \tag{10a}$$

$$Q(z,D): = \frac{1}{2}|\gamma_D|\ (T_D - z)^{-1} \left(\frac{1}{2}|\gamma_D|\right) \tag{10b}$$

Hence the resonances are also points where the analytically continued function (10b) is not invertible. Now the following assertion is valid

THEOREM 4: Suppose that $\gamma \in C_0^{\infty}(\mathbb{R})$ is non-zero, fulfills $|\gamma(s)| \leq 2\lambda_1^{1/2}$, and choose

$$\sum = \mathbb{C} \setminus \left[\mathbb{R}_- \cup (z:\ \mathrm{Re}\ z \geq \lambda_1 - \eta^2 + (\mathrm{Im}\ z)^2)\right] \tag{11}$$

for some $\eta < (\lambda_1 - \varepsilon_n)^{1/2}$. Then $Q(\cdot,D)$ can be continued from the upper complex halfplane to $\sum$. For all large enough D, the function $Q_1(\cdot,D)$

has just n simple poles in the lower-halfplane of $\sum$ at $z_j(D) = \xi_j(D) - \frac{i}{2}\Gamma_j(D)$ and

$$\lim_{D\to\infty} z_j(D) = \varepsilon_j \tag{12}$$

for $j = 1,\dots,n$. Moreover, the resonances $z_j(D)$ exhibit spectral concentration: choose a one-parameter family $\{\delta_j(D):\, D>0\}$ of positive numbers such that

$$\lim_{D\to\infty} \delta_j(D) = \lim_{D\to\infty} \frac{\Gamma_j(D)}{\delta_j(D)} = 0 \tag{13a}$$

and denote $\Delta_j(D) = (\xi_j(D) - \delta_j(D),\, \xi_j(D) + d_j(D))$, then the corresponding spectral measure converges strongly,

$$\text{s-}\lim_{D\to\infty} E_{H_D}(\Delta_j(D)) = P_j \quad , \tag{13b}$$

to the projection on the eigenspace referring to ε_j.

To prove the theorem, one can use a general factorization-technique theorem by Howland (1971). In order to express $Q_1(z,D)$ from (10), one has to know the *free* resolvent $(T_D - z)^{-1}$. We introduce the auxiliary operator $T_D^{(0)}$ which differs from T_D by the boundary conditions which are changed to the separated ones, $f_2(\mp D) = 0$ and $L_0(u_j) = 0$ for $j=1,3$. The resolvent $(T_D^{(0)} - z)^{-1}$ is a known integral operator; in particular, its *inner* part has the kernel

$$K(x,y;z) = \frac{\mathrm{ch}(\kappa(2D - |x-y|)) - \mathrm{ch}(\kappa|x+y|)}{2\kappa\ \mathrm{sh}(2\kappa D)} \tag{14}$$

for $-D < x,y < D$, where $\kappa = (\lambda_1 - z)^{1/2}$. The sought resolvent is then given by the Krein formula,

$$(T_D - z)^{-1} = (T_D^{(0)} - z)^{-1} + \sum_{j,k=1}^{4} \mu_{jk}(z)\ |g_j(z)\rangle\langle g_k(\bar{z})| \quad , \tag{15}$$

where the coefficients $\mu_{jk}(z)$ and the vectors $g_j(z)$ can be found from the requirement that $(T_D - z)^{-1}$ maps κ into the domain $D(T_D)$. This yields a set of linear equations which can be solved explicitly - cf. Exner (1990).

The most important step in application of the Howland's theorem is to check that $Q(z,D)$ tends to $Q(z,\infty)$ corresponding to the operator (3) in the operator norm uniformly in $\sum$. Since (14) gives

$$-\frac{1}{2}|\gamma_D(x)|\; K(x,y;z)\left(\frac{1}{2}|\gamma_D(y)|\right) \longrightarrow -\frac{1}{4}|\gamma(x)|\;\frac{e^{-\kappa|x-y|}}{2\kappa}|\gamma(y)|$$

as $D \to \infty$, where the *rhs* is just the kernel of $Q(z,\infty)$, it remains to check the contribution from the finite-rank part in (15) is zero. It follows from the explicit form of $\mu_{jk}(z)$ and $g_j(z)$ mentioned above. The formulae (14) and (15) allow to verify also the remaining hypotheses of the Howland's theorem, in particular, compactness of the operator $-\frac{1}{2}|\gamma_D|\;(T_D-z)^{-1}\left(\frac{1}{2}|\gamma_D|\right)$ for z from the resolvent set $\rho(T_D)$, existence of analytic continuation of $Q(\cdot,D)$ from the upper complex halfplane to Σ and the strong convergence $(T_D-z)^{-1} \longrightarrow (T_\infty-z)^{-1}$ as $D \to \infty$ for $\mathrm{Im}\, z \neq 0$, where T_∞ is the *free* counterpart of the operator (3); details can be found in (Exner, 1990).

3. CONCLUDING REMARKS

There are many ways in which the results presented here can be extended but we leave this to a future publication. Here we limit ourselves with two brief comments.

First of all, we have not mentiond in Theorem 4 the rate of spectral concentration. A semiclassical estimate shows that it should be exponential, $\Gamma_j(D) \simeq \exp(-\mathrm{const.}\,(\lambda_1 - \varepsilon_j)^{1/2})$. This is a positive feature from the experimentalist point of view because it gives hope for observation of sharp resonances on not very long quantum wires.

The second remark concerns the ways in which the resonance effect can be manifested. It certainly leads to a sheer variation in the transmission coefficient between the halfplanes which is related to the conductance by well-known formula

$$G = \frac{2e^2}{h}\;\frac{T(\varepsilon)}{1-T(\varepsilon)} \tag{16}$$

(Landauer, 1981). Tuning the applied voltage, one can change the electron energy, and by (15) the conductance of the whole structure. Let us mention an indirect but clear indication for curvature-induced resonances in quantum wires in a recent experiment by Timp et al. (1981): they have found that the resistance of a many-problem junction depends on the number of right-angle turns the electron must pass on its way between a pair of electrodes.

ACKNOWLEDGEMENT

I would like to thank my Romanian friends, especially G.Nenciu, for inviting me to the Poiana Brasov summer school. Remembering the sheer contrast between its creative atmosphere and the very depressive environment, I wish them a quick healing of the wounds and lasting freedom for the future.

REFERENCES

Baumgaertel, H. and Demuth, M. (1976) *Perturbations of unstable eigenvalues of finite multiplicity,* J.Funct.Anal. **22**, 187-203.

Bulla, W. and Gesztesy, F. (1985) *Deficiency indices and singular boundary conditions*, J.Math.Phys. **26**, 2520-2528.

Duclos, P. and Exner, P. (1990) *Bound states in curved quantum waveguides in two and three dimensions*, in preparation.

Exner, P. (1990) *A model of resonance scattering on curved quantum wires*, Annalen der Physik, to appear.

Exner, P. and Seba, P. (1987) *Quantum motion on a halfline connected to a plane*, J.Math.Phys. **28**, 386-391, 2304.

Exner, P. and Seba, P. (1989) *Bound states in curved quantum waveguides*, J.Math.Phys. **30**, 2574-2580.

Exner, P., Seba., P. and Stovicek, P. (1989) *On existence of a bound state in an L-shapes waveguide*, Czech.J.Phys. **B30**, 1181-1191.

Howland, J. (1971) *Spectral concentration and virtual poles II,* Trans.Amer.Math.Soc. **162**, 141-156.

Kato, T. (1966) *Wave operators and similarity for some non-selfadjoint operators*, Math.Annal. **162**, 258-279.

Klaus, M. (1977) *On the bound state of Schrödinger operators in one dimension*, Ann.Phys. **108**, 288-300.

Landauer, R. (1981) *Can a length of perfect conductor have a resistance ?*, Phys.Lett. **A85**, 91-93.

Newton, R.G. (1983) *Bounds on the number of bound states for the Schrödinger equation in one or two dimensions*, J.Oper.Theory **10**, 119-125.

Temkin, H., et al. (1987) *Low temperature photoluminiscence from InGaAs/InP quantum wires and boxes*, Appl.Phys.Lett. **50**, 413-415.

Timp, G., et al. (1988) *Propagation around a bend in a multichannel electron waveguide*, Phys.Rev.Lett. **60**, 2081-2084.

EIGENFUNCTION EXPANSIONS FOR HYPERBOLIC LAPLACIANS

R. FROESE
Department of Mathematics
University of British Columbia
Vancouver, B.C.
Canada V6T 1Y4

1 Introduction

The geometry near infinity of a non-compact complete Riemannian manifold is reflected in the essential spectrum its Laplace operator. The nature of this correspondence, however, is subtle, and the manifolds whose Laplace operators have a well understood spectral theory are ones whose structure near infinity is simple. In this lecture I would like to explain some recent work of P. Hislop, P. Perry and myself on the Laplacian for three-dimensional non-compact hyperbolic manifolds. These manifolds are are simple near infinty, in the sense of being geometrically finite, but somewhat less simple than the ones that had been considered previously, in that they have cusps of non-maximal rank.

We will construct continuum eigenfunctions, $E(u; b; s)$, which, as functions of u, are eigenfunctions of the Laplace operator with eigenvalue $s(2-s)$. The variable b ranges over the boundary at infinity. These functions provide a generalized eigenfunction expansion for the continuous spectral subspace of the Laplacian. The goal of this work is to show that they have meromorphic continuations, in the variable s, to the entire complex plane.

The study of continuum eigenfunctions of Laplace operators on hyperbolic manifolds dates back to the work of the number theorists Maass [M] and Roelke [R]. They realized that these eigenfunctions are Eisenstein series for an associated discrete group of hyperbolic isometries. They considered two-dimensional manifolds with finite volume. These are the most interesting examples to number theorists, and their study, and the study of the associated Selberg trace formula and Zeta

A. Boutet de Monvel et al. (eds.), Recent Developments in Quantum Mechanics, 265–277.

function, continues to the present. For an overview, with many references, see Terras [T]. Continuum eigenfunctions can be thought of as scattering states and Fadeev [F] introduced methods from quantum scattering theory to their study. Lax and Phillips, in their book [LP1], showed how the Lax-Phillips scattering theory could be applied and have continued to study increasingly more general hyperbolic manifolds, in higher dimensions and with possibly infinite volume [LP2-5]. More general manifolds have also been studied by many other authors, including Patterson [Pa], Mandouvalos [Ma1-3], Agmon [A1], Mazzeo and Melrose [MM] and Perry [Pe].

What is new in our work on Eisenstein series, is that we allow the manifold to have cusps of non-maximal rank. This means that the boundary at infinty has singular directions. We use a partition of unity at infinity and a geometric resolvent formula to localize the analysis. An important ingredient in our method is a Mourre estimate which provides the needed control over the resolvent of the Laplace operator near the spectrum.

2 Schrödinger Operators with Short Range Potentials

Since Schrödinger operators are more likely to be familiar to the participants at this summer school than hyperbolic Laplacians, I want to remind you about eigenfunction expansions for Schrödinger operators with short range potentials. These are similar in many ways to the Eisenstein series. This analogy was developed by Agmon in his article [A2], which should be consulted for further information.

Consider the Schrödinger operator acting in $L^2(\mathbf{R}^n)$ given by

$$H = -\Delta + V,$$

where V is a potential which decreases rapidly enough at infinity, say, $V(x) = O(|x|^{-(1+\epsilon)})$. There exist continuum eigenfunctions, $e(x;\omega;k)$, where $x \in \mathbf{R}^n$, $\omega \in S^{n-1}$ and $k \in \mathbf{R}$, satisfying

$$He(x;\omega;k) = k^2 e(x;\omega;k)$$

with asymptotic behaviour

$$e(x;\omega;k) \sim \exp(ik\, x \cdot \omega)$$

for large $|x|$. These eigenfunctions define two distorted Fourier transforms on the continuous spectral subspace of H,

$$\mathcal{F}_\pm : E_c(H)L^2(\mathbf{R}^n) \to L^2(\mathbf{R} \times S^{n-1}, k^{n-1}dkds(\omega))$$

by the formula

$$\mathcal{F}_{\pm}f(\omega,k) = (2\pi)^{-n/2}\int_{\mathbf{R}^n}\overline{e(x;\omega;\pm k)}f(x)d^n x.$$

These are unitary maps which diagonalize H.

The scattering operator $S(k)$ maps $e(x;\omega;k)$ to $e(x;-\omega;-k)$, providing the connection between the two expansions.

Often, one is interested in the analytic properties of the eigenfunctions in the spectral parameter, k. In general, one knows that $e(x;\omega;k)$ is analytic in k for $\mathrm{Im}(k) > 0$ and continuous onto $\mathrm{Im}(k) = 0$, which are the values of k for which k^2 lies in the spectrum. In favourable situations, the eigenfunctions might have meromorphic continuations onto the lower half plane. Poles in the lower half plane of such a continuation are called resonances, and are expected to co-incide with the poles in the continuation of $S(k)$. The purpose of our work is to show that in the hyperbolic setting, these meromorphic extensions exist.

The eigenfunctions $e(x;\omega;k)$ are, of course, the analogues of the Eisenstein series $E(u;b;k)$. The spectral parameter k corresponds to s, with the upper half plane $\mathrm{Im}(k) \geq 0$ corresponding to the half plane $\mathrm{Re}(s) \geq 1$. The boundary at infinity in the Schrödinger case is the sphere S^{n-1}. It labels the directions in which scattering can occur. For fixed k, each point on the boundary at infinity corresponds to a continuum eigenfunction which looks asymptotically like the plane wave, $\exp(ik\, x\cdot\omega)$.

3 Some Hyperbolic Geometry

We will work with the upper half plane model of hyperbolic space. In this model, hyperbolic space in three dimensions is represented by

$$\mathbf{H}^3 = \{(x_1,x_2,x_3) : x_3 > 0\}.$$

The Riemannian metric (of constant negative sectional curvature) is given by

$$ds^2 = x_3^{-2}(dx_1^2 + dx_2^2 + dx_3^2).$$

Geodesics are either straight lines parallel to the x_3 axis, or semicircles making a right angle with plane $\{x_3 = 0\}$. Geodesic hyperplanes are either planes parallel to the x_3 axis, or hemispheres sitting on the plane $\{x_3 = 0\}$. The plane $\{x_3 = 0\}$, together with the point at infinity, is the boundary at infinty for $\mathbf{H}^3$. It can be identified with the Riemann sphere, $\hat{\mathbf{C}}$.

The isometries of $\mathbf{H}^3$ are given by $SL(2,\mathbf{C})$. To describe the action, we first identify the point $(x_1,x_2,x_3) \in \mathbf{H}^3$ with the quaternion (with zero k component)

$w = x_1 + x_2 i + x_3 j$. Then $SL(2,\mathbf{C})$ acts via linear fractional transformations as follows. If

$$\gamma = \begin{pmatrix} a & b \\ c & d \end{pmatrix} \in SL(2,\mathbf{C}),$$

then

$$\gamma w = (aw+b)(cw+d)^{-1}.$$

Notice that this action extends to the boundary at infinity, where it becomes the usual action of $SL(2,\mathbf{C})$ on the Riemann sphere via fractional linear transformations.

We will consider infinite volume manifolds, M, which are quotients of $\mathbf{H}^3$ by the action of a discrete subgroup Γ of isometries. Of course, if we allow any subgroup Γ, then the resulting space $M = \mathbf{H}^3/\Gamma$ can be very wild at infinty. We will restrict our attention to geometrically finite groups.

Definition: A *fundamental domain*, F, for Γ is an open connected subset of $\mathbf{H}^3$ satisfying

1. no two elements of F are congruent under the action of Γ,

2. translates of $\overline{F}$ cover $\mathbf{H}^3$, i.e., $\bigcup_{\gamma\in\Gamma} \gamma\overline{F} = \mathbf{H}^3$.

Definition: A discrete group of isometries is *geometrically finite* if it has a fundamental domain bounded by finitely many geodesic hyperplanes.

From now on we will assume that $M = \mathbf{H}^3/\Gamma$, where Γ is geometrically finite. A model for M is given by a fundamental domain whose bounding hyperplanes are identified by the action of Γ. These bounding hyperplanes extend to the boundary at infinity for $\mathbf{H}^3$. The boundary at infinity for M, which we denote B, can be thought of as the part of the boundary of infinity for $\mathbf{H}^3$ enclosed by its intersection with these hyperplanes, with the bounding segments identified by the action of the group extended to the boundary at infinity (which is just the usual action of $SL(2,\mathbf{C})$ on the Riemann sphere).

The behaviour of M at infinity is determined by the way in which the bounding segments for B come together. This is related to the presence or absence of parabolic elements in Γ. A parabolic element of Γ has, by definition, exactly one fixed point when acting on the boundary at infinity.

If there are no parabolic elements, then the bounding segments all meet non-tangentially, and B is a smooth compact manifold. Near infinity, M always looks like a piece of $\mathbf{H}^3$, namely, $\{(x_1,x_2,x_3) \in \mathbf{H}^3 : x_1^2+x_2^2+x_3^2 < 1\}$.

If there is a parabolic element of maximal rank, then its fixed point is the point of tangency for two or more pairs of bounding segments. Such a fixed point is disconnected from the rest of the boundary at infinity, and its presence results in a finite rank change in the scattering matrix. So, for simplicity, we will assume that there are no parabolic elments of maximal rank.

If there is a parabolic element, γ, of non-maximal rank, then a single pair of the bounding segments of B meet tangentially at the fixed point of γ. The important point to notice is that the fixed point is not disconnected from the rest of the boundary at infinty. To simplify matters, we can change co-ordinates so that the fixed point of γ is the point at infinity in our upper half space model, and γ itself acts by translations, i.e., $\gamma(x_1, x_2, x_3) = (x_1 + 1, x_2, x_3)$. In these co-ordinates, the bounding hypersurfaces corresponding to γ are the planes parallel to the x_3 axis given by $x_1 = 1/2$ and $x_1 = -1/2$. Identifying these by the action of γ, we obtain a model for $\mathbf{H}^3/T$, where T is the group of integer translations generated by γ. Near the parabolic fixed point, M is isometric to a piece of $\mathbf{H}^3/T$ near the point at infinty, namely $\{(x_1, x_2, x_3) \in \mathbf{H}^3/T : x_2^2 + x_3^2 > 1\}$. We do not include the parabolic fixed point itself in the definition of B. With this understanding, B is a smooth, but non-compact, manifold.

It will be convenient to use another co-ordinate system near the parabolic fixed point. Let

$$\begin{aligned} b_1 &= x_1 \\ b_2 &= -x_2/(x_2^2 + x_3^2) \\ b_3 &= x_3/(x_2^2 + x_3^2) \end{aligned}$$

$$\begin{aligned} x_1 &= b_1 \\ x_2 &= -b_2/(b_2^2 + b_3^2) \\ x_3 &= b_3/(b_2^2 + b_3^2). \end{aligned} \tag{1}$$

These new co-ordinates can be extended to the boundary at infinty yielding

$$\begin{aligned} b_1 &= x_1 \\ b_2 &= -1/x_2 \end{aligned}$$

Of course, the circle $b_2 = 0$ is not in the range of this map. However, if we add this circle (at each parabolic fixed point), we obtain a smooth compactification of $\bar{B}$ of B.

The assumptions that we have made on M can be summed up as follows. There exists a finite open cover, $\{\mathcal{U}_0, \mathcal{U}_1, \ldots, \mathcal{U}_k\}$, of M, where $\mathcal{U}_0$ has compact closure and all the other $\mathcal{U}_i$'s are isometric either to the subset $\{(x_1, x_2, x_3) \in \mathbf{H}^3 : x_1^2+x_2^2+x_3^2 < 1\}$ of $\mathbf{H}^3$ or the subset $\{(x_1, x_2, x_3) \in \mathbf{H}^3/T : x_2^2 + x_3^2 > 1\}$ of $\mathbf{H}^3/T$.

The sets $\mathcal{U}_i$, for $i \geq 1$, can be extended to the boundary at infinity, to give a finite open cover for B consisting of discs and pairs of half cylinders. If each pair of cylinders is joined by adding a circle at infinity, we obtain an open cover for $\bar{B}$.

4 The Laplace Operator

The Laplacian on $\mathbf{H}^3$, in the co-ordinates of the upper half space model, is given by

$$\Delta = (x_3 D_3)^2 - 2x_3 D_3 + x_3^2(D_1^2 + D_2^2). \tag{2}$$

The spectrum of the Laplacian, acting in $L^2(\mathbf{H}^3, d\mu)$, is $[1, \infty)$ and is purely absolutely continuous. The Green's function, $G_0(u; v; s)$, which is the integral kernel of the resolvent $(\Delta - s(2-s))^{-1}$, can be computed explicitly. The result is

$$G_0(x; y; s) = \frac{\Gamma(s)}{(2s - s)\pi\Gamma(s-1)} \sigma^{-s} {}_2F_1(s, s-1, 2s+1, \sigma^{-1}),$$

where

$$\sigma(x, y) = 1 + \frac{|x-y|^2}{4x_3 y_3}$$

and ${}_2F_1$ is the hypergeometric function.

The Laplacian on $\mathbf{H}^3/T$ is also given by (2), except that the variable x_1 now ranges over the circle. Again, the spectrum is $[1, \infty)$ and purely absolutely continuous. The Green's function, $G_1(x, y, s)$, can be computed explicitly in terms of Bessel functions. The expression is given in [FHP2]. What is important for our purposes is that the asymptotics of $G_1(x, y, s)$ as x and y approach infinity can be computed.

The Laplacian in which we are interested is the Laplacian on $M = \mathbf{H}^3/\Gamma$. For this Laplacian, it is known [LP2-5] that there is absolutely continuous spectrum in the interval $[1, \infty)$ with no imbedded eigenvalues, and possibly discrete spectrum between 0 and 1. We want to find eigenfunctions corresponding to the continuous part of the spectrum.

The Laplacian on $\mathbf{H}^3/\Gamma$ can be thought of as acting on functions defined on $\mathbf{H}^3$ which are automorphic with respect to Γ and square integrable over any fundamental domain. This suggests that one might try to define automorphic continuum eigenfunctions for the Laplacian on $\mathbf{H}^3/\Gamma$ by starting with the explicitly computable eigenfunctions of the Laplacian on $\mathbf{H}^3$, and averaging them over the group. For example, if we take the eigenfunction, x_3^s, for the Laplacian on $\mathbf{H}^3$, we might expect that

$$\sum_{\gamma\in\Gamma}(\gamma x)_3^s \tag{3}$$

should be a corresponding eigenfunction for $\mathbf{H}^3/\Gamma$. This sum is an Eisenstein series. The trouble with it is that for the values of s corresponding to the continuous spectrum, namely $\mathrm{Re}(s) = 1$, the series does not converge. It does converge (in the presence of parabolics the sum is slightly modified) if $\mathrm{Re}(s)$ is large enough , and it is not hard to see that the eigenfunctions that we construct are equal to such a sum for large $\mathrm{Re}(s)$. So we will be providing a meromorphic continuation of such a sum to the entire complex plane. Actually, the question that we are able to avoid, namely, what is the borderline between convergence and divergence of the Eisenstein series, is interesting and has an been studied. It turns out that the borderline depends on the the position of the lowest eigenvalue of the Laplacian.

5 Green's Function Asymptotics and Eigenfunctions

We will obtain the continuum eigenfunctions from the asymptotic behaviour of the Green's function, $G(x; w; s)$ as w tends to infinity. The needed estimates are given in the following theorem. In a cusp neighbourhood of infinity there is an extra set of estimates which describes the behaviour at the singularity. This theorem, and the following ones, are not stated with complete precision here, since the details tend to obscure the essential ideas. Precise statements and proofs can be found in [FHP2].

Theorem 5.1 *The Green's function has the following asymptotic expansion, for* $\mathrm{Re}(s) \geq 1$ *and* $s \neq 1$.

1. *For w in any neighbourhood of infinity*

$$G(x; w; s) = w_3^s E(x; w_1, w_2; s) + O(w_3^{s+1})$$

$$\frac{\partial G}{\partial w_3}(x; w; s) = s w_3^{s-1} E(x; w_1, w_2; s) + O(w_3^s)$$

2. *In a cusp neighbourhood of infinity*

$$G(x; w(b); s) = w_3^{1/2} b_3^{s-1/2} F(x; b_1, b_2; s) + O(w_3^{1/2} b_3^{s+1/2})$$

$$\frac{\partial G}{\partial b_3}(x; w(b); s) = \frac{\partial w_3^{1/2} b_3^{s-1/2}}{\partial b_3} F(x; b_1, b_2; s) + O(w_3^{1/2} b_3^{s-1/2})$$

Note that unless $b_2 = 0$ the two sets of estimates really say the same thing, because then $b_3 \sim x_3$. In fact, it is not hard to see that

$$F(x; b_1, b_2; s) = |b_2|^{-2s+1} E(x; b_1, -1/b_2; s)$$

as long as $b_2 \neq 0$. The function F has a smooth extension over this singular point. However, this is not the whole story, as we will see.

The proof of Theorem 5.1 has two main ingredients. The first of these is the Mourre estimate, which says that for a suitable conjugate operator A,

$$E[\Delta, A]E \geq \alpha E + K$$

where E denotes the spectral projector for Δ corresponding to some interval contained in $(0, \infty)$, $\alpha > 0$ and K is compact. The Mourre estimate together with bounds on the commutator, $[\Delta, A]$, and the double commutator, $[[\Delta, A], A]$, implies that the operator $(A+1)^{-1}R(s)(A+1)^{-1}$ (here $R(s)$ denotes the resolvent $(\Delta - s(2-s))^{-1}$) is analytic in the half plane $\mathrm{Re}(s) > 1$, except for poles at the values of s for which $s(2-s)$ is an eigenvalue of Δ, and continuous onto the line $\mathrm{Re}(s) = 1$, except at $s = 1$ (see, e.g., [CFKS]). This is what is actually needed in the proof of Theorem 5.1. The Mourre estimate for geometrically finite hyperbolic manifolds (in any dimension) was proven in [FHP1].

The second ingredient in the proof is a geometric resolvent formula. This formula is derived from an analogue of the familiar Schrödinger operator resolvent equation

$$(H + V - z)^{-1} = (H - z)^{-1} - (H - z)^{-1}V(H + V - z)^{-1}.$$

It says that if x and w are in a neighbourhood of infinity, then the Greens function $G(x; w; s)$ can be written in the form

$$G(x; w; s) = G_0(x; w; s) + \langle K_{xs}, (A+1)^{-1}R(s)(A+1)^{-1}K_{ws}\rangle.$$

Here $G_0(x; w; s)$ is the Green's function for the model space for the neighbourhood of infinity under consideration. The second term contains the functions K_{xs}, which depend parametrically on x and s. Although the exact expressions for these somewhat messy, they are given in terms of the free Green's function. They can be shown to have L^2 asymptotic expansions

$$K_{ws} = w_3^s k_{w_1 w_2 s} + O(w_3^{s+1}).$$

where $k_{w_1 w_2 s}$ is an L^2 valued function, smooth in w_1, w_2 and analytic in s. In fact, it is not difficult to obtain an asymptotic expansion to any order, but this is not needed for our purposes. Thus, the geometric resolvent formula can be thought of as a way of turning operator-valued analyticity and continuity information for the operator $(A+1)^{-1}R(s)(A+1)^{-1}$ into pointwise information for the Green's function.

The functions $E(x;w_1,w_2;s)$ of Theorem 5.1, as functions of x, are P.D.E. eigenfunctions of the Laplace operator. To see this, start with the equation satisfied by the Green's function, namely,

$$(\Delta_x - s(2-s))G(x;w;s) = \delta(x,w).$$

Now multiply the equation by w_3^{-s} and take the limit as w_3 tends to zero. On the left side, we obtain $(\Delta_x - s(2-s))E(x;w_1,w_2;s)$, while on the right side, the singularity of the delta function gets pushed out to infinity, and equals zero in the limit.

As functions of w_1 and w_2, the functions $E(x;w_1,w_2;s)$ are defined locally on B—but not globally. This is because if we take two points on the boundary at in infinity for $\mathbf{H}^3$ which are related by an element in Γ, and thus represent the same element of B, and compute the limits which define E at these points, the results differ by a factor. This identifies $E(x;w_1,w_2;s)$, for fixed x and s, as a section of a line bundle, $\mathcal{M}_s$. A computation shows that if $e_s(w_1,w_2)$ (resp. $e_{2-s}(w_1,w_2)$) is a local expression for a section of $\mathcal{M}_s$ (resp. $\mathcal{M}_{2-s}$) then $e_s(w_1,w_2)e_{2-s}(w_1,w_2)dw_1 \wedge dw_2$ is a globally defined 2-form on B. Thus there is an L^2 pairing of sections of $\mathcal{M}_s$ and of $\mathcal{M}_{2-s}$. In particular, when $\mathrm{Re}(s)=1$, complex conjugation provides a natural map from sections of $\mathcal{M}_s$ to sections of $\mathcal{M}_{2-s}$, so we have an L^2 inner product naturally defined.

Theorem 5.1 can be used to prove the following functional equation for the Green's function.

Theorem 5.2 *For* $\mathrm{Re}(s)=1$,

$$G(x;y;s)-G(x;y;2-s) = (2-2s)\int_{\bar{B}} E(y;w_1,w_2;s)E(x;w_1,w_2;2-s)dw_1\wedge dw_2. \quad (4)$$

This identity is proven using Green's formula for a region bounded by a surface which is pushed out to infinity. The asymptotics in Theorem 5.1 are used to show that the boundary term in the limit is equal to the right side of (4).

This formula is very suggestive, since for $\mathrm{Re}(s)=1$, the left side is the limit onto the spectrum of the integral kernel of the imaginary part of the resolvent $R(s)=R(1+ik)=(\Delta-1-k^2)^{-1}$. This limit is related to the spectral projections of Δ through Stone's formula. Therefore, it is not surprising that eigenfunctions $E(x;w_1,w_2;1+ik)$ can be used to produce a continuum eigenfunction expansion for Δ. As is the case for Schrödinger operators, there are two expansion, corresponding to positive and negative k. In the following theorem, $\tilde{\Gamma}_2(\mathcal{M}_{1\pm ik})$ denotes a space of L^2 sections of $\mathcal{M}_{1\pm ik}$. For more details, see [FHP2].

Theorem 5.3 *The maps*

$$\mathcal{F}_\pm : E_c(\Delta)L^2(M, d\mu) \to \int_{(0,\infty)}^{\oplus} \tilde{\Gamma}_2(\mathcal{M}_{1\pm ik})dk$$

defined on the continuous subspace of Δ *in* $L^2(M, d\mu)$ *by*

$$(\mathcal{F}_\pm\psi)(k, w_1, w_2) = \sqrt{2/\pi k}\int_M E(u; w_1, w_2; 1 \pm ik)\psi(u)d\mu(u)$$

are isometries diagonalizing the Laplacian: for $\psi \in \mathcal{D}(\Delta) \cap E_c(\Delta)L^2(M, d\mu)$,

$$(\mathcal{F}_\pm\Delta\psi)(k, w_1, w_2) = (1 + k^2)(\mathcal{F}_\pm\psi)(k, w_1, w_2).$$

6 The Scattering Operator and Meromorphic Continuation

If we start with equation (4), multiply by x_3^{-s} and take the limit $x_3 \downarrow 0$, then, using the asymptotics of Theorem 5.1, we find that

$$E(u; x_1, x_2; s) = \lim_{x_3 \downarrow 0}\Big\{ x_3^{2-2s}E(u; x_1, x_2; 2 - s)$$
$$+(2 - 2s)\int_{\bar{B}} x_3^{-s}E(x; w_1, w_2; s)E(u; w_1, w_2; 2 - s)dw_1 \wedge dw_2\Big\}. \qquad (5)$$

Here is the idea behind the definition of the scattering operator. Replace $E(u; \cdot; 2-s)$ with an "arbitrary" section $e(\cdot\,; 2 - s)$ of $\mathcal{M}_{2-s}$ in the right side of (5), and call the resulting left side $S(s)e$. This defines the operator $S(s)$. Equation (5) can now be written

$$E(u; \cdot; s) = S(s)E(u; \cdot; 2 - s).$$

With a suitable definition of analyticity, we can show that $E(u; \cdot\,; s)$ is analytic for $\mathrm{Re}(s) > 1$, except for s such that $s(2-s)$ is an eigenvalue of Δ, and continuous onto the line $\mathrm{Re}(s) = 1$, except at $s = 1$. Furthermore, it is at least formally apparent that the operator $S(s)$ has an analytic continuation to $\mathrm{Re}(s) > 1$. If we can find a meromorphic inverse for $S(s)$ which is continuous onto the line $\mathrm{Re}(s) = 1$ then we will have that

$$S(s)^{-1}E(u; \cdot\,; s) = E(u; \cdot\,; 2 - s).$$

The left side will be meromorphic to the right of, and continuous onto the line $\mathrm{Re}(s) = 1$, except at $s = 1$, while the right side will be analytic to the left of the same line, except for s such that $s(2-s)$ is an eigenvalue for Δ, and continuous onto the line, except at $s=1$. This gives the desired meromorphic continuation.

Thus, to carry out the meromorphic continuation, we must study the scattering operator. Since the scattering operator is defined as the limit of integral operators whose integral kernels are constructed from Eisenstein series, $E(x;w_1,w_2;s)$, it is the limit of these functions, as x_3 tends to zero, which we must control. The limit will be singular on the diagonal $(x_1,x_2)=(w_1,w_2)$. It is the nature of this singularity which determines the crucial properties of the scattering operator. The geometric resolvent formula gives us a decomposition of $E(x;w_1,w_2;s)$ into a free part and a smooth part. The free part has the same singularity as the Eisenstein series for the appropriate model space, and can be determined explicitly.

In the absence of cusps, there is only one model space, and the singularity looks the same in every direction at infinity. In this case, the scattering operator turns out to be essentially Δ_{w_1,w_2}^{s-1}, where Δ_{w_1,w_2} is the Euclidean Laplacian in the w_1,w_2 variables. This operator can be inverted using the standard calculus of pseudo-differential operators.

When cusps of non-maximal rank are present, the situation is a bit more complicated. We will not give details here, but refer the reader to the paper [FHP2]. Here is a sketch of the main idea. Near the cusp, the Eisenstein series $E(x;w_1,w_2;s)$ is a periodic function of w_1. Make a Fourier decomposition with respect to this variable. Let $E^0(x;w_2;s)$ denote the zero Fourier mode and let $E^{\perp}(x;w_1,w_2;s)$ denotes the sum of the non-zero components. Acting on $E^{\perp}(x;w_1,w_2;s)$, the scattering operator has the Fourier representation $((2\pi n)^2+\xi^2)^{s-1}$ for $n\neq 0$. This is easy to invert, since for $n\neq 0$ it is strictly positive. To handle the zero mode, we use the fact that $E^0(x;w_2;s)=|w_2|^{-2s+1}F(x;1/w_2;s)$ for large w_2, where $F(x;b;s)$ is smooth in b. We then compute what the effect of the scattering operator has on F in this representation. It turns out that applying the scattering operator to E amounts to applying the operator $(d/db)^{s-1}$ to F.

With this information about the action of the scattering operator, we can go back and make a precise definition of what we mean by an "arbitrary" section. This is a space of sections whose non-zero Fourier modes decay near each cusp, and whose zero Fourier modes have the representation above. Acting on this space, it can be shown that the scattering operator can be inverted for $\mathrm{Re}(s)=1$ and that the inverse has a meromorphic continuation. Then the meromorphic continuation of the Eisenstein series follows, and we obtain the following result.

Theorem 6.1 *The Eisenstein series $E(u;x_1,x_2;s)$, for fixed u, x_1 and x_2 have meromorphic continuations from the half plane* $\mathrm{Re}(s)>1$ *to* $\mathbb{C}\backslash\{1\}$.

7 References

[A1] S. Agmon, *On the spectral theory of the Laplacian on non-compact hyperbolic manifolds,* Journées 'Équations aux derivées partielles' (Saint-Jean de Monts, 1987), Exp. No. XVII. École Polytechnique, Palaiseau, 1987.

[A2] S. Agmon, *Spectral Theory of Schrödinger Operators on Euclidean and Non-Euclidean Spaces* C.P.A.M. **39** (1986), Number S, Supplement.

[CFKS] H.L. Cycon, R.G. Froese, W. Kirsch and B. Simon, *Schrödinger Operators, with Application to Quantum Mechanics and Global Geometry,* Springer Texts and Monographs in Physics, Springer-Verlag, New York, 1986.

[F] L. Faddeev, *Expansion in eigenfunctions of the Laplace operator in the fundamental domain of a discrete group on the Labacevskii plane,* Trudy Moscow. Mat. Obshch. **17**, (1967), 323-350.

[FHP1] R. Froese, P. Hislop, P. Perry, *A Mourre Estimate and Related Bounds for Hyperbolic Manifolds with Cusps of Non-maximal Rank,* to appear in J. Funct. Anal.

[FHP2] R. Froese, P. Hislop, P. Perry, *The Laplace Operator on a Hyperbolic Three Manifold with Cusps of Non-Maximal Rank*, preprint.

[LP1] P. Lax and R.S. Phillips, *Scattering Theory for Automorphic Functions*, Annals of Math Studies **87**, Princeton University Press, 1976. [LP2-LP4] P. Lax and R.S. Phillips, *Translation representation for automorphic solutions of the non-Euclidean wave equation, I, II and III,* Comm. Pure Appl. Math. **37**, 303–328, 1984, **37**, 779–813, 1984, **38**, 179–208, 1985.

[LP5] P. Lax and R.S. Phillips, *Translation representation for automorphic solutions of the non-Euclidean wave equation, IV,* Preprint

[M] H. Maass, *Über eine neue Art von nichtanalytischen automorphen Functionen und die Bestimmumng Dirichletscher Reihen durch Functionalgleichungen,* Math. Ann. **121**, (1949), 141-183.

[Ma1] N. Mandouvalos, *The Theory of Eisenstein Series for Kleinian Groups,* In *The Selberg Trace Formula and Related Topics,* Hejhal, Sarnak and Terras eds., (Contemporary Mathematics **53**, 1986) 357–370.

[Ma2] N. Mandouvalos, *Scattering operator, inner product formula, and "Maass-Selberg" relations for Kleinian groups,* AMS Memoir **400**, 1989.

[Ma3] N. Mandouvalos, *Spectral theory and Eisenstein series for Kleinian groups,* Cambridge Unversity preprint, 1986.

[MM] R. Mazzeo and R. Melrose, *Meromorphic extension of the resolvent on complete spaces with asymptotically constant negative curvature,* J. Funct. Anal. **75**, 260–310,

1987.

[P] S.J. Patterson, *The Laplacian Operator on a Riemann Surface*, I, II and III, Compositio Math. **31** (1975), 83-107; **32** (1976),71-112, **33** (1976), 227-259.

[P] P. Perry, *The Laplace operator on a hyperbolic manifold, II. Eisenstein series and the scattering matrix,* J. Reine Angew. Math. **398**, 67-91, 1989.

[R] W. Roelcke, *Das Eigenwertproblem der automorphen Formen in der hyperbolischen Ebene,* Teil I, Math. Ann. **167** (1966), 292-337: Teil II, Math. Ann. **168**, (1967), 261-324.

[T] A. Terras, *Harmonic Analysis on Symmetric Spaces and Applications I*, Springer-Verlag, 1985.

THE METHOD OF DIFFERENTIAL INEQUALITIES

Anne Boutet de Monvel and Vladimir Georgescu *)

Laboratoire de Physique Mathématique et Géométrie
Université Paris VII - 2, place Jussieu, 75251 Paris Cedex 05

1. In this paper we shall present a variant of the method of differential inequalities which, although much simpler than the original method of Mourre, allows one in the case of the Laplace operator (for example) to get the limiting absorption principle in the Besov space of Agmon and Hörmander and to obtain a large class of locally smooth operators. Both the local behaviour and the behaviour at infinity of the operators in this class seems to be optimal. We insist in this paper on the simplicity of the method, which explains why we consider a very restricted class of hamiltonians. In particular, the Mourre estimate is not explicitely needed. A much more general variant of the method of differential inequalities (representing a direct development of Mourre's ideas), which enables us to obtain essentially optimal results in the N-body case, is presented in our paper *Locally Smooth Operators and Limiting Absorption Principle for N-Body Hamiltonians* (joint work with M.Mantoiu), which also contains references to other works. The method of the present note can be extended such as to give Hardy type inequalities for the Laplace operator in a very natural way (see our paper *Mourre Estimates and Hardy Type Inequalities* in collaboration with M.Mantoiu).

2. Throughout the abstract part of this paper we shall fix a triplet $(\mathcal{G}, \mathcal{H}, W)$ consisting of a (complex, separable) Hilbert space $\mathcal{H}$ (with norm denoted $\|\cdot\|$), a dense subspace $\mathcal{G}$ provided with a new, finner, scalar product which makes it a Hilbert space (the corresponding norm is $\|\cdot\|_1$) and a strongly continuous unitary group $W(\alpha) = e^{iA\alpha}$ in $\mathcal{H}$ which leaves $\mathcal{G}$ invariant, i.e. $W(\alpha)\mathcal{G} \subset \mathcal{G}$ for all $\alpha \in \mathbb{R}$.

*) Supported by C N R S
Permanent address: Institute of Mathematics of the Romanian Academy, Bucharest, ROMANIA

A. Boutet de Monvel et al. (eds.), Recent Developments in Quantum Mechanics, 279–298.

One can show that the restriction of $W(\alpha)$ to $\mathcal{G}$ is a C_0-group in $\mathcal{G}$; we shall not prove it here, because this fact is obvious in the applications we have in mind.

Let H be a selfadjoint operator in $\mathcal{H}$ with domain equal to $\mathcal{G}$ (by closed graph theorem $\|\cdot\|_1$ is equivalent to the graph-norm associated to H). We shall say that H *is A-homogeneous of degree* m for some $m \in \mathbb{R}$ if

$$W(\alpha)\, H\, W(-\alpha) = e^{m\alpha}H \quad \text{for all } \alpha \in \mathbb{R} \quad . \tag{1}$$

Obviously, this is formally (in fact rigorously, if suitably interpreted) equivalent to $[iA,H] = mH$. If $m \neq 0$, then $a = \frac{i}{m} A$ and $b = H$ may be thought as the generators of a solvable two-dimensional Lie algebra, the only nontrivial commutation relation being $[a,b] = b$. We shall prove below that in this case ($m \neq 0$) the spectrum of H is purely absolutely continuous outside zero. A natural question may be raised now: can one generalize the results and techniques which follow to a wider class of Lie algebras ? We shall restrain ourselves to the case (1) although the method which we present below extends easily to larger classes of Lie algebras (an example will be given later).

3. Let H be a selfadjoint operator in $\mathcal{H}$ with domain $\mathcal{G}$. Assume H is A-homogeneous of degree $m \neq 0$. For pedagogical reasons, we shall begin with the most elementary form of our method. Denote $R(z) = (H-z)^{-1}$ for z a complex number not in the spectrum of H. Then:

$$\frac{d}{dz} R(z) = R(z)^2 \quad . \tag{2}$$

On the other hand, a formal calculation gives:

$$\begin{aligned}[iA,\ R(z)] &= -\ R(z)[iA,H]R(z) = -\ mR(z)HR(z) = \\ &= -\ mz\ R(z)^2 - mR(z) \quad .\end{aligned} \tag{3}$$

For a rigorous proof, observe that (1) is equivalent to:

$$W(\alpha)\ R(z)\ W(-\alpha) = e^{-m\alpha}\ R(e^{-m\alpha}z) \tag{4}$$

and then take the derivative with respect to α at $\alpha = 0$. From (2) and (3) we get, assuming $z \neq 0$:

$$\frac{d}{dz} R(z) = \frac{1}{imz}\ [A,\ R(z)] - \frac{1}{z} R(z) \quad . \tag{5}$$

Let us consider now any bounded operator L in $\mathcal{H}$ such that $L(\mathcal{H}) \subset D(A)$ (domain of A in $\mathcal{H}$). Then AL and L*A are bounded operators in $\mathcal{H}$ and (5) implies:

$$\frac{d}{dz} L^*R(z)L = \frac{1}{imz} (L^*A\cdot R(z)L - L^*R(z)\cdot AL) - \frac{1}{z} L^*R(z)L \quad . \tag{6}$$

Assume $|z| \geq a > 0$; since $\|L^*A\| = \|AL\|$ we get:

$$\|\frac{d}{dz} L^*R(z)L\| \leq \frac{\|AL\|}{a|m|} (\|R(z)L\| + \|L^*R(z)\|) + \frac{1}{a} \|L^*R(z)L\| \quad . \tag{7}$$

The next step consists in using the so-called *quadratic estimates* of Mourre. Namely, for any bounded operator L and for any $z \in \mathbb{C}$ with $\mathrm{Im} z \neq 0$:

$$\|R(z)L\| = \|L^*R(z)\| \leq \left[\frac{\|L^*R(z)L\|}{|\mathrm{Im} z|}\right]^{1/2} \quad . \tag{8}$$

The proof of this is trivial: taking $z = \lambda + i\mu$, $\mu > 0$, we have

$$\|R(\lambda\pm i\mu)L\|^2 = \|L^*R(\lambda\pm i\mu)^*R(\lambda\pm i\mu)L\| =$$

$$= \frac{1}{2\mu} \|L^*\left[R(\lambda+i\mu) - R(\lambda-i\mu)\right]L\| \leq \frac{1}{\mu} \|L^*R(\lambda+i\mu)L\| \quad .$$

Using (8) into (7) we get a so-called *differential inequality*: if $|\mathrm{Re} z| \geq a > 0$ then:

$$\|\frac{d}{dz} L^*R(z)L\| \leq \frac{2\|AL\|}{a|m|\sqrt{|\mathrm{Im} z|}} \|L^*R(z)L\|^{1/2} + \frac{1}{a} \|L^*R(z)L\| \quad . \tag{9}$$

We shall use this estimate in order to prove:

Proposition 1: *Assume that the selfadjoint operator* H *with domain* $\mathcal{G}$ *in* $\mathcal{H}$ *is* A*-homogeneous of degree* $m \neq 0$. *Then for any bounded operator* L *in* $\mathcal{H}$ *such that* $L(\mathcal{H})$ *is included in the domain of* A *in* $\mathcal{H}$ *and for any* $a > 0$ *there is a constant* $C < \infty$ *such that for* $|\mathrm{Re} z| \geq a$, $\mathrm{Im} z \neq 0$:

$$\|L^*R(z)L\| \leq C \quad . \tag{10}$$

In particular L *is* H*-smooth on* $\mathbb{R}\backslash(-a,+a)$ *for any* $\varepsilon > 0$ *and* H *has no singularly continuous spectrum and no non-zero eigenvalues.*

Proof: Let $z = \lambda + i\mu$, $|\lambda| \geq a$, $\mu > 0$ and $S_\mu = L^*R(\lambda+i\mu)L$. Then $S'_\mu \equiv \frac{d}{d\mu} S_\mu = i \frac{d}{dz} L^*R(z)L\Big|_{z=\lambda+i\mu}$, hence there is a constant $b < \infty$ such that

$$\|S'_\mu\| \leq \frac{b}{\sqrt{\mu}} \|S_\mu\|^{1/2} + b\|S_\mu\| \quad . \tag{11}$$

It is clear that $\|S_\mu\| \leq \|L\|^2\mu^{-1}$. Replacing this in the right-hand side of (11), we get after integration $\|S_\mu\| \leq$ const $|\ln\mu|$. Now we repeat the procedure and get $\|S_\mu\| \leq$ const after integration.

Assume that φ is a vector in the range of L. Using (10) and (8) we get $\|R(\lambda+i\mu)\varphi\| \leq c\,\mu^{-1/2}$ for some constant c and all $\lambda \in \mathbb{R}\setminus(-a,+a)$, all $\mu > 0$. This may be written:

$$\int \frac{\mu}{(t-\lambda)^2+\mu^2} \|E(dt)\psi\|^2 \leq c \quad .$$

It is well-known that this implies the absolute continuity of the measure $\|E(dt)\psi\|^2$ on $\mathbb{R}\setminus(-a,+a)$. 𝔔 𝔈 𝔇

In order to make a *smooth* transition to section **4**, it is convenient to present the estimate (7) in a slightly different way. Let us denote $|\varphi\rangle\langle\psi|$ the operator of rank one in $\mathcal{H}$ defined by $u \mapsto \varphi\langle\psi,u\rangle$, where φ,ψ are arbitrary vectors in $\mathcal{H}$. Observe that $\||\varphi\rangle\langle\psi|\| = \|\varphi\|\cdot\|\psi\|$ and $(|\varphi\rangle\langle\psi|)^* = |\psi\rangle\langle\varphi|$. If S is an operator in $\mathcal{H}$ whose domain contains φ then $S|\varphi\rangle\langle\psi| = |S\varphi\rangle\langle\psi|$. Now let L be as before and ψ be an arbitrary vector in $\mathcal{H}$. Then $|L\psi\rangle\langle\psi|$ is a bounded operator in $\mathcal{H}$ whose image is contained in D(A) (because $L\psi \in D(A)$). Hence we may replace the operator L in (7) by $|L\psi\rangle\langle\psi|$. Then $L^*R(z)L$ will become $\langle L\psi, R(z)L\psi\rangle \cdot |\psi\rangle\langle\psi|$ whose norm is $|\langle L\psi, R(z)L\psi\rangle|\cdot\|\psi\|^2$, etc. We shall obtain:

$$\left|\frac{d}{dz} \langle L\psi, R(z)L\psi\rangle\right| \leq \frac{\|AL\psi\|}{a|m|} (\|R(z)L\psi\| + \|R(z)^*L\psi\|) + \frac{1}{a} |\langle L\psi, R(z)L\psi\rangle| \quad . \tag{12}$$

If we want to get a version of (9) in this context it is better to use in place of (8) the estimate:

$$\|R(z)\varphi\| = \|R(z)^*\varphi\| \leq \left|\frac{\langle\varphi, R(z)\varphi\rangle}{\mathrm{Im} z}\right|^{1/2} \tag{13}$$

for all $\varphi \in \mathcal{H}$ (this follows from (8) by taking $L = |\varphi\rangle\langle\varphi|$). In section 6. we shall follow such a procedure, but we shall use a consequence of (13) which we state below. Let κ be a finite constant such that $\|u\|_1 \le \kappa(\|u\| + \|Hu\|)$ (recall that $\|\cdot\|_1$ is equivalent to the graph-norm associated to H). Observe that $R(z)\varphi \in \mathcal{G}$ if $\varphi \in \mathcal{H}$ and $\text{Im}z \neq 0$. We have then:

$$\|R(z)\varphi\|_1 + \|R(\bar{z})\varphi\|_1 \le 2\kappa\|\varphi\| + 2\kappa(1+|z|) \left|\frac{\langle\varphi, R(z)\varphi\rangle}{\text{Im}z}\right|^{1/2} \tag{14}$$

In fact:

$$\|R(z)\varphi\|_1 \le \kappa\|R(z)\varphi\| + \kappa\|HR(z)\varphi\| \le \kappa\|R(z)\varphi\| + \kappa\|(H-z)R(z)\varphi\| +$$

$$+ \kappa|z|\|R(z)\varphi\| = \kappa\|\varphi\| + \kappa(1 + |z|) \|R(z)\varphi\|$$

and then we use (13).

4. We have proved the above proposition only in order to explain the main idea of the method of differential inequalities in a simple although non-trivial situation. We have found many H-smooth operators and as a consequence we get the limiting absorption principle in $B(\mathcal{E},\mathcal{E}^*)$ if $\mathcal{E}$ is the Hilbert space defined by: $\mathcal{E} = L(\mathcal{H})$, $\|u\|_{\mathcal{E}} = \|L^{-1}u\|_{\mathcal{H}}$, where L is assumed injective with dense range (observe that we shall have $\mathcal{E} \subset \mathcal{H}$ continuously and densely so that, if $\mathcal{E}^*$ is the adjoint Hilbert space of $\mathcal{E}$ and if we identify $\mathcal{H} = \mathcal{H}^*$ by Riesz lemma, then $\mathcal{E} \subset \mathcal{H} \subset \mathcal{E}^*$). Now we shall consider a variant of the above proof which allows us to get Besov space estimates and an optimal class of locally smooth operators.

Let us fix $z = \lambda+i\mu$ with λ in a compact subset of $\mathbb{R}\setminus\{0\}$ and $0 < \mu < 1$. We shall consider then an auxiliary parameter $\varepsilon > 0$ and denote

$$G_\varepsilon = R(z + i\varepsilon) = R(\lambda + i(\varepsilon+\mu)) \quad .$$

(5) clearly implies

$$\frac{d}{d\varepsilon} G_\varepsilon = \frac{1}{m(z+i\varepsilon)} [A,G_\varepsilon] - \frac{i}{z+i\varepsilon} \quad G_\varepsilon \quad . \tag{15}$$

It is obvious that the arguments of section 3 give $\|L^*G_\varepsilon L\| \le$ const for $\varepsilon \ge 0$, the constant being independent of λ and μ. Since $G_o = R(z)$, we get (10).

The advantage of this new variant is that we can replace L by an ε-dependent family of bounded operators L_ε with divergent norm as $\varepsilon \to 0$. Then we could consider, as at the end of section 3., the quantity $\langle\psi, L_\varepsilon^* R(z+i\varepsilon)L_\varepsilon\psi\rangle = \langle L_\varepsilon\psi, G_\varepsilon L_\varepsilon\psi\rangle$ and try to establish a differential inequality (generalising (12)) in order to study its behaviour as $\varepsilon \to 0$. Here ψ is an arbitrary vector in $\mathcal{H}$. It is, however, simpler to consider in place of $L_\varepsilon\psi$ a more general family ψ_ε of vectors in $\mathcal{H}$. So, let us consider a function $(0,1] \ni \varepsilon \mapsto \psi_\varepsilon \in \mathcal{H}$ which is weakly C^1. Denoting G'_ε, ψ'_ε derivatives with respect to ε, we obtain using (15):

$$\frac{d}{d\varepsilon}\langle\psi_\varepsilon, G_\varepsilon\psi_\varepsilon\rangle = \langle\psi'_\varepsilon, G_\varepsilon\psi_\varepsilon\rangle + \langle G_\varepsilon^*\psi_\varepsilon, \psi'_\varepsilon\rangle -$$

$$- \frac{i}{z+i\varepsilon}\langle\psi_\varepsilon, G_\varepsilon\psi_\varepsilon\rangle + \frac{1}{m(z+i\varepsilon)}\langle\psi_\varepsilon, [A, G_\varepsilon]\psi_\varepsilon\rangle \quad . \tag{16}$$

(observe that, due to (3), $[A, G_\varepsilon]$ is a bounded operator). Assume for a moment that ψ_ε belongs to the domain of A (in $\mathcal{H}$). Since $[A, G_\varepsilon]$ is the derivative of $-i\ W(\alpha)G_\varepsilon\ W(-\alpha)$ with respect to α at $\alpha=0$, it is clear that:

$$\langle\psi_\varepsilon, [A, G_\varepsilon]\psi_\varepsilon\rangle = -i\frac{d}{d\alpha}\langle W(-\alpha)\psi_\varepsilon, G_\varepsilon W(-\alpha)\psi_\varepsilon\rangle\Big|_{\alpha=0} =$$

$$= \lim_{\alpha\to 0}\left\langle\frac{W(-\alpha)-1}{-i\alpha}\psi_\varepsilon, G_\varepsilon W(-\alpha)\psi_\varepsilon\right\rangle + \tag{17}$$

$$+ \lim_{\alpha\to 0}\left\langle G_\varepsilon^*\psi_\varepsilon, \frac{W(-\alpha)-1}{i\alpha}\psi_\varepsilon\right\rangle = \langle A\psi_\varepsilon, G_\varepsilon\psi_\varepsilon\rangle - \langle G_\varepsilon^*\psi_\varepsilon, A\psi_\varepsilon\rangle .$$

However, such an identity may be established under more general conditions on ψ_ε using a more sophisticated formalism. Since this is very important in applications, we shall stop now the argument in order to introduce some new objects.

5. For any Hilbert space $\mathcal{F}$ we denote $\mathcal{F}^*$ the *adjoint space*, namely the Hilbert space defined as follows: it is the vector space of all anti-linear continuous mappings φ: $\mathcal{F} \to \mathbb{C}$ provided with the dual norm $\|\varphi\|_{\mathcal{F}^*} = \sup\{|\varphi(u)| \mid u \in \mathcal{F}, \|u\|_{\mathcal{F}} \le 1\}$. We put $\varphi(u) = \langle u,\varphi\rangle = \overline{\langle\varphi, u\rangle}$ if $u \in \mathcal{F}$ and $\varphi \in \mathcal{F}^*$. If $(\cdot,\cdot)_{\mathcal{F}}$ is the scalar product of $\mathcal{F}$, then $u \longmapsto (\cdot,u)_{\mathcal{F}}$ is a linear, bijective, isometric mapping of $\mathcal{F}$ onto $\mathcal{F}^*$

(this is Riesz lemma). If this map is used to identify $\mathcal{F}$ and $\mathcal{F}^*$, then $\langle u,\varphi\rangle = (u,\varphi)_{\mathcal{F}}$. But we shall not make this identification in general. On the other hand, we always identify $\mathcal{F}^{**} = (\mathcal{F}^*)^* = \mathcal{F}$ (this is implicit in the notation $\langle\varphi,u\rangle = \overline{\langle u,\varphi\rangle}$ if $u \in \mathcal{F}$, $\varphi \in \mathcal{F}^*$); in fact, each continuous, anti-linear functional on $\mathcal{F}^*$ can be written as $\varphi \longmapsto \overline{\langle u,\varphi\rangle}$ for some $u \in \mathcal{F}$ and $\|u\|_{\mathcal{F}}$ equals the norm of this functional.

Recall that $\mathcal{G}$ is a dense subspace of $\mathcal{H}$ provided with a stronger norm $\|\cdot\|_1$ which makes it a Hilbert space. We have denoted $\langle\cdot,\cdot\rangle$, $\|\cdot\|$ the scalar product and the norm of $\mathcal{H}$. We shall identify $\mathcal{H} = \mathcal{H}^*$ by the procedure described above. Then, if $i\colon \mathcal{G} \longrightarrow \mathcal{H}$ is the embedding, then i is linear, continuous and with dense range; so the adjoint mapping $i^*\colon \mathcal{H}^* = \mathcal{H} \longrightarrow \mathcal{G}^*$ has the same properties. We shall use i^* in order to identify $\mathcal{H}$ with a dense subspace of $\mathcal{G}^*$. Let $\|\cdot\|_{-1}$ be the norm of $\mathcal{G}^*$, hence $\|\cdot\|$ is a norm on the dense subspace $\mathcal{H}$ of $\mathcal{G}^*$, stronger than $\|\cdot\|_{-1}$. We have obtained a triplet of spaces: $\mathcal{G} \subset \mathcal{H} \subset \mathcal{G}^*$, each space being a Hilbert space continuously and densely embedded in the next one. The second embedding may be explicitely described as follows: each $\varphi \in \mathcal{H}$ determines a continuous anti-linear functional $u \mapsto \langle u,\varphi\rangle$ on $\mathcal{G}$, hence an element of $\mathcal{G}^*$ which is denoted by the same symbol φ. Observe that now the bracket $\langle u,\varphi\rangle$ may be interpreted as scalar product in $\mathcal{H}$ if $u,\varphi \in \mathcal{H}$, or as the action $\varphi(u)$ of φ on u if $\varphi \in \mathcal{G}^*$ and $u \in \mathcal{G}$, or as $\overline{\langle\varphi,u\rangle} = \overline{u(\varphi)}$ if $u \in \mathcal{G}^*$ and $\varphi \in \mathcal{G}$. We shall then have $|\langle u,\varphi\rangle| \le \|u\|\cdot\|\varphi\|$, or $|\langle u,\varphi\rangle| \le \|u\|_1\cdot\|\varphi\|_{-1}$, or $|\langle u,\varphi\rangle| \le \|u\|_{-1}\cdot\|\varphi\|_1$ respectively.

Let us consider the strongly continuous unitary group $W(\alpha) = e^{iA\alpha}$ in $\mathcal{H}$, cf. beginning of section 2. We already said that the assumption $W(\alpha)\mathcal{G} \subset \mathcal{G}$ for all α implies $W(\alpha) \in B(\mathcal{G})$ (i.e. the restriction of $W(\alpha)$ to $\mathcal{G}$ is a continuous linear operator in $\mathcal{G}$; this is trivial by closed graph theorem) and also assures the continuity of the function $\mathbb{R} \ni \alpha \mapsto W(\alpha)u \in \mathcal{G}$ for each $u \in \mathcal{G}$ (this is not quite trivial). Hence W induces a C_0-group in $\mathcal{G}$ by restriction. Denote for the moment $W_{\mathcal{G}}(\alpha) = W(\alpha)|\mathcal{G}$ considered as operator in $\mathcal{G}$. So $W_{\mathcal{G}}(\alpha) \in B(\mathcal{G})$ and we may consider its adjoint $W_{\mathcal{G}}(\alpha)^* \in B(\mathcal{G}^*)$. It is easily shown that $W_{\mathcal{G}}(\alpha)^*|\mathcal{H} = W(-\alpha)$ and that $\alpha \mapsto W_{\mathcal{G}}(\alpha)^* \in B(\mathcal{G}^*)$ is weakly (hence strongly) continuous. It follows that $W_{\mathcal{G}^*}(\alpha) \overset{d}{=} W_{\mathcal{G}}(-\alpha)^*$ is a C_o-group in $\mathcal{G}^*$. The groups W and $W_{\mathcal{G}}$ may now be obtained by restriction of $W_{\mathcal{G}^*}$ to $\mathcal{H}$, respectively $\mathcal{G}$. For this reason we shall denote these three groups by the same symbol W.

Each C_o-semigroup in a Banach space has a uniquely defined infinitesimal generator, which is a closed, linear, densely defined operator. For W in $\mathcal{H}$, this generator is the selfadjoint operator A in $\mathcal{H}$ (with domain D(A)) defined by the condition $W(\alpha) = e^{iA\alpha}$. Let $A_{\mathcal{G}}$ (respectively $A_{\mathcal{G}*}$) be the generator of $W_{\mathcal{G}}$ (respectively $W_{\mathcal{G}*}$), hence $A_{\mathcal{G}}$ (respectively $A_{\mathcal{G}*}$) is a densely defined, closed operator in $\mathcal{G}$ (respectively $\mathcal{G}^*$) such that $W_{\mathcal{G}}(\alpha)= e^{i\alpha A_{\mathcal{G}}}$ (respectively $W_{\mathcal{G}*}(\alpha)= e^{i\alpha A_{\mathcal{G}*}}$) in the sense of semigroup theory. One can then show:

1) $D(A_{\mathcal{G}}) = \{u \in \mathcal{G} \mid u \in D(A) \text{ and } Au \in \mathcal{G}\}$,

$$u \in D(A_{\mathcal{G}}) \Longrightarrow A_{\mathcal{G}}u = Au \quad ;$$

2) $D(A) = \{u \in \mathcal{H} \mid u \in D(A_{\mathcal{G}*}) \text{ and } A_{\mathcal{G}*}u \in \mathcal{H}\}$,

$$u \in D(A) \Longrightarrow Au = A_{\mathcal{G}*}u \quad .$$

Hence A and $A_{\mathcal{G}}$ are restrictions of the same operator $A_{\mathcal{G}*}$ acting in $\mathcal{G}^*$. We shall denote by the same symbol A these three operators and we shall denote $D(A;\mathcal{G}^*)$ (respectively $D(A;\mathcal{H})$, $D(A;\mathcal{G})$) its domain in $\mathcal{G}^*$ (respectively $\mathcal{H}$, $\mathcal{G}$). Clearly:

$$D(A;\mathcal{H}) = \{u \in \mathcal{H} \cap D(A;\mathcal{G}^*) \mid Au \in \mathcal{H}\} \quad ,$$

$$D(A;\mathcal{G}) = \{u \in \mathcal{G} \cap D(A;\mathcal{G}^*) \mid Au \in \mathcal{G}\} \quad .$$

An explicit definition is:

$$D(A;\mathcal{G}^*) = \left\{ u \in \mathcal{G}^* \mid \begin{array}{l} \text{the function } \alpha \longmapsto W(\alpha)u \in \mathcal{G}^* \\ \text{is derivable at } \alpha = 0 \end{array} \right\} \qquad (18)$$

$$Au = -i \frac{d}{d\alpha} W(\alpha)u \Big|_{\alpha=0} \quad ,$$

and similarly for $\mathcal{G}^*$ replaced by $\mathcal{H}$ or $\mathcal{G}$.

6. Let us go back to the context of section **4**. Let us show that *the identity (17) is still valid if the vector ψ_ε belongs* to $D(A;\mathcal{G}^*)$, the domain of A in $\mathcal{G}^*$ (from now on we assume that this is fulfilled). In fact, since $\psi_\varepsilon \in D(A;\mathcal{G}^*)$ and $G^*_\varepsilon \psi_\varepsilon \in \mathcal{G}$ (because $\psi_\varepsilon \in \mathcal{H}$ also) we shall have (see (18)):

$$\lim_{\alpha\to 0} \langle G^*_\varepsilon \psi_\varepsilon, \frac{W(-\alpha) - 1}{i\alpha} \psi_\varepsilon\rangle = - \langle G^*_\varepsilon\psi_\varepsilon, A\psi_\varepsilon\rangle$$

where the bracket has to be interpreted as the action of the element $A\psi_\varepsilon \in \mathcal{G}^*$ onto $G^*_\varepsilon\psi_\varepsilon \in \mathcal{G}$. Similarly, observe that $G_\varepsilon W(-\alpha)\, \psi_\varepsilon$ is convergent in $\mathcal{G}$ as $\alpha \to 0$ to $G_\varepsilon\psi_\varepsilon$ hence (18) implies again:

$$\lim_{\alpha\to 0} \langle\frac{W(-\alpha) - 1}{- i\alpha} \psi_\varepsilon , G_\varepsilon W(-\alpha)\psi_\varepsilon\rangle = \langle A\psi_\varepsilon, G_\varepsilon\psi_\varepsilon\rangle .$$

This clearly proves (17), the brackets in the last member being interpreted as anti-dualities between $\mathcal{G}$ and $\mathcal{G}^*$, cf. the explanations from section 5.

Replacing (17) in (16) we obtain then:

$$\frac{d}{d\varepsilon} \langle\psi_\varepsilon, G_\varepsilon\psi_\varepsilon\rangle = \langle\psi'_\varepsilon, G_\varepsilon\psi_\varepsilon\rangle + \langle G^*_\varepsilon\psi_\varepsilon, \psi'_\varepsilon\rangle$$

$$- \frac{i}{z+i\varepsilon} \langle\psi_\varepsilon, G_\varepsilon\psi_\varepsilon\rangle + \frac{1}{m(z+i\varepsilon)} [\langle A\psi_\varepsilon, G_\varepsilon\psi_\varepsilon\rangle - \langle G^*_\varepsilon \psi_\varepsilon, A\psi_\varepsilon\rangle] . \quad (19)$$

Since $|z+i\varepsilon| \geq |z|$ for $\varepsilon > 0$ (recall $\mathrm{Im} z = \mu > 0$), we may estimate the right-hand side such as to get:

$$|\frac{d}{d\varepsilon} \langle\psi_\varepsilon, G_\varepsilon\psi_\varepsilon\rangle| \leq \frac{1}{|z|} |\langle\psi_\varepsilon, G_\varepsilon\psi_\varepsilon\rangle| +$$

$$+ (\|\psi'_\varepsilon\|_{-1} + \frac{1}{|mz|} \|A\psi_\varepsilon\|_{-1})\cdot(\|G_\varepsilon\psi_\varepsilon\|_1 + \|G^*_\varepsilon\psi_\varepsilon\|_1) . \quad (20)$$

Now we shall use the quadratic estimate under the form (14). Since $G_\varepsilon = R(z+i\varepsilon)$ we get:

$$\|G_\varepsilon\psi_\varepsilon\|_1 + \|G^*_\varepsilon\psi_\varepsilon\|_1 \leq 2\kappa\|\psi_\varepsilon\| + 2\kappa(1+|z|) \left|\frac{\langle\psi_\varepsilon , G_\varepsilon\psi_\varepsilon\rangle}{\varepsilon+\mu}\right|^{1/2} .$$

Replacing in (20) we obtain (recall $\mu > 0$):

$$|\frac{d}{d\varepsilon} \langle\psi_\varepsilon, G_\varepsilon\psi_\varepsilon\rangle| \leq 2\kappa\|\psi_\varepsilon\| \left(\|\psi'_\varepsilon\|_{-1} + \frac{1}{|mz|} \|A\psi_\varepsilon\|_{-1}\right) +$$

$$+ 2\kappa(1+|z|)\, \varepsilon^{-1/2}\left(\|\psi'_\varepsilon\|_{-1} + \frac{1}{|mz|}\|A\psi_\varepsilon\|_{-1}\right) |\langle\psi_\varepsilon, G_\varepsilon\psi_\varepsilon\rangle|^{1/2} +$$

$$+ \frac{1}{|z|} |\langle\psi_\varepsilon, G_\varepsilon\psi_\varepsilon\rangle| . \quad (21)$$

It is convenient to use this estimate in a less precise form. Since $z = \lambda + i\mu$ with λ in a compact subset of $\mathbb{R}\setminus\{0\}$ and $0 < \mu < 1$ there is a constant $c > 0$ such that for all $\varepsilon \in (0,1)$:

$$c\left|\frac{d}{d\varepsilon} \langle\psi_\varepsilon, G_\varepsilon\psi_\varepsilon\rangle\right| \le \|\psi_\varepsilon\|(\|\psi'_\varepsilon\|_{-1} + \|A\psi_\varepsilon\|_{-1}) + $$

$$+ \varepsilon^{1/2}(\|\psi'_\varepsilon\|_{-1} + \|A\psi_\varepsilon\|_{-1}) \; |\langle\psi_\varepsilon, G_\varepsilon\psi_\varepsilon\rangle|^{1/2} + |\langle\psi_\varepsilon, G_\varepsilon\psi_\varepsilon\rangle| \quad . \tag{22}$$

Let us stress the fact that c is independent of the family $\{\psi_\varepsilon\}$. (22) *is our main differential inequality.*

An implicit estimate of the function $\langle\psi_\varepsilon, G_\varepsilon\psi_\varepsilon\rangle$ such as (22) can be transformed into an explicit estimate by using a version of Gronwall lemma which we state below.

Proposition 2: *Let* $f\colon (0,1] \to \mathbb{C}$ *be a continuously differentiable function such that:*

$$|f'(\varepsilon)| \le a(\varepsilon) + b(\varepsilon) \; |f(\varepsilon)|^{1/2} + c(\varepsilon)|f(\varepsilon)|$$

for some positive, locally integrable functions a, b, c *on* [0,1]. *Then for all* $\varepsilon \in (0,1)$:

$$|f(\varepsilon)| \le 2\left[|f(1)| + \int_\varepsilon^1 a(t)dt + \left(\int_\varepsilon^1 b(t)dt\right)^2\right] \exp \int_\varepsilon^1 c(t)dt \; . \tag{23}$$

In particular, if a, b, c *are integrable on* (0,1), *then* $f(0) = \lim_{\varepsilon\to 0} f(\varepsilon)$ *exists and the preceding inequality is valid for* $\varepsilon = 0$ *also.*

This result is a consequence of corollary 4.4 from chapter 3 of *Ordinary Differential Equations* by Ph. Hartman (second edition). For a simple, direct proof, see our first paper cited in section 1 above. The main point of estimate (23) is that the coefficient b enters only quadratically (the usual form of Gronwall lemma would give an exponential dependence on b).

If we use proposition 2 in connection with (22) we obtain our main technical result:

Proposition 3: *Assume that the self-adjoint operator* H *with domain* $\mathcal{G}$ *in* $\mathcal{H}$ *is* A*-homogeneous of degree* $m \ne 0$. *Then for each compact subset* Λ *of* $\mathbb{R}\setminus\{0\}$ *there is a constant* $c > 0$ *such that for* $z = \lambda + i\mu$, $\lambda \in \Lambda$, $0 < \mu < 1$, *and for all families* $\{\psi_\varepsilon\}_{0<\varepsilon\le 1}$ *of vectors in* $\mathcal{H}$ *such that*

$\varepsilon \mapsto \psi_\varepsilon \in \mathcal{H}$ *is (weakly)* C^1 *and* $\psi_\varepsilon \in D(A;\mathcal{G}^*)$ (see section 5) *the following estimate is satisfied* $(0 < \varepsilon < 1)$:

$$c|\langle\psi_\varepsilon, R(z+i\varepsilon)\psi_\varepsilon\rangle| \le \|\psi_1\|^2 + \int_\varepsilon^1 \|\psi_\tau\| \left[\|\psi'_\tau\|_{-1} + \|A\psi_\tau\|_{-1}\right] d\tau$$

$$+ \left(\int_\varepsilon^1 \left[\|\psi'_\tau\|_{-1} + \|A\psi_\tau\|_{-1}\right] \frac{d\tau}{\sqrt{\tau}} \right)^2 \quad . \tag{24}$$

7. In the context of proposition 3, we denote Ψ the family $\{\psi_\varepsilon\}_{0<\varepsilon\le 1}$ and:

$$N(\Psi) = \sup_{0<\varepsilon<1} \sqrt{\varepsilon}\|\psi_\varepsilon\| + \int_0^1 \left[\|\psi'_\varepsilon\|_{-1} + \|A\psi_\varepsilon\|_{-1}\right] \frac{d\varepsilon}{\sqrt{\varepsilon}} \quad . \tag{25}$$

We then see that (24) implies for some constant c:

$$|\langle\psi_\varepsilon, R(z+i\varepsilon)\psi_\varepsilon\rangle| \le c\, N(\Psi)^2 \quad . \tag{26}$$

We would like to have such an estimate for the more general object $\langle\psi^1_\varepsilon, R(z+i\varepsilon)\psi^2_\varepsilon)$ where the families $\{\psi^j_\varepsilon\}_{0<\varepsilon\le 1}$ have the same properties as $\{\psi_\varepsilon\}$. For this we use the simple:

Lemma 1: *Let* V *be a complex vector space provided with a seminorm* $|\cdot|$ *which may take the value* $+\infty$. *Let* $Q\colon V \times V \to \mathbb{C}$ *be a sesquilinear form such that* $|Q(v,v)| \le |v|^2$ *for all* $v \in V$. *Then* $|Q(v_1,v_2| \le 4|v_1||v_2|$ *for all* $v_1, v_2 \in V$.

Proof: Let $q(v) = Q(v,v)$. Then we have the polarisation formula:

$$Q(v_1,v_2) = \frac{1}{4}\sum_{k=0}^{3} i^k q(i^k v_1 + v_2)$$

(Q is linear in v_2). For any strictly positive number λ we shall have:

$$|Q(v_1, v_2)| = |Q(\lambda v_1, \lambda^{-1}v_2)| \le \frac{1}{4}\sum_{k=0}^{3}|q(i^k\lambda v_1 + \lambda^{-1}v_2)| \le$$

$$\le \frac{1}{4}\sum_{k=0}^{3} |i^k\lambda v_1 + \lambda^{-1}v_2|^2 \le (\lambda|v_1| + \lambda^{-1}|v_2|)^2 \quad .$$

It is enough to take $\lambda = |v_2|^{1/2}|v_1|^{-1/2}$ if $|v_1| \ne 0$ and $|v_1| \ne 0$. If, for example, $|v_2| = 0$, let $\lambda \to 0$ above; we obtain $Q(v_1,v_2) = 0$. Q.E.D.

Let us denote V the space of functions Ψ: $(0,1] \to \mathcal{H}$, $\Psi = \{\psi_\varepsilon\}_{0<\varepsilon\le 1}$, which are weakly C^1 and are such that $\psi_\varepsilon \in D(A;\mathcal{G}^*)$ for all ε. Then $N(\Psi)$ is a seminorm on V (which takes the value $+\infty$) and we may take $Q(\Psi^1,\Psi^2) = \langle\psi_\varepsilon^1, R(z+i\varepsilon)\psi_\varepsilon^2\rangle$ for some fixed $\varepsilon \in (0,1)$. Hence:

Proposition 4: *Under the conditions of proposition 3, the constant* c *may be chosen such that*

$$|\langle\psi_\varepsilon^1, R(z+i\varepsilon)\psi_\varepsilon^2\rangle| \le c\, N(\Psi^1)N(\Psi^2) \tag{27}$$

for all families $\{\psi_\varepsilon^1\}$, $\{\psi_\varepsilon^2\}$ *with the same properties as* $\{\psi_\varepsilon\}$ *from proposition 3.*

8. We would like now to interpret the relation $N(\Psi) < \infty$ in a more *intrinsic* way. It is clear that only the limit $\varepsilon \to 0$ is of interest for us in (27) and that the families $\{\psi_\varepsilon^j\}$ play only an auxiliary role in the argument: they should be thought as constructed in some way from vectors $\psi^j = \lim_{\varepsilon\to 0} \psi_\varepsilon^j$, assuming that the limit exists in some space. In fact, observe that from $N(\Psi) < \infty$ it follows $\int_0^1 \|\psi_\varepsilon'\|_{-1} d\varepsilon < \infty$, hence $\lim_{\varepsilon\to 0} \psi_\varepsilon \stackrel{d}{=} \psi$ exists (strongly) in $\mathcal{G}^*$.

Let us introduce a new Hilbert space $\mathcal{A}$: we take $\mathcal{A} = D(A;\mathcal{G}^*)$ (see (18)) provided with the Hilbert norm:

$$\|\varphi\|_{\mathcal{A}} = \left[\|\varphi\|_{-1}^2 + \|A\varphi\|_{-1}^2\right]^{1/2} \tag{28}$$

(i.e. we consider on $\mathcal{A}$ the graph-topology associated to the operator A in $\mathcal{G}^*$). Observe that:

$$\|\psi_\varepsilon\|_{-1} \le \|\psi_1\|_{-1} + \int_\varepsilon^1 \|\psi_\tau'\|_{-1}\, d\tau$$

hence:

$$\int_0^1 \|\psi_\varepsilon\|_{-1} \frac{d\varepsilon}{\sqrt{\varepsilon}} \le 2\|\psi_1\|_{-1} + 2\int_0^1 \|\psi_\tau'\|_{-1} \sqrt{\tau}\, d\tau .$$

This shows that the seminorm (on functions Ψ) given by:

$$\tilde{N}(\Psi) = \sup_{0<\varepsilon<1} \sqrt{\varepsilon}\, \|\psi_\varepsilon\| + \int_0^1 \left[\|\psi_\varepsilon'\|_{-1} + \|\psi_\varepsilon\|_{\mathcal{A}}\right] \frac{d\varepsilon}{\sqrt{\varepsilon}} \tag{29}$$

is equivalent to $N(\Psi)$. Now, taking into account section 1.8.1 from *Interpolation Theory* by H. Triebel, we see that the finiteness of the integral from (29) implies $\psi \in (\mathcal{G}^*, \mathcal{A})_{1/2,1}$, where $(E,F)_{\theta,p}$ denotes the interpolation space obtained by the real method from the Banach spaces E, F, with $0 < \theta < 1$, $1 \le p \le \infty$. Unfortunately, the first term in (29) puts new conditions on ψ. In order to avoid the use of some complicated spaces, we shall restrict ourselves to a space smaller that $\mathcal{A}$.

Let $\mathcal{K}$ be any Hilbert space such that $\mathcal{K} \subset \mathcal{H}$ continuously and densely and $\mathcal{K} \subset \mathcal{A}$ continuously. Then $\|\psi_\varepsilon\|_{\mathcal{A}} \le c\|\psi_\varepsilon\|_{\mathcal{K}}$ so that $N(\Psi)$ is dominated by the seminorm:

$$N_{\mathcal{K}}(\Psi) = \sup_{0<\varepsilon<1} \sqrt{\varepsilon}\, \|\psi_\varepsilon\| + \int_0^1 \left[\|\psi'_\varepsilon\|_{-1} + \|\psi_\varepsilon\|_{\mathcal{K}}\right] \frac{d\varepsilon}{\sqrt{\varepsilon}} \quad . \tag{30}$$

As before we see that the finiteness of the last integral implies $\psi \in (\mathcal{G}^*, \mathcal{K})_{1/2,1}$. Reciprocally, for each such ψ, there is a family $\{\psi_\varepsilon\}_{0<\varepsilon\le1}$ which makes finite the last integral and $\psi_\varepsilon \to \psi$ in $\mathcal{G}^*$ as $\varepsilon\to0$. We shall see now that the first term in the right-hand side of (30) does not require any new assumption on ψ.

Lemma 2: *For each $\psi \in (\mathcal{G}^*, \mathcal{K})_{1/2,1}$ there is a function $\Psi = \{\psi_\varepsilon\}_{0<\varepsilon\le1}$ with $\psi_\varepsilon \in \mathcal{K}$ such that Ψ: $(0,1] \longrightarrow \mathcal{G}^*$ is of class C^1 and:*

$$N_{\mathcal{K}}(\Psi) \le c\, \|\psi\| \tag{31}$$

for some constant c independent of ψ; here $\|\cdot\|$ is the norm in $(\mathcal{G}^*, \mathcal{K})_{1/2,1}$. Moreover, $\psi_\varepsilon \to \psi$ strongly in $\mathcal{G}^{-1/2} \equiv (\mathcal{G}^*, \mathcal{H})_{1/2,2}$.

Proof: Since $\mathcal{K} \subset \mathcal{H}$, it will be enough to consider in place of $N_{\mathcal{K}}$ the seminorm $\tilde{N}_{\mathcal{K}}$ defined as in (30) but with $\|\psi_\varepsilon\|$ replaced by $\|\psi_\varepsilon\|_{\mathcal{K}}$ (we have $N_{\mathcal{K}} \le c\, \tilde{N}_{\mathcal{K}}$). We shall also prove the last statement with $\mathcal{G}^{-1/2}$ replaced by the smaller space $(\mathcal{G}^*, \mathcal{K})_{1/2,2}$. Observe that $\mathcal{K} \subset \mathcal{G}^*$ continuously and densely (because $\mathcal{H} \subset \mathcal{G}^*$ has these properties).

More generally, let X, Y be two Hilbert spaces such that $X \subset Y$ continuously and densely. Then there is a unique positive selfadjoint operator Λ in Y with domain X such that $\|u\|_X = \|\Lambda u\|_Y$ for all $u \in Y$. It is known (see section 1.14 in the book of Triebel cited above) that the norm in $(X,Y)_{\theta,p}$ can be taken equal to:

$$\|u\|_{\theta,p} = \|u\|_Y + \left[\int_0^1 [\varepsilon^{1-\theta}\|(1+\varepsilon\Lambda)^{-1}u\|_X]^p \frac{d\varepsilon}{\varepsilon}\right]^{1/p} . \tag{32}$$

Let us define $(0 < \varepsilon \le 1)$:

$$\psi_\varepsilon = (1 + \varepsilon\Lambda)^{-1}\psi .$$

Clearly $\psi_\varepsilon \in X$ for any $\psi \in Y$ and as an element of Y it depends strongly C^1 on ε. Moreover:

$$\psi'_\varepsilon = -(1 + \varepsilon\Lambda)^{-1}\Lambda(1+\varepsilon\Lambda)^{-1}\psi .$$

Let us estimate the first term in the expression of $\tilde{N}_{\mathcal{K}}(\Psi)$:

$$\sup_{0<\varepsilon<1} \sqrt{\varepsilon}\,\|\psi_\varepsilon\|_X = \sup_{0<\varepsilon<1} \sqrt{\varepsilon}\,\|(1 + \varepsilon\Lambda)^{-1}\psi\|_X \le \|\psi\|_{1/2,\infty} .$$

But $(Y,X)_{\theta,p} \subset (Y,X)_{\theta,q}$ if $1 \le p \le q \le \infty$. Hence the last term above is dominated by $\|\psi\|_{1/2,1}$. Then

$$\|\psi'_\varepsilon\|_Y \le \|\Lambda(1 + \varepsilon\Lambda)^{-1}\psi\|_Y = \|(1 + \varepsilon\Lambda)^{-1}\psi\|_X .$$

It follows immediately from (32) that $\tilde{N}_{\mathcal{K}}(\psi) \le c\,\|\psi\|_{1/2,1}$. Finally, let us prove $\psi_\varepsilon \to \psi$ strongly in $(Y,X)_{1/2,2}$. An equivalent norm in this space is $\|\Lambda^{1/2}u\|_Y$. So, it is enough to prove that $\Lambda^{1/2}\psi_\varepsilon \to \Lambda^{1/2}\psi$ in Y if $\psi \in (Y,X)_{1/2,2}$; but this is obvious. Q.E.D.

9. Let us go back to proposition 4. Take there families $\{\psi^1_\varepsilon\}$ and $\{\psi^2_\varepsilon\}$ associated to some vectors $\psi^1, \psi^2 \in (\mathcal{G}^*,\mathcal{K})_{1/2,1}$ as in lemma 2. Observe that $R(z + i\varepsilon) \to R(z)$ as operators $\mathcal{G}^{-1/2} \to \mathcal{G}^{+1/2}$, for each fixed z. Since the constants in (27) and (31) are independent of z (whose real part belongs to a compact in $\mathbb{R}\setminus\{0\}$) we get:

$$|\langle\psi^1, R(z)\psi^2\rangle| \le c\,\|\psi^1\|\cdot\|\psi^2\| \tag{33}$$

for a constant c and all z as in proposition 3, all $\psi^j \in (\mathcal{G},\mathcal{K})_{1/2,1}$. We can now prove our main result:

Theorem: *Let* H *be a selfadjoint operator in* $\mathcal{H}$ *with domain* $\mathcal{G}$. *Assume* H *is* A*-homogeneous of degree* $m\neq 0$. *Let* $\mathcal{K}$ *be any Hilbert space such that* $\mathcal{K} \subset \mathcal{H}$ *continuously and densely and* $\mathcal{K} \subset D(A;\mathcal{G}^*)$ (cf. (18)). *Denote*

$\mathcal{E} = (\mathcal{G}^*, \mathcal{K})_{1/2,1}$ *(Banach space defined by real interpolation),* $\mathcal{E}^* = (\mathcal{G}, \mathcal{K}^*)_{1/2,\infty}$ *its dual and* $\mathcal{G}^{-1/2} = (\mathcal{G}^*, \mathcal{H})_{1/2,2}$, $\mathcal{G}^{+1/2} = (\mathcal{G}^{-1/2})^* = (\mathcal{G}, \mathcal{H})_{1/2,2}$. *Then* $\mathcal{E} \subset \mathcal{G}^{-1/2}$ *continuously and densely and* $\mathcal{G}^{1/2} \subset \mathcal{E}^*$ *continuously (but not densely); in particular* $B(\mathcal{G}^{-1/2}, \mathcal{G}^{1/2}) \subset B(\mathcal{E}, \mathcal{E}^*)$ *continuously. Let*

$$\mathbb{C}^*_\pm = \{z \in \mathbb{C} \mid z \neq 0, \pm \operatorname{Im} z \geq 0\}$$

and $\mathring{\mathbb{C}}^*_\pm$ *its interior. Then* $\mathring{\mathbb{C}}^*_\pm \ni z \longmapsto R(z) \in B(\mathcal{G}^{-1/2}, \mathcal{G}^{1/2})$ *is an holomorphic function which extends to a *-weakly continuous function on* $\mathbb{C}^*_\pm$ *when considered with values in* $B(\mathcal{E}, \mathcal{E}^*)$. *In particular:*

$$R(\lambda \pm io) \stackrel{d}{=} \lim_{\mu \to \pm o} R(\lambda + i\mu) \in B(\mathcal{E}, \mathcal{E}^*)$$

*exists *-weakly in* $B(\mathcal{E}, \mathcal{E}^*)$ *for each* $\lambda \neq 0$ *and depends *-weakly continuously on* λ.

Proof: Observe first that $\mathcal{K} \subset \mathcal{A}$ continuously by closed graph theorem. Then (33) implies:

$$\|R(z)\|_{B(\mathcal{E}, \mathcal{E}^*)} \leq c \tag{34}$$

for all $z = \lambda + i\mu$, λ in a compact subset of $\mathbb{R}\setminus\{0\}$ and $0 < \mu < 1$. Hence the limiting absorption principle is valid in $B(\mathcal{E}, \mathcal{E}^*)$. It remains to prove the continuity assertions: we must show that for $\psi^1, \psi^2 \in \mathcal{E}$, the function $z \longrightarrow \langle\psi^1, R(z)\psi^2\rangle$ extends to a continuous function on $\mathbb{C}^*_+$. Because of (34) and of the density of $\mathcal{K}$ in $\mathcal{E}$ it is enough to assume $\psi^j \in \mathcal{K}$. By polarisation, we may take $\psi^1 = \psi^2 \equiv \psi$. Construct the ψ_ε as in lemma 6 for example and observe that $\langle\psi, R(z)\psi\rangle = \langle\psi_\varepsilon, G_\varepsilon\psi_\varepsilon\rangle\big|_{\varepsilon=0}$. Hence:

$$\langle\psi, R(z)\psi\rangle = \langle\psi_1, G_1\psi_1\rangle - \int_0^1 \frac{d}{d\varepsilon} \langle\psi_\varepsilon, G_\varepsilon\psi_\varepsilon\rangle \, d\varepsilon \quad . \tag{35}$$

Taking into account (22 and (26) we get an estimate of the form:

$$\left|\frac{d}{d\varepsilon} \langle\psi_\varepsilon, G_\varepsilon\psi_\varepsilon\rangle\right| \leq f(\varepsilon)$$

with $\int_0^1 f(\varepsilon)d\varepsilon < \infty$ and f independent of z if $z = \lambda + i\mu$, λ being in a compact Λ from $\mathbb{R}\setminus\{0\}$ and $0 < \mu < 1$. On the other hand, for each $\varepsilon > 0$

the function $\frac{d}{d\varepsilon}\langle\psi_\varepsilon, G_\varepsilon\psi_\varepsilon\rangle = +i\ \langle\psi_\varepsilon,\ R(z+i\varepsilon)^2\psi_\varepsilon\rangle$ depends continuously on $z = \lambda + i\mu$ for $\lambda \in \Lambda$ and $\mu \in [0,1]$. By dominated convergence theorem applied in (35) we get the same property for $\langle\psi, R(z)\psi\rangle$. This finishes the proof. Q.E.D.

It is trivial to deduce from the theorem the next corollary. We denote $\mathring{\mathcal{E}}^*$ the closure of $\mathcal{G}$ in $\mathcal{E}^*$; one can show that $(\mathring{\mathcal{E}}^*)^* = \mathcal{E}$.

Corollary: *The continuous linear operators* T: $\mathring{\mathcal{E}}^* \to \mathcal{H}$ *are locally* H-*smooth on* $\mathbb{R}\setminus\{0\}$.

10. We shall consider now several examples. Let $\mathcal{H} = L^2(\mathbb{R}^n)$ and $W(\alpha)$ the unitary group in $\mathcal{H}$ associated to dilations in $\mathbb{R}^n$: $(W(\alpha)\psi)(x) = e^{n\alpha/2}\ \psi(e^\alpha x)$. We shall denote Q_j the operator of multiplication by x_j and $P_j = -i\frac{\partial}{\partial x_j}$; then $Q = (Q_1,\dots,Q_n)$ and $P = (P_1,\dots,P_n)$ are vector-operators. It is easily shown that the infinitesimal generator of W is:

$$A = \frac{1}{2}(P\cdot Q + Q\cdot P) = \frac{1}{2}\sum_{j=1}^{n}(P_jQ_j + Q_jP_j) \qquad .$$

If $\mathcal{F}$ is the operator of Fourier transformation (unitary in $\mathcal{H}$) then $P_j = \mathcal{F}^*Q_j\mathcal{F}$, $Q_j = -\mathcal{F}^*P_j\mathcal{F}$ and $\mathcal{F}^*A\mathcal{F} = -A$. Also $\mathcal{F}^*W(\alpha)\mathcal{F} = W(-\alpha)$.

Think about P_j, Q_j as selfadjoint operators in $\mathcal{H}$. Then P_j is A-homogeneous of degree -1 and Q_j is A-homogeneous of degree +1 in the sense of section 1, i.e. $W(\alpha)$ leaves invariant their domains and:

$$W(\alpha)\ P_jW(-\alpha) = e^{-\alpha}P_j$$

$$W(\alpha)\ Q_jW(-\alpha) = e^{\alpha}Q_j \qquad .$$

A much more general assertion is easily proved. Let $F: \mathbb{R}^n \to \mathbb{R}$ be a homogeneous polynomial of degree $m = 1,2,3,\dots$, i.e. $F(x) = \sum_{|\alpha|=m} a_\alpha x^\alpha$ for some $a_\alpha \in \mathbb{R}$, with $\alpha = (\alpha_1,\dots,\alpha_n)$, $\alpha_j \in \{0,1,2,\dots\}$, $|\alpha| = \alpha_1+\dots+\alpha_n$, $x^\alpha = x_1^{\alpha_1}\dots x_n^{\alpha_n}$. Then we may define self-adjoint operators $F(Q)$ and $F(P)$ in $\mathcal{H}$ in an obvious way: the domain of $F(Q)$ is the set of $\psi \in \mathcal{H}$ such that the function $F(x)\psi(x)$ belongs to $\mathcal{H}$ etc; then $F(P) = \mathcal{F}^*F(Q)\mathcal{F}$.

It is cleaar that $W(\alpha)$ leaves the domain of $F(Q)$ invariant and $W(\alpha)F(Q)W(-\alpha) = e^{m\alpha}F(Q)$, i.e. $F(Q)$ *is A-homogeneous of degree* m. Making a Fourier transformation one sees that $F(P)$ *if A-homogeneous of degree* $-m$.

In the above argument the fact that F is a polynomial has not been used: F may be any real Borel function on $\mathbb{R}^n$ such that $F(\lambda x) = \lambda^m F(x)$ for $x \in \mathbb{R}^n$ and $\lambda > 0$ (in particular $F(0) = 0$). Proposition 1 implies that *the spectrum of* $F(Q)$ *and* $F(P)$ *is purely absolutely continuous outside zero if* $m \neq 0$.

Observe that F has the above property if and only if $F(x) = f(\frac{x}{|x|})|x|^m$ where $f: S \to \mathbb{R}$ is a Borel function and S is the unit sphere in $\mathbb{R}^n$. Working in polar coordinates one can identify $\mathcal{H} \cong L^2(S) \otimes L^2(\mathbb{R}_+)$ $(\mathbb{R}_+ = (0,\infty))$ which transforms $F(Q)$ into the operator $f(\omega) \otimes r^m$ and $W(\alpha)$ into $1 \otimes \omega(\alpha)$ where $(\omega(\alpha)u)(r) = e^{\alpha/2}\ u(e^{\alpha}r)$ in $L^2(\mathbb{R}_+)$. Any operator of the form $T \otimes r^m$ with $m \neq 0$ and T selfadjoint in $L^2(S)$ may be shown to have purely absolutely continuous spectrum outside zero by this method. However this fact is rather trivial and may be shown directly for operators of the form $T \otimes R$ whenever R has a purely absolutely continuous spectrum.

On the other hand, the assertions of the theorem and of the corollary are not at all trivial in this context. Let us denote $\mathcal{H}^s_t$ the weighted Sobolev space defined as the completion of $\mathcal{S}(\mathbb{R}^n)$ (Schwartz test functions) under the norm $\|\langle P\rangle^s\langle Q\rangle^t u\|$ (where $\langle a\rangle = (1 + |a|^2)^{1/2}$). Put $\mathcal{H}^s = \mathcal{H}^s_0$, $\mathcal{H}_t = \mathcal{H}^0_t$. Observe that $A \in B(\mathcal{H}_{+1}, \mathcal{H}^{-1})$. If $\mathcal{G}$ is a Hilbert space such that $\mathcal{G} \subset \mathcal{H}^{+1}$ continuously and densely, then $\mathcal{H} \subset \mathcal{H}^{-1} \subset \mathcal{G}^*$, hence $A \in B(\mathcal{H}_1, \mathcal{G}^*)$ which easily implies $\mathcal{H}_1 \subset D(A;\mathcal{G}^*)$. Then clearly we may take in the theorem: $\mathcal{K} = \mathcal{H}_1$.

Let $F: \mathbb{R}^n \to \mathbb{R}$ be a homogeneous functions of degree $m \geq 1$ and such that $|F(\omega)| \geq \text{const} > 0$ if $|\omega| = 1$. Take $\mathcal{G} = D(F(P))$ provided with graph norm. Then $\mathcal{G} \subset \mathcal{H}^1$, $F(P)$ is A-homogeneous of degree $-m$ and we may take $\mathcal{K} = \mathcal{H}_1$. Observe that $\mathcal{G} = \mathcal{H}^m$ if F is also bounded on the unit sphere. In this case

$$\mathcal{E} = \left[\mathcal{H}^{-m},\ \mathcal{H}_{+1}\right]_{1/2,1} = \mathcal{H}^{(-m/2,1)}_{(1/2,1)}$$

$$\mathcal{E}^* = \left[\mathcal{H}^{m}\ \ \mathcal{H}_{-1}\right]_{1/2,\infty} = \mathcal{H}^{(m/2,\infty)}_{(-1/2,\infty)} \quad .$$

The theorem says then that the limiting absorption principle is valid for such F(P) in $B(\mathcal{E},\mathcal{E}^*)$. The space $\mathcal{E}$ obtained in this way is probably optimal (from this point of view) in the scale of spaces obtained from $\mathcal{H}^s_t$ by real interpolation (this is a class of weighted Besov spaces). In fact, for all $\varepsilon > 0$:

$$\mathcal{H}^{-m/2+\varepsilon}_{1/2+\varepsilon} \subset \mathcal{H}^{(-m/2,1)}_{(1/2,1)} \subset \mathcal{H}^{-m/2}_{1/2} \quad ,$$

$$\mathcal{H}^{m/2}_{-1/2} \subset \mathcal{H}^{(m/2,\infty)}_{(-1/2,\infty)} \subset \mathcal{H}^{m/2-\varepsilon}_{-1/2-\varepsilon} \quad ,$$

and one can show, for example, that the limiting absorption principle is *not* valid for $F(P) = P^2$ in $B(\mathcal{H}^{-1}_{1/2},\ \mathcal{H}^{+1}_{-1/2})$ (m = 2 here).

11. We would like to present in this section some examples which show the importance of the assumption $W(\alpha)\mathcal{G} \subset \mathcal{G}$ for all α. Let $\mathcal{H} = L^2(\mathbb{R}_+)$ and $P = -i\frac{d}{dx}$ considered only formally for the moment. Let $A = \frac{1}{2}(QP + PQ)$ as before, i.e. $(W(\alpha)\psi)(x) = e^{\alpha/2}\psi(e^{\alpha}x)$. Then *formally* P is A-homogeneous of degree -1, hence P^2 is of degree -2. However, in order to define P^2 as a selfadjoint operator in $\mathcal{H}$, one has to chose some boundary conditions at zero. These are of the following forms;

1) Dirichlet: $\psi(0) = 0$;
2) Neumann: $\psi'(0) = 0$;
3) Mixed: $a\psi(0) + b\psi'(0) = 0$; here $a,b \in \mathbb{R}\setminus\{0\}$.

It is easily shown that in the first two cases the domain of P^2 is W-invariant and P^2 is A-homogeneous of degree - 2 in the sense of section **1**; in particular, in both cases the spectrum of P^2 is purely absolutely continuous outside zero, which is a well-known fact. On the other hand, in the third case the domain is obviously not W-invariant. In fact, if $\psi_\alpha(x) = e^{\alpha/2}\psi(e^{\alpha}x)$, then $\psi_\alpha(0) = e^{\alpha/2}\psi(0)$ and $\psi'_\alpha(0) =$

$= e^{3\alpha/2}\psi'(0)$ hence $a\psi_\alpha(0) + b\psi'_\alpha(0) = 0$ for all α if and only if $\psi(0) = \psi'(0) = 0$. So in this case P^2 is *not* A-homogeneous. Besides, P^2 with mixed boundary conditins has a strictly negative eigenvalue if a,b have the same sign, namely $\lambda = -(\frac{a}{b})^2$ with corresponding eigenfunction $u(x) = \exp(-\frac{a}{b}x)$, hence even the conclusion of proposition 1 is not true.

As another example, let $H = P^2 + aQ^{-2}$in the same Hilbert space, with $a \in \mathbb{R}\backslash\{0\}$. *Formally* H is A-homogeneous of dgree - 2. If $a > -\frac{1}{4}$, then H is essentially selfadjoint on $C_0^\infty((0,\infty))$ and positive. It is easily shown that in this case H is rigorously A-homogeneous of degree -2, so proposition 1 implies it has a purely absolutely continuous spectrum. If $a \le -\frac{1}{4}$, then H has many selfadjoint extensions, but all have an unbounded negative spectrum consisting of eigenvalues, hence it has no selfadjoint extension with domain invariant under the dilations.

One may consider a generalisation of the last example which is of some interest. Let $\mathcal{H}_0$ be a Hilbert space, Λ a selfadjoint operator in $\mathcal{H}_0$ such that $\Lambda \ge -\frac{1}{4} + \varepsilon$ for some $\varepsilon > 0$, and $H = P^2 + \Lambda Q^{-2} \cong 1 \otimes P^2 + \Lambda \otimes Q^{-2}$in $\mathcal{H} = L^2(\mathbb{R}_+; \mathcal{H}_0) \cong \mathcal{H}_0 \otimes L^2(\mathbb{R}_+)$. Then H is essentially selfadjoint on $C_0^\infty((0,\infty); D(\Lambda))$ and is A-homogeneous of degree -2, so that H is positive and has a purely absolutely continuous spectrum. To this category belong operators of the form $P^2 + V(Q)$ in $\mathcal{H} = L^2(\mathbb{R}^n)$ if $V\colon \mathbb{R}^n \to \mathbb{R}$ is Borel, homogeneous of degree -2, and $V(x) \ge -\frac{1}{4} + \varepsilon$ if $|x| = 1$. For example, many-body hamiltonians of the form:

$$H = -\Delta + \sum_{1\le i<j\le N} \frac{c_{ij}}{|x_i - x_j|^2}$$

may be treated.

12. Finally, we would like to indicate a class of hamiltonians, larger than the A-homogeneous ones, for which the preceding arguments may be extended without any effort. Assume H is a selfadjoint operator in $\mathcal{H}$ with domain $\mathcal{G}$ and $[iA,H] = mH + n$ for some real numbers m,n not both zero (hence, the Lie algebra has the generators A, H and the identity I). Then, as in (3) we obtain

$$[iA, R(z)] = (n - mz)\, R(z)^2 - m\, R(z) \quad .$$

Assuming $n - mz \neq 0$, we get in place of (5)

$$\frac{d}{dz} R(z) = \frac{i}{n - mz} [A, R(z)] + \frac{m}{n - mz} R(z)$$

Starting from this identity we can repeat the same arguments as before. The role of the critical point $\lambda = 0$ is now played by the point $\lambda = \frac{n}{m}$ if $m \neq 0$. If $m = 0$, remark that we are considering the canonical commutation relations ($n \neq 0$) and the theorem gives a non-trivial result even in such a simple case. We would also like to note that the case of A-homogeneous H with $H > 0$ is also *formally* related to the canonical commutation relations: if $\tilde{A} = H^{-1/2}AH^{-1/2}$, then $\tilde{A}$ is (formally) selfadjoint and $[i\tilde{A}, H] = m \neq 0$.

SUPERSYMMETRIC QUANTUM MECHANICS[1]

H. GROSSE
Institut für Theoretische Physik
Universität Wien
Boltzmanngasse 5, A-1090 Vienna, Austria

ABSTRACT. We summarize recent developments of supersymmetric quantum mechanics. We start from the susy oscillator, mention the factorization schemes and discuss the order of levels of Schrödinger operators as an example. We mention soliton equations and the inverse scattering problem and discuss susy breaking and index problems for Dirac operators. The construction of Lie-supergroups suggests a generalization of the well-known theorems of von Neumann and Wigner to superspace. We mention finally studies of the general structure of susy models. A number of relations between the operator formulation and the stochastic formulation result.

1 Introduction: From the Susy Oscillator to Lie-Superalgebras

In these lectures we review on the one hand *super*s*y*mmetric quantum mechanics and we shall mention on the other hand our own developments in collaboration with a number of colleagues. From the technical point of view susy quantum mechanics is much simpler than susy field theory. Nevertheless, certain characteristic properties of supersymmetry can be studied and analyzed already in this simple setting.

Supersymmetry results from the combination of continuous symmetries and discrete symmetries. Although the structure which we obtain is extremely simple and some roots go back to the last century, the history of the subject turns out to be rather surprising.

In the early seventieth people tried to explain hadron physics with the help of strings. Besides the bosonic one, the "old" fermionic string model was formulated. In 1974 Wess and Zumino [1] replaced the embedding space by Minkowski space time, "erased" the world sheet index, combined bosons and fermions and developed and formulated space-time supersymmetry. Afterwards a large number of physicists treated susy models and supergravity (for reviews see e.g. [2,3]). Nevertheless, it was not before 1981 when Witten formulated susy quantum mechanics [4]. We should also mention that the observation of anomaly cancellations in 1984 led to a revival of string models. The new superstring models are believed to give a description of all fundamental interactions including gravity.

---

[1]Part of project P5588 of the "Fonds zur Förderung der wissenschaftlichen Forschung in Österreich".

A. Boutet de Monvel et al. (eds.), Recent Developments in Quantum Mechanics, 299–327.

After the work of Witten a large number of papers appeared treating various aspects of susy quantum mechanics [5,6,7]. It is surprising that historically nobody tried to combine a bosonic oscillator with a fermionic one just after the invention of quantum mechanics. The cancellation of the zero point energy gives already a hint towards the fact that certain infinites do not show up in susy field theory. In addition, the problem of exponentiating a super-Lie-algebra shows up already at this level.

The first application concerns the factorization schemes which was used already by Schrödinger and developed further by Infeld and Hull, by Green and many others. The method goes back to a theorem of Darboux from the last century. We have used such a scheme to show how levels are ordered as a function of the angular momentum and as a function of the number of nodes of the reduced wave function. At the beginning we have reinvented this scheme without knowing its connection to supersymmetry, like Monsieur Jourdain "Qui a fait la prose, sans savoir". We describe the level order results in chapter three. Afterwards we summarize the inverse scattering transform method and show that it gives a general scheme to relate two Hamiltonians; susy turns out to be a special case. Solitons of the modified KdV equation can be related to those of the KdV equation by the so-called Miura transformation. The reason behind that is given by "commutation relations" which are related to susy.

Next we treat the Dirac operator. It is given by a kind of square root of the Schrödinger operator. Whether susy is realized within a model or broken is connected to an index problem. There exist a number of regularization schemes for the spectral asymmetry. In order to calculate this index, susy is of great help. In chapter five we shall also mention results on the two-dimensional susy quantum field theory models. Recently relations between the operator formalism and the stochastic formalism for susy models have been worked out. The special role of coordinates which lead to total invariance of the Lagrangian is mentioned. They enter the Nicolai mapping too. This transformation is seen to be a similarity transformation. We end the last chapter with a short discussion of supersymmetric functional integrals in which a generalization of the Wiener measure to superspace is obtained.

We shall start from the simplest system which we know in physics, namely the *bosonic* harmonic oscillator with Hamiltonian

$$H_B = -\frac{\omega}{2}\frac{d^2}{dx^2} + \frac{\omega}{2}x^2 \quad \text{on } L^2(\mathbf{R}, dx). \tag{1.1}$$

For simplicity reasons we have put $m = 1/\omega$ and rewrite (1.1) in a "factorized" form

$$H_B = \omega(a^\dagger a + \frac{1}{2}), \qquad a = (\frac{d}{dx} + \omega)\frac{1}{\sqrt{2}}, \qquad a^\dagger = (-\frac{d}{dx} + \omega)\frac{1}{\sqrt{2}}, \tag{1.2}$$

where $a^\dagger$ and a are creation and annihilation operators of quanta of our oscillator. Their commutator equals one:

$$[a, a^\dagger] = 1, \qquad [H_B, a^\dagger] = \omega a^\dagger. \tag{1.3}$$

(1.3) shows the "bosonic" nature of our system. The spectrum and associated eigenfunctions of H_B are given by

$$E_n = \omega(n + \frac{1}{2}), \qquad n \in \mathbf{N}_0, \qquad a|0\rangle = 0, \qquad \frac{(a^\dagger)^n}{\sqrt{n!}}|0\rangle = |n\rangle, \tag{1.4}$$

where $|0\rangle$ denotes the ground state vector and $\omega/2$ is the zero point energy.

Next we "double" the system by adding a second *bosonic* oscillator and introduce

$$h = \frac{1}{2}(p_1^2 + \omega_1^2 q_1^2) + \frac{1}{2}(p_2^2 + \omega_2^2 q_2^2) \quad \text{on } L^2(\mathbf{R}^2, dq_1 dq_2), \tag{1.5}$$

This two-dimensional oscillator has an additional symmetry if $\omega_1 = \omega_2$. h is then invariant under rotations which are generated by the angular momentum operator

$$\ell = q_1 p_2 - q_2 p_1, \qquad [\ell, h] = 0. \tag{1.6}$$

We may replace the two real coordinates q_1, q_2 by a single complex one $q = q_1 + iq_2$. The above mentioned rotation then becomes a phase transformation $q \to e^{i\varphi}q$. Such an invariance implies a conserved charge. In our case it is the angular momentum which serves as a charge. We note that ℓ acts in the space of two bosonic degress of freedom and it is easy to exponentiate ℓ to obtain the global unitary group of transformations.

We intend next to add a "fermionic" degree of freedom to the bosonic oscillator of equ. (1.2) and obtain the susy oscillator

$$H = \omega(a^\dagger a + c^\dagger c) \quad \text{on } L^2(\mathbf{R}, dx) \otimes \mathbf{C}^2, \tag{1.7}$$

where $c^\dagger$ and c denote fermionic creation and annihilation operators obeying

$$\{c, c^\dagger\} = 1, \qquad c^{\dagger 2} = c^2 = 0. \tag{1.8}$$

We could start from the Hamiltonian $\omega_1(a^\dagger a + 1/2) + \omega_2(c^\dagger c - 1/2)$ instead of (1.7); then no symmetry would have been implied. We note that the zero point energy of the fermionic oscillator is negative. For the case of equal frequencies $\omega_1 = \omega_2 = \omega$, the zero point energies of the bosonic and the fermionic oscillator cancel.

We note that the algebra (1.8) can well be represented by σ-matrices, since $\{\sigma^-, \sigma^+\} = 1$, $\sigma^{+2} = \sigma^{-2} = 0$. This representation is meant in equ. (1.7) when we wrote that H acts in $L^2(\mathbf{R}, dx) \otimes \mathbf{C}^2$. For more degrees of freedom we have to introduce the Klein-Jordan-Wigner transformation in order to represent the algebra of operators fulfilling the canonical anticommutation relations by the algebra obeyed by the matrices $\mathbf{1} \otimes \ldots \mathbf{1} \otimes \sigma^j \otimes \mathbf{1} \ldots \otimes \mathbf{1}$.

We may rewrite (1.7) as

$$H = \omega\{a^\dagger a(1 - c^\dagger c) + (a^\dagger a + 1)c^\dagger c\} = \omega(a^\dagger a c c^\dagger + a a^\dagger c^\dagger c) = \{Q, Q^\dagger\}, \tag{1.9}$$

where we introduced operators Q and $Q^\dagger$ through

$$Q = \sqrt{\omega}\, a\sigma^+ = \sqrt{\omega}\, a c^\dagger, \qquad Q^\dagger = \sqrt{\omega}\, a^\dagger \sigma^- = \sqrt{\omega}\, a^\dagger c, \tag{1.10}$$

which turn out to be a kind of charges, called supercharges, as we shall see soon. We observe that Q and $Q^\dagger$ can be written as

$$Q = \sqrt{\omega}\begin{pmatrix} 0 & a \\ 0 & 0 \end{pmatrix}, \quad Q^\dagger = \sqrt{\omega}\begin{pmatrix} 0 & 0 \\ a^\dagger & 0 \end{pmatrix}, \quad \{Q, Q^\dagger\} = \omega\begin{pmatrix} aa^\dagger & 0 \\ 0 & a^\dagger a \end{pmatrix} = H. \tag{1.11}$$

The supercharges represent a kind of square root of the Hamiltonian H. They commute with H as a simple calculation shows:

$$[Q, H] = \omega^{3/2}[ac^\dagger, a^\dagger a + c^\dagger c] = \omega^{3/2}(c^\dagger a - ac^\dagger) = 0. \tag{1.12}$$

They generate therefore a symmetry of the Hamiltonian which is called a *super*symmetry. We observe already at this stage that $a^\dagger a$ and $aa^\dagger$ are parts of H and are "essentially" isospectral, which should mean that their spectra coincide except for zero modes

$$\text{spect}\,(a^\dagger a) \setminus \{0\} = \text{spec}\,(aa^\dagger) \setminus \{0\}. \tag{1.13}$$

We prove (1.13) for the pure point spectrum by observing that

$$aa^\dagger\psi = \varepsilon\psi \Rightarrow a^\dagger a(a^\dagger\psi) = \varepsilon(a^\dagger\psi), \tag{1.14}$$

and $a^\dagger\psi$ is an eigensolution for $a^\dagger a$ except if $a^\dagger\psi = 0$. Clearly, for the continuous spectrum we have to be more careful and use, for example, the resolvents of the operators and commutation relations à la Deift [8]. Zero modes will play an essential role later on, when we discuss index problems and susy breaking.

We introduce the ground state $|0,0\rangle = |0\rangle \otimes \begin{pmatrix} 0 \\ 1 \end{pmatrix}$ for H by requiring that $a|0,0\rangle = c|0,0\rangle = 0$. All eigensolutions of H are given by

$$|n,m\rangle = \frac{(a^\dagger)^n}{\sqrt{n!}}(c^\dagger)^m|0,0\rangle, \qquad n = 0,1,2,\ldots, \qquad m = 0,1, \tag{1.15}$$

and the spectrum is pure point and all excited states are double degenerate. Only the ground state is not degenerate. We observe that

$$Q|n,0\rangle \propto |n-1,1\rangle, \qquad Q^\dagger|n,1\rangle \propto |n+1,0\rangle; \tag{1.16}$$

Q and $Q^\dagger$ map from a "bosonic" state $|n,0\rangle$ to a "fermionic" one $|n-1,1\rangle$ with same energy and vice versa. We may introduce the fermion number operator N_F and the Klein operator K acting in $\mathcal{H} = L^2(\mathbf{R}) \otimes \mathbf{C}^2$ as

$$N_F = \begin{pmatrix} 1 & 0 \\ 0 & 0 \end{pmatrix}, \qquad K = (-1)^{N_F} = \begin{pmatrix} -1 & 0 \\ 0 & 1 \end{pmatrix}, \tag{1.17}$$

and observe that $\mathcal{H}$ is decomposed into $\mathcal{H}_B \oplus \mathcal{H}_F = \mathcal{H}$, where $K\mathcal{H}_B = \mathcal{H}_B$ and $K\mathcal{H}_F = -\mathcal{H}_F$. We obtain on the one hand a Z_2-grading of states, but on the other hand a Z_2-grading for operators is obtained too, according to whether they commute or anticommute with K. For example, we obtain $\{K, Q\} = 0$ and $[H, K] = 0$; more generally

$$\{K, O\} = 0 \text{ iff } O = \begin{pmatrix} 0 & x_1 \\ x_2 & 0 \end{pmatrix}, \qquad [K, E] = 0 \text{ iff } E = \begin{pmatrix} x_3 & 0 \\ 0 & x_4 \end{pmatrix}, \tag{1.18}$$

where x_i denote some possible nonzero operator, and we assign grad zero to the even operators E but grad one to the odd ones O.

We remark that H in equ. (1.7) is already normal ordered. Zero point energies for bosons and fermions cancel. For infinite number of degrees of freedom no normal ordering is necessary either. The Hamiltonian is therefore automatically lower bounded. This is the first indication that in susy models divergences are cancelled. It is surprising that history did not follow such a simple line of reasoning as we have indicated it above.

Q generates a symmetry and was called a supercharge. Indeed we obtain

$$\delta_Q a^\dagger := [Q, a^\dagger] = \sqrt{\omega}\, c^\dagger, \qquad \delta_Q c := \{Q, c\} = \sqrt{\omega}\, a, \tag{1.19}$$

for the infinitesimal transformations. $a^\dagger$ is mapped under Q into $c^\dagger$ and c into a. We deduce from (1.19) an essential property. Since Q is an odd operator but $a^\dagger$ is even we have to take the commutator of Q with $a^\dagger$. On the other hand, c is odd so that we have to take the anticommutator of Q and c. The algebraic structure which we obtained is that of a Lie-superalgebra or graded Lie-algebra where for general fermionic operators F and bosonic ones B the brackets

$$\{F, F\} = B, \qquad [F, B] = F, \qquad [B, B] = B \tag{1.20}$$

close. The introduction of Lie-superalgebras was the key which allowed to overcome no-go-theorems prohibiting the extension of Poincaré symmetry in order to combine it with internal symmetries. All Lie-superalgebras have been classified.

(1.19) is the infinitesimal version of a supersymmetry. It is obviously not possible to exponentiate directly Q and to deal at the same time with commutators and anticommutators. In order to formulate a supergroup transformation we shall introduce parameters θ_i which anticommute with Q and more generally with all odd operators but commute with all even ones and are anticommuting among themselves and nilpotent:

$$\{\theta_i, F\} = 0, \qquad [\theta_i, B] = 0, \qquad \theta_i\theta_j + \theta_j\theta_i = 0, \qquad \theta_i^2 = 0. \tag{1.21}$$

This allows to unit equs. (1.19) into

$$\delta_Q R := [\theta Q, R] = \begin{cases} \theta[Q, R] & \text{for } R = B \\ \theta\{Q, R\} & \text{for } R = F. \end{cases} \tag{1.22}$$

The anticommuting parameters together with real parameters like the time (or space coordinates in field theory) span the *superspace*. In chapter 6, we shall start from (1.21) - (1.22) and introduce supergroup transformations.

2 First Application

2.1 Factorization

It is evident that the structure, which we have found in chapter one, is not restricted to the harmonic oscillator. In fact, a simple reformulation allows for a generalization, which is called factorization. We shall add a few examples, some historical notes and explain relations between different methods which are suitable to generate susy quantum mechanical models. They run under the names of factorization, shape independence condition, and Riccati equation. The Darboux transformation originates from the last century and is closely connected to the susy structure. The Crum and Klein transformations indicate the connection to the inverse scattering transform method, which we shall mention in chapter four.

From the first part we deduce for the harmonic oscillator that

$$a^\dagger|n\rangle = \bar{g}_{n+1}|n+1\rangle, \qquad a|n+1\rangle = g_{n+1}|n\rangle, \qquad \bar{g}_n = g_n^*, \tag{2.1}$$

where $\bar{g}_n$ and g_n denote the normalization constants and are given by $g_n = \sqrt{n}$. It follows that

$$\langle n+1|n+1\rangle = \langle n|n\rangle \quad \text{and} \quad |n\rangle = \prod_{r=1}^{n} \frac{a^\dagger}{\bar{g}_r}|0\rangle, \qquad a|0\rangle = 0. \tag{2.2}$$

Cross multiplication of (2.1) yields a consistency condition

$$aa^\dagger - a^\dagger a = g_{n+1}\bar{g}_{n+1} - \bar{g}_n g_n, \tag{2.3}$$

and the spectrum of the Hamiltonian can be obtained as

$$H_B|n\rangle = \omega(\bar{g}_n g_n + \frac{1}{2})|n\rangle. \tag{2.4}$$

Next we give two examples: A look to the spectra of the Schrödinger equation for the Coulomb problem as well as for the three-dimensional oscillator reveals degeneracies for different angular momentum. In the first case all states corresponding to angular momentum ℓ and $\ell+1$ are degenerate except for the lowest ℓ-state. A similar situation occurs, if one compares the spectra of angular momentum ℓ and $\ell+2$ for the harmonic oscillator.

Let us denote by h_ℓ^C the radial Schrödinger equation for the Coulomb problem

$$h_\ell^C \varphi_\ell(r) = (-\frac{d^2}{dr^2} - \frac{1}{r} + \frac{\ell(\ell+1)}{r^2})\varphi_\ell(r) = E_\ell \varphi_\ell(r). \tag{2.5}$$

As remarked before the spectra of h_ℓ^C and $h_{\ell+1}^C$ are "essentially" isospectral. h_ℓ^C can easily be factorized since

$$h_\ell^C = (\frac{d}{dr} + \frac{\ell+1}{r} - \frac{1}{2(\ell+1)})(-\frac{d}{dr} + \frac{\ell+1}{r} - \frac{1}{2(\ell+1)}) - \frac{1}{4(\ell+1)^2} =: a_\ell^{C\dagger} a_\ell^C - \frac{1}{4(\ell+1)^2}. \tag{2.6}$$

This was known to Schrödinger. The ℓ-th ground state wave function can be obtained from equating $a_\ell^C \varphi_\ell^0(r) = 0$, which yields $\varphi_\ell^0(r) = r^{\ell+1}\exp(-r/2(\ell+1))$ and yields eigenvalues $E_\ell^0 = -1/4(\ell+1)^2$.

From the definition (2.6) of a_ℓ^C and $a_\ell^{C\dagger}$ we obtain the algebra

$$[a_\ell^C, a_\ell^{C\dagger}] = -2\frac{\ell+1}{r^2}, \qquad h_{\ell+1}a_\ell^C = a_\ell^C h_\ell, \tag{2.7}$$

which indicates that a_ℓ^C and $a_\ell^{C\dagger}$ map from angular momentum ℓ-states to $(\ell+1)$-states and vice versa. We obtain the remaining eigenfunctions corresponding to eigenvalue E_L^0 by

$$\varphi_L^\ell(r) \propto (\prod_{\lambda=L}^{\ell+1} a_\lambda^C)\varphi_L^0(r) \tag{2.8}$$

as long as $0 < \ell < L$. L plays the role of the principal quantum number.

The factorization of the radial Schrödinger operator for the three-dimensional harmonic oscillator is obtained, for example, by writing

$$H_\ell^0 = -\frac{d^2}{dr^2} + \frac{\ell(\ell+1)}{r^2} + \frac{r^2}{4} = a_\ell^{0\dagger} a_\ell^0 + \ell + \frac{3}{2}, \qquad a_\ell^{0\dagger} = -\frac{d}{dr} - \frac{\ell+1}{r} + \frac{r}{2}, \tag{2.9}$$

and the algebra of operators a_ℓ^0, $a_\ell^{0\dagger}$, the spectrum and eigenfunctions are easily obtained.

The general formulation of the factorization scheme starts from the system of first order equations [9]:

$$A_{n+1}^\dagger \psi_{n,n_0} = \bar{G}_{n+1} \psi_{n+1,n_0}, \qquad A_{n+1} \psi_{n+1,n_0} = G_{n+1} \psi_{n,n_0}, \tag{2.10}$$

where $\bar{G}_n$ and G_n may depend on a real parameter E. The label n_0 is fixed by the vacuum condition

$$A_{n_0} \psi_{n_0,n_0} = 0. \tag{2.11}$$

Cross multiplication of (2.10) yields as a consistency condition

$$A_{n+1} A_{n+1}^\dagger - A_n^\dagger A_n = G_{n+1} \bar{G}_{n+1} - \bar{G}_n G_n \tag{2.12}$$

for any value of the parameter n. From the above scheme we obtain a sequence of "essentially" isospectral factorized Hamiltonians

$$H_n = A_n^\dagger A_n + \varepsilon_n, \tag{2.13}$$

where ε_n are constants. From (2.11) and (2.10) we obtain the condition that the sequence breaks off

$$G_{n_0}(E) = 0, \tag{2.14}$$

which determines the value of E. The parameter remains constant for the whole sequence labelled by n_0. All solutions for such a sequence are obtained from (2.10) by

$$H_n \psi_{n,n_0} = (G_n(E)\bar{G}_n(E) + \varepsilon_n)\psi_{n,n_0}, \qquad \psi_{n,n_0} = \prod_{r=n_0}^{n} \frac{A_{r+1}^\dagger}{\bar{G}_{r+1}} \psi_{n_0,n_0}. \tag{2.15}$$

Remarks: Infeld and Hull [10] based their formulation on the replacement of the Sturm-Liouville eigenvalue problems

$$(\frac{d^2}{dx^2} + v(x,m))\psi(x,m) = \lambda\psi(x,m) \tag{2.16}$$

by a pair of eigenvalue problems of the type

$$\begin{aligned} A_{m+1} A_{m+1}^\dagger \psi(x,m) &= (\lambda - L(m+1))\psi(x,m), \\ A_m^\dagger A_m \psi(x,m) &= (\lambda - L(m))\psi(x,m), \end{aligned} \tag{2.17}$$

where $L(m)$ is an unknown function of the parameter m and A_m is of the form

$$A_m = \frac{d}{dx} + w(x,m) \tag{2.18}$$

and maps solutions of (2.16) into each other

$$A_m\psi(x,m) \propto \psi(x,m-1), \qquad A^{\dagger}_{m+1}\psi(x,m) \propto \psi(x,m+1). \tag{2.19}$$

(2.12) results as consistency condition. The above scheme is obviously identical to the factorization method presented before.

In order to illustrate the above scheme we discuss the relativistic Coulomb problem: We start from the Dirac operator $H_D = \vec{\alpha}\vec{p} + \beta m - \alpha/r$ on $L^2(\mathbf{R}^3, d^3x) \otimes \mathbf{C}^4$ and introduce large and small spinor components. On $L^2(\mathbf{R}^+, dr) \otimes \mathbf{C}^2$ we obtain two coupled radial equations of the form

$$\begin{aligned} (\frac{d}{dr} - \frac{k}{r})G(r) &= (1 - \frac{\alpha}{r} - E)F(r) \\ (\frac{d}{dr} + \frac{k}{r})F(r) &= (1 + \frac{\alpha}{r} + E)G(r), \end{aligned} \tag{2.20}$$

where k denotes the eigenvalue of the operator $K = \tau_3 \otimes (\vec{\sigma}\vec{L} + 1)$ acting on $L^2(\mathbf{R}^3, d^3x) \otimes \mathbf{C}^4$. τ_3 denotes the third Pauli matrix. The eigenvalues of K are $k = \pm(j + 1/2)$ and the total angular momentum j equals $\ell \pm 1/2$. For fixed j the spectrum is always double degenerate except for the lowest state of the ladder. A transformation of the type

$$\begin{aligned} \bar{G} &= (\sqrt{k-\alpha} + \sqrt{k+\alpha})G + (\sqrt{k-\alpha} - \sqrt{k+\alpha})F \\ \bar{F} &= (\sqrt{k-\alpha} - \sqrt{k+\alpha})G + (\sqrt{k-\alpha} + \sqrt{k+\alpha})F \end{aligned} \tag{2.21}$$

has been introduced by Infeld and Hull and studied further by Biedenharn and more recently by Thaller. (2.21) allows to obtain a form suitable for the factorization procedure. Introducing $\rho = Er$, $\gamma = \sqrt{k^2 - \alpha^2}$ we obtain from (2.20) and (2.21)

$$\begin{aligned} (-\frac{d}{d\rho} + \frac{\gamma}{\rho} - \frac{\alpha}{\gamma})\bar{G} &= (\frac{k}{\gamma} - \frac{1}{E})\bar{F} \\ (\frac{d}{d\rho} + \frac{\gamma}{\rho} - \frac{\alpha}{\gamma})\bar{F} &= (\frac{k}{\gamma} + \frac{1}{E})\bar{G}. \end{aligned} \tag{2.22}$$

γ can be interpreted as the (noninteger) parameter for the factorization scheme and therefore (2.22) fits precisely into it, where the operators A_γ, $A^{\dagger}_\gamma$ are given by

$$A_\gamma = \frac{d}{d\rho} + \frac{\gamma}{\rho} - \frac{\alpha}{\gamma}, \qquad A^{\dagger}_\gamma = -\frac{d}{d\rho} + \frac{\gamma}{\rho} - \frac{\alpha}{\gamma} \tag{2.23}$$

and $g_\gamma = \frac{k}{\gamma} - \frac{1}{E_\gamma}$, $\bar{g}_\gamma = \frac{k}{\gamma} + \frac{1}{E_\gamma}$. The ladder operators for the relativistic case are obtained from the nonrelativistic ones by replacing the angular momentum ℓ by the noninteger valued parameter γ. Equs. (2.21) can be written in the form

$$A^{\dagger}_\gamma \psi_{\gamma-1} = \bar{g}_\gamma \psi_\gamma, \qquad A_\gamma \psi_\gamma = g_\gamma \psi_{\gamma-1} \tag{2.24}$$

where $(\bar{G}, \bar{F}) = (\psi_{\gamma-1}, \psi_\gamma)$. In order to obtain the whole spectrum of H_D we need a sequence of radial equations [11]. This is obtained by replacing γ in (2.23) and (2.24) by $\gamma_n = \gamma + n$, with $n = 0, 1, 2, \ldots$. It is easy to check that

$$A^\dagger_{\gamma+n} = -\frac{d}{d\rho} + \frac{\gamma+n}{\rho} - \frac{\alpha}{\gamma+n}, \qquad A_{\gamma+n} = \frac{d}{d\rho} + \frac{\gamma+n}{\rho} - \frac{\alpha}{\gamma+n} \tag{2.25}$$

together with

$$g_{\gamma+n} = \frac{k_n}{\gamma+n} - \frac{1}{E_n}, \qquad \bar{g}_{\gamma+n} = \frac{k_n}{\gamma+n} + \frac{1}{E_n}, \qquad k_n = \sqrt{\gamma_n^2 + \alpha^2} \tag{2.26}$$

fulfil the consistency condition. The spectrum is obtained from the vacuum condition $g_{\gamma+n}(E_n) = 0$ and leads to the Sommerfeld eigenvalue formulae after we eliminate γ:

$$E_{n,k} = \frac{1}{\sqrt{1 + \dfrac{\alpha^2}{(n + \sqrt{k^2 - \alpha^2})^2}}}.$$

The eigenfunctions can also be obtained by the algebraic procedure.

Remarks: In the nonrelativistic Coulomb problem there exists a conserved quantity, the Runge-Lenz vector, which stabilizes the ellipses

$$\vec{F} = \vec{L} \times \vec{p} - \vec{p} \times \vec{L} + \alpha \frac{\vec{x}}{|\vec{x}|}, \qquad [\vec{F}, H_C] = 0, \tag{2.28}$$

where $H_C = \frac{p^2}{2} - \frac{\alpha}{r}$ on $L^2(\mathbf{R}^3, d^3x)$. This implies the $O(4)$ invariance group for the bound state problem and the $O(3,1)$ symmetry for the scattering problem. If one applies $\vec{F}$ to vectors of the form $Y_\ell^m(\vartheta, \varphi)\varphi_\ell(r)$ with $m = \ell$, the resulting radial operator is identical to the ladder operator of equ. (2.6). This implies a relationship between the factorization and the group theoretical structure of the Coulomb problem.

It is not so well-known that an additional conserved quantity besides the total angular momentum $\vec{J}$ and the spin-orbit operator K exists also for the relativistic Coulomb problem. This was introduced by Johnson and Lippman and further studied by Biedenharn, and is given by

$$R = \vec{\sigma} \otimes \left(\frac{1}{2\alpha}\tau_3(\vec{L} \times \vec{p} - \vec{p} \times \vec{L}) + \frac{\vec{x}}{|\vec{x}|}\right) + \frac{1}{|\vec{x}|}\tau_2 \cdot K. \tag{2.29}$$

This integral of motion $[H_D, R] = 0$ explains the degeneracy for $\pm k$. In the nonrelativistic limit $R \to \vec{\sigma} \cdot \vec{F}$. R anticommutes with K, $\{R, K\} = 0$, which implies that R maps eigenstates of K to eigenvalue k into those with value $-k$.

The connection of R to the factorization scheme is seen if we write R as

$$R = \tau_3 \otimes \frac{\vec{\sigma} \cdot \vec{x}}{|\vec{x}|}\gamma\left(\frac{d}{d\rho} - \frac{\gamma}{\rho} + \frac{\alpha}{\gamma}\right). \tag{2.30}$$

It might be that there exists a supersymmetry for the relativistic Coulomb problem which incorporates R as well.

2.2 Factorization – Susy QM

The $N = 2$ superalgebra $S(2)$ was introduced by Witten and is given in terms of the supercharge Q and its adjoint $Q^\dagger$ by

$$Q^2 = Q^{\dagger 2} = 0, \qquad \{Q, Q^\dagger\} = H, \qquad [H, Q] = [H, Q^\dagger] = 0. \tag{2.31}$$

We realize this algebra by putting

$$Q = \sigma^+ A, \qquad Q^\dagger = \sigma^- A^\dagger, \qquad A = \frac{d}{dx} + W(x), \tag{2.32}$$

and obtain

$$H = -\frac{d^2}{dx^2} + W^2(x) + \sigma_3 W'(x). \tag{2.33}$$

The Hamiltonian of susy QM consists therefore of a pair of Schrödinger Hamiltonians $H_+ = AA^\dagger$ and $H_- = A^\dagger A$ which are "essentially" isospectral and the eigenfunctions to nonzero energies are double degenerate:

$$H_\pm = p^2 + V_\pm(x), \qquad V_\pm = W^2(x) \pm W'(x). \tag{2.34}$$

In addition, we observe that the spectrum of H is nonnegative.

Iff $\varepsilon = \inf \operatorname{spec} H$ equals zero we call the supersymmetry to be unbroken. This means that $Q|0\rangle = 0$ and $H|0\rangle = 0$, where $|0\rangle$ denotes the ground state wave function. Q annihilates the "vacuum", the symmetry is realized. Broken supersymmetry means $\varepsilon > 0$. We do not call this case spontaneously broken symmetry, since the representations of the superalgebra are still superunitarily equivalent.

2.3 Connection of Facorization and Shape Invariance Conditions

If we consider a one-parameter family of problems and assume that $V_+(a, x)$ and $V_-(a, x)$ are related to each other by [12]

$$V_+(a, x) = V_-(\bar{a}, x) + L(\bar{a}) \quad \text{for } \bar{a} = f(a), \tag{2.35}$$

where $L(\bar{a})$ is assumed to be independent of x, we obtain solvable model Hamiltonians. (2.35) implies for the superpotential the condition

$$W^2(a, x) + W'(a, x) - W^2(\bar{a}, x) + W'(\bar{a}, x) = 2L(\bar{a}). \tag{2.36}$$

The spectrum of $H = p^2 + V_-(a, x) = H_-$ is determined by constructing a series of Hamiltonians H_n, $n = 0, 1, 2, \ldots$ with $H_0 = H_-$, $H_1 = H_+$ and

$$H_n = p^2 + V_-(a_n, x) + \sum_{k=1}^{n} L(a_k), \tag{2.37}$$

where a_n is determined by n applications of the function $f : a_n = f^{(n)}(a)$.

Historical Notes: The connection of the structure of susy quantum mechanics to the Riccati equation (1724) is obtained, if we consider the differential equation $u'' = a(x)u$ for continuous $a(x)$ on an interval. If $u(x)$ is nowhere vanishing, we may introduce $W(x) = \frac{d}{dx} \ln u(x)$, which solves the Riccati equation $W' + W^2 = a$.

Bernoulli solved already in 1702 the nonlinear equation $w' + w^2 + x^2 = 0$ by transforming it to the second order equation $y'' + x^2 y = 0$.

In 1827 Cauchy asked the question about factorization of differential equations in analogy to the fundamental theorem of algebra. Jacobi proved that any self-adjoint differential operator L of order $2m$ can be written as product of a differential operator p of order m times its adjoint $L = p\bar{p}$. Cayley (1868) transformed $u'' = x^{2q-2}u$ into $y' + y^2 = x^{2q-2}$. Frobenius proved first the complete factorization which implies the existence of $v_i(x)$ such that

$$L(y) := y^{(n)} + p_1(x)y^{(n-1)} + \ldots = \prod_{i=1}^{n}\left(\frac{d}{dx} - \frac{v_i'(x)}{v_i(x)}\right)y. \tag{2.38}$$

Darboux's theorem dates back to 1882 and starts from $L\phi = (-\frac{d^2}{dx^2} + v(x))\phi = 0$, uses the factorization $L = A^\dagger A$ with $A = \frac{d}{dx} - (\ln \phi_0)'$ assuming $\phi_0 \neq 0$ and asserts that

$$\tilde{L} = AA^\dagger = -\frac{d^2}{dx^2} + v(x) - 2\frac{d}{dx}\left(\frac{\phi_0'}{\phi_0}\right) \quad \text{solves} \quad \tilde{L}(A\phi) = 0. \tag{2.39}$$

This result is related to Crum's and Krein's transformation, which are closely connected to the inverse scattering problem.

3 Second Application: Order of Levels of Schrödinger Operators

Assume, somebody asks the following simple question: How are the levels of a three-dimensional Schrödinger problem with rotation symmetric potential distributed? We consider therefore

$$H_\ell u_{n,\ell}(r) = \left(-\frac{d}{dr^2} + \frac{\ell(\ell+1)}{r^2} + V(r)\right)u_{n,\ell}(r) = E_{n,\ell}u_{n,\ell}(r) \quad \text{on } L^2(\mathbf{R}^+, dr), \tag{3.1}$$

where n denotes the number of nodes the reduced wave function $u_{n,\ell}$ has within $[0,\infty)$.

Let us start with a few historical remarks. A. Martin asked first the question concerning low lying states in charmonium. A potential of the form $-\alpha/r + (1-\alpha)r$ shows a number of convenient features and between the ground state and the first excited S-state three P-states were found. In addition, a D-state above the first excited S-state exists. The question is, which class of potentials shows such a level ordering. If we study a combination of a Coulomb and an oscillator potential, we expect the ordering

$$E_{1S} \leq E_{2P} \leq E_{2S} \leq E_{3D} \leq E_{3P} \leq E_{3S} \ldots \tag{3.2}$$

and, in addition, if we introduce the notation of equ. (3.1), we expect that $E_{n,\ell} > E_{n-1,\ell+1}$.

The first results of A. Martin and myself concerned the low lying levels [13]. We found local conditions on the potential such that the ordering (3.2) is implied. For example, $\frac{d^3}{dr^3}(r^2V) \gtrless 0 \Rightarrow E_{2S} \gtrless E_{2P}$. We started from the case of degeneracy and used the nodal structure of the wave functions involved. As a next step, it was shown that within the WKB approximation

$$n + \frac{1}{2} = \oint_{E=E_n} dr p_r, \qquad N = n + \ell + 1, \qquad \Delta V \gtrless 0 \Rightarrow \left.\frac{\partial E}{\partial \ell}\right|_N \gtrless 0 \tag{3.3}$$

the splitting of levels is controlled by the Laplacian of the potential. In addition, we found within first order perturbation theory [14] the

Theorem: Let $V(r) = -\frac{1}{r} + \lambda v(r)$, then

$$\Delta v(r) \gtrless 0 \Rightarrow \delta = \frac{d}{d\lambda}(E_{n,\ell} - E_{n-1,\ell+1})|_{\lambda=0} \gtrless 0. \tag{3.4}$$

For the proof we found out the sign of

$$\delta = \int_0^\infty dr v(r)(u_{n,\ell}^2 - u_{n-1,\ell+1}^2). \tag{3.5}$$

But observe that there are ladder operators $a_\ell^C, a_\ell^{C\dagger}$ connecting $u_{n,\ell}$ and $u_{n-1,\ell+1}$. A similar result was obtained by starting from the harmonic oscillator.

The generalization to the nonperturbative case has been obtained by B. Baumgartner, A. Martin and myself [15]:

Theorem:

$$\text{If } \Delta V(r) \gtrless 0 \quad \forall r \neq 0 \Rightarrow E_{n,\ell} \gtrless E_{n-1,\ell+1}. \tag{3.6}$$

The ideas of the proof are the following

a) We used the factorization: Let u_ℓ denote the ground state wave function for fixed ℓ and $g_\ell = -u_\ell'/u_\ell$. H_ℓ can be expressed as

$$H_\ell = A_\ell^\dagger A_\ell + E_\ell, \qquad A_\ell^\dagger = -\frac{d}{dr} + g_\ell(r), \tag{3.7}$$

which implies that

$$(H_\ell + 2g_\ell')A_\ell = A_\ell H_\ell. \tag{3.8}$$

We start from H_ℓ and obtain an operator $H_\ell + 2g_\ell'$ which has "essentially" the same spectrum like H_ℓ.

b) Next we have to compare $H_\ell + 2g_\ell'$ with $H_{\ell+1}$ or g_ℓ' with $(\ell+1)/r^2$:

Lemma:

$$\text{If } \Delta V \gtrless 0 \quad \forall r \neq 0 \Rightarrow g'_\ell \gtrless \frac{\ell+1}{r^2}. \tag{3.9}$$

This is the essential step in the proof. We used the Coulomb potential a few times as comparison potential. In order to conclude that $g'_\ell < (\ell+1)/r^2$ it is actually sufficient to assume that $\Delta V < 0$ for $r < R$ for some R fixed, and $dV/dr < 0$ for $r > R$.

c) From the Min-Max principle we finally conclude that

$$H_\ell + 2g'_\ell \gtrless H_{\ell+1}, \quad \text{if } g'_\ell \gtrless \frac{\ell+1}{r^2}, \tag{3.10}$$

and therefore $E_{n,\ell} \gtrless E_{n-1,\ell+1}$.

Remarks: Since it is possible to transform the Coulomb problem to the oscillator, a similar result as above holds starting from the latter. More generally we introduced r^α as a new variable and obtained

Theorem: Let $V \geq 0$, then

$$\begin{aligned} D_\alpha V > 0, \qquad & 1 < \alpha < 2 \;\Rightarrow\; E_{n,\ell} > E_{n-1,\ell+\alpha} \\ D_\alpha V < 0, \qquad & \alpha > 2 \text{ or } \alpha < 1 \;\Rightarrow\; E_{n,\ell} < E_{n-1,\ell+\alpha} \end{aligned} \tag{3.11}$$

where

$$D_\alpha = \frac{d^2}{dr^2} + (5-3\alpha)\frac{1}{r}\frac{d}{dr} + 2(1-\alpha)(2-\alpha)\frac{1}{r^2}.$$

It would be surprising, if the above techniques could not be applied to the continuous spectrum as well. It leads to

Theorem: Let

$$V(r) = -\frac{Z}{r} + v(r), \qquad \Delta V(r) \gtrless 0 \,\forall r \neq 0 \Rightarrow \delta_{\ell+1}(E) \gtrless \delta_\ell(E), \tag{3.12}$$

where $\delta_\ell(E)$ denotes the ℓ-th phase shift for energy E.

We quote also a perturbative result for the level ordering of the Dirac operator [16]. Let $E_{n,k}(\lambda)$ and $E_{n,-k}(\lambda)$ be energies of the operator $h = \vec{\alpha}\vec{p} + \beta m - \alpha/r + \lambda v(r)$, which become degenerate for $\lambda = 0$, then

$$\Delta V(r) \gtrless 0 \,\forall r \neq 0 \Rightarrow \lim_{\lambda \to 0} \frac{E_{n,-k}(\lambda) - E_{n,k}(\lambda)}{\lambda} \gtrless 0. \tag{3.13}$$

The sign of the Laplacian of the potential plays again a crucial role for the level ordering.

Applications: As we have mentioned already, our main motivation was connected to the quarkonia spectra. Two features occur: $E_{n,\ell} > E_{n-1,\ell+1}$ is implied by $\Delta V > 0$. In order to motivate this condition we write the force as $-V' = -Z(r)/r^2$, and introduce an effective charge $Z(r)$. $Z(r)$ should therefore decrease as $r \to 0$ which is in accordance with asymptotic freedom. $E_{n,\ell} < E_{n-1,\ell+2}$ is implied by $\frac{d}{dr}\frac{1}{r}\frac{d}{dr}V < 0$. The potential should therefore be increasing and concave. This follows from lattice QCD and reflection positivity.

It turned out that a further application of our inequalities can be found in atomic physics. If we consider atoms with closed shells plus one outer electron within the Hartree approximation

$$(-\Delta - \frac{Z}{r} + \int \frac{d\vec{y}\rho(\vec{y})}{|\vec{x}-\vec{y}|})\psi_{n,\ell} = E_{n,\ell}\psi_{n,\ell}, \tag{3.14}$$

the density $\rho(\vec{y})$ is not known. Nevertheless, we deduce from our level ordering theorem that $E_{n,\ell} < E_{n-1,\ell+1}$. If we look at the periodic table and check how nature fills the shells for atoms, we realize that after the $1S$ shell the $2S$ shell is filled before the $2P$ shell, after the $2P$ shell comes the $3S$ shell, and then the $3P$ shell, in accordance with our theorem. The $3D$ state actually overcomes also the $4S$ level, which makes irregularities in the filling scheme and goes beyond our results.

4 Inverse Scattering Method and Solitons – Relation to Susy Quantum Mechanics

We start with an example and determine the susy partner of a constant potential:

Example: In order to factorize the equation $-\frac{d^2}{dx^2}\phi = -\kappa^2\phi$, we solve $f^2 - f' = \kappa^2$ for f. We put first $f = -u'/u$, solve $u'' = \kappa^2 u$ for $u = \alpha \cosh \kappa(x - x_0)$. We require symmetry and take $x_0 = 0$. $f(x) = -\kappa \tanh \kappa x$ and the new potential becomes

$$V(x) = f^2 + f' = \kappa^2 - \frac{2\kappa^2}{\cosh^2 \kappa x}. \tag{4.1}$$

V has one bound state and is reflectionless. The Schrödinger equation for the new potential is solved for energy $E = q^2$ by

$$\psi = A\phi = (iq - \kappa \tanh \kappa x)e^{iqx} \stackrel{x\to\pm\infty}{\longrightarrow} (iq \mp \kappa)e^{iqx}. \tag{4.2}$$

We recognize that no term proportional to e^{-iqx} occurs, the potential is reflectionless, and the S-matrix equals $S = (iq - \kappa)/(iq + \kappa)$.

The procedure of adding one bound state to a given potential can be solved in general. With the help of the inverse scattering method [17] one may add or subtract as many bound states as one likes. In order to determine uniquely a potential with N bound states one has to give the scattering data $\{R(k), \varepsilon_\ell, c_\ell\}$, where $R(k)$ denotes the reflection coefficient to energy $E = k^2$, ε_ℓ are the bound state energies and c_ℓ normalization constants for the ℓ-th bound state wave function. The potential is recovered by defining a kernel

$$G(x,y) := \int_{-\infty}^{\infty} \frac{dk}{2\pi} e^{ik(x+y)}R(k) + \sum_{\ell=1}^{N} c_\ell^2 e^{-\kappa_\ell(x+y)}, \qquad \kappa_\ell^2 = -\varepsilon_\ell, \tag{4.3}$$

and solving the Gelfand-Levitan-Marchenko equation for the kernel $K(x,y)$:

$$K(x,y) + G(x,y) + \int_x^\infty dz K(x,z) G(z,y) = 0 \text{ for } y > x. \tag{4.4}$$

The potential is recovered by taking $-2\frac{d}{dx}K(x,x) = V(x)$. (4.4) can be solved easily if $R(k) \equiv 0$ since it becomes a Fredholm equation with separable kernel. All reflectionless potentials are then given by

$$V(x) = -2\frac{d^2}{dx^2} \ln \det(1 + C(x)), \qquad C_{\ell m}(x) = \frac{c_\ell c_m}{\kappa_\ell + \kappa_m} e^{(\kappa_\ell + \kappa_m)x}. \tag{4.5}$$

This solves the problem of adding or subtracting bound states or changing normalization constants starting from a constant potential. The two potential problem, starting from a particular one, is similar. We quote the result for adding one bound state:

Theorem: Let $H = -\frac{d^2}{dx^2} + V(x)$ and denote by $f_1(k,x)$, $f_2(k,x)$ Jost solutions with asymptotic behaviour

$$f_1(k,x) \overset{x\to\infty}{\simeq} e^{ikx} \qquad \text{and} \qquad f_2(k,x) \overset{x\to-\infty}{\simeq} e^{-ikx}.$$

Assume that $\kappa > \kappa_n > \kappa_{n-1} > \ldots > \kappa_1$ and take $\alpha > 0$. Define $g_\alpha(x) = f_1(i\kappa, x) + \alpha f_2(i\kappa, x)$. The new Hamiltonian

$$\bar{H} = -\frac{d^2}{dx^2} + V(x) - 2\frac{d^2}{dx^2} \ln g_\alpha(x) \tag{4.6}$$

has $n+1$ bound states at energies $-\kappa^2 < -\kappa_n^2 < \ldots < -\kappa_1^2$. The new normalization constants are given by

$$\bar{c}_j = \frac{\kappa + \kappa_j}{\kappa - \kappa_j} c_j, \qquad j = 1, \ldots, n, \qquad \bar{c}_{n+1} = \frac{2\kappa T(i\kappa)}{\alpha}. \tag{4.7}$$

The new reflection- and transmission coefficients become

$$\bar{T}(k) = \frac{k + i\kappa}{k - i\kappa} T(k), \qquad \bar{R}(k) = -\frac{k + i\kappa}{k - i\kappa} R(k). \tag{4.8}$$

The idea of the proof follows the factorization method and shows the connection to susy. From the construction it follows that $g_\alpha^{-1} \in L^2$. We have

$$H + \kappa^2 = A^\dagger A, \qquad \bar{H} + \kappa^2 = AA^\dagger, \qquad A = g_\alpha \frac{d}{dx} g_\alpha^{-1}, \qquad A^\dagger = -g_\alpha^{-1} \frac{d}{dx} g_\alpha, \tag{4.9}$$

and the theorem follows from studying the asymptotic behaviour of solutions of $\bar{H}$.

We mention that the direct step from a potential $V(x)$ to scattering data $\{R(k), \varepsilon_\ell, c_\ell\}$ and the inverse step are the essential tools to solve a large class of nonlinear partial differential equations. If a "weak" nonlinearity and a "weak" dispersion allows the formation of solitons the inverse transform method may be applied. Examples are given by the KdV, m-KdV, Sine-Gordon, nonlinear Schrödinger equation, Toda lattice and all its generalizations. We consider, as an example, the KdV equation for $v(t,x)$:

$$v_t = 6vv_x - v_{xxx}. \tag{4.10}$$

Starting from initial values $v(0,x)$, we obtain scattering data at time zero $\{R_0(k), \varepsilon_\ell^0, c_\ell^0\}$ in the direct step. The time evolution of the latter is simple:

$$c_\ell^t = e^{4\kappa_\ell^3 t} c_\ell^0, \qquad R_t(k) = e^{-8ik^3 t} R_0(k), \qquad T_t(k) = T_0(k), \qquad \varepsilon_\ell^t = \varepsilon_\ell^0. \tag{4.11}$$

(4.11) can be obtained with the help of the Lax-pair, which is behind the integrability of that equation. It is very simple, using the Feynman-Hellmann theorem to check the isospectral property of the KdV flow. The spectrum of the Schrödinger operator $-d^2/dx^2 + v(t,x)$ is independent of t if $v(t,x)$ evolves according to the KdV equation. If we take the time evolution (4.11) for c_ℓ^t in equ. (4.5) we obtain all solitons of that equation.

The above remarks suggest that there should be connections between the susy structure and nonlinear equations. We note that if $V = f^2 - f_x - \kappa^2$ solves the KdV-equation $\bar{V} = f^2 + f_x - \kappa^2$ gives a solution too. In addition, f solves then the m-KdV equation:

$$f_t + 6(\kappa^2 - f^2)f_x + f_{xxx} = 0. \tag{4.12}$$

The analogous relation between Dirac operator Q and Schrödinger operator H for the nonlinear equations is given by the Miura transformation

$$w_t^\pm - 6w^\pm w_x^\pm + w_{xxx}^\pm = (2\phi \pm \frac{\partial}{\partial x})(\phi_t - 6\phi^2\phi_x + \phi_{xxx}), \qquad w^\pm = \phi^2 \pm \phi_x, \tag{4.13}$$

which relates solutions of the KdV equation to those of the m-KdV equation. The connection of the above transformation to commutation rules of susy quantum mechanics has recently been used to investigate all soliton solutions of both equations. A generalization to the KP and m-KP hierarchies are possible. These are three-dimensional integrable nonlinear equations. Recently, the structure of these models became more clear. They result from an orbit of the Lie-group GL_∞ represented within the fermionic Fock space. This connects to second quantization methods. The vertex operator generates solitons in the sense that the N-soliton solutions are obtained by applying N times vertex operators to the vacuum. A connection to conformal quantum field theory results.

5 Dirac Operator and Index Problems: Susy Breaking, Second Quantization

Years ago we studied the nonrelativistic approximation for the three-dimensional Dirac operator with the help of rigorous resolvent techniques. The analytic expansion for energy levels as a function of $1/c^2$ was obtained with commutation relations similar to the algebra of susy [18].

Here we would like to study the one-dimensional Dirac operator of the form

$$h_D = \alpha p + \beta v + \gamma w, \qquad \alpha = \begin{pmatrix} 0 & -i \\ i & 0 \end{pmatrix}, \qquad \beta = \begin{pmatrix} 0 & 1 \\ 1 & 0 \end{pmatrix}, \qquad \gamma = \alpha\beta, \tag{5.1}$$

on $L^2(\mathbf{R}, dx) \otimes \mathbf{C}^2$, and show that susy helps in calculating the index. For simplicity reasons we take $w = const = 0$. h can therefore be identified with a self-adjoint supercharge

$$Q = \begin{pmatrix} 0 & A^\dagger \\ A & 0 \end{pmatrix}, \qquad A = \frac{d}{dx} + \phi(x). \tag{5.2}$$

We assume $\phi \in L^\infty$, $\phi' \in L^1$ and $\int dx(1 + |x|^2)|\phi^2 \pm \phi' - \phi_+^2\theta(x) - \phi_-^2\theta(-x)| < \infty$, where $\phi_\pm$ denote the asymptotic values of the potential ϕ for $x \to \pm\infty$.

If A is a Fredholm operator, which means that it is closed with closed range and $\dim\ker A < \infty$ and $\dim\ker A^\dagger < \infty$ the index is defined by

$$i(A) = \dim\ker A - \dim\ker A^\dagger \in \mathbf{N}. \tag{5.3}$$

If A is not Fredholm, we have to use a certain regularization. The heat kernel regularization has been used by a number of people. We found it very convenient [19] to use the resolvent regularization and to define

$$\begin{aligned} \Delta_z(A) &= (-z)\mathrm{Tr}\{(H_1 - z)^{-1} - (H_2 - z)^{-1}\} \\ \Delta(A) &\doteq \lim_{z\to 0} \Delta_z(A), \end{aligned} \tag{5.4}$$

where H_1 and H_2 are the two Schrödinger operators connected to Q

$$Q^2 = \begin{pmatrix} A^\dagger A & 0 \\ 0 & AA^\dagger \end{pmatrix} = \begin{pmatrix} H_1 & 0 \\ 0 & H_2 \end{pmatrix}, \qquad H_j = -\frac{d^2}{dx^2} + \phi^2 + (-)^j\phi'. \tag{5.5}$$

Remark: If A is Fredholm an expansion of the form

$$(-z)\{(H_1 - z)^{-1} - (H_2 - z)^{-1}\} = P_1 - P_2 - z\sum_{n=0}^{\infty} z^n(T_1^{n+1} - T_2^{n+1}),$$

$$T_j = n\text{-}\lim_{z\to 0}(H_j - z)^{-1}(1 - P_j) \tag{5.6}$$

is convergent in trace norm and the two indices coincide $\Delta(A) = i(A)$.

Our idea was to use relative scattering theory between H_1 and H_2 in order to calculate $\Delta(A)$. We remark that a nonzero index $i(A) \neq 0$ is only sufficient for unbroken supersymmetry. Only if we know, for example, that $\dim\ker A = 0$, a nonvanishing index indicates unbroken susy and vice versa.

Next we summarize the results for the index calculation in the model defined by equ. (5.2). It turns out that

$$\Delta(A) = \begin{cases} (\operatorname{sgn}\phi_+ - \operatorname{sgn}\phi_-)/2 & \text{if } \phi_+ \neq 0, \phi_- \neq 0 \\ (\operatorname{sgn}\phi_+)/2 & \text{if } \phi_+ \neq 0, \phi_- = 0 \\ 0 & \text{if } \phi_+ = \phi_- = 0. \end{cases} \tag{5.7}$$

Half-integer values occur if either H_1 or H_2 (but not both) has a zero energy resonance. We summarize these findings in Table 1.

	# of $E=0$ res		# of $E=0$ bd. st			
	H_1	H_2	H_1	H_2	$\Delta(A)$	$i(A)$
$\phi_+ > 0 > \phi_-$	0	0	1	0	1	1
$\phi_+, \phi_- > 0$	0	0	0	0	0	0
$\phi_- = 0,\ \phi_+ > 0$	1	0	0	0	1/2	–
$\phi_+ = \phi_- = 0$	1	1	0	0	0	–

Table 1

The determination of $\dim\ker A$ and $\dim\ker A^\dagger$ is easy since

$$\begin{aligned} Af = 0 \quad &\Longrightarrow \quad f(x) = f(0)\exp[-\textstyle\int_0^x dt\phi(t)] \overset{x\to\pm\infty}{\sim} O(e^{-\phi_\pm x}) \\ A^\dagger g = 0 \quad &\Longrightarrow \quad g(x) = g(0)\exp[\textstyle\int_0^x dt\phi(t)] \overset{x\to\pm\infty}{\sim} O(e^{\phi_\pm x}). \end{aligned} \tag{5.8}$$

That a resonance contributes one half to $\Delta(A)$, is similar as in Levinson's theorem (which is actually an index theorem).

We remark that the indices are topologically invariant. They depend only on the asymptotic values of $\phi(x)$. In field theory one defines an effective ground state charge $\langle q\rangle$ and obtains with the help of a regularization scheme a relation to the spectral asymmetry or the index: $\langle q\rangle = \Delta(A)/2$.

In order to show how scattering theory helps in calculating the index, we first state our assumptions: We assume that

a) $\mathrm{Tr}((H_1 - z)^{-1} - (H_2 - z)^{-1}) < \infty$ for $z \in \rho(H_1) \cap \rho(H_2)$;

b) let $H_1 - H_2 = V_{12} = u \cdot w$ denote the relative potential, then $u(H_2 - z)^{-1}w$ should be analytic in z in trace topology and $u(H_2 - z)^{-1}$ and $(H_2 - z)^{-1}w$ should be Hilbert-Schmidt operators.

c) Finally, at high energy we require that $\lim_{|z|\to\infty, \mathrm{Im}\, z\neq 0} \det(1 + u(H_2 - z)^{-1}w) = 1$. We deduce the existence of Krein's spectral shift function $\xi_{12}(\lambda)$, such that

$$\mathrm{Tr}\{(H_1 - z)^{-1} - (H_2 - z)^{-1}\} = -\int \frac{d\lambda \xi_{12}(\lambda)}{(\lambda - z)^2} \tag{5.9}$$

for $\lambda \in \rho(H_1) \cap \rho(H_2)$. We quote the relation of ξ_{12} to the relative scattering S-matrix S_{12}, $\det S_{12} = \exp(-2i\pi i \xi_{12}(\lambda))$.

Lemma: By differentiating the logarithmic determinant and using cyclicity under the trace we show that

$$\frac{d}{dz}\ln\det\{1 + u(H_2 - z)^{-1}w\} = \frac{d}{dz}\mathrm{Tr}\ln\{1 + u(H_2 - z)^{-1}w\} = -\mathrm{Tr}\{(H_1 - z)^{-1} - (H_2 - z)^{-1}\}. \tag{5.10}$$

Combining (5.9) and (5.10) gives a formula for the index

$$\Delta_z(A) = z\int_{-\varepsilon}^{\varepsilon} d\lambda \frac{\xi_{12}(\lambda)}{(\lambda - z)^2} + O(z) \overset{z\to 0}{\longrightarrow} \xi_{12}(0_-) - \xi_{12}(0_+). \tag{5.11}$$

For the explicit calculation of the index we consider relative scattering from potential $\phi^2+\phi'$ to $\phi^2 - \phi'$. We define Jost functions $f_j^{\pm}$ for $j = 1, 2$ with asymptotic behaviour

$$f_j^{\pm}(E,x) \overset{x\to\pm\infty}{\longrightarrow} e^{\pm ik_\pm x}, \qquad k_\pm = \sqrt{E - \phi_\pm^2}. \tag{5.12}$$

Scattering data for Hamiltonians H_j can be expressed in terms of Wronskians. The transmission coefficient for the relative scattering is then given by

$$T_{12} = \frac{T_1}{T_2}, \qquad \sqrt{k_-/k_+}\, T_j = \frac{W(f_j^-, \bar{f}_j^-)}{W(f_j^-, f_j^+)}. \tag{5.13}$$

If we study the Volterra integral equation for this scattering problem we identify T_{12}^{-1} with the Fredholm-determinant

$$T_{12}^{-1} = \det(1 - 2|\phi'|^{1/2} \operatorname{sgn} \phi'(H_2 - z)^{-1}|\phi'|^{1/2}). \tag{5.14}$$

Combining with (5.10) yields finally

$$\Delta_z(A) = z\frac{d}{dz}\ln\frac{W(f_1^-, f_1^+)}{W(f_2^-, f_2^+)} \qquad \text{for } z \in \mathbf{C}\setminus\{\sigma_p(H_1)\cup\sigma_p(H_2)\}. \tag{5.15}$$

At this stage we use susy relations for distributional solutions. From

$$f_2^{\pm} = (\pm ik_\pm + \phi_\pm)^{-1} A f_1^{\pm} \Rightarrow W(Af_1^-(z), Af_1^+(z)) = zW(f_1^-(z), f_1^+(z)), \tag{5.16}$$

we deduce our explicit expression for the r.h.s. of (5.15)

$$\frac{W(f_1^-, f_1^+)}{W(f_2^-, f_2^+)} = \frac{ik_+ + \phi_+}{ik_- + \phi_-}. \tag{5.17}$$

Usually one calculates the index for a finite volume and takes limits. Scattering methods allow to work all the time on the whole line.

Example: For the half line problem with $A = \dfrac{d}{dr} + \phi(r)$ on $H_0^{2,1}(0,\infty)$ functions with $\mathcal{H} = L^2([0,\infty), dr)\otimes \mathbf{C}^2$ we obtain results which are summarized in Table 2.

	# of $E=0$ res		# of $E=0$ bd. st			
	H_1	H_2	H_1	H_2	$\Delta(A)$	$i(A)$
$\phi_+ > 0$	0	0	0	0	0	0
$\phi_+ < 0$	0	0	0	1	-1	-1
$\phi_+ = 0$	0	1	0	0	$-1/2$	–

Table 2

A zero energy resonance yields again a fractional index.

Example: Using the inverse scattering method for the Dirac operator h_D of equ. (5.1), we determined all reflectionless potentials of the coupled m-KdV equations. If we take bound state energies $\varepsilon_1, \ldots, \varepsilon_N$ within the gap $(-m, m)$ the calculation of the index yields a continuously varying effective charge

$$\langle q \rangle = \frac{1}{\pi} \sum_{\ell=1}^{N} \alpha_\ell, \qquad \varepsilon_\ell = m \cdot \cos \alpha_\ell. \tag{5.18}$$

Example: Magnetic field model on $L^2(\mathbf{R}^2, d^2x) \otimes \mathbf{C}^2$. There exist two-dimensional susy models too. We define the operator

$$A = (-i\frac{\partial}{\partial x_1} - a_1(x)) + i(i\frac{\partial}{\partial x_2} + a_2(x)), \tag{5.19}$$

where $\vec{a} = (a_1, a_2)$ denotes the vector potential, and choose the gauge where $\vec{a} = (\partial_2\phi, -\partial_1\phi)$ with ϕ behaving asymptotically like

$$\phi(x) = -F \ln |\vec{x}| + C + O(|x|^{-\epsilon}). \tag{5.20}$$

F denotes the total magnetic flux, $F = \int \frac{d^2q}{2\pi} b(q)$, and $b = \partial_1 a_2 - \partial_2 a_1 = -\Delta\phi$ the magnetic field. A is not a Fredholm operator. The pair of Hamiltonians obtained from factorization are

$$H_j = \frac{1}{2}\{(\frac{1}{i}\vec{\nabla} - \vec{a})^2 - (-)^j b\}, \qquad j = 1, 2. \tag{5.21}$$

Since $\Delta(A)$ is topologically invariant we may choose

$$\phi(r, R) = \begin{cases} -\dfrac{F}{2}\dfrac{r^2}{R^2} & \text{if } |x| = r \leq R \\[2ex] -\dfrac{F}{2}(1 + \ln \dfrac{r^2}{R^2}) & \text{if } r \geq R. \end{cases} \tag{5.22}$$

We have explicitly indicated the dependence on R. Let U_ε be the unitary group of dilations which transforms $g(x) \to U_\varepsilon g(x) = \varepsilon^{-1} g(x/\varepsilon)$ and therefore $H_j(R)$ into $U_\varepsilon H_j(R) U_\varepsilon^{-1} = \varepsilon^2 H_j(\varepsilon R)$. Let $S_{12}(R)$ be the scattering operator. After decomposition according to the spectral parameter λ we obtain $S_{12}(\lambda, R)$. Scaling yields

$$S_{12}(\lambda, R) = S_{12}(\varepsilon^2\lambda, R/\varepsilon), \qquad \xi_{12}(\lambda, R) = \xi_{12}(\varepsilon^2\lambda, R/\varepsilon), \tag{5.23}$$

and finally that ξ_{12} is independent of R. An explicit model calculation yields as a result $\Delta(A) = -F$.

Example: Wess-Zumino Model Jaffe, Lesniewski and Lewenstein [20] have treated the interaction of one complex boson $z(t)$ with two fermions $\psi_1(t)$ and $\psi_2(t)$. They realized the algebra

$$\{\bar{\psi}_1, \psi_2\} = \{\bar{\psi}_2, \psi_1\} = 1, \qquad \{\psi_i, \psi_j\} = \{\bar{\psi}_i, \bar{\psi}_j\} = 0 \tag{5.24}$$

by Euclidean γ-matrices

$$\psi_1 = \frac{\gamma_0 - i\gamma_3}{2}, \qquad \bar{\psi}_1 = \frac{\gamma_1 + i\gamma_2}{2}, \qquad \psi_2^* = \bar{\psi}_1, \qquad \psi_1^* = \bar{\psi}_2, \tag{5.25}$$

$$\gamma_0 = \begin{pmatrix} 0 & 1 \\ 1 & 0 \end{pmatrix}, \qquad \gamma_j = \begin{pmatrix} 0 & i\sigma_j \\ -i\sigma_j & 0 \end{pmatrix}, \qquad \gamma = \begin{pmatrix} 1 & 0 \\ 0 & -1 \end{pmatrix},$$

where $\gamma = \gamma_0\gamma_1\gamma_2\gamma_3$ denotes the grading operator. The two-dimensional model is called holomorphic quantum mechanics. There are two supercharges Q_1 and Q_2:

$$Q_1 = i\bar{\psi}_1\partial - i\bar{\psi}_2\partial V^*, \qquad Q_2 = i\psi_2\bar{\partial} + i\psi_1\partial V, \qquad Q = Q_1 + Q_2, \tag{5.26}$$

which yield explicitly

$$Q = \begin{pmatrix} 0 & Q_- \\ Q_+ & 0 \end{pmatrix}, \qquad Q_- = \begin{pmatrix} \partial V & i\partial \\ i\bar{\partial} & -\partial V^* \end{pmatrix}, \qquad Q_+ = \begin{pmatrix} \partial V^* & i\partial \\ i\bar{\partial} & -\partial V \end{pmatrix}. \tag{5.27}$$

The Hamiltonian $H = Q^2$ has the following matrix representation

$$H = \begin{pmatrix} h_1 & 0 \\ 0 & h_2 \end{pmatrix}, \tag{5.28}$$

$$\begin{aligned} h_1 &= Q_-Q_+ = h_2 + \begin{pmatrix} 0 & -i\partial^2 V \\ i\partial^2 V^* & 0 \end{pmatrix}, \\ h_2 &= Q_+Q_- = (-\partial\bar{\partial} + |\partial V|^2)\begin{pmatrix} 1 & 0 \\ 0 & 1 \end{pmatrix}. \end{aligned}$$

As examples one may take $V(z) = \lambda z^n$, $n \in \mathbf{N}_+$. It is then easy to show that $\dim\ker h_2 = 0$; there are no zero modes of Q_-, since $\partial V\Omega = \bar{\partial}\Omega = 0$ yield $\Omega = 0$. The calculation of $i(Q_+) = \dim\ker Q_+ = n - 1$ is tedious. There results an interesting conclusion for $n \geq 2$: Although the ground state is degenerate, susy is unbroken.

Remark on second quantization of susy problems: The above mentioned model has been treated also by second quantization methods. We shall make here only a remark concerning the general setting and mention one result. One starts from the one-particle Hilbert space $\mathcal{H} = L^2(T^1) \oplus L^2(T^1)$ over the torus and realizes the CAR on the fermionic Fock space $\mathcal{H}_F$ over $\mathcal{H}$. Time zero fermion fields $\psi_j(x)$, $j = 1, 2$, are defined by a mode expansion. Similarly, the CCR are realized on the bosonic Fock space $\mathcal{H}_B$ and time zero fields $\varphi(x)$ and $\pi(x)$ are obtained. The total Hilbert space is given by $\mathcal{H} = \mathcal{H}_F \otimes \mathcal{H}_B$.

The free supercharge Q_0 is defined by

$$Q_0 = \int_{T_1} \frac{dx}{\sqrt{2}}\{\psi_1(\pi - \partial_1\varphi^* - im\varphi) + \psi_2(\pi^* - \partial_1\varphi - im\varphi^*) + h.c.\}, \tag{5.29}$$

and yields the free Hamiltonian

$$H_0 = Q_0^2 = \int_{T_1} dx\{|\pi(x)|^2 + |\partial_1\varphi(x)|^2 + m^2|\varphi(x)|^2 + \bar{\psi}(x)(i\gamma_1\partial_1 - m)\psi(x)\}, \tag{5.30}$$

on the domain of vectors having only finitely many particles. H_0 is equal to its normal ordered form.

Together with E. Langmann we have recently studied the susy external field problem. The interaction is given by adding

$$Q_I^e = -i\int_{T_1} dx(\psi_1 W\varphi + \psi_2 W^*\varphi^\dagger) + h.c. \tag{5.31}$$

to $Q_0 : Q = Q_0 + Q_I^e$. This yields interactions for the bosons of the form $W^2 \pm W'$ and for the fermions of the form $\sigma_1 W$. We studied the quantization map, the susy generalization of the Schwinger term and the question of equivalent and inequivalent representation of the algebra of CCR and CAR.

The nonlinear field theoretic model on the cylinder has been studied in [21]. The total supercharge is constructed by introducing a regularization κ

$$Q(\kappa) = Q_0 + Q_\kappa^I, \qquad H(\kappa) = Q^2(\kappa), \tag{5.32}$$

$$Q_\kappa^I = -i\int_{T_1} dx\{\psi_{1\kappa}(\partial P)(\varphi_\kappa) + \psi_{2\kappa}(\partial P)(\varphi_\kappa)\} + h.c.,$$

where $\varphi_\kappa = \chi_\kappa * \varphi$ and $\psi_{j,\kappa} = \chi_\kappa * \psi_j$ and χ_κ denotes a cutoff function. It turns out that the resolvents of $Q(\kappa)$ and $H(\kappa)$ converge in operator norm to limits and the index $i(Q(\kappa)) = constant$ and given by the zero momentum contribution, which is the quantum mechanical index.

6 Mathematical Formulation of Susy Quantum Mechanics: Superunitary Supergroups, von Neumann Theorem, Wigner Theorem

Together with L. Pittner we formulated a kind of axioms for susy quantum mechanics and generalized theorems of quantum mechanics to superspace [22].

The Hilbert space $\mathcal{H}$ is supposed to be the sum of even and odd parts, $\mathcal{H} = \mathcal{H}_0 \oplus \mathcal{H}_1$. N_0 and N_1 are projection operators onto $\mathcal{H}_0$ and $\mathcal{H}_1$ such that $N_0 + N_1 = 1$. $K = N_0 - N_1$ denotes the Klein operator.

We assume that there are given self-adjoint operators $Q_n = Q_n^\dagger$, $n = 1, \ldots, N$, with $Q_n^2 = H$ and $\mathcal{D} = \text{dom}\, Q_n = \text{dom}\, H^{1/2}$, such that the anticommutation relations hold as quadratic forms for all $\psi, \phi \in \mathcal{D}$:

$$\langle Q_n\psi|Q_m\phi\rangle + \langle Q_m\psi|Q_n\phi\rangle = 0, \qquad n \neq m,$$
$$\langle Q_n\psi|K\phi\rangle + \langle K\psi|Q_n\phi\rangle = 0. \tag{6.1}$$

This implies that

$$\langle H\psi|K\phi\rangle = \langle K\psi|H\phi\rangle, \tag{6.2}$$

and HN_0 and HN_1 are "essentially" isospectral. From (6.1) and (6.2) it follows that we study representations of the Z_2 graded Lie superalgebra $S(N)$. The representations are obtained by combining bosons with fermions: For f degrees of freedom

$$B_k = (x_k + \frac{\partial}{\partial x_k})\frac{1}{\sqrt{2}}, \qquad B_k^\dagger = (x_k - \frac{\partial}{\partial x_k})\frac{1}{\sqrt{2}} \tag{6.3}$$

on $L^2(\mathbf{R}^f, d^f x)$ obey the canonical commutation relations (CCR):

$$[B_k, B_\ell^\dagger] = \delta_{k\ell} I, \qquad [B_k, B_\ell] = 0, \qquad k, \ell = 1, \ldots, f, \quad \text{on dom } x_k \cap \text{ dom } p_k. \tag{6.4}$$

For the fermions we work on $\mathbf{C}^{2^f}$ and realize the canonical anticommutation relations (CAR):

$$\{c_k, c_\ell^\dagger\} = \delta_{k\ell} I, \qquad \{c_k, c_\ell\} = 0, \qquad k, \ell = 1, \ldots, f, \tag{6.5}$$

either by operators or by σ-matrices (Klein-Jordan-Wigner transformation) or by introducing a Grassmann alsgebra G_f of polynomials in f anticommuting variables $\varepsilon_1, \ldots, \varepsilon_f$. A general element of G_f is written as

$$\xi = c_0 I + \sum_{1 \leq i_1 \leq \ldots \leq f} c_{i_1 \ldots i_p} \varepsilon_{i_1} \ldots \varepsilon_{i_p}, \qquad \{\varepsilon_i, \varepsilon_j\} = 0, \qquad i_j = 1, \ldots, f. \tag{6.6}$$

If we introduce the left derivatives $\partial_k = \partial / \partial \varepsilon_k$, we obtain the algebra

$$\{\varepsilon_k, \partial_\ell\} = \delta_{k\ell} I, \qquad \{\varepsilon_k, \varepsilon_\ell\} = 0 = \{\partial_k, \partial_\ell\}. \tag{6.7}$$

With the help of the natural scalar product on G_f we have given the adjoint operator to ε_k and obtain $\varepsilon_k^\dagger = \partial_k$. From (6.7) we realize that there exists an isomorphism between $c_k \leftrightarrow \partial_k$ and $c_k^\dagger \leftrightarrow \varepsilon_k$.

We require that

$$[B_\ell, \varepsilon_k] = [B_\ell^\dagger, \varepsilon_k] = 0 \tag{6.8}$$

and realize the algebra (6.1) by forming supercharges $Q = \sqrt{2} \sum_{k=1}^f B_k c_k^\dagger$. In order to transform bosons into fermions we define superunitary transformations and construct Lie-supergroups. This can be done by introducing anticommuting parameters $\theta_1, \ldots, \theta_M$ which span the Grassmann algebra $\mathcal{D}_M$. $\{t, \theta_\ell\}$ form the superspace. We take the skewsymmetric tensor product of the Clifford algebra K_{2f} spanned by ε_k, ∂_k, with the Grassmann algebra $\mathcal{D}_M$ and obtain the rules

$$\{\theta_\ell, \varepsilon_k\} = \{\theta_\ell, \theta_k\} = 0, \qquad \{\theta_\ell, \theta_k\} = 0. \tag{6.9}$$

If we take the above structure seriously we have to extend the Hilbert space $\mathcal{H}_f$ of quantum mechanics too. We go over to the $\mathcal{D}_M$-modul $\mathcal{H}_f \oplus \theta \mathcal{H}_f$ whose elements are of the form $\bar{\psi} = \psi + \theta \phi$ with $\psi, \phi \in \mathcal{H}_f$. For simplicity reasons we took $M = 1$. The scalar product of $\mathcal{H}_f$ turns into a sesquilinear form. We have to be careful in establishing the correct rules for this form. For *bosonic* ϕ and ψ we have, for example,

$$\langle \varepsilon \phi | \theta \varepsilon \psi \rangle = \langle \phi | \partial \theta \varepsilon \psi \rangle = -\langle \phi | \theta \psi \rangle = -\theta \langle \phi | \psi \rangle. \tag{6.10}$$

If ϕ and ψ are general elements of $\mathcal{H}_f$, we are allowed to take out θ only if we absorb the minus sign of (6.10). The correct rule turns out to be given by

$$\langle \theta \phi | \psi \rangle = \langle \phi | \theta \psi \rangle = \theta \langle \phi | K \psi \rangle, \tag{6.11}$$

where K denotes the Klein operator. This sesquilinear form takes values in $\mathcal{D}_M$, and no norm can be defined. As for adjointness a consistent definition is given by $(\theta A)^\dagger = A^\dagger \theta$. A physical interpretation of the space $\mathcal{H}_f \overset{\text{gr.t.pr.}}{\otimes} \mathcal{D}_M$ is not clear. In a way $\mathcal{H}_f$ is the space in which we live, while $\mathcal{H}_f \overset{\text{gr.t.pr.}}{\otimes} \mathcal{D}_M$ is the space from which we live.

Definition: The transformation $g(t) = 1 + \theta t A$ is called superunitary iff A is odd and self-adjoint.

From this definition we obtain the group multiplication law $g(t)g(s) = g(t+s)$. There exists a converse, a kind of generalization of Stone's theorem. If $g(t) = 1 + \theta A(t)$ fulfills $g(t)g(s) = g(t+s)$ and the mapping $t \to \langle\phi|A(t)\psi\rangle$ is continuous, $A(t) = tA$ generates a superunitary transformation.

In the following we shall turn to two anticommuting parameters $\theta = \theta_1 + i\theta_2$, $\theta^* = \theta_1 - i\theta_2$, and write down the most general group of superunitary transformations [22], which depends on four parameters:

$$g(t,s,r) = \exp[itH + isQ\theta + is^*\theta^*Q^\dagger + irH\theta\theta^*], \qquad t,r \in \mathbf{R},\ s \in \mathbf{C}. \tag{6.12}$$

The composition law reads

$$g(t,s,r)g(t',s',r') = g(t+t', s+s', r+r' + 2\,\mathrm{Im}(s's^*)). \tag{6.13}$$

g is superunitary in the sense that

$$g(t,s,r)^\dagger = g(t,s,r)^{-1} = g(-t,-s,-r). \tag{6.14}$$

We generalized two standard theorems of quantum mechanics. The first one concerns uniqueness of representations of the CCR and CAR: Let $F = \{c_k, c_\ell^\dagger, B_k, B_\ell^\dagger\}$ be an operator family which represents the C(A)CR irreducible on $\mathcal{H}_f$. There exists then a unitary operator U such that

$$UB_kU^\dagger = (x_k + ip_k)/\sqrt{2}, \qquad Uc_kU^\dagger = (-\sigma^3)\otimes\ldots\otimes(-\sigma^3)\otimes \underbrace{\sigma^-}_{k\text{-th position}} \otimes 1\otimes\ldots\otimes 1. \tag{6.15}$$

All representations are the direct sum of countable many "fermionic oscillator" representations (6.15).

We studied representations in the $\mathcal{D}_1$-modul. Let F be as before and let us identify $B = (x+ip)/\sqrt{2}$ and $c^\dagger = \varepsilon$ in $\mathcal{H}_{f=1}$. Let D and G be densely defined closed linear operators and define

$$\begin{aligned} \bar{B} &= B + \theta D, & D &= D_{10}\varepsilon + D_{01}\partial, \\ \bar{c} &= \partial + \theta G, & G &= G_{00}1 + G_{11}\varepsilon\partial, \end{aligned} \tag{6.16}$$

on a suitable domain. We have put $f = 1$ for simplicity reasons. Assume that $\bar{F} = \{\bar{B}, \bar{B}^\dagger, \bar{c}, \bar{c}^\dagger\}$ fulfills the C(A)CR too on a suitable domain of definition.

Theorem: Under the above conditions there exists a uniquely determined operator A, such that

$$G = \{A, \partial\}, \qquad D = [A, B], \qquad A = G_{00}\varepsilon + G_{00}^\dagger\partial, \tag{6.17}$$

such that the transformation $B \to \bar{B}$, $c \to \bar{c}$ is implemented by the superunitary operator $\exp(\theta A)$ and

$$e^{\theta A}Be^{-\theta A} = B + \theta[A, B], \qquad e^{\theta A}\partial e^{-\theta A} = \partial + \theta\{A, \partial\}, \tag{6.18}$$

hold.

This shows a kind of super-von Neumann theorem [23]. It holds for f degrees of freedom too. It is also true for more θ-variables. The proof for the last case turns out to be tricky.

Our second result concerns a generalization of the Wigner theorem on implementation of symmetries: Let $\mathcal{P}(\mathcal{H})$ be the projective Hilbert space of $\mathcal{H}$, where we identify $\psi, \eta \in \mathcal{H}$ with $\|\psi\| = \|\eta\| = 1$ if $\psi = e^{i\theta}\eta$. For such classes $[\psi]$ and $[\eta]$ we define a product by taking

$$\langle[\psi]|[\eta]\rangle = |\langle\psi|\eta\rangle|. \tag{6.19}$$

A bijection α such that

$$\mathcal{P}(\mathcal{H}) \stackrel{\alpha}{\rightarrow} \mathcal{P}(\mathcal{H}) \quad \text{with} \quad \langle\alpha[\psi]|\alpha[\eta]\rangle = \langle[\psi]|[\eta]\rangle \tag{6.20}$$

is called a Wigner automorphism. The theorem says that every Wigner automorphism is of the form $\alpha[\psi] = [U\psi]$ with U being unitary or antiunitary and essentially unique.

Since our sesquilinear form maps into $\mathcal{D}_M$ it is impossible to generalize directly (6.19). But $\mathcal{P}(\mathcal{H})$ can also be described as space of projections P_ψ of rank one. We obtain

$$|\langle[\psi]|[\eta]\rangle|^2 = \mathrm{Tr}\ P_\psi P_\eta, \tag{6.21}$$

and (6.21) can be generalized. We consider "coded states" $\Psi = \psi_0 + \theta_1\psi_1 + \theta_2\psi_2 + \theta_1\theta_2\psi_{12}$, where ψ_0 denotes the "body" and $\Psi - \psi_0$ the "soul", and normalize $\langle\Psi|\Psi\rangle = 1$. We define "coded" projection operators

$$P_\Psi\Phi = \Psi\langle\Psi|\Phi\rangle, \qquad P_\Psi^2 = P_\Psi = P_\Psi^\dagger, \tag{6.22}$$

such that equivalence classes $[\Psi]$ are defined and

$$P_\Psi = P_\Phi \Longleftrightarrow \Psi = e^{ir_0}(1 + \theta_1\theta_2 r_{12})\Phi, \qquad r_0, r_{12} \in \mathbf{R}. \tag{6.23}$$

We call τ a supersymmetry if it is a bijection from the set of "coded" projection operators onto itself, such that

$$\mathrm{str}\ P_\Phi P_\Psi = \mathrm{str}\ \bar{P}_\Phi \bar{P}_\Psi, \qquad \bar{P}_\Phi = \tau(P_\Phi), \qquad \bar{P}_\Psi = \tau(P_\Psi), \tag{6.24}$$

where str denotes the supertrace on $\mathcal{H}_f$. From Wigner's theorem we deduce the existence of a unitary operator which maps bodies onto each other. After imposing a condition if bodies are orthogonal we obtain the existence of an essentially unique superunitary operator implementing the supersymmetry [24]. The proof is tricky.

7 General Structure of Susy Models

Although we used up to now always the Hamiltonian approach an equivalent Lagrangian formalism exists. We may realize the superalgebra on superspace spanned by $\{t, \theta, \bar{\theta}\}$, for example. For this simplest case a real scalar superfield has the expansion

$$\phi(t, \theta, \bar{\theta}) = x(t) + i\bar{\psi}(t)\theta + i\psi(t)\bar{\theta} + F\theta\bar{\theta}. \tag{7.1}$$

The generators of supersymmetry are represented by

$$Q_1 = -\frac{\partial}{\partial\bar\theta} - i\theta\frac{\partial}{\partial t}, \qquad Q_2 = \frac{\partial}{\partial\theta} + i\bar\theta\frac{\partial}{\partial t}, \tag{7.2}$$

fulfilling the correct algebra. Infinitesimal transformations of ϕ are given by

$$\delta\phi = -i(\varepsilon_1\xi_1 Q_1 + \varepsilon_2\xi_2 Q_2)\phi := \delta_1\phi + \delta_2\phi, \tag{7.3}$$

where ξ_i denotes anticommuting parameters. (7.3) yields the transformation laws for components. There exist covariant derivatives

$$D_1 = -\frac{\partial}{\partial\bar\theta} + i\theta\frac{\partial}{\partial t}, \qquad D_2 = \frac{\partial}{\partial\theta} - i\bar\theta\frac{\partial}{\partial t}, \tag{7.4}$$

such that $\delta D_j\phi = D_j(\delta\phi)$ results. One can introduce supervector fields and generalize to superspace the standard differential geometric formalism.

For the Witten model the invariant Lagrangian is given by

$$\mathcal{L}_{\text{susy}} = \frac{1}{2}D_1\phi D_2\phi - w(\phi) = A + iS_1\theta + iS_2\bar\theta + L\theta\bar\theta. \tag{7.5}$$

It is easy to calculate the variation δL and we obtain a total derivative term. By expanding $D_1\phi D_2\phi$ in (7.5) we obtain

$$\begin{aligned} A &= \frac{1}{2}\psi\bar\psi - w(x) \\ L &= \frac{\dot x^2}{2} + \frac{i}{2}(\bar\psi\dot\psi - \dot{\bar\psi}\psi) + \frac{1}{2}F^2 - w'F + w''\psi\bar\psi. \end{aligned} \tag{7.6}$$

(7.6) yields an algebraic equation for the auxiliary field $F = w'$. The susy-invariant action

$$S = \int dt \int d\theta d\bar\theta \mathcal{L}_{\text{susy}} = \int dt L, \tag{7.7}$$

is given by L. A Legendre transformation yields the Hamiltonian from (7.6)

$$H = \frac{p^2}{2} + \frac{w'^2}{2} + \frac{1}{2}w''[\bar\psi, \psi], \tag{7.8}$$

which is well-known to us. We quote next how components of $\mathcal{L}_{\text{susy}}$ transform under susy transformation. (7.3) and (7.5) yield

$$\begin{aligned} \delta_1 A &= \varepsilon_1\xi_1 S_2, & \delta_2 A &= -\varepsilon_2\xi_2 S_1, \\ \delta_1 L &= i\varepsilon_1\xi_1\dot S_2, & \delta_2 L &= i\varepsilon_2\xi_2\dot S_1, \end{aligned} \tag{7.9}$$

and the remarkable result that special combinations are invariant

$$\delta_1(L - i\dot A) = 0, \qquad \delta_2(L + i\dot A) = 0. \tag{7.10}$$

We can start the canonical formalism with either taking $L_1 = L - i\dot A$ or $L_2 = L + i\dot A$ as a new Lagrangian. In the first case we change from p to the new momentum $\pi_1 = p - iw'$, in the second case to $\pi_2 = p + iw'$. These changes are given by a similarity transformation on superspace [25]

$$\exp[-i\theta\bar{\theta}\frac{\partial}{\partial t}]\mathcal{L}_{\text{susy}}\exp[i\theta\bar{\theta}\frac{\partial}{\partial t}] = \mathcal{L}_{\text{susy}} - i\theta\bar{\theta}\frac{\partial A}{\partial t}, \tag{7.11}$$

and singles out special coordinates. What is the meaning of these changes for the quantum theory? We represent the heat kernel for the Hamiltonians

$$H_{\pm} = \frac{p^2}{2} + w'^2(x) \pm w''(x) \tag{7.12}$$

in stochastic language. Let $b(s)$, $0 \leq s \leq t$, be a family of Gaussian distributed random variables and introduce the Wiener process with expectation value $E(\cdot)$ and covariance $E(b(s)b(t)) = \min(s,t)$. The heat kernel representations

$$(e^{-tH_{\pm}}f)(x) = E(\exp[-\frac{1}{2}\int_0^t d\tau(w'^2 \pm w'')(x + b(\tau))]f(x + b(t))) \tag{7.13}$$

can be changed with the help of Ito's Lemma:

$$\int_0^t db(s)w'(b(s)) = w(b(t)) - w(b(0)) - \frac{1}{2}\int_o^t ds w''(b(s)). \tag{7.14}$$

This gives for (7.13)

$$(e^{-tH_{\pm}}f)(x) = e^{\pm w(x)}E(\exp[-\frac{1}{2}\int_0^t d\tau w'^2(x + b(\tau)) \pm \int_0^t db(s)w'(b(s))]$$
$$\cdot \exp[\mp w(x + b(t))]f(x + b(t))). \tag{7.15}$$

Intuitively speaking we have formed complete squares $\dot{x}^2 \mp 2\dot{x}w' + w'^2$ in the exponent. Nevertheless, the model is not a free one. (7.15) reads in operator language

$$(e^{-tH_{\pm}})(x,y) = e^{\pm w(x)}e^{-tH^{\pm}_{FP}}(x,y)e^{\mp w(y)}. \tag{7.16}$$

The above similarity transformations on superspace were introduced from the demand that the new Lagrangians L_j are totally invariant. $H^{\pm}_{FP}$ are Fokker-Planck Hamiltonians of the form

$$H^{\pm}_{FP} = \frac{p^2}{2} \mp iw'p, \tag{7.17}$$

which generate contraction semigroups on $L^2(\mathbf{R}, \Omega^2_{\pm}dx)$, where $\Omega_{\pm}$ are ground state wave functions of $H_{\pm}\Omega_{\pm} = 0$ expressed in terms of the potential w by $\Omega_{\pm} = \exp(\mp w)$. This mapping from the Schrödinger equation to Fokker-Planck equations is well-known. For the Wiener process the change of variables corresponds to a Langevin equation

$$b(s) \to B(s), \qquad db(s) = w'(b(s)) + dB(s). \tag{7.18}$$

If we introduce a new expectation

$$\bar{E}(\cdot) = E(\exp[\int_0^t db(s)w'(b(s)) - \frac{1}{2}\int_0^t ds w'^2(b(s))]\cdot), \tag{7.19}$$

we realize that $\bar{E}(B(s)) = 0$, $\bar{E}(B(s)B(t)) = \min(s,t)$ etc. These relations are called stochastic identities. They can be translated into the operator language and yield constraints on wave functions.

We have sketched above a number of relations connected to the introduction of new variables. There is one more connection. They are the correct variables for the Nicolai transformation. If we consider a functional integral for bosons and fermions, we may integrate out the latter and obtain a determinant. Changing to new variables yields a Jacobian which might cancel the fermionic one. This program has been realized for a number of susy quantum mechanics models.

For simplicity reasons we have restricted ourselves to the Witten model. The above interrelations have been worked out for the two-dimensional magnetic field model as well as for the Wess-Zumino model. In all models the mentioned role of special coordinates occurs.

Together with D. Bollé and R. Corns we have recently extended the work of Rogers on susy functional integrals. We have found a kernel representation of the superunitary transformation introduced in chapter six. In this way we generalized the Wiener measure to a measure for superpaths.

Final Remark: Although we have treated only these aspects, which are close to our own research we hope that the generality and great applicability of susy quantum mechanics became apparent. In this sense it should also enter standard quantum mechanics textbooks. Roots of this structure can be found in mathematics of the last century. But the many connections have been seen only recently. This explains also the enormously growing literature from which we quoted only some parts.

Acknowledgement: We thank the organizers of this school, Profs. P. Dita, G. Nenciu and R. Purice for all their efforts.

References

[1] J. Wess and B. Zumino, Nucl. Phys. B70 (1974) 39.

[2] V.A. Kostelecky and D.K. Campbell, Physica 15D (1985) 1.

[3] M.F. Sohnius, Phys. Rep. 128 (1985) 39.

[4] E. Witten, Nucl. Phys. B185 (1981) 513.

[5] P. Salomonson and J.W. van Holten, Nucl. Phys. B196 (1982) 509.

[6] M. de Crombrugghe and V. Rittenberg, Ann. Phys. NY 151 (1983) 99.

[7] L.E. Gendenshtein and I.V. Krive, Usp. Fiz. Nauk 146 (1986) 645.

[8] P.A. Deift, Duke Math. Journal 45 (1978) 267.

[9] A. Stahlhofen and K. Bleuler, "An Algebraic Form of the Factorization Method", Duke Univ./Univ. of Bonn preprint (1988).

[10] L. Infeld and T.E. Hull, Rev. Mod. Phys. 23 (1951) 21.

[11] C.V. Sukumar, J. Phys. A18 (1985) L 697.

[12] A. Stahlhofen, J. Phys. A: Math. Gen. 22 (1989) 1053.

[13] H. Grosse and A. Martin, Phys. Rep. 60 (1980) 341.

[14] H. Grosse and A. Martin, Phys. Lett. 134B (1984) 368.

[15] B. Baumgartner, H. Grosse and A. Martin, Phys. Lett. 146B (1984) 363.

[16] H. Grosse, Phys. Lett. 197 (1987) 413.

[17] K. Chadan and P.C. Sabatier, "Inverse Problems in Quantum Scattering Theory", Springer 1989.

[18] F. Gesztesy, H. Grosse and B. Thaller, Phys. Lett. 116B (1982) 155.

[19] D. Bollé, F. Gesztesy and B. Simon, Lett. Math. Phys. 13 (1987) 127; and same authors with W. Schweiger, Jour. Math. Phys. 28 (1987) 1512.

[20] A. Jaffe, A. Lesniewski and M. Lewenstein, Ann. of Phys. NY 178 (1987) 313.

[21] A. Jaffe, A. Lesniewski and J. Weitsman, Commun. Math. Phys. 112 (1987) 75.

[22] H. Grosse and L. Pittner, Jour. of Phys. A: Math. Gen. 20 (1987) 4265 and 21 (1988) 3239.

[23] H. Grosse and L. Pittner, Jour. Math. Phys. 29 (1988) 110.

[24] H. Grosse and L. Pittner, Jour. of Phys. A: Math. Gen.

[25] D. Bollé, P. Dupont and H. Grosse, "On the General Structure of Quantum Mechanical Susy Models", Univ. Leuwen preprint 1989.

PROPAGATION DES SINGULARITES GEVREY POUR LA DIFFRACTION

Bernard Lascar Richard Lascar
Université Pierre et Marie Curie,UA 213
4, Place Jussieu. Paris 75231 France

ABSTRACT. We describe in these notes a result of propagation of Gevrey singularities for diffractive waves.

Dans ces notes nous allons décrire la propagation des singularités Gevrey pour le problème de Dirichlet dans le cas d'une onde diffractive. Ce problème a été étudié dans le cas C^∞ par Melrose [8], Taylor [9] , Eskin [3] ,dans le cas analytique par Sjöstrand [10], dans le cas Gevrey par Lebeau [7] . Notre travail consiste à étendre le résultat de Lebeau au cas d'obstacles non analytiques, notre approche est cependant totalement différente et c'est la méthode que nous utilisons qui constitue l'essentiel de l'intérêt de ce travail.

Le résultat peut être énoncé de la façon suivante : si l'obstacle a la régularité $G^{s'}$ une singularité G^s portée par une onde diffractive ne se propage pas le long de l'obstacle tant que $s > 2s'+1$ avec égalité quand $s'=1$. Le résultat de Lebeau est le cas $s'=1$ des obstacles analytiques.

Par un changement de variables on étudie plutôt une équation aux dérivées partielles du second ordre $P= D_{x_1}{}^2-R(x,D_{x'})$ où les variables $x=(x_1,x')$ sont dans $\mathbf{R}^n$ On travaillera dans $x_1 \geqq 0$.

On dit qu'un opérateur pseudodifférentiel de degré m est de classe $G^{s'}$ si son symbole de Weyl $\sigma(x,\xi)$ satisfait des estimations :
$|D_x{}^\alpha D_\xi{}^\beta \sigma(x,\xi)| \leqq CA^{|\alpha|+|\beta|}\alpha!^{s'}\beta!^{s'}<\xi>^{m-|\beta|}$ où $<\xi>=(1+|\xi|)$, C et A étant des constantes positives.

On supposera ici que R est un opérateur pseudodifférentiel de degré 2 en (x,ξ') de classe $G^{s'}$ et qu'il a un symbole principal homogène de degré 2 $r(x,\xi')$ réel,

On désigne par $r_0(x',\xi')$ la restriction de r à $x_1=0$. la région glancing est $G=\{r_0(x,\xi')=0\}$. On supposera que r_0 est de type principal et que les points glancing sont diffractifs c'est à dire $\dfrac{\partial r}{\partial x_1} > 0$ près des points de G. L'exemple canonique est le cas de $r=x_1\xi_n{}^2+\xi_{n-1}\xi_n$ près d'un point où $\xi_{n-1}=0$ et $\xi_n>0$.

Soit ρ_0 un point de G, on note $\gamma_-(\rho_0)$ resp $\gamma_+(\rho_0)$ la demi bicaractéristique rentrante resp sortante issue de ρ_0.

On désigne pour un nombre $s>1$ par $WF^{(s)}(u)$ et $WF^{(s)}{}_b(u)$ le front d'onde et le front d'onde au bord Gevrey s.

On dira qu'une distribution u sur $\mathbf{R}^{n+1}{}_+$ est s régulière si elle s'étend en une distribution $C^\infty(\mathbf{R}_{x_1},D'(\mathbf{R}^n))$ et si $WF^{(s)}(u) \cap N^*{}_{x_1=c}=\emptyset$, $N^*{}_{x_1=c}$ est le conormal à l'hypersurface $x_1=c$, pour tout $c>0$. Une solution u de $Pu=f$ où f est G^s est s régulière quand P est un opérateur différentiel.

Le résultat s'énonce comme:

A. Boutet de Monvel et al. (eds.), Recent Developments in Quantum Mechanics, 329–339.

Théorème.

Soit P un opérateur de classe $G^{s'}$, $s'>1$, ρ_0 un point de G, s satisfaisant $s>2s'+1$. Soit u une distribution s régulière telle que $\rho_0 \notin WF^{(s)}{}_b(Pu)$. On suppose que la demi bicaractéristique $\gamma_-(\rho_0)\backslash\{\rho_0\}$ resp $\gamma_+(\rho_0)\backslash\{\rho_0\}$ ne rencontre pas $WF^{(s)}(u)$. On suppose que $\rho_0 \notin WF^{(s)}(u|_{x_1=0})$.

Alors $\rho_0 \notin WF^{(s)}(\frac{\partial u}{\partial x_1}|_{x_1=0})$.

Corollaire.

Si Ω est un obstacle strictement convexe, $\partial\Omega$ est $G^{s'}$, les résonances du problème extérieur pour l'équation des ondes sont contenues dans une région $\{\mu \in \mathbf{C}^n, \operatorname{Re}\mu \leqq -C|\mu|^{1/s}\}$ pour une certaine constante C, pour tout nombre $s>2s'+1$.

On fait maintenant quelques commentaires sur ce résultat et on décrit la structure de la preuve.

C'est le cas des points diffractifs qui est le plus intéressant mais aussi le plus difficile pour la propagation des singularités Gevrey. C'est en effet là qu'un indice critique s=3 pour le cas analytique apparait, ceci est déjà évident dans le modèle de Friedlander. L'étude du cas analytique par Lebeau est fait par une méthode de paramétrixe utilisant les fonctions d'Airy. Outre leur complexité les méthodes de paramétrixes nous paraissent moins adaptées que les méthodes d'estimation d'énergie pour discuter de la meilleure régularité Gevrey que l'on obtient pour un opérateur pseudodifférentiel Gevrey, cela apparait déjà pour la propagation des singularités Gevrey à l'intérieur. Nous avons pris le parti de faire une estimation d'énergie Gevrey. Notre point de départ a été celle du cas C^∞ due à Ivrïi telle qu'on la trouve dans Hörmander [4]. La grande différence entre une méthode d'énergie C^∞ et une méthode Gevrey est que cette dernière utilise des estimations avec poids. Cependant la particularité du cas diffractif est que les fonctions de poids ne peuvent plus être tangentielles. Ceci fait que le formalisme de [6], utilisant une transformation canonique complexe convenable, n'est plus adaptable. Il reste cependant l'idée que l'on doit utiliser non plus des opérateurs Fourier intégraux à phase complexe d'ordres infinis mais leurs reste dans la division par P. Ces opérateurs seront seulement décrits comme des opérateurs pseudodifférentiels irréguliers et d'ordres infinis. On se contentera d'adapter dans ce cas le calcul de Weyl de [4], décrivant les propriétés des opérateurs à l'aide de poids et de métriques. La condition pour que les métriques vérifient le principe d'incertitude est précisément $s \geqq 3$. Notre condition $s>2s'+1$ apparait plus technique, mais elle se manifeste clairement dans le processus de la division par P par exemple. Notre opinion est que $s \geqq 2s'+1$ devrait être nécessaire.

Une difficulté que nous rencontrerons est que du fait de l'absence de deux commutateurs suffisamment indépendants toutes les opérations – division et majorations par exemple – devront se faire sur les opérateurs et non plus au niveau d'un certain symbole principal.

Le plan de ces notes est le suivant : au paragraphe 1 on décrit un calcul de Weyl adapté à notre situation, au paragraphe 2 il s'agit de la division par P, au paragraphe 3 on décrit l'estimation d'énergie obtenue et on finit la preuve.

1. Calcul symbolique.

On va décrire ici les propriétés du calcul symbolique des classes d'opérateurs pseudo-différentiels irréguliers Gevrey utilisés dans la suite de l'article. On ne cherche pas ici la plus grande généralité possible.

On utilise systèmatiquement la quantifiquation avec un grand paramètre λ, $op_{1/2}(a)=(\lambda/2\pi)^n\int a((x+y)/2,\xi)e^{i\lambda(x-y)\xi}d\xi$. On considèrera aussi un paramètre μ, $\mu=\mu_0\lambda^{-1+1/s}$, μ_0 est un petit paramétre qui sera fixé à la fin de la preuve, on supposera toujours que $\lambda\mu\geqq 1$.

On décompose les variables en (ξ,y,ζ), ξ est la variable duale de x, x=0 représente le bord du domaine, on a pris des coordonnées symplectiques de sorte que par une transformation canonique tangentielle r est transformée en y.

Les symboles seront Gevrey s' en (y,ζ) et s_0 en ξ, $s_0>1$ sera fixé plus tard aussi proche de 1 qu'il sera nécessaire.

On définit S(m,G) comme la classe des fonctions satisfaisant des estimations :

$$|D_\xi^\alpha D_y^\beta D_\zeta^\gamma a(x,\xi)| \leqq C\, A^{\alpha+\beta+\gamma}(\lambda\mu+\alpha^{s_0})^\alpha(\lambda\mu)^{2\beta}\beta!^{s'}(\lambda\mu+|\gamma|)^\gamma\gamma!^{s'-1}\, m(X). \quad (1.\ 1).$$

m(X) est le poids, la métrique est $G(t)=(\lambda\mu)^2|t_\xi|^2+(\lambda\mu)^4|t_y|^2+(\lambda\mu)^2|t_\zeta|^2$.

On utilisera aussi la classe S'(m,G) des fonctions satisfaisant à :

$$|D_\xi^\alpha D_y^\beta D_\zeta^\gamma a(x,\xi)| \leqq C\, A^{\alpha+\beta+\gamma}(\lambda\mu+\alpha^{s_0})^\alpha(\lambda\mu)^{2\beta}\beta!^{s'}\gamma!^{s'}\, m(X). \quad (1.\ 2).$$

On voit que la métrique G est constante mais dépend de paramètres, les poids eux seront variables et d"ordre infini".

La métrique duale G^σ est définie par $G^\sigma(t)=\sup_\theta\lambda^2(\sigma(t,\theta))^2/G(\theta)$. $h^2=\sup_t G(t)/G^\sigma(t)$, $h=(\lambda\mu)^3/\lambda$, la condition $s\geqq 3$ assure que $G\leqq G^\sigma$. Les conditions sur les poids sont :

(i) Il existe c et C>0 tels que pour $G(Y)\leqq c$, $1/C\leqq m(X+Y)/m(X)\leqq C$.

(ii) Il existe r>0 tel que pour tout $\varepsilon>0$ il existe $C_\varepsilon>0$ tel que si $|Y|\leqq r$

$m(X+Y)/m(X)\leqq C_\varepsilon\exp(\varepsilon(G^\sigma(Y))^{1/2s'})$.

(iii) Il existe C>0 tel que $m(X+Y)/m(X)\leqq C\exp(C\lambda\mu)$.

L'interprétation de ces conditions est claire (i) et (ii) signifient que m est à variation lente et σ tempérée (la condition de croissance polynomiale a été remplacée par une exponentielle fractionnaire).

On vérifie que $m(X)=\exp(\lambda\mu(\phi_0+\inf(1,\max(\xi-\alpha,y_+^{1/2}))))$ où ϕ_0 et α sont des fonctions Gevrey bornées $y_+=\max(0,y)$, satisfait ces conditions si $s>2s'+1$.

Les éléments f de S(m,G) ont une extension presque analytique à $\mathbf{C}^{2n}$ notée encore f qui est dans la classe S(m(ReX),G), ceci se prouve en adaptant l'argument de Borel de [2].

$$|\bar\partial f(X)|\leqq C'm(\mathrm{Re}X)\exp(-A'(G(\mathrm{Im}X))^{-1/2(s'-1)}) \quad (1.\ 3).$$

On peut énoncer le résultat principal de ce paragraphe.

Proposition 1. 1.

Soient $f\in S(m,G)$ et $g\in S(m',G)$ où m et m' satisfont (i)–(iii), $f\#g\in S(mm',G)$ et

$$f\#g-\sum_{0\leqq j\leqq N-1}1/j!(i/2\lambda^{-1}\sigma(D_x,D_\xi;D_y,D_\eta))^j(f(x,\xi)g(y,\eta))|_{(x,\xi)=(y,\eta)}\in S(mm'((\lambda\mu)^3/\lambda)^N,G). \quad (1.\ 4).$$

2.Théorème de division.

On va décrire dans ce paragraphe un résultat de division d'un opérateur peudo-différentiel de la classe S(m, g) par P.

On écrit ici $p=(\xi-\alpha(x,\xi'))^2-y$, on note les variables sous la forme (ξ,y,ζ), ζ dénote donc toutes les autres variables.

Dans $y\geqq 0$, on a $f(x,\xi)=q(f)(x,\xi)+L(\partial_\xi^2 f)\,p$ où $L(f)=\int_0^1 L_t(f)\,dt$ avec

$L_t(f)=\int_t^1 f(t\xi+(1-t)\alpha+(1-2s+t)y^{1/2},y,\zeta)\,ds$, $q(f)(x,\xi)=f_0(x,\xi')+(\xi-\alpha)f_1(x,\xi')$,

$f_0(x,\xi')=1/2(f(\alpha+y^{1/2},y,\zeta)+f(\alpha-y^{1/2},y,\zeta))$,

$f_1(x,\xi')=1/2(f(\alpha+y^{1/2},y,\zeta)-f(\alpha-y^{1/2},y,\zeta))/y^{1/2}$.

Il s' agit maintenant d'étudier comment ces opérateurs opérent dans les classes de Gevrey. Pour ce faire on introduit des normes formelles adéquates.

$$N(f;T)(X)=\sum_{\alpha,\beta,\gamma}|D_\xi^\alpha D_y^\beta D_\zeta^\gamma f|/\alpha!\beta!\gamma!(\alpha+2\beta+|\gamma|)!^{s_0-1}|\gamma|!^{s'-s_0}\beta!^{s''}(\lambda\mu)^\alpha(\lambda\mu)^{2\beta}(\lambda\mu)^\gamma T_\xi^\alpha T_y^\beta T_\zeta^\gamma$$

On a noté $s''=s'-(2s_0-1)$.

Soit $g=L_t^0(f)=\int_t^1 f(t\xi+(1-2s+t)\,y^{1/2},\,y,\,\zeta)\,ds$ qui est L_t dans lequel on a fait $\alpha=0$.

Il existe une constante $C>0$ telle que

$$N(L_t^0(f);\,T)\,(X)\leq(1-t)\sup_{-1\leq\theta\leq 1}N(f;\,\phi_t(T))(X_{t,\theta}) \qquad (2.\ 1)$$

avec $\phi_t(T)=(T_\xi+CT_y{}^{1/2},T_y,T_\zeta)$ et $X_{t,\theta}=(t\xi+\theta(1-t)y^{1/2},y,\zeta)$.

Il faut maintenant effectuer la division des opérateurs pseudo-differentiels. Procédons d'abord formellement, on définit par induction une suite de symboles $a_{\alpha,q}$ par $a_{0,0}=a$, puis $a_{\alpha,j}=L(\partial_\xi{}^2 b_{\alpha,j})$ et $b_{\alpha,j}=-\sum_{\beta+\gamma=\alpha,\ \beta\neq 0}R_\beta(a_{\gamma,j})-p_1 a_{\alpha,j-1}$ où R_β est le terme d'ordre β dans le calcul de Weyl, donc $R_\beta(f)=c_\beta\partial^{\beta'}\sigma\partial^\beta f/\beta!$ β' est un multi-indice de longueur $|\beta|$, c_β est une constante majorée par $\lambda^{-|\beta|}2^{-|\beta|}$, σ est le symbole total de P $\sigma=p+p_1$, p_1 est de degré -1. Il nous faut tenir compte maintenant de la fonction α qui intervient dans le symbole de p. Pour cela on supposera que $a_{\gamma,j}=e^{\lambda\mu\phi_0}\tau_\alpha a'_{\gamma,j}$ où τ_α est l'opérateur de translation $\tau_\alpha f(\xi,y,\zeta)=f(\xi-\alpha(x,\xi'),y,\zeta)$.

On peut maintenant énoncer le résultat principal de ce paragraphe.

Proposition 2.1

Pour chaque α, j on peut écrire $a'_{\alpha,j}$ sous la forme d'une somme $a'_{\alpha,j}=\sum_{l,q,r\,l\leq 2|\alpha|,q\leq|\alpha|,r\leq 2j,N\leq|\alpha|+j}a'_{\alpha,j,l,q,r,N}$ où chaque $a'_{\alpha,j,l,q,r,N}$ s'exprime comme une intégrale $a'_{\alpha,j,l,q,r,N}=\int_{[0,1]^N}a'_{\alpha,j,l,q,r,N}(t_1,..,t_N)\,dt_1...dt_N$ et on a : pour $T_y,T_\zeta\leq c(\lambda\mu)$

$$N(a'_{\alpha,j,l,q,r,N}(t_1,...,t_N),T)\leq M^{\alpha+j}\max_{0\leq t\leq 1}(1+\phi_t(T)^{-1})^\alpha(1+\phi_t(T)_\xi{}^{-1})^\alpha\alpha!^{s'-1}F(\alpha,j,l,q,r,\eta,\rho,\varepsilon,\varepsilon_1,\varepsilon_2)$$
$$c_{\alpha,j,l,q,r,N}(t_1,\,...,\,t_N)N(a';\,\mu_1(\phi_{t_1...t_N,\varepsilon}(T)_\xi+\rho),\mu_2T_y,\mu_3T_\zeta) \qquad (2.\ 2)$$

Où M est une constante convenable, η, ρ, ε, ε_1, ε_2 sont des nombres positifs tous plus petits que 1 sauf ρ,

$$F(\alpha,j,l,q,r,\eta,\rho,\varepsilon,\varepsilon_1,\varepsilon_2)=\rho^{-(n+r)}\eta^{-(s_0-1)(n+r)}\varepsilon^{-(s_0\alpha_1+(2s_0-1)\alpha_2+s_0\alpha_3)}\varepsilon_1{}^{-s''\alpha_2}\varepsilon_2{}^{-(s'-s_0)\alpha_3}$$
$$\phi(n+r)^{s_0}\phi(\gamma)^{2s_0-1}\phi(\gamma_2)^{s''}\phi(\gamma_3)^{s'-s_0}(\lambda\mu)^{n+r+\alpha_1+2\alpha_2+\alpha_3}\lambda^{-(|\alpha|+q+j)}$$

$\phi(n)$ est la fonction n^n, ici $n=2|\alpha|-1$.

$c_{\alpha,j,l,q,r,N}(t_1,\,...,\,t_N)$ est une fonction telle que pour tout θ dans $[0,\,1]$ l'intégrale $\int_{t_1...t_N=\theta}c_{\alpha,j,l,q,r,N}(t_1,\,...,\,t_N)\,dt_1...\,dt_N\leq C^{\alpha+j}(1-\theta)^{2|\alpha|-1+r}/(2|\alpha|-1+r)!$;

$\mu_1=(1+\eta)^{s_0-1}(1+\varepsilon)^{s_0}$, $\mu_2=(1+\eta)^{2(s_0-1)}(1+\varepsilon)^{2s_0-1}(1+\varepsilon_1)^{s''}$, $\mu_3=(1+\eta)^{s_0-1}(1+\varepsilon)^{s_0}(1+\varepsilon_2)^{s'-s_0}$;

$\phi_{t,\varepsilon}(T)=(tT_\xi+C(1-t)(1+\varepsilon_1)^{s''/2}T_y{}^{1/2},T_y,T_\zeta)$; il reste enfin à préciser que α_1 ,α_2, α_3 désignent respectivement les composantes en ξ, y, ζ de α.

La proposition 2. 1 ne permet pas d'obtenir directement les estimations voulues. Pour faire converger les itérations il faut partir d'une fonction a' qui vérifie de meilleures estimations des dérivées en ξ et ζ que ce que donne la norme N. C'est à dire :

$$|D_\xi{}^\alpha D_y{}^\beta D_\zeta{}^\gamma a'|\leq CA^{\alpha+\beta+\gamma}\,(\lambda\mu+\alpha^{s_0})^\alpha\,(\lambda\mu)^{2\beta}\beta!^{s'}\,(\lambda\mu+|\gamma|)^\gamma\gamma!^{s'-1}\,m(X). \qquad (2.\ 3).$$

optimisant le choix de ρ, on trouve que les $a'_{\alpha,j,l,q,r,N}$ satisfont à :

$$|D_\xi{}^\delta D_y{}^\beta D_\zeta{}^\gamma a'_{\alpha,j,l,q,r,N}|\leq CM^{\alpha+j}(\alpha!^{2s'-2+s_0-1}(\lambda\mu)^{q+3|\alpha|+2j}+\alpha!^{2s'-1+4(s_0-1)}j!^{2(s_0-1)}(\lambda\mu)^{2|\alpha|})\lambda^{-(|\alpha|+q+j)}\,A^{\delta+\beta+\gamma}$$
$$(\lambda\mu+\delta^{s_0})^\delta\,(\lambda\mu)^{2\beta}\beta!^{s'}\,(\lambda\mu+|\gamma|)^\gamma\gamma!^{s'-1}\,m(X). \qquad (2.\ 4).$$

Si on multiplie m(X) par $\exp(\lambda\mu\phi_0)$ on a la même estimation pour les a_α.

Sous la condition $s>2s'+1$, on peut donc sommer les a'_α et les a_α modulo un symbole Gevrey à décroissance exponentielle 1/s.

On obtient alors.

Proposition 2.2

Si $a\in S(m,\,G)$, il existe $a_0(x,\xi')$ et $a_1(x,\xi')$ avec $a_0\in S(m_0,\,g)$ et $a_1\in S(m_0\delta^{-1},\,g)$ et $j\in S(m(\lambda\mu)^2,\,G)$, $r\in S(m(\lambda\mu)^2\,,G)$ à décroissance exponentielle 1/s dans $y\geq 0$ de sorte que :

$a=a_0+(\xi-\alpha)a_1+p\#j+r$. Avec $m=\exp(\lambda\mu(\phi_0+\inf(1,\max(\xi-\alpha,y_+{}^{1/2}))))$, $m_0=\exp(\lambda\mu(\phi_0+\inf(1,y_+{}^{1/2})))$, $\delta=(\lambda\mu)^{-1}+|y|^{1/2}$.

Si $a=e^{\lambda\mu\phi_0}\tau_\alpha a'$ où $a'\in S'(m', G)$, alors $a_0=e^{\lambda\mu\phi_0}a'_0$ et $a_1=e^{\lambda\mu\phi_0}a'_1$ avec $a'_0\in S'(m'_0, g)$ et $a'_1\in S'(m'_0\delta^{-1}, g)$ avec $m'=\exp(\lambda\mu\inf(1,\max(\xi,y_+{}^{1/2})))$, $m'_0=\exp(\lambda\mu\inf(1,y_+{}^{1/2}))$.

3. L'inégalité d'énergie.

On transforme le problème de propagation de singularités en un problème semi classique et on réduit r à y, en conjuguant l'opérateur P par des O.I.F. semiclassiques dont les symboles sont supportés dans $\xi\neq0$.

Dans ce paragraphe on va utiliser le calcul symbolique et le théorème de division prouvés précédemment pour établir une inégalité d'énergie. On procède de la façon suivante : on trouve un opérateur Q dont le commutateur avec P est positif, puis on fait des divisions par P pour obtenir des opérateurs linéaires en D_x dont le commutateur avec P sera encore positif modulo P. Il y aura en fait quelques complications techniques par rapport à ce schéma.

P est un opérateur pseudodifférentiel de degré zéro quadratique en D_x, par une transformation canonique tangentielle (dépendant de x comme paramètre) qui transforme r en y on se ramène au cas où $p=(\xi-\alpha(x,y,\zeta))^2-y$, la condition $r'_x>0$ se traduit donc par $\alpha'_\eta<0$. On localisera plus loin dans $\{(x,\xi')\in W\}$. Soit W_0 un ouvert tel que $\overline{W}\subset W_0$. On cherche $Q=op_{1/2}(q)$ où q contient une exponentielle $e^{\lambda\mu\phi}$, pour une phase analytique ϕ. On introduit donc la notation $e_\phi=op_{1/2}(e\, e^{\lambda\mu\phi})$, e est un symbole Gevrey s_0 supporté dans un voisinage W'_0 de l'ensemble de $(x,\xi',\alpha+\tau y^{1/2})$ où $(x,\xi')\in W_0$ et $-1\leqq\tau\leqq1$. $e'_{-\phi}$ est une paramétrixe de e_ϕ, $e'_{-\phi}=op_{1/2}(e^{-\lambda\mu\phi}\, e')$ où e' est un symbole réel Gevrey s_0 supporté dans W_0, $e'_{-\phi}\#e_\phi=1+\zeta_0$, ζ_0 est à décroissance exponentielle 1/s dans un voisinage W' de l'ensemble de $(x,\xi',\alpha+\tau y^{1/2})$ où $(x,\xi')\in W$, $-1\leqq\tau\leqq1$, (ceci est dû au fait que $\lambda\mu^2\leqq1$ donc $e^{-\lambda\mu\phi}$ inverse suffisamment e_ϕ). On cherche Q sous la forme $Q=e_{\phi/2}Q'e_{\phi/2}$, où Q' est autoadjoint, donc $i(P^*Q-QP)=e_{\phi/2}i(P_\phi{}^*Q'-Q'P_\phi)e_{\phi/2}+\zeta_1+R$ où $P_\phi=e_{\phi/2}Pe'_{-\phi/2}$, $\zeta_1=op_{1/2}(e^{\lambda\mu\phi}\zeta'_1)$ où ζ'_1 est un symbole à support compact Gevrey s_0 en ξ et s' en (y,ζ) de degré μ^{-1} à décroissance exponentielle 1/s dans W'. R est à décroissance exponentielle 1/s. P_ϕ est donc un opérateur pseudo-différentiel Gevrey s_0 en ξ et s' en (y,ζ) dont le symbole p_ϕ vérifie $p_\phi=p+i\mu/2\{p,\phi\}+O(\lambda^{-1}+\mu^2)$ dans W'. Si q' est le symbole de Q', q' est de degré μ^{-1}
$i(P_\phi{}^*Q'-Q'P_\phi)=-2\mathrm{Im}(\overline{p}_\phi\#q')=2\mathrm{Im}p_\phi q'+1/\lambda\{\mathrm{Re}p_\phi,q'\}+O(\lambda^{-2}\mu^{-1})=\mu\{p,\phi\}q'+O((\lambda\mu)^{-1})$.

Il faut maintenant préciser le choix de ϕ, $\phi=\xi-\alpha+\phi_0(x,y,\zeta)$, donc $\{p,\phi\}=-\alpha'_\eta+\phi_0{}'_\eta+2(\xi-\alpha)(\phi_0{}'_x+\{\phi_0,\alpha\})$. Dans la suite on adoptera $\phi_0=-Cx-|(y,\zeta)|^2+\alpha-\alpha*\nu_{\varepsilon'}$, où $\nu_{\varepsilon'}$ est un noyau Gaussien, ε' une petite constante et C une grande constante seront fixées en fonction de la géométrie de l'ouvert W dans lequel on microlocalise, $\phi_0-\alpha$ est analytique. On a $-\alpha'_\eta+\phi_0{}'_\eta>0$ et $\phi_0{}'_x+\{\phi_0,\alpha\}<0$, donc $\{p,\phi\}>0$ sauf si $\xi-\alpha\geqq-(-\alpha'_\eta+\phi_0{}'_\eta)/\phi_0{}'_x+\{\phi_0,\alpha\}\geqq c$. Soit $\chi\in C^\infty(\mathbb{R})$, $\chi\equiv1$ pour $t\leqq c/2$ supportée dans $t<c$. On introduit la troncature $\chi_1=\chi(\xi-\alpha)$, sur le support de χ_1 $\{p,\phi\}>0$, sur le support de $1-\chi_1$ $\xi-\alpha\geqq c/2$. On posera $q'=\omega\chi_1/\mu\{p,\phi\}+\varepsilon\omega/\mu$, où ω est supportée par W $\mu\{p,\phi\}q'=1+\varepsilon\omega\{p,\phi\}+\omega(\chi_1-1)+\omega-1$; ε est une constante assez petite. Soit $\chi_2=\omega(1-\chi_1)$. La fonction $\chi_2-\chi_2(\alpha+y^{1/2})/2y^{1/2}(\xi-\alpha+y^{1/2})$ s'annule sur $p^{-1}(0)$, elle s'écrit donc pq'_2. La fonction $\chi_2(\alpha+y^{1/2})$ a son support contenu dans $y\geqq(c/2)^2$, qui représente une zone dans laquelle on peut factoriser P sous la forme $P\equiv(D_x-\Lambda_-)(D_x-\Lambda_+)$ modulo R à décroissance exponentielle 1/s dans $y\geqq c$; le symbole λ_+ de Λ_+ commence par $\alpha+y^{1/2}$, celui de Λ_- par $\alpha-y^{1/2}$. On note

$P_+=D_x-\Lambda_+$, $P_-=D_x-\Lambda_-$. Si $P_{+,\phi}=e_{\phi/2}P_+e'_{-\phi/2}$, on écrit $\chi_2=P_{+,\phi}^*Q'_1+Q'_1P_{+,\phi}+P_\phi^*Q'_2+Q'_2P_\phi+\rho+\zeta_2$, ρ et ζ_2 sont des symboles Gevrey s_0 en ξ et s' en (y,ζ), ρ est de degré μ, ζ_2 est à décroissance exponentielle 1/s hors de W'. Q'_1 est autoadjoint de degré 0 et supporté par $y\geqq c_1$, Q'_2 est autoadjoint de degré 0. Donc

$i(P_\phi^*Q'-Q'P_\phi)=1+\varepsilon\omega\{p,\phi\}+\rho_1+P_{+,\phi}^*Q'_1+Q'_1P_{+,\phi}+P_\phi^*Q'_2+Q'_2P_\phi+\zeta_3$, ρ_1 est un symbole réel Gevrey s_0 en ξ et s' en (y,ζ) de degré $(\lambda\mu)^{-1}$, ζ_3 est à décroissance exponentielle 1/s hors de W'. On peut donc écrire $1+\varepsilon\omega\{p,\phi\}+\rho_1=C_1^*C_1$ où C_1 est obtenu par la série géométrique de $(1+\varepsilon\omega\{p,\phi\}+\rho_1)^{1/2}$ dont on prouve à l'aide d'arguments analogues à ceux de Beals qu'il s'agit bien d'un opérateur pseudo-différentiel de degré 0, Gevrey s_0 en ξ et s' en (y,ζ) si ε est assez petit. Donc

$i(P^*Q-QP)\equiv C^*C+P_+^*Q_1+Q_1P_++P^*Q_2+Q_2P+\zeta_4$ avec $Q_1\equiv e_{\phi/2}Q'_1e_{\phi/2}$, $Q_2\equiv e_{\phi/2}Q'_2e_{\phi/2}$, $C\equiv C_1e_{\phi/2}$. Q_1 est modulo un opérateur à décroissance exponentielle 1/s encore supporté par une zone $y\geqq c_2$, est le produit par $e^{\lambda\mu\phi}$ d'un symbole à support compact de degré 0 Gevrey s_0 en ξ et s' en (y,ζ) (il n'est plus tangentiel). ζ_4 se décrit comme ζ_1. R est à décroissance exponentielle 1/s On divise Q_1 par P_-, $Q_1\equiv Q_3+Q_4P_-$, Q_3 est tangentiel et supporté par $y\geqq c_2$, $Q_4\in S(m(\lambda\mu),G)$ avec $m=\exp(\lambda\mu(\phi_0+\inf(1,\max(\xi-\alpha,y_+^{1/2}))))$ Q_4 est supporté dans $y\geqq c_2$. $\equiv$ signifie modulo un opérateur de $S(m\mu^{-1},G)$ à décroissance exponentielle 1/s. Soit :

$i(P^*Q-QP)\equiv C^*C+P_+^*Q_3^*+Q_3P_++P^*Q_2+Q_2P+P^*Q_4^*+Q_4P+\zeta_5$. ζ_5 est analogue à ζ_1. On décompose $Q_4=Q_5+iQ_6$ selon les parties autoadjointes et anti autoadjointes, Q_5 et Q_6 appartiennent à $S(m(\lambda\mu),G)$. Donc

$$i(P^*(Q-Q_6)-(Q-Q_6)P)\equiv C^*C+P_+^*Q_3^*+Q_3P_++P^*(Q_2+Q_5)+(Q_2+Q_5)P+\zeta_5 \quad (3.\ 1)$$

avec $Q=e_{\phi/2}(\omega\chi_1/\mu\{p,\phi\}+\varepsilon\omega/\mu)e_{\phi/2}$, $Q_6\in S(m(\lambda\mu),G)$, $Q_3\in S(m_0,g)$ est tangentiel et supporté par $y\geqq c_2$, $Q_7=Q_2+Q_5\in S(m(\lambda\mu),G)$ est autoadjoint. On note que $Q-Q_6$ est elliptique dans W' car $\lambda\mu<1/\mu$. Pour simplifier les notations on note ζ_5 par ζ.

Q s'écrit $Q=op_{1/2}(q)$ avec $q=e^{\lambda\mu\phi}(\omega\chi_1/\mu\{p,\phi\}+\varepsilon\omega/\mu+O(1))$.

L'étape suivante consiste à faire des divisions par P pour obtenir un opérateur linéaire en D_x dont le commutateur avec P sera positif.

Utilisant la proposition 2. 2, on écrit $C=C_0+KP+R$, où $K\in S((\lambda\mu)^2m_1,G)$, $R\in S((\lambda\mu)^2m_1,G)$ où $m_1=m^{1/2}$, R est à décroissance exponentielle 1/s dans $y\geqq 0$, de plus $c_0=c_0^0(x,\xi')+(\xi-\alpha)c_0^1(x,\xi')$, où les les c_0^j s'écrivent $e^{\lambda\mu\phi_0}c'^j_0$, les c'^j_0 vérifient l'estimation (2.14) avec le poids $m_1^0=\exp(\lambda\mu/2\inf(1,y_+^{1/2}))$. On effectue de même les divisions de Q et Q_6, sous une forme cette fois plus autoadjointe : $Q=Q_0+P^*J_1+J_1P+R_1$, J_1 et R_1 sont autoadjoints, dans la classe $S(m\mu^{-1}(\lambda\mu)^2,G)$, R_1 est décroissance exponentielle 1/s dans $y\geqq 0$. $Q_0=op_{1/2}(q_0+(\xi-\alpha)q_1)$, $q_1=1/2y^{1/2}(q(\xi-\alpha+y^{1/2},y,\zeta)-q(\xi-\alpha-y^{1/2},y,\zeta))+S(m_0\mu^{-1}\delta^{-1}(\lambda\mu)^3/\lambda,g)$ dans $y\geqq 0$. Or $1/2y^{1/2}(q(\xi-\alpha+y^{1/2},y,\zeta)-q(\xi-\alpha-y^{1/2},y,\zeta))\geqq c\mu^{-1}\delta^{-1}m_0$ pour $(x,\xi')\in W$ et $y\geqq 0$; on choisit un prolongement, par un théorème de Borel, à $y\leqq 0$ de sorte que q_1 est elliptique positif dans W.

$Q_6=Q_6^0+P^*J_2+J_2P+R_2$, cette fois J_2 et R_2 sont dans $S(m(\lambda\mu)^3,G)$. On procède de même pour $\zeta=\zeta_0+P^*\zeta'+\zeta'P+R_3$, ζ_0 sera à décroissance exponentielle 1/s dans W. On obtient donc à partir de (3. 1) :

$$i(P^*(Q_0+Q_6^0)-(Q_0+Q_6^0)P)=C_0^*C_0+\zeta_0+P_+^*Q_3^*+Q_3P_++$$
$$P^*(Q_2+Q_5+\zeta'+K^*C_0+1/2K^*KP)+(Q_2+Q_5+\zeta'+C_0^*K+1/2P^*K^*K)P+1/i(P^{*2}J-JP^2)+R \quad (3.\ 2),$$

$R\in S(m\mu^{-1}(\lambda\mu)^2,G)$ est à décroissance exponentielle 1/s dans $y\geqq 0$, avec la notation $J=J_1+J_2$. Un terme de la forme P^*L+LP n'est pas gênant, comme on le verra plus loin. On remarque que $i(P^{*2}J-JP^2)=P^*A+AP$ avec $A=1/i(JP-P^*J)$. Le cas de

$P^*(C_0K+1/P^*2K^*K)^*+(C_0^*K+1/2P^*K^*K)P$ demande plus de soin. On écrit $C_0^*K+1/2P^*K^*K=1/2(C^*+C_0^*)K$. On traite d'abord le cas du terme $P^*K^*C_0+C_0^*KP$. On décompose C_0^*K sous la forme $C_0^*K=A+iB$, A et B sont autoadjoints. Il faut étudier $i(PB-BP)$, on divise B par P sous la forme $B\equiv B_0+B'P+P^*B'$, on va soustraire à Q_0 le terme B_0, il faut voir que celà ne change pas l'ellipticité du terme en D_x. On calcule $B=1/2i(C_0^*K-K^*C_0)=op_{1/2}(b)$,. Calculant précisèment B on peut montrer qu'il s'exprime sous la forme $B\equiv B_0+B'P+P^*B'$, où le facteur contenant D_x de B_0 est $o(\mu^{-1})$ dans $S(m_0,g)$. Le terme $i(P^*(B'P+P^*B')-(B'P+P^*B')P)=i(P^{*2}B'-B'P^2)$ s'écrit lui sous la forme P^*L+LP comme on l'a vu plus haut. La contribution de $B_1=C^*K$ se traite plus facilement car $B_1\in S(m(\lambda\mu)^2,G)$ donc il donnera après division un terme de $S(m\delta^{-1}(\lambda\mu)^2,g)$ comme reste et $(\lambda\mu)^2=o(\mu^{-1})$.

On a obtenu :

$$i(P^*(Q_0+B_0)-(Q_0+B_0)P)=C_0^*C_0+\zeta_0+P_+^*Q_3^*+Q_3P_++P^*L+LP+R, \tag{3. 3}$$

où $B_0=op_{1/2}(b_0+(\xi-\alpha)b_1)$, $b_1\in S(m_0\delta^{-1}o(\mu^{-1}),g)$, $b_0\in S(m_0o(\mu^{-1}),g)$, $L\in S(m(\lambda\mu)^k,G)$, $R\in S(m(\lambda\mu)^k,G)$ est à décroissance exponentielle 1/s dans $y\geqq 0$. Il est clair que $i(P^*(Q_0+B_0)-(Q_0+B_0)P)-(C_0^*C_0+P_+^*Q_3^*+Q_3P_+)$ est quadratique en D_x, il s'écrit donc sous la forme $op_{1/2}(a_0+(\xi-\alpha)a_1)+P^*A_2+A_2P$, où A_2 est tangentiel. On va calculer plus précisèment les degrés de a_0 et a_1. $a_0\in S((\lambda\mu)\delta m_0,g)$, $a_1\in S((\lambda\mu)m_0,g)$, $a_2\in S((\lambda\mu)\delta^{-1}m_0,g)$ les calculs donnent un résultat moins bon que ce qu'il devrait être d'un facteur $(\lambda\mu)\delta$, ceci provient uniquement de l'estimation de $q_0{}'_y$ et $q_1{}'_y$, en choisissant correctement les prolongements à $y\leqq 0$, on peut supposer que dans $-y(\lambda\mu)^2$ grand q_0 et q_1 sont réguliers, donc dans $y\leqq 0$ $a_0\in S(m_0,g)$, $a_1\in S(\delta^{-1}m_0,g)$, $a_2\in S(\delta^{-2}m_0,g)$.

On a donc écrit

On écrit $\zeta_0=op_{1/2}(a'_0+(\xi-\alpha)a'_1)$, $a'_0\in S(\mu^{-1}m_0,g)$, $a'_1\in S(\mu^{-1}\delta^{-2}m_0,g)$, a'_0 et a'_1 sont à décroissance exponentielle 1/s dans W.

$i(P^*(Q_0+B_0)-(Q_0+B_0)P)-(C_0^*C_0+\zeta_0+P_+^*Q_3^*+Q_3P_+)=P^*L+LP+R=op_{1/2}(a''_0+(\xi-\alpha)a''_1)+A_2P+P^*A_2$. Avec $a''_0=a'_0+a_0$, $a''_1=a'_1+a_1$. Donc $P^*J+JP+R=op_{1/2}(a''_0+(\xi-\alpha)a''_1)$, avec $J=L-A_2$, $A_2\in S(\mu^{-1}\delta^{-2}m,G)$, $J\in S(m(\lambda\mu)^k,G)$. On va prouver que $J=op_{1/2}(j)$ est à décroissance exponentielle dans $y\geqq 0$. On développe $P^*J+JP=pj+\sum_{\beta\neq 0,|\beta|<N}R_\beta(j)+\bar{R}_N(j)$ en utilisant le calcul symbolique. On écrit $\sum_{\beta\neq 0,|\beta|<N}R_\beta(j)+\bar{R}_N(j)+R=r_0+(\xi-\alpha)r_1+pL(\partial_\xi{}^2(\sum_{\beta\neq 0,|\beta|<N}R_\beta(j)+\bar{R}_N(j)))$. Donc dans $y\geqq 0$, $j=-L(\partial_\xi{}^2(\sum_{\beta\neq 0,|\beta|<N}R_\beta(j)+\bar{R}_N(j)+R)$. On note $R'_\beta(j)=-L(\partial_\xi{}^2R_\beta(j))$, $r_N=-L(\partial_\xi{}^2(\bar{R}_N(j)+R))$. Soit pour tout N, $j=\sum_{\beta\neq 0,|\beta|<N}R'_\beta(j)+r_N$ dans $y\geqq 0$. Donc : dans $y\geqq 0$ $j=\sum_{0\leqq k\leqq N-1,k\leqq|\beta_1|+\ldots+|\beta_k|<N}R'_{\beta_1}\ldots R'_{\beta_k}(r_{N-(|\beta_1|+\ldots+|\beta_k|)})$. La méthode utilisée pour la récurrence au paragraphe 1 prouve que j est à décroissance exponentielle 1/s dans $y\geqq 0$. a''_0 et a''_1 sont donc à décroissance exponentielle 1/s dans $y\geqq 0$.

On a obtenu que :

$$i(P^*(Q_0+B_0)-(Q_0+B_0)P)-(C_0^*C_0+P_+^*Q_3^*+Q_3P_+)-(P^*A_2+PA_2)=op_{1/2}(a_0+a_1(\xi-\alpha)) \tag{3. 4}$$

$a_0\in S(m_0,g)$, $a_1\in S(\delta^{-1}m_0,g)$, $a_2\in S(\delta^{-2}m_0,g)$ dans $y(\lambda\mu)^2\leqq C$, a_0 et a_1 sont à décroissance exponentielle 1/s dans $W\cap\{y\geqq 0\}$. On peut assurer que a_0, a_1 et a_2 sont réels. On écrit donc $a_0=j_0+k_0$ et $a_1=j_1+k_1$, où j_0 et j_1 sont supportées dans $W\cap\{y\leqq 0\}$, $j_0\in S(m_0,g)$, $j_1\in S(\delta^{-1}m_0,g)$, k_0 et k_1 sont à décroissance exponentielle dans W.

On va exploiter le facteur $C_0^*C_0$ pour majorer $op_{1/2}(j_0+j_1(\xi-\alpha))$. Il faudra prendre deux multiplicateurs Q_0 et Q'_0 de façon à obtenir des opérateurs C_0 et C'_0 "indépendants" dans $y\leqq 0$. On va étudier C_0 un peu plus précisèment. $C=op_{1/2}(ce^{\lambda\mu\phi/2})$ où c est un symbole à support compact Gevrey s_0 en ξ, s' en (y,ζ) elliptique dans W'. On peut appliquer la proposition 2.2 et donc $C_0=op_{1/2}(c_0)+op_{1/2}(c_1)(D_x-\alpha)$, $c_0=e^{\lambda\mu\phi_0/2}\tilde{c}_0$, $c_1=e^{\lambda\mu\phi_0/2}\tilde{c}_1$, $\tilde{c}_0$ et $\tilde{c}_1$ vérifient les inégalités (1.2). $\tilde{c}_0|_{y=0}=c(\alpha,0,\zeta)+o(1)$, $\tilde{c}_1|_{y=0}=\lambda\mu c(\alpha,0,\zeta)+o(\lambda\mu)$. $c(\alpha,0,\zeta)>0$ partout, à condition de modifier ζ_0 et A_2.

C' est construit de la même façon que C mais avec la phase $\phi'=2(\xi-\alpha)+\phi_0$. ϕ' n'est plus analytique mais est linéaire en ξ, ceci fait que dans la composition de deux opérateurs $Q_i=op_{1/2}(q_i\, e^{\lambda\mu\phi'/2})$ où les q_i sont des symboles à support compact Gevrey s_0 en ξ et s' en (y,ζ) on obtient encore un opérateur $Q=Q_1Q_2=op_{1/2}(qe^{\lambda\mu\phi'})+R$ avec un symbole q à support compact Gevrey s_0 en ξ et s' en (y,ζ), R est à décroissance exponentielle 1/s, car le point critique en (Y,Z) de la phase $\sigma(Y,Z)+i\mu/2(\phi'(X+Y)+\phi'(X+Z))$ est une fonction qui ne dépend pas de ξ.

Par division on obtient $C'_0=op_{1/2}(c'_0)+op_{1/2}(c'_1)(D_x-\alpha)$, $c'_0=e^{\lambda\mu\phi_0/2}\tilde{c}'_0$ $c'_1=e^{\lambda\mu\phi_0/2}\tilde{c}'_1$, $\tilde{c}'_0$ et $\tilde{c}'_1$ vérifient les inégalités (1.2). $\tilde{c}'_0|_{y=0}=c'(\alpha,0,\zeta)+o(1)$, $\tilde{c}'_1|_{y=0}=2\lambda\mu c'(\alpha,0,\zeta)+o(\lambda\mu)$.

On posera $U=(u_0,u_1)$, $u_0=u$, $u_1=(D_x-\alpha)u$. On a écrit C_0 et C'_0 de sorte que : $|C_0u|^2=|c_0u_0+c_1u_1|^2$ et $|C'_0u|^2=|c'_0u_0+c'_1u_1|^2$. On écrit cela sous forme de système.

Le système $M_0=\begin{pmatrix} c_0 & c_1 \\ c'_0 & c'_1 \end{pmatrix}$ est elliptique sur y=0, ce sera encore vrai dans une zone $0\leq y\leq c(\lambda\mu)^{-2}$, on prolongera dans $y\leq 0$ $\tilde{c}_0$, $\tilde{c}'_0\in S(1,g)$, $\tilde{c}_1$, $\tilde{c}'_1\in S(\delta^{-1},g)$ de sorte que le système soit elliptique en mettant le poids δ^{-1} sur la deuxième composante dans une zone $(\lambda\mu)^2y\leq c$, où c est une petite constante positive.

Dans une zone $(\lambda\mu)^2y\leq c$, le symbole $\sigma(x_0,x_1)$ de $M_0^*M_0$ vérifie $\sigma(x_0,x_1)\geq ce^{\lambda\mu\phi_0}(|x_0|^2+\delta^{-2}|x_1|^2)$ dans $y(\lambda\mu)^2\leq c$, où c est assez petit.

Soit $n=\begin{pmatrix} n_0 & 0 \\ 0 & n_1 \end{pmatrix}$ où $n_0=C(1-\chi)(y(\lambda\mu)^2)e^{\lambda\mu\phi_0}ch(2\lambda\mu y^{1/2})$, $n_1=C(1-\chi)(y(\lambda\mu)^2)\delta^{-2}e^{\lambda\mu\phi_0}ch(2\lambda\mu y^{1/2})$, $\chi(t)=1$ pour $t\leq c/2$, $supp\chi\subset\{t\leq c\}$, C est une grande constante. $n'=n+\sigma=(n'_{ij})$, $0\leq i,j\leq 1$, $n'_{ij}\in S(\delta^{-(i+j)}m'_0,g)$ avec $m'_0=\exp(\lambda\mu(\phi_0+2\inf(1,y_+{}^{1/2})))$. Les n'_{ij} s'écrivent $n'_{ij}=e^{\lambda\mu\phi_0}\tilde{n}'_{ij}$ où les $\tilde{n}'_{ij}$ satisfont à des estimations comme (1.2) – c'est à dire ont des dérivées régulières en ζ. La matrice n' est partout définie positive $n'(x_0,x_1)\geq cm'_0(|x_0|^2+\delta^{-2}|x_1|^2)$.

Soit $\mu_0=\begin{pmatrix} \mu_0^0 & 0 \\ 0 & \mu_0^1 \end{pmatrix}$ une matrice où $\mu_0^0=op_{1/2}(e^{\lambda\mu\phi_0/2}\widetilde{\mu}_0^0)$, $\mu_0^1=op_{1/2}(e^{\lambda\mu\phi_0/2}\widetilde{\mu}_0^1)$, où $\widetilde{\mu}_0^0\in S(m'_0{}^{1/2},g)$, $\widetilde{\mu}_0^1\in S(m'_0{}^{1/2}\delta^{-1},g)$ sont des symboles elliptiques vérifiant (1.2). $\mu_0{}^0$ et $\mu_0{}^1$ ont des inverses (pour λ grand) $(\mu_0^0)^{-1}=op_{1/2}(e^{-\lambda\mu\phi_0/2}\tilde{\mu}_0)$, et $(\mu_0^1)^{-1}=op_{1/2}(e^{-\lambda\mu\phi_0/2}\tilde{\mu}_1)$ où $\tilde{\mu}_0\in S(m'_0{}^{-1/2},g)$ $\tilde{\mu}_1\in S(\delta m'_0{}^{-1/2},g)$ et vérifient les estimations (1.2). $\mu_0{}^{-1}$ designe l'inverse de l'opérateur μ_0.

On pose $\sigma_0=\mu_0^{*-1}\sigma\mu_0^{-1}$, $n'_0=\mu_0^{*-1}n'\mu_0^{-1}$, $\sigma_0\geq 0$ et $n'_0>0$ sont de degré zéro et vérifient (1.2).

Soit $j=op_{1/2}(e^{\lambda\mu\phi_0}\tilde{j})$ où $\tilde{j}$ est une matrice 2×2 autoadjointe dont les entrées $\tilde{j}_{ij}\in S(m'_0\delta^{-(i+j)},g)$ vérifient (1.2) et sont supportées dans $y\leq 0$. On pose $j_0=n'^{-1/2}_0\mu_0^{*-1}j\mu_0^{-1}n'^{-1/2}_0$. j_0 est de degré zéro et son symbole vérifie (1.2).

Soit $\chi_1=op_{1/2}(\chi_1(y(\lambda\mu)^2)$ où χ_1 vaut 1 dans $t\leq c_1/2$ et est supportée dans $t\leq c_1$, c_1 est une constante positive plus petite que c. $(1-\chi_1)j_0$ et $j_0(1-\chi_1)$ sont à décroissance exponentielle 1/s quand s>s'+2.

On en déduit que $(j_0U,U)\leq C|\chi_1U|^2+(\rho U,U)$ où ρ est à décroissance exponentielle 1/s. Donc

$(jU,U)\leq C|\chi_1 n'_0{}^{1/2}\mu_0U|^2+(\rho_1U,U)$. (3. 5).

$\chi_1(n'_0{}^{1/2}-\sigma_0{}^{1/2})$ est un opérateur dont la norme dans L^2 est à décroissance exponentielle 1/s. Donc (3. 5) donne :

$(jU,U)\leq C_1|\chi_1\sigma_0{}^{1/2}\mu_0U|^2+C\exp(-1/C\lambda^{1/s})|U|^2$, soit

$(jU,U)\leq C_1\sigma(U,U)+C\exp(-1/C\lambda^{1/s})|U|^2$. (3. 6).

On va utiliser (2. 6) pour majorer le terme $op_{1/2}(j_0+(\xi-\alpha)j_1)$ où j_0 et j_1 ont été déterminées en (2. 4). On écrit :

$j=op_{1/2}(j_0+(\xi-\alpha)j_1)=op_{1/2}(j_0+r_0)+1/2((D_x-\alpha)op_{1/2}(j_1)+op_{1/2}(j_1)(D_x-\alpha))$, $r_0\in S(m_0(\lambda\mu)^3/\lambda,g)$ est à décroissance exponentielle 1/s dans $y\geq\varepsilon(\lambda\mu)^{-2}$. On n'utilise pas (3. 6) directement car on souhaite avoir une constante C_1 arbitrairement petite.

Si $j_0\in S(m_0,g)$ et $j_1\in S(m_0\delta^{-1},g)$ sont supportées par $y\leq 0$, pour tout $\varepsilon>0$ il existe $j_2\in S(m_0\delta^{-2},g)$ supporté par $y\leq\varepsilon(\lambda\mu)^{-2}$de sorte que : $j_0+\xi j_1\leq(\xi^2-y+\varepsilon(\lambda\mu)^{-2})j_2$. Le maximum de j_2 étant indépendant de ε.

On exprime $op_{1/2}(j_0+(\xi-\alpha)j_1-((\xi-\alpha)^2-y+\varepsilon(\lambda\mu)^{-2})j_2)=j_0+(y-\varepsilon(\lambda\mu)^{-2})j_2+1/2((D_x-\alpha)j_1+j_1(D_x-\alpha)j_1-(D_x-\alpha)j_2(D_x-\alpha)+r_0+r_1(D_x-\alpha)+(D_x-\alpha)r_1$, où $r_0\in S(m_0o(1),g)$ $r_1\in S(m_0\delta^{-1}o(1),g)$, $r_0$et r_1 sont réels et à décroissance exponentielle 1/s dans $y\geq 2\varepsilon(\lambda\mu)^{-2}$. On considére le système $A=(a_{ij})$, $a_{00}=j_0+(y-\varepsilon(\lambda\mu)^{-2})j_2+r_0$, $a_{10}=a_{01}=1/2j_1+r_1$, $a_{11}=-j_2$, $a_{ij}\in S(m_0\delta^{-(i+j)},g)$. Ce système est autoadjoint, son symbole $\sum_{i,j}a_{ij}\xi^{i+j}\leq o(1)e^{\lambda\mu\phi_0}(1+\xi^2\delta^{-2})$ à cause de l'inégalité prouvé au lemme 1. On considérera le système $A'=\mu_0{}^{-1}A\mu_0{}^{-1}$, où μ_0 est l'opérateur utilisé dans la preuve de (3. 6) dont on peut assurer qu'il est autoadjoint, $A'=(a'_{ij})$ $a'_{ij}\in S(1,g)$ et cette fois $\sum_{i,j}a'_{ij}\xi^{i+j}\leq o(1)(1+\xi^2)$. On peut lui appliquer l'inégalité de Gärding, donc $(A'U,U)\leq o(1)|U|^2$. Si χ a son support dans t<c on trouve que

$(A'U,U)\leq o(1)|\chi U|^2+C\exp(-1/C\lambda^{1/s})|U|^2$. Donc $(AU,U)\leq o(1)|\chi\mu_0U|^2+C\exp(-1/C\lambda^{1/s})|U|^2$.

On applique maintenant (3. 6) à $j=\mu_0\chi^2\mu_0$ et on obtient

$(AU,U)\leq o(1)(\sigma U,U)+C\exp(-1/C\lambda^{1/s})|U|^2$ (3. 7).

Avec des constantes o(1) et C dépendant de ε.

Ceci signifie que $j=op_{1/2}(j_0+(\xi-\alpha)j_1)$ satisfait à :

$(ju,u)\leq o(1)(\sigma U,U)+(op_{1/2}((\xi-\alpha)^2-y+\varepsilon(\lambda\mu)^{-2})j_2)u,u)+C\exp(-1/C\lambda^{1/s})|U|^2$ avec $U=(u,(D_x-\alpha)u)$.

$\varepsilon((\lambda\mu)^{-2}j_2u,u)\leq\varepsilon(C'+o(1))C_1\sigma(U,U)+C\exp(-1/C\lambda^{1/s})|U|^2$. On choisira ε de sorte que $\varepsilon C'C_1$ soit petit.

$op_{1/2}((\xi-\alpha)^2-y)j_2)=P^*j_2+j_2P+op_{1/2}(r'_0+(\xi-\alpha)r'_1)$, $r'_0\in S(m_0o(1),g)$ $r'_1\in S(m_0\delta^{-1}o(1),g)$, r'_0 et r'_1 sont réels et à décroissance exponentielle 1/s dans $y\geq 2\varepsilon(\lambda\mu)^{-2}$. On appliquera directement l'argument de (3. 6) à $op_{1/2}(r'_0+(\xi-\alpha)r'_1)$.

Résumant la discussion ci dessus on a majoré :

$(ju,u)\leq 1/2(\ |C_0u|^2+|C'_0u|^2)+((P^*j_2+j_2P)u,u)\ +C\exp(-1/C\lambda^{1/s})(|u|^2+|(D_xu|^2)$. (3. 8).

On applique (3. 8) à $-j$, (3. 4) donne

$((i(P^*(Q_0+B_0)-(Q_0+B_0)P)-(P_+{}^*Q_3{}^*+Q_3P_+)-(P^*\tilde{A}_2+P\tilde{A}_2)+i(P^*(Q'_0+B'_0)-(Q'_0+B'_0)P)-(P_+{}^*Q'_3{}^*+Q'_3P_+)-(P^*\tilde{A}'_2+P\tilde{A}'_2))u,u)\geqq 1/2(\ |C_0u|^2+|C'_0u|^2)-C\exp(-1/C\lambda^{1/s})(|u|^2+|(D_xu|^2)+(ku,u)+(k'u,u).$ (3. 9).

On a noté avec un ' les opérateurs construits avec la phase ϕ', k et k' sont linéaires en D_x, dans des classes $S(m'_0,g)$ et à décroissance exponentielle 1/s dans W, $\tilde{A}_2=A_2+j_2$, $\tilde{A}'_2=A'_2+j'_2$.

On veut une inégalité dans le demi espace $\mathbf{R}^n{}_+=(x\geqq 0)$, on note $(u,v)_+$ le produit scalaire dans $L^2(\mathbf{R}^n{}_+)$, et $(u,v)_\partial$ le produit scalaire sur le bord. On remarque que si Q_1 est tangentiel et autoadjoint on a

$1/i((D_x{}^2u,Q_1D_xu)_+-(Q_1D_xu,D_x{}^2u)_+)=(Q_1D_xu,D_xu)_\partial+\sum_{0\leq i,j\leq 1}(Q_{ij}D_x{}^ju,D_x{}^iu)_+.$

Si $u\in H^1{}_0(\mathbf{R}^n{}_+)$ on déduit de (3. 9) que :

$$2\operatorname{Im}(Pu,(Q_0+Q'_0+B_0+B'_0)u)_+-2\operatorname{Re}(P_+u,(Q^*{}_3+Q'_3{}^*)u)_+-2\operatorname{Re}(Pu,(\tilde{A}_2+\tilde{A}'_2)u)_+\geqq$$
$$((Q_1+Q'_1)D_xu,D_xu)_\partial+1/2(\ |C_0u|_+{}^2+|C'_0u|_+{}^2)-C\exp(-1/C\lambda^{1/s})(|u|_+{}^2+|(D_xu|_+{}^2)+(ku,u)_++(k'u,u)_+. \quad (3.\ 10).$$

On a noté par Q_1 la partie en D_x de Q_0+B_0. On a vu que $Q_1\in S(m_0\delta^{-1}\mu^{-1},g)$ est elliptique positif dans W.

Il n'est pas restrictif de se ramener à $H^1{}_0(\mathbf{R}^n{}_+)$ car si on sait seulement que $\gamma_0u=u|_{x_1=0}$ est à décroissance exponentielle 1/s dans W, on appliquera (3. 10) à $v=\psi(x,D')u-\psi(0,x',D')\gamma_0u$ et $\psi(0,x',D')\gamma_0u$ est à décroissance exponentielle 1/s, au lieu d'appliquer (3. 10) à $v=\psi(x,D')u$.

$W=\{(y,\zeta);\ |x|\leqq\varepsilon,\ |(x',\xi')|\leqq\delta\}$, ε et δ sont petits et fixés plus loin, on aura fait en sorte que tous les points de $\{(x,\xi',\alpha\pm y^{1/2});\ y\geqq 0\ (x,\xi')\in W\}$ sont atteints après éventuellement une réfléxion sur le bord par une bicaractéristique issue d'un voisinage d'un point $\rho'_0\in\gamma_-(\rho_0)\setminus\rho_0\}$. $\psi(x,\xi')$ est supportée dans W, le support de $d\psi$ est proche de ∂W. On étudie un terme comme $(P\psi u,Q_0u)$ où $Q_0\in S(m_0,g)$. ψPu est à décroissance exponentielle 1/s par hypothèse, il reste donc $([P,\psi]u,Q_0u)$. $[P,\psi]$ est à décroissance exponentielle 1/s hors d'un voisinage de ∂W, il reste donc ψ_1Q_0u où ψ_1 est supportée près de ∂W. Pour un point $(x,\xi)\in T^*\mathbf{R}^n{}_+$ on note par $y_0(x,\xi)$ la valeur de y au point où la bicaractéristique arrivant en (x,ξ) touche le bord, si ça ne se produit pas ou bien le point (x,ξ) est non caractéristique ou bien on sait déjà qu'il n'est pas dans le $WF^{(s)}(u)$ par la propagation à l'intérieur. Dans les deux derniers cas le résultat est connu ou facile. La condition $\alpha'_\eta<0$ donne $y_0(x,\xi)\geqq y-kx$ si $(\xi-\alpha)^2\geqq c_2x$, où c_2 et k sont des constantes positives. Le poids $m_0=\exp(\lambda\mu(\phi_0+y_+{}^{1/2}))$ est à décroissance exponentielle (avec un taux dépendant de μ_0) quand $\phi_0+y_+{}^{1/2}<0$. On réalise la condition $\partial W\cap\{\phi_0+y_+{}^{1/2}\geqq 0\}\subset\{(x,\xi');\ y_0(x,\xi',\alpha\pm y^{1/2})>0\}$ en choisissant ε, δ petits et la constante C intervenant dans la définition de ϕ_0 grande par rapport à ε. Par le théorème sur la réfléxion transverse les points $\{(x,\xi',\alpha\pm y^{1/2}),\ y_0(x,\xi',\alpha\pm y^{1/2})>0\}$ ne sont pas dans le $WF^{(s)}{}_b(u)$. Donc $\partial W\cap\{\phi_0+y_+{}^{1/2}\geqq 0\}\cap\pi(WF^{(s)}{}_b(u))=\emptyset$, $\pi(WF^{(s)}{}_b(u))$ est un compact puisque u est s régulière. Donc ψ_1Q_0u est à décroissance exponentielle 1/s. Pour les termes faisant intervenir P_+u, on fait remarquer que si χ a son support dans $y>0$ χP_+u est décroissance exponentielle 1/s car $v=P_+u$ satisfait $P_-v=0$ dans $y\geqq c'$ mais dans les caractéristiques de P_- on a $\xi-\alpha<0$, ces points sont donc joints (sans réflexion) à un point voisin de ρ'_0 par une bicaractéristique contenue dans $x\geqq 0$, on conclut par la propagation des singularités à l'intérieur si $x>0$, sinon il s'agit d'un point hyperbolique.

On conclut donc que $((Q_1+Q'_1)\psi D_x u,\psi D_x u)_\partial$ est à décroissance exponentielle 1/s. $Q''_1=Q_1+Q'_1$ est elliptique positif dans la classe $S(m'_0\delta^{-1}\mu^{-1},g)$ dans W, W est un voisinage du support de ψ, on en déduit que dans $\phi_0+2y_+^{1/2}\geq-\varepsilon$, pour un ε petit, $D_x u|_{x_1=0}$ est à décroissance exponentielle 1/s, donc $\rho_0\notin WF^{(s)}(D_x u|_{x_1=0})$.

Bibliographie.

[1] C. Bardos- G. Lebeau- J. Rauch. Scattering frequencies and Gevrey 3 singularities. *Inventiones math.* 90 (1987). 77-114.

[2] L. Carleson . On universal moment problems. *Math. Scand.* 9. (1961). 197-206.

[3] G. Eskin. General initial boundary problems for second order hyperbolic equations. D. Reidel. Co. Dordrecht, Boston, London. (1981). 19-54.

[4] L. Hörmander. The Analysis of linear differential operators III. Springer-Verlag 1985.

[5] V. Ivrii. Wave front sets of solutions of boundary value problems for a class of symetric hyperbolic systems. *Sibirian Math. Journal.* 21. (1980). 527-534.

[6] B. Lascar. Propagation des singularités Gevrey pour des opérateurs hyperboliques. *American Journal of Maths.* 110. (1988). 413-449.

[7] G. Lebeau. Régularité Gevrey 3 pour la diffraction. *Comm. in Partial Differential Equations.* 9. 15. (1985). 1437-1494.

[8] R. Melrose. Microlocal parametrices for diffractive boundary value problems. *Duke Math. Journal.* 42. (1975). 605-635.

[9] M. Taylor. Grazing rays and reflection of singularities to wave equations. *Comm. Pure Appl. Math.* 29. (1976). 1-38.

[10] J. Sjöstrand. Propagation of analytic singularities for second order Dirichlet problems. I, II, III. *Comm. in Partial Differential Equations.* 5.1. (1980). 41-94. 5. 2. (1980). 187-207. 6. 5. (1981). 499-567.

[11] J. Sjöstrand. Singularités analytiques microlocales. *Astérisque* 95. (1984). S. M.F.

REDUCTION AND GEOMETRIC PREQUANTIZATION AT THE COTANGENT LEVEL

Mircea PUTA
University of Timişoara
Department of Geometry-Topology
1900 Timişoara - Romania

1. INTRODUCTION

Let (M, ω) be a symplectic manifold (possibly infinite dimensional), G a Lie group (possibly infinite dimensional) with Lie algebra $\mathcal{G}$ and $\phi: G \times M \to M$ a symplectic action of G on M, with and Ad^*-equivariant momentum map $J: M \to \mathcal{G}^*$, i.e.

$$
\begin{cases}
\phi_g^* \omega = \omega \\[2mm]
X_{\hat{J}(\xi)} = \xi_M \\[2mm]
J(\phi_g(x)) = \mathrm{Ad}^*_{g^{-1}}(J(x)) \quad ,
\end{cases}
$$

for each $x \in M$, $g \in G$ and $\xi \in \mathcal{G}$, where $\hat{J}(\xi): M \longrightarrow \mathbb{R}$ is a smooth real function on M given by:

$$
\hat{J}(\xi)(x) \overset{def}{=} J(x)(\xi) \quad ,
$$

ξ_M is the infinitesimal generator of the action ϕ i.e.

$$
\xi_m(x) \overset{def}{=} \left. \frac{d}{dt} \right|_{t=0} \phi_{\exp(t\xi)}(x) \quad ,
$$

and $\mathrm{Ad}^*: G \times \mathcal{G}^* \longrightarrow \mathcal{G}^*$ given by

$$
\mathrm{Ad}^*_{g^{-1}} = (\mathrm{Ad}_{g^{-1}})^* = (T_e(R_g \circ L_{g^{-1}}))^*
$$

is the coadjoint action of G on $\mathcal{G}^*$.

For each $\mu \in \mathcal{G}^*$, let us denote by G_μ the isotropy group of μ under the coadjoint action Ad^*, i.e.

A. Boutet de Monvel et al. (eds.), Recent Developments in Quantum Mechanics, 341–350.
© 1991 Kluwer Academic Publishers.

$$G_\mu = \left\{ g \in G \mid Ad^*_{g^{-1}} \mu = \mu \right\} .$$

Then due to the fact that H is Ad^*-equivariant under G_μ the orbit space $M_\mu \overset{def}{=} J^{-1}(\mu)/G_\mu$ is well defined and it is called the reduced phase space of M.

If $\mu \in \mathcal{G}^*$ is a regular value of J and G_μ acts free and properly on $J^{-1}(\mu)$, then M_μ is a smooth manifold and the projection $\pi_\mu : J^{-1}(\mu) \to M_\mu$ is a submersion. Moreover there exists an unique (weak) symplectic structure ω_μ on M_μ which is consistent with the symplectic structure ω on M, i.e.

$$\pi^*_\mu \omega_\mu = i^*_\mu \omega ,$$

where $i_\mu : J^{-1}(\mu) \longrightarrow M$ is the inclusion.

In this form the reduced phase space has appeared for the first time in a paper of J. Marsden and A. Weinstein [5], but the general idea on the reduction goes back to E. Cartan.

Let $H: M \to \mathbb{R}$ be an G-invariant Hamiltonian on M and H_μ its induced Hamiltonian on M_μ, i.e.

$$H_\mu \circ \pi_\mu = H \circ i_\mu .$$

As real valued functions H and H_μ define the Hamiltonian vector fields X_H and X_{H_μ} on $J^{-1}(\mu)$ and respectively M_μ. Then it can be proved that the integral curves of X_H project onto the integral curves of X_{H_μ}, or equivalently that the Hamiltonian vector fields X_H and X_{H_μ} are π_μ-related, that is:

$$T\pi_\mu \circ X_H = X_{H_\mu} \circ \pi_\mu , \qquad [1].$$

Therefore we have two simplectiv manifolds (M, ω) and (M_μ, ω_μ) such that the dynamics of the first one projects onto the dynamics of the second one. The problem is now to determine the prequantum equivalent of this property from the geometric prequantization point of view. We remined that geometric prequantization associates in a geometrical way to each quantizable manifold (M, ω), and Hilbert space $\mathcal{H}$ of quantum states and to each smooth function (classical observable) on M a prequantum operator δ_f on $\mathcal{H}$.

Since it is very difficult to obtain results in a completelly general setting we shall concentrate here only to the particular case $M = T^*Q$, $\omega = d\theta$, the canonical symplectic structure on T^*Q, G acts by point transformations on Q and $\mu = 0$. This choice has the following physical motivation: $(T^*Q,\ \omega = d\theta)$ can be viewed as the extended phase space of a constrained mechanical system, $((T^*Q)_o, \omega_o)$ as the corresponding reduced phase space, $J = 0$ as the constrained set, and G as the gauge group.

For this configuration we shall prove that reduction and geometric prequantization are interchangeable processes. In the particular case when Q and Q_o are simple-connected manifolds it follows that the unique geometric prequantization on T^*Q projects onto the unique geometric prequantization of $(T^*Q)_o$. Also we shall point out that some examples from theoretical physics as: the planar n-body problem, Einstein's equations in vacuum are particular cases of the above picture.

2. GENERALITIES ON THE COTANGENT BUNDLE

Let Q be a smooth n-dimensional manifold and T^*Q its cotangent bundle with the canonical symplectic structure $\omega = d\theta$, locally in canonical coordinates we have:

$$\theta = \sum_{j=1}^{n} p_j\, dq^j \quad ; \qquad \omega = \sum_{j=1}^{n} dp_j \wedge dq^j \quad .$$

The cotangent bundle T^*Q is a 2n-dimensional manifold and it is also an oriented one with respect to the volume form ε_ω given by the following relation:

$$\varepsilon_\omega \overset{def}{=} (-1)^{\frac{n(n-1)}{2}} \frac{1}{n!}\,\omega^n \quad .$$

In physical applications Q is the configuration space of some mechanical system and T^*Q is its corresponding phase space.

If $H: T^*Q \to \mathbb{R}$ is the energy of the Hamiltonian of the system then its behaviour in time is given by the integral curves of the Hamiltonian vector field X_H, where X_H is uniquely detemined by the relation:

$$X_H \lrcorner\ \omega + dH = 0.$$

Since $\omega = d\theta$ is an exact form, the cotangent bundle (T^*Q, ω) is a quantizable manifold in the sense of geometric prequantization.

Indeed, its prequantum bundle L^{ω} is a trivial one, i.e. $L^{\omega} = (T^{*}Q \times \mathbb{C}, pr_1, T^{*}Q)$; the Hermitian structure on L^{ω} is given by:

$$((q,\alpha_q,z_1),(q,\alpha_q,z_2)) \stackrel{def}{=} z_1\bar{z}_2 \quad ,$$

for each $(q,\alpha_q) \in T^{*}Q$; $z_1, z_2 \in \mathbb{C}$, and if we identify the space $\Gamma(L^{\omega})$ with the space $C^{\infty}(T^{*}Q, \mathbb{C})$ of smooth, complex valued functions defined on $T^{*}Q$, then the connection ∇^{ω} of L^{ω} whose curvature is $(1/\hbar)\omega$ and which is compatible with the above Hermitian structure is globally defined by the following relation:

$$\nabla^{\omega}_{X}(f) = X(f) - (i/\hbar)\theta(X)\cdot f \quad ,$$

for each $X \in \mathcal{X}(T^{*}Q)$, $f \in C^{\infty}(T^{*}Q, \mathbb{C})$.

The Hilbert representation space $\mathcal{H}$ of quantum states can be identified with the space $L^2(T^{*}Q, \mathbb{C})$ of square integrable, complex valued functions on $T^{*}Q$, where the integrability is taken with respect to the volume form $(1/2\pi\hbar)^n \varepsilon_{\omega}$, $\hbar$ being the Plack constant divided by 2π.

Finally, for each $f \in C^{\infty}(T^{*}Q, \mathbb{R})$ the prequantum operator, δ_f

$$\delta_f : \mathcal{H} \longrightarrow \mathcal{H}$$

is given by:

$$\delta_f \stackrel{def}{=} -i\hbar\nabla^{\omega}_{X_f} + f \quad .$$

Some properties of the geometric prequantization are given in the following theorem:

THEOREM 2.1. ([1], [2]). Let $I_{\mathcal{H}} : \mathcal{H} \to \mathcal{H}$ be the identity operator on $\mathcal{H}$ and $\ell_{T^{*}Q}$ the constant function on $T^{*}Q$ given by

$$\ell_{T^{*}Q} : x \in T^{*}Q \longmapsto \ell \in \mathbb{R} \quad .$$

Then for each $f, g \in C^{\infty}(T^{*}Q, \mathbb{R})$ and $\lambda \in \mathbb{R}$ we have:

(i) $\delta_{f+g} = \delta_f + \delta_g$

(ii) $\delta_{\lambda f} = \lambda\cdot\delta_f$

(iii) $[\delta_f, \delta_g] \stackrel{def}{=} \delta_f\cdot\delta_g - \delta_g\cdot\delta_f = -i\hbar\delta_{\{f,g\}}$

(iv) $\delta_{\ell_{T^*Q}} = I_{\mathcal{H}}$

(v) $\delta_{fg} = f\cdot\delta_g + g\cdot\delta_f - fg$.

Let us observe now that the relations (i)-(iv) of the above theorem give us an extension of the Van Hove theorem concerning the Dirac problem, from the Euclidean case to the case of the cotangent bundle. On the other hand the last relation of the theorem has deep implications in geometric prequantization of the constrained mechanical systems, [2],[14].

3. REDUCTION AND GEOMETRIC PREQUANTIZATION AT THE COTANGENT LEVEL (case $\mu = 0$)

Let G be a Lie group which acts free and proper on Q by transformations $\phi_g: Q \to Q$ and define the lifted action on the cotangent bundle T^*Q by pushing forward the ℓ-forms:

$$\phi_g^*(\alpha_q)(v) \overset{def}{=} \alpha_q(T^{-1}\phi_g(v))$$

for each $q \in Q$, $\alpha_q \in T_q^*Q$ and $v \in (T\phi_g)_q Q$.

The lifted action preserves the canonical ℓ-form θ on T^*Q, it is free and proper and has a momentum map $J\colon T^*Q \to \mathcal{G}^*$ given by:

$$J(\alpha_q)(\xi) \overset{def}{=} \alpha_q((\xi_Q)_q) \quad ,$$

for each $q \in Q$, $\alpha_q \in T_q^*Q$ and $\xi \in \mathcal{G}$.

It is easy to see that J is an Ad^* - equivariant momentum map, $\mu = 0 \in \mathcal{G}^*$ is a regular value of J and the isotropy group G_o of $\mu = 0$ under the coadjoint action is the whole G. Then we have:

THEOREM 3.1. ([15], [1]. [3], [4]) (i) The reduced phase space $((T^*Q,\omega)$ of the cotangent bundle $((T^*Q,= d\theta)$ is symplectodiffeomorphic with the cotangent bundle $(T^*Q_o, d\theta_o)$, where Q_o is the orbit space Q/G and θ_o is the canonical ℓ-form on $T^{*o}Q$.

(ii) The following equality holds:

$$\pi_o^*\theta_o = i_o^*\theta \quad .$$

Using the above mentioned symplectodiffeomorphism we can conclude that the reduced phase space $(T^*Q_o, d\theta_o)$ is a quantizable manifold. Let $(\mathcal{H}_o, \delta^o)$ be its Hilbert representation space and the prequantum operator given by geometric prequantization. Then we can prove:

THEOREM 3.2. ([16], [7], [3]). Let $\mathcal{H}_G$ be the Hilbert space of G-invariant states of the cotangent bundle $(T^*Q, \omega = d\theta)$ and $C^\infty_G(T^*G, \mathbb{R})$ the space of smooth, G-invariant, real valued functions on T^*G. Then for each $f \in C^\infty_G(T^*Q, \mathbb{R})$, and $g \in \mathcal{H}_G$ we have:

(i) $\delta_f(g) \in \mathcal{H}_G$;

(ii) $[\delta_f(g)]_o = \delta^o_{f_o}(g_o)$, or in other words reduction and geometric prequantization are interchangeable processes;

(iii) If Q and Q_o are simple-connected manifolds then the unique prequantization on the cotangent bundle $(T^*Q, \omega = d\theta)$ projects onto the unique prequantization of the reduced phase space $(T^*Q_o, d\theta_o)$.

REMARK 3.1. All the above results remain also true if Q is an infinite dimensional Banach manifold and this is very important for applications.

REMARK 3.2. The general case $\mu \neq 0$ is more complicated and are known only some partial results and several examples [3], [7], [8].

REMARK 3.3. All the above results remain also true if instead of geometric prequantization we consider the half-form correction of geometric quantization in vertical polarization, [3], [9].

4. APPLICATIONS

In the last section I want to point out some examples which are particular cases of the above picture.

EXAMPLE 4.1. (the planar n-body problem [3], [11]). Consider n particles of masses $m_1, \ldots, m_n$ moving in $\mathbb{R}^2$ subject to Newton's gravitational law. If we remove the collisions from the model, the configuration space is the 2n-dimensional manifold:

$$Q = \{q \in (\mathbb{R}^2)^n \,|\, q \notin \Delta\}$$

where

$$\Delta = \bigcup_{i,j} \{\Delta_{ij} \mid 1 \le i < j \le n\}$$

and

$$\Delta_{ij} = \{q \in (\mathbb{R}^2)^n \mid q^i = q^j\} \qquad ,$$

and the phase space is its cotangent bundle $T^*Q = Q \times (\mathbb{R}^2)^n$ with the canonical symplectic structure $\omega = d\theta$.

The additive group $(\mathbb{R}^2,+)$ acts freely and properly on Q by translations:

$$\phi(g,(q^1,\ldots,q^n)) \overset{\text{def}}{=} (g+q^1,\ldots,g+q^n)$$

and the lifted action to T^*Q:

$$\phi^*(g,(q,p)) = (g+q^1,\ldots,g+q^n,p_1,\ldots,p_n)$$

has the moment map given by:

$$J(q,p) = \sum_{j=1}^{n} p_j \;.$$

Therefore the reduced phase space $((T^*Q)_o,\omega_o)$ is symplectodiffeomorphic via the Theorem 4.1. with $(T^*Q_o,\, d\theta_o)$ where:

$$T^*Q = \left\{(q,p) \in T^*Q \,\middle|\, \sum_{j=1}^{n} m_j q^j = 0 \;\; , \;\; \sum_{j=1}^{n} p_j = 0\right\}$$

$d\theta_o$ is the canonical symplectic structure on T^*Q_o and the following relation holds:

$$\pi^*_o\theta_o = i^*_o\theta \qquad .$$

Physically speaking the reduced phase space can be identified with the classical phase space which is obtained if we fix the center of mass at the origin.

It follows that all conditions of our Theorem 4.2. hold and therefore we can conclude that reduction and geometric prequantization of the planar n-body problem are interchangeable processes. Moreover the unique geometric prequantization of the cotangent bundle $(T^*Q,\omega = d\theta)$ projects onto the unique geometric prequantization of the reduced phase space $(T^*Q)_o,\omega_o)$.

EXAMPLE 4.2. (Einstein's vacuum field equations, [3]). Let V be an 4-dimensional manifold with a Lorentzian metric g, M a compact 3-dimensional oriented manifold and $i: M \to V$ an embedding of M such that the embedded manifold $\Sigma = i(M)$ is space-like, that is the pullback $i^*(g) = g$ is a Riemannian metric on M.

Einstein's vacuum field equations state thet the Ricci tensor vanishes:

$$\mathrm{Ricc}(g) = 0 \quad .$$

These equations can be written in an Hamiltonian form if we take as the configuration space $\mathcal{M}$ the space of all C^∞-Riemannian metrics on M and as the phase space its cotangent bundle $T^*\mathcal{M}$ with the canonical symplectic structure $\omega = d\theta$, [1].

The group of C^∞-diffeomorphisms of M denoted by Diff(M) acts in a canonically way on $\mathcal{M}$ by pull-back:

$$\phi: (\eta, g)\mathrm{Diff}(M) \times \mathcal{M} \to \phi(\eta, g) \overset{def}{=} (\eta^{-1})^*(g)$$

and the lifted action on $T^*\mathcal{M}$ has the momentum map J given by:

$$J(g, \pi)(X) \overset{def}{=} \int_M \langle \pi, L_X g \rangle$$

where $L_X g$ is the Lie derivative of g with respect to the vector field X.

Of particular interest in the case $J^{-1}(0)$ referred to as the divergence constraint in general relativity. Then the reduced phase space $((T^*\mathcal{M})_o, \omega_o)$ is symplectodiffeomorphic via the Theorem 3.1. and the Remark 3.1. to $(T^*\mathcal{M})_o, \omega_o)$ and the relation

$$\pi_o^*\theta_o = i_o^*\theta$$

also holds. Using now the Theorem 3.2. and the Remark 3.1. we can conclude that reduction and geometric prequantization of the Einstein's vacuum field equations are interchangeable processes.

EXAMPLE 4.3. (Maxwell's equations in vacuum, [10]). It is well known that Maxwell's equations in vacuum can be written in an Hamiltonian form if we take as the configuration space $\mathcal{M}$ the space of vector fields on $\mathbb{R}^3$ and as the phase space its corresponding cotangent bundle with the canonical symplectic structure $\omega = d\theta$, [6].

The additive group of smooth real valued functions on $\mathbb{R}^3$ acts in a naturally way on by:

$$(A, \varphi) \longmapsto A + \nabla\varphi$$

and then the lifted action on $T^*\mathcal{M}$ has the momentum map J given by:

$$J(A, Y) = -\operatorname{div} Y.$$

Then the reduced phase space $((T^*\mathcal{M})_o, \omega_o)$ is symplectodiffeomorphic via the Theorem 3.1. and the Remark 3.1. to $(T^*\mathcal{M}_o, d\theta_o)$ and the relation:

$$\pi_o^* \theta_o = i_o^* \theta$$

also holds. Using the Theorem 3.2. and the Remark 3.1. we can conclude that reduction and geometric prequantization of the Maxwell's equations in vacuum are interchangeable processes, and moreover that the unique prequantization on $T^*\mathcal{M}$ projects onto the unique prequantization on $T^*\mathcal{M}_o$.

I want to finish my considerations with the observation that for the general case $\mu \neq 0$ there are also some examples where we have the interchangeability between reduction and geometric prequantization. This is the case of the Kaluza-Klein theory of a relativistic charged particle [3], the case of the heavy top [8] and the case of spherical pendulum [2].

REFERENCES

[1] Abraham, R., Marsden, J.: *Foundations of mechanics,* Second Edition, Addison Wesley 1978.

[2] Blau, M.: On the geometric quantization of constrained systems, Class.Quantum Grav. **5** (1988) 1033-1044.

[3] Gotay, M.J.: Constraints. Reduction and Quantization, J.Math.Phys. **27**, (1986), 2051-2067.

[4] Kummer, M.: On the construction of the reduced phase space of a Hamiltonian system with symmetry, Indiana Univ.Math.Journ. **30** (1981), 281-291.

[5] Marsden, J., Weinstein, A.: Reduction of symplectic manifolds with symmetry, Raports on Math.Phys. **5** (1974), 121-130.

[6] Marsden, J., Weinsten, A.: The Hamiltonian structure of the Maxwell-Vlasov equations, Physica **4D** (1982), 394-406.

[7] Puta, M.: On the reduced phase space of a cotangent bundle, Lett.Math.Phys. **8** (1984), 189-194.

[8] Puta, M.: Geometric quantization of the Heavy Top, Lett.Math.Phys. **11** (1986), 105-112, Erratum and Addendum, **12** (1986) 169.

[9] Puta, M.: Geometric quantization of the reduced phase space of the cotangent bundle, Proceedings of the Conference 24-30 August (1986) Brno, Czechoslovakia, Brno (1987), 273-282.

[10] Puta, M.: On the geometric prequantization of Maxwell-equations, Lett.Math.Phys. **13** (1987) 99-103.

[11] Puta, M.: The planar n-body problem and geometric quantization, The XVIII-th National Conference on Geometry and Topology, Oradea-Felix, October 4-7, 1987, Preprint No.2 (1988), 151-154.

[12] Puta, M.: Geometric quantization of the sperical pendulum, Serdica Bulgaricae Math.Publ. **14** (1988) 198-201.

[13] Puta, M.: Geometric prequantization of the Einstein's vacuum field equations (to appear).

[14] Puta, M.: Dirac constrained mechanical systems and geometric prequantization (to appear).

[15] Satzer, M.J.: Canonical reduction of mechanical systems invariant under abelian group actions with an application to celestical mechanics, Indiana Univ.Math.Journ. **26** (1977), 951-976.

[16] Sniatycki, J.: Constraints and quantization, Lect.Notes in Math., vol. 1037 (1983) 301-334.

DIRAC PARTICLES IN MAGNETIC FIELDS

B. THALLER
Institute of Mathematics
Hans-Sachs-Gasse 3
A-8010 Graz
Austria

ABSTRACT. We give a review of spectral and scattering theory for spin-1/2 particles in an external magnetic field. The supersymmetric point of view is strongly emphasized. Recent results on Foldy-Wouthuysen transformations, properties of the resolvent, threshold eigenvalues and scattering theory are presented.

1. Magnetic fields

In any space dimension $\nu \geq 2$ the magnetic field strength B is given by a 2-form

$$B(x) = \sum_{\substack{i,k=1 \\ i<k}}^{\nu} F_{ik}(x) dx_i \wedge dx_k \tag{1}$$

satisfying $dB = 0$ (exterior derivative). We can write $B = dA$ with a magnetic "vector" potential (1-form) A and obtain

$$F_{ik}(x) = \frac{\partial A_i(x)}{\partial x_k} - \frac{\partial A_k(x)}{\partial x_i}, \qquad A(x) = \sum_{i=1}^{\nu} A_i(x) dx_i. \tag{2}$$

We want to stress that the vector potential is not directly observable. Therefore, one should formulate all results under assumptions on the field strengths. Eventually, we shall use the gauge freedom (see below) to make the description as simple as possible. Throughout this lecture we assume that each component of B is a smooth function in $C^\infty(\mathbb{R}^\nu)$. We refer to the cited literature for possible generalizations (e.g., [1]). We are only interested in results for dimensions $\nu = 2, 3$. For $\nu = 3$ we identify B and A with vector fields $\mathbf{B}(\mathbf{x}) = \big(F_{23}(\mathbf{x}), F_{31}(\mathbf{x}), F_{12}(\mathbf{x})\big)$ and $\mathbf{A}(\mathbf{x}) = \big(A_1(\mathbf{x}), A_2(\mathbf{x}), A_3(\mathbf{x})\big)$ satisfying $\mathbf{B} = \mathrm{rot}\mathbf{A}$, i.e., $\mathrm{div}\mathbf{B} = 0$. In two dimensions, $B = \partial_1 A_2 - \partial_2 A_1$ is simply a scalar field. This corresponds to the three dimensional situation $\mathbf{B}(\mathbf{x}) = \big(0, 0, B(x_1, x_2)\big)$.

A. Boutet de Monvel et al. (eds.), Recent Developments in Quantum Mechanics, 351–366.

2. Dirac and Pauli operators

We recall the Dirac operator with a magnetic field,

$$H(A) \equiv \begin{cases} c\sum_{i=1}^{3} \alpha_i(p_i - A_i) + \beta mc^2 & \text{if } \nu = 3, \\ c\sum_{i=1}^{2} \sigma_i(p_i - A_i) + \sigma_3 mc^2 & \text{if } \nu = 2, \end{cases} \tag{3}$$

where $p_i = -i\partial/\partial x_i$, and where $\vec{\alpha} = (\alpha_1, \alpha_2, \alpha_3)$ and β are the famous 4×4 "Dirac matrices", which in the standard representation are defined by

$$\beta = \begin{pmatrix} 1 & 0 \\ 0 & -1 \end{pmatrix}, \quad \alpha_i = \begin{pmatrix} 0 & \sigma_i \\ \sigma_i & 0 \end{pmatrix}, \quad i = 1, 2, 3. \tag{4}$$

Here σ_i denote the 2×2 "Pauli matrices"

$$\sigma_1 = \begin{pmatrix} 0 & 1 \\ 1 & 0 \end{pmatrix}, \quad \sigma_2 = \begin{pmatrix} 0 & -i \\ i & 0 \end{pmatrix}, \quad \sigma_3 = \begin{pmatrix} 1 & 0 \\ 0 & -1 \end{pmatrix}. \tag{5}$$

The Dirac operator (3) is essentially self-adjoint on $C_0^\infty(\mathbb{R}^3)^4$, resp. $C_0^\infty(\mathbb{R}^2)^2$, even without restriction on the growth of B or A at infinity [2].

In the standard representation $H(A)$ has the abstract "supersymmetric" form

$$H(A) = \begin{pmatrix} mc^2 & cD^* \\ cD & -mc^2 \end{pmatrix} = cQ + mc^2\tau, \tag{6}$$

where

$$Q = \begin{pmatrix} 0 & D^* \\ D & 0 \end{pmatrix}, \quad \tau = \begin{pmatrix} 1 & 0 \\ 0 & -1 \end{pmatrix} \tag{7}$$

and

$$D \equiv \begin{cases} \sum_{i=1}^{3} \sigma_i(p_i - A_i) & \text{if } \nu = 3, \\ (p_1 - A_1) + i(p_2 - A_2) & \text{if } \nu = 2, \end{cases}. \tag{8}$$

Note that $D \neq D^*$ in two dimensions. The Pauli operator

$$H_P(A) \equiv \frac{1}{2m} D^* D = \begin{cases} H_S(\mathbf{A})1 - (1/2m)\vec{\sigma}\cdot\mathbf{B}, & \text{if } \nu = 3, \\ H_S(A) - (1/2m)B, & \text{if } \nu = 2 \end{cases} \tag{9}$$

can be obtained as the nonrelativistic limit of $H(A)$, see Refs. [3], [4], [5]. Here, $H_S(A) = \frac{1}{2m}(-i\nabla - A)^2$ is the nonrelativistic Schrödinger operator for a spinless particle. See Chapter 6 of Ref. [6] for a review of its properties and references.

3. Vector potentials and gauge freedom

The 1-form A is not uniquely determined by (1). Replacing A by $A + dg$, where $g \in C^\infty(\mathbb{R}^\nu)$ (see [7] for more general g's), does not change the magnetic field strength and leads to an equivalent mathematical description of this physical system ("gauge invariance"). Indeed, if $A^{(2)} = A^{(72)} + \nabla g$, then the unitary transformation $\Psi(\mathbf{x}) \to e^{ig(\mathbf{x})}\Psi(\mathbf{x})$ transforms $H(A^{(72)})$ into $H(A^{(2)})$

$$e^{ig}\, H(A^{(72)})\, e^{-ig} = H(A^{(2)}) \tag{10}$$

(similar for H_P and H_S).

The use of the vector potential in quantum mechanics is sometimes counterintuitive, because the vector potential is usually nonzero in regions where the magnetic field strength vanishes. Consider, e.g., a magnetic field $B(x)$ in two dimensions with compact support and nonvanishing flux $\int B(x)d^2x$. Clearly in this case we expect the particle to move freely once it has left the support of B. But using Stokes law $\oint A ds = \int B d^2x$ we see that $A(x)$ cannot decay faster than $|x|^{-1}$, as $|x| \to \infty$. Hence the vector potential keeps influencing the wavefunction of the particle also at large distances from the support of B. There have been attempts to formulate quantum mechanics entirely in terms of B [8].

4.Supersymmetry

Now we turn to the proof of some spectral properties of $H(A)$. Any operator H of the form (6) will be called a "Dirac operator with supersymmetry".

Theorem. (Thaller, [9]). *Let $H = cQ + mc^2\tau$ be a Dirac operator with supersymmetry. Then there is a unitary transformation U, which brings H to the diagonal form*

$$UHU^{-1} = \begin{pmatrix} \sqrt{c^2D^*D + m^2c^4} & 0 \\ 0 & -\sqrt{c^2DD^* + m^2c^4} \end{pmatrix} = \tau|H|. \tag{11}$$

The unitary transformation is given by

$$U = a_+ + \tau(\mathrm{sgn}Q)a_-, \quad a_\pm = \frac{1}{\sqrt{2}}\sqrt{1 \pm \frac{mc^2}{|H|}}. \tag{12}$$

Proof: For simplicity, we set $c = 1$. It is easy to verify the following formulas for the bounded operators $a_\pm$

$$a_+^2 + a_-^2 = 1, \quad a_+^2 - a_-^2 = m|H|^{-1}, \quad 2a_+a_- = |Q|\,|H|^{-1}. \tag{13}$$

Furthermore we note that $|H| = (Q^2 + m^2)^{1/2}$ commutes with τ and Q, and the following commutation relations hold on $\mathcal{D}(H) = \mathcal{D}(Q)$

$$[H, a_\pm] = [Q, a_\pm], \quad H\tau\,(\mathrm{sgn\,Q}) = -\tau\,(\mathrm{sgn\,Q})\,\mathrm{H}. \tag{14}$$

Now we can verify Eq. (14) in the following way

$$\begin{aligned} UHU^{-1} &= (a_+ + \tau\,(\operatorname{sgn} Q)\,a_-)H(a_+ - \tau\,(\operatorname{sgn} Q)\,a_-) = \\ &= (a_+^2 + 2\tau\,(\operatorname{sgn} Q)\,a_+a_- - a_-^1)H \\ &= (m|H|^{-1} + \tau\,(\operatorname{sgn} Q)\,|Q|\,|H|^{-1})H \\ &= (m + \tau Q)|H|^{-1}H = \tau(m\tau + Q)|H|^{-1}H \\ &= \tau H^2|H|^{-1} = \tau|H| \end{aligned} \tag{15}$$

The matrix form of $\tau|H|$ immediately follows from (12), (13) and $|H| = \sqrt{H^2}$. *QED.*

A generalization of this result has been obtained recently in Ref. [5].

Corollary. *The spectrum of $H(A)$ is symmetric with respect to 0 except possibly at $\pm mc^2$. The open interval $(-mc^2, +mc^2)$ does not belong to the spectrum.*

Proof: We use (11) and the fact that $\sigma(D^*D) = \sigma(DD^*)$ except at 0. The Theorem now follows immediately from the spectral mapping theorem. *QED.*

We have $H(\mathbf{A})\Psi = mc^2\Psi$ if and only if $\Psi = \begin{pmatrix} \psi_1 \\ 0 \end{pmatrix}$ with $D^*D\psi_1 = 0$, or equivalently $D\psi_1 = 0$. On the other hand we see that $H(\mathbf{A})\Psi = -mc^2\Psi$ if and only if $\Psi = \begin{pmatrix} 0 \\ \psi_2 \end{pmatrix}$ with $DD^*\psi_2 = 0$, or equivalently $D^*\psi_2 = 0$. Hence in three dimensions, where $D^* = D$, the spectrum is always symmetric, even at 0. The situation is completely different in two dimensions, as we learn from the following theorem.

6. Eigenvectors belonging to $\pm mc^2$

Theorem. (Aharonov-Casher, [10]). *In two dimensions, let $B(x)$ be a magnetic field with compact support, and denote*

$$F = \frac{1}{2\pi}\int_{\mathbb{R}^2} B(x)d^2x \tag{16}$$

a) If $F = n + \epsilon$ (where n is a positive integer, and $0 \leq \epsilon < 1$), then $+mc^2$ (but not $-mc^2$) is an eigenvalue of the Dirac operator $H(A)$ defined in (3).
b) If $F = -n - \epsilon$, then $-mc^2$ (but not $+mc^2$) is an eigenvalue of $H(A)$.
In both cases the multiplicity of the eigenvalue is n, if $\epsilon > 0$, and $n-1$, if $\epsilon = 0$.

Proof: The Green function of Δ in two dimensions is $\frac{1}{2\pi}\ln|x|$, therefore

$$\phi(x) = \frac{1}{2\pi}\int_{\mathbb{R}^2} \ln|x-y|B(y)d^2y \tag{17}$$

satisfies $\Delta\phi(x) = B(x)$, and

$$\phi(x) - F\ln|x| = O\Big(\frac{1}{|x|}\Big), \quad \text{as } |x| \to \infty. \tag{18}$$

We choose the vector potential $A = (-\partial_2\phi, \partial_1\phi)$, and look for a solution of

$$c\vec{\sigma}\cdot(p - A)\psi = 0, \quad \vec{\sigma} = (\sigma_1, \sigma_2) \tag{19}$$

Writing

$$\omega = e^{\sigma_3\phi}\psi \tag{20}$$

we find that (19) is equivalent to

$$\vec{\sigma}\cdot p\omega = 0 \quad \text{or} \quad \begin{aligned} \left(\frac{\partial}{\partial x_1} + i\frac{\partial}{\partial x_2}\right)\omega_1(x) = 0, \\ \left(\frac{\partial}{\partial x_1} - i\frac{\partial}{\partial x_2}\right)\omega_2(x) = 0. \end{aligned} \tag{21}$$

These equations are equivalent to the Cauchy Riemann equations. Hence ω_1 (resp. ω_2) has to be an entire analytic function in the variable $z = x_1 + ix_2$ (resp. $\overline{z} = x_1 - ix_2$). For large $|z| = |x|$ these functions behave as

$$\omega_1(x) \approx e^{+F\ln|x|}\psi_1(x) = |x|^{+F}\psi_1(x), \tag{22}$$
$$\omega_2(x) \approx e^{-F\ln|x|}\psi_2(x) = |x|^{-F}\psi_2(x). \tag{23}$$

If $F > 0$ then ω_2 is square integrabel at infinity and hence zero, because an analytic function cannot vanish in all directions, as $|z| \to \infty$. This shows that $\psi_2 = 0$ and therefore only $+mc^2$ can be an eigenvalue of $H(A)$. But for this we have to fulfill the condition

$$\psi_1 = e^{-\phi}\omega_1 \in L^2(\mathbb{R}^2) \tag{24}$$

which requires that ω_1 should not increase faster than $|x|^{F-1-\delta}$, for some $\delta > 0$. Since ω_1 is an entire function, it must be a polynomial in $x_1 + ix_2$ of degree $\leq n-1$ (resp. $n-2$, if $\epsilon = 0$). Hence there are n linearly independent solutions ψ_1 of $D\psi_1 = 0$, namely (for $\epsilon \neq 0$)

$$e^{-\phi},\ e^{-\phi}(x_1 + ix_2),\ e^{-\phi}(x_1 + ix_2)^2,\ \ldots\ ,\ e^{-\phi}(x_1 + ix_2)^{n-1}. \tag{25}$$

An analogous reasoning applies to the case $F < 0$. *QED.*

The proof of the Aharonov-Casher Theorem shows that if one can find a solution ϕ of $\Delta\phi(x) = B(x)$, such that $e^{-\phi}$ (or $e^{+\phi}$) is a rapidly decreasing function in $\mathcal{S}(\mathbb{R}^3)$, then the eigenvalue $+mc^2$ (or $-mc^2$) is infinitely degenerate and hence in $\sigma_{ess}(H(A))$ (see also Ref. [11]). This is indeed the case, e.g., for a homogeneous magnetic field.

In case of a spherically symmetric magnetic field in two dimensions ($B(x) = B(r)$, $r = |x|$) a solution of $\Delta\phi(r) = (\partial^2/\partial r^2 + (1/r)\partial/\partial r)\phi(r) = B(r)$ is given by

$$\phi(r) = \int_0^r ds \frac{1}{s}\int_0^s dt B(t)t \tag{26}$$

Hence for a magnetic field with infinite flux like

$$B(r) \approx r^{\delta-2}, \quad \text{for some } \delta > 0,\ r \text{ large}, \tag{27}$$

we find

$$\phi(r) \approx \frac{1}{\delta} r^{\delta}, \quad \text{for } r \text{ large}, \tag{28}$$

i.e., $e^{-\phi}$ decreases faster than any polynomial in $|x|$. Therefore $+mc^2$ is infinitely degenerate in this case. If even $B(r) \to \infty$, as $r \to \infty$, then the next Theorem shows, that $+mc^2$ is the only possible element in the essential spectrum of the Dirac operator.

Theorem. (Helffer-Nourrigat-Wang, [2]). *If in two dimensions $B(x) \to \infty$ (resp. $B(x) \to -\infty$), as $|x| \to \infty$, then λ with $\lambda \neq +mc^2$ (resp. $\lambda \neq -mc^2$) is not in the essential spectrum of the Dirac operator.*

Proof: We assume $B(x) \to +\infty$, the other case can be treated analogously. We show that λ with $\lambda \neq mc^2$ is not in $\sigma_{ess}(H(A))$, because for all $\Psi = \binom{\psi_1}{\psi_2} \in C_0^\infty(\mathbb{R}^2)^2$ with support outside a ball with sufficiently large radius R there is a constant $C(\lambda)$ such that

$$\|(H(A) - \lambda)\Psi\| \geq C(\lambda)\|\Psi\| \tag{29}$$

In order to prove this, we choose R so large, that

$$B(x) \geq \frac{1}{c^2}|\lambda - mc^2|(3 + 2|\lambda + mc^2|), \quad \text{for all } |x| \geq R. \tag{30}$$

Denoting $\Phi = \binom{\phi_1}{\phi_2} = (H(A) - \lambda)\Psi$, i.e.,

$$\begin{aligned} \phi_1 &= cD^*\psi_2 - (\lambda - mc^2)\psi_1 \\ \phi_2 &= cD\psi_1 - (\lambda + mc^2)\psi_2 \end{aligned} \tag{31}$$

we find

$$\begin{aligned} \|D^*\psi_2\|^2 &= (\psi_2, D^*D\psi_2) = \|(p-A)^2\psi_2\|^2 + (\psi_2, B(x)\psi_2) \\ &\geq \frac{1}{c^2}|\lambda - mc^2|(3 + 2|\lambda + mc^2|)\|\psi_2\|^2, \end{aligned} \tag{32}$$

provided suppψ_2 is outside the ball with radius R. Hence

$$\begin{aligned} (3 + 2|\lambda + mc^2|)\|\psi_2\|^2 &\leq \frac{1}{|\lambda - mc^2|}\|\phi_1 + (\lambda - mc^2)\psi_1\|^2 \\ &= \frac{1}{|\lambda - mc^2|}\|\phi_1\|^2 + 2(\text{sgn}\lambda)\text{Re}(\phi_1, \psi_1) + |\lambda - mc^2|\|\psi_1\|^2. \end{aligned} \tag{33}$$

Since

$$
\begin{aligned}
(\operatorname{sgn}\lambda)\operatorname{Re}(\phi_1,\psi_1) &= (\operatorname{sgn}\lambda)\operatorname{Re}(cD^*\psi_2,\psi_1) - |\lambda - mc^2|\|\psi_1\|^2 \\
&\leq \|\psi_2\|\|cD\psi_1\| - |\lambda - mc^2|\|\psi_1\|^2 \\
&\leq \|\psi_2\|\|\phi_2\| + |\lambda + mc^2|\|\psi_2\|^2 - |\lambda - mc^2|\|\psi_1\|^2 \qquad (34)
\end{aligned}
$$

we find

$$
\begin{aligned}
&3\|\psi_2\|^2 + |\lambda - mc^2|\|\psi_1\|^2 \\
&\leq \frac{1}{|\lambda - mc^2|}\|\phi_1\|^2 + 2\|\psi_2\|\|\phi_2\| \qquad (35)
\end{aligned}
$$

Now we have either $\|\psi_2\| \leq \|\phi_2\|$ or $\|\psi_2\| \geq \|\phi_2\|$. In each case

$$
\|\Psi\|^2 = \|\psi_1\|^2 + \|\psi_2\|^2 \leq 2\max\left\{1, \frac{1}{(\lambda - mc^2)^2}\right\}\|\Phi\|^2, \qquad (36)
$$

which proves the Theorem. *QED.*

7. Three space dimensions

There is no analogue of the Aharonov-Casher result for $\nu = 3$. Also the theorem of Helffer, Nourrigat, and Wang is very specific to two dimensions. Concerning eigenvalues at $\pm mc^2$ in three dimensions, so far only some examples are known.

Example. (Loss-Yau, [12]). If, for some real valued λ, we had a solution of

$$
\vec{\sigma}\cdot\mathbf{p}\Psi(\mathbf{x}) = \lambda(\mathbf{x})\Psi(\mathbf{x}), \qquad (37)
$$

which satisfies

$$
\langle\Psi,\Psi\rangle(\mathbf{x}) \equiv \sum_{i=1}^{2}\overline{\psi_i(\mathbf{x})}\psi_i(\mathbf{x}) \neq 0, \qquad (38)
$$

then we can find a solution of $\vec{\sigma}\cdot(\mathbf{p}-\mathbf{A})\Psi = 0$. First note that

$$
\langle\Psi,\vec{\sigma}\Psi\rangle\langle\Psi,\vec{\sigma}\Psi\rangle = \langle\Psi,\Psi\rangle^2, \qquad (39)
$$

implies

$$
\vec{\sigma}\cdot\frac{\langle\Psi,\vec{\sigma}\Psi\rangle}{\langle\Psi,\Psi\rangle}\Psi = \Psi, \qquad (40)
$$

and hence

$$
\vec{\sigma}\cdot\mathbf{A}(\mathbf{x})\Psi(\mathbf{x}) = \lambda(\mathbf{x})\Psi(\mathbf{x}), \qquad (41)
$$

if we choose

$$
\mathbf{A}(\mathbf{x}) = \lambda(\mathbf{x})\frac{\langle\Psi,\vec{\sigma}\Psi\rangle}{\langle\Psi,\Psi\rangle}. \qquad (42)
$$

But a solution of (37) is easy to find. Choose, for example,

$$\Psi(\mathbf{x}) = \frac{1 + i\vec{\sigma}\cdot\mathbf{x}}{(1+x^2)^{3/2}}\Phi_0, \tag{43}$$

where $\Phi_0 \in \mathbb{C}^2$, with $\langle\Phi_0, \Phi_0\rangle = 1$. Note that

$$0 \neq \langle\Psi, \vec{\sigma}\Psi\rangle(\mathbf{x}) = \frac{1}{(1+x^2)^3}\{(1-x^2)\mathbf{w} + 2(\mathbf{w}\cdot\mathbf{x})\mathbf{x} + 2\mathbf{w}\wedge\mathbf{x}\}, \tag{44}$$

where $w = \langle\Phi_0, \vec{\sigma}\Phi_0\rangle$ is a unit vector in $\mathbb{R}^3$. We obtain

$$\vec{\sigma}\cdot\mathbf{p}\Psi(\mathbf{x}) = \frac{3}{1+x^2}\Psi(\mathbf{x}), \tag{45}$$

and finally

$$\mathbf{A}(\mathbf{x}) = 3(1+x^2)\langle\Psi, \vec{\sigma}\Psi\rangle, \qquad \mathbf{B}(\mathbf{x}) = 12\langle\Psi, \vec{\sigma}\Psi\rangle. \tag{46}$$

The vector field A can be obtained by stereographic projection from a parallel basis vector field on the three dimensional sphere. Hence the flow lines are circles on the Hopf tori.

A characterisation of the essential spectrum in two or three dimensions is given by the next Theorem. This result is not so typical for Dirac operators, because similar statements are true for the nonrelativistic Schrödinger operator without spin (see Leinfelder [6], Miller and Simon [13], [14]).

Theorem. (Leinfelder-Miller-Simon). *In two or three dimensions, if* $|B(x)| \to 0$, *as* $|x| \to \infty$, *then*

$$\sigma_{ess}(H(A)) = (-\infty, -mc^2] \cup [mc^2, \infty). \tag{47}$$

If $B(x)$ *is bounded, then the distance from an arbitrary* $\lambda \notin (-mc^2, mc^2)$ *to* $\sigma_{ess}(H(A))$ *is less than* $2\sqrt{12}\sup\sqrt{|B(x)|}$.

Proof: It is sufficient to consider the essential spectrum of the operator

$$D_\nu = c\sum_{i=1}^{\nu}\sigma_i(p_i - A_i) \equiv c\vec{\sigma}\cdot(p - A) \tag{48}$$

in dimensions $\nu = 2, 3$, because $H(A)$ is unitarily equivalent to

$$\sigma_3\sqrt{(D_2)^2 + m^2c^4} \quad \text{for } \nu = 2, \tag{49}$$

and

$$\begin{pmatrix} \sqrt{(D_3)^2 + m^2c^4} & 0 \\ 0 & -\sqrt{(D_3)^2 + m^2c^4} \end{pmatrix} \quad \text{for } \nu = 3. \tag{50}$$

In order to prove $k \in \sigma_{ess}(D_\nu)$ it is sufficient to find an orthonormal sequence of vectors $\Psi^{(n)}$ in the domain of D_ν, such that

$$\lim_{n\to\infty} \|(D_\nu - k)\Psi^{(n)}\| = 0 \tag{51}$$

(Weyl's criterion). Moreover, the distance between k and $\sigma_{ess}(D_\nu)$ is less than d, if for a suitable orthonormal sequence $\Psi^{(n)}$

$$\|(D_\nu - k)\Psi^{(n)}\| \leq d. \tag{52}$$

We are going to construct suitable vectors $\Psi^{(n)}$ as follows. Let $B_n = B_{\rho_n}(x^{(n)})$ be a sequence of disjoint balls with centers $x^{(n)}$ and radii ρ_n. Any two L^2-functions with support in different balls are orthogonal. We use the gauge freedom to define within these balls vector potentials $A^{(n)}$ which are determined by the local properties of B in that region (unlike the original A-field). For each n we define

$$A^{(n)}(x) = \int_0^1 B\big(x^{(n)} + (x - x^{(n)})s\big) \wedge (x - x^{(n)})s ds, \tag{53}$$

or, written in components

$$A_i^{(n)}(x) = \int_0^1 \sum_{i=1}^{\nu} F_{ki}(x^{(n)} + (x - x^{(n)})s)(x_k - x_k^{(n)})s ds, \quad i = 1, \ldots, \nu. \tag{54}$$

It is easy to see that

$$\sup_{x\in B_n} |A^{(n)}(x)| \leq \rho_n \sup_{x\in B_n} |B(x)|. \tag{55}$$

Furthermore, if A is the vector potential we started with, then

$$A - A^{(n)} = \nabla g^{(n)}, \quad \text{with } g^{(n)} \in C^\infty(\mathbb{R}^\nu). \tag{56}$$

Finally, we choose

$$\Psi^{(n)}(x) = \frac{1}{\sqrt{2}} \begin{pmatrix} 1 \\ 1 \end{pmatrix} \frac{1}{\rho_n^{\nu/2}} \, j\left(\frac{x - x^{(n)}}{\rho_n}\right) \exp\big(ig^{(n)}(x) - ikx_1\big), \tag{57}$$

where j is a localization function with the following properties:

$$j \in C_0^\infty(\mathbb{R}^\nu), \quad \text{supp} j \subset \{x \mid |x| \leq 1\}, \quad \int |j(x)|^2 d^\nu x = 1. \tag{58}$$

It is easily verified that $\mathrm{supp}\Psi^{(n)} \subset B_n$, and $\|\Psi^{(n)}\| = 1$. A little calculation gives for all $k \in \mathbb{R}$

$$\begin{aligned}
&\{\vec{\sigma}\cdot(p - A(x)) - k\}\Psi^{(n)}(x) \\
&= -i\frac{1}{\rho_n}\vec{\sigma}\cdot(\nabla j)\left(\frac{x - x^{(n)}}{\rho_n}\right)\frac{1}{\sqrt{2}}\begin{pmatrix}1\\1\end{pmatrix}\frac{1}{\rho_n^{\nu/2}}\exp\left(ig^{(n)}(x) - ikx_1\right) \\
&- \vec{\sigma}\cdot A^{(n)}(x)\Psi^{(n)}(x).
\end{aligned} \tag{59}$$

Using (55) we obtain the estimate

$$\|\{\vec{\sigma}\cdot(p - A) - k\}\Psi^{(n)}\| \le \frac{1}{\rho_n}\int |\nabla j(x)|^2 d^\nu x + \rho_n \sup_{x\in B_n}|B(x)| \tag{60}$$

Assume first, that $|B(x)| \le M$ for all x. Choosing $\rho_n = \rho$, all n, we find that (60) is bounded, i.e., $\sigma_{ess}(D_\nu)$ is not empty. Moreover, if we choose $j(x) = \mathrm{const.}\cos^2(\pi|x|/2)$, then it is easy to see that $\int |\nabla j(x)|^2 d^3x \approx 11.62$. Setting $\rho = (12/M)^{1/2}$ we obtain the bound $2(12M)^{1/2}$ for (60). Hence the distance of an arbitrary $k \in \mathbb{R}$ to $\sigma_{ess}(D_\nu)$ is less than this constant. The distance from an arbitrary $\lambda \notin (-mc^2, mc^2)$ to the next point in $\sigma_{ess}(H(A))$ is bounded by the same constant.

If $|B(x)| \to 0$, as $|x| \to \infty$, we can proceed as follows. There is a sequence of disjoint balls B_n with increasing radius ρ_n, such that

$$\sup_{x\in B_n}|B(x)| \le \frac{1}{\rho_n^2} \tag{61}$$

But then (60) is bounded by constant$/\rho_n \to 0$, as $n \to \infty$. Hence any $k \in \mathbb{R}^\nu$ is in the essential spectrum of D_ν. *QED.*

In case of electric or scalar potentials which decay at infinity, the essential spectrum mainly consists of a continuous spectrum associated to scattering states. This is not necessarily the case for magnetic fields. This can be seen most clearly by looking at spherically symmetric examples.

8. Cylindrical symmetry

In two dimensions, if the magnetic field strength is cylindrically symmetric we can pass to coordinates $r = |x|$, $\phi = \arctan(x_2/x_1)$. We denote the coordinate unit vectors by $\mathrm{e}_r = \frac{1}{r}(x_1, x_2)$, $\mathrm{e}_\phi = \frac{1}{r}(-x_2, x_1)$, write $B(x) \equiv B(r)$, and choose $A(x) = A_\phi(r)\mathrm{e}_\phi$, where

$$A_\phi(r) = \frac{1}{r}\int_0^r B(s)s\,ds, \quad B(r) = \left(\frac{d}{dr} + \frac{1}{r}\right)A_\phi(r) = \frac{1}{r}\frac{d}{dr}(A_\phi(r)r). \tag{62}$$

In this notation the flux of B is given by $F = \lim_{r\to\infty}(A_\phi(r)r)$. The Dirac operator in cylindrical coordinates can be written as

$$\begin{aligned} H(A) &\equiv c\vec{\sigma}\cdot(p-A)+\sigma_3 mc^2 \\ &= c(\vec{\sigma}\cdot \mathrm{e}_r)\mathrm{e}_r\cdot(p-A)+c(\vec{\sigma}\cdot \mathrm{e}_\phi)\mathrm{e}_\phi\cdot(p-A)+\sigma_3 mc^2 \\ &= c(\vec{\sigma}\cdot \mathrm{e}_r)\left\{-i\left(\frac{\partial}{\partial r}+\frac{1}{2r}\right)+i\frac{1}{r}\sigma_3 J_3 - i\sigma_3 A_\phi(r)\right\}+\sigma_3 mc^2. \end{aligned} \tag{63}$$

The angular momentum operator $J_3 = L_3 + \sigma_3/2$ commutes with $H(A)$ and the spinors

$$\chi_{m_j} = \begin{pmatrix} a\, e^{i(m_j-1/2)\phi} \\ b\, e^{i(m_j+1/2)\phi} \end{pmatrix}, \quad a, b \in \mathbb{C}, \quad m_j = \pm\frac{1}{2}, \pm\frac{3}{2}, \pm\frac{5}{2}, \ldots \tag{64}$$

form a complete set of orthogonal eigenvectors of J_3 in $L^2(S^1)^2$ with the properties

$$J_3\chi_{m_j} = m_j\chi_{m_j}, \tag{65}$$

$$(\vec{\sigma}\cdot \mathrm{e}_r)\chi_{m_j} = \begin{pmatrix} b\, e^{i(m_j-1/2)\phi} \\ a\, e^{i(m_j+1/2)\phi} \end{pmatrix}. \tag{66}$$

Any function $\Psi(r,\phi)$ in $L^2(\mathbb{R}^2)^2$ can be written as a sum

$$\Psi(r,\phi) = \sum_{m_j} \begin{pmatrix} \dfrac{1}{\sqrt{r}} f_{m_j}(r) e^{i(m_j-1/2)\phi} \\ -i\dfrac{1}{\sqrt{r}} g_{m_j}(r) e^{i(m_j+1/2)\phi} \end{pmatrix}, \tag{67}$$

with suitable functions f_{m_j} and g_{m_j} in $L^2([0,\infty), dr)$. The action of $H(A)$ on Ψ can be described on each angular momentum subspace as the action of a "radial Dirac operator" h_{m_j} defined in $L^2([0,\infty), dr)^2$

$$h_{m_j}\begin{pmatrix} f_{m_j} \\ g_{m_j} \end{pmatrix} = \begin{pmatrix} mc^2 & cD^* \\ cD & -mc^2 \end{pmatrix}\begin{pmatrix} f_{m_j} \\ g_{m_j} \end{pmatrix} \tag{68}$$

with

$$D = \frac{d}{dr} - \frac{m_j}{r} + A_\phi(r), \tag{69}$$

and $H(A)$ is unitarily equivalent to a direct sum of the operators h_{m_j}. A little calculation shows

$$\left.\begin{matrix} D^*D \\ DD^* \end{matrix}\right\} = -\frac{d^2}{dr^2} + \frac{\left(m_j \mp \frac{1}{2}\right)^2 - \frac{1}{4}}{r^2} - 2\frac{m_j \mp \frac{1}{2}}{r} A_\phi(r) + A_\phi^2(r) \mp B(r). \tag{70}$$

From (62) we see that if $B(r)r \to \infty$ then also $A_\phi(r) \to \infty$, as $r \to \infty$. In this case the term A_ϕ^2 dominates in (70) the interaction at large values of r and clearly the Schrödinger operator D^*D (resp. DD^*) has a pure point spectrum. By (11) the same is true for the Dirac operator on each angular momentum subspace and hence for $H(A)$. Let us summarize these observations in the following Theorem.

Theorem. (Miller-Simon [13], [14]). *In two dimensions, if B is spherically symmetric and $B(r)r \to \infty$, as $r \to \infty$, then the Dirac operator $H(A)$ has a pure point spectrum.*

In addition, if $B(r) \to 0$, then $B(r)r \to \infty$ implies that there is a complete set of orthonormal eigenvectors of $H(A)$ belonging to eigenvalues which are dense in the the union of the intervals $(-\infty, -mc^2]$ and $[mc^2, \infty)$.

9. A review of further results

It is immediately clear from the proof of the Theorem of Leinfelder-Miller-Simon that the condition $|B(x)| \to 0$ can be weakened considerably. It is sufficient to require that there is a sequence of balls with increasing radius on which B tends to zero. These balls can be widely separated and it does not matter how B behaves elsewhere [7,12]. It is even sufficient to require a similar behaviour of the derivatives of B [2]. More, precisely, one defines functions

$$\epsilon_r(x) = \frac{\sum_{|\alpha|=r} |D^\alpha B|}{1 + \sum_{|\alpha|<r} |D^\alpha B|}, \quad \text{if } r \geq 1, \text{ and } \epsilon_0(x) = |B(x)|. \tag{71}$$

and introduces the assumption

(A_r): There exist a sequence of disjoint balls B_n of radii r_n with $r_n \to \infty$ such that the function $\epsilon_r(x)$ restricted to the union of these balls tends to zero at infinity.

It can be seen that if (A_r) holds for some $r \geq 2$, then one of the assumptions (A_0) or (A_1) is true [2]. If the components of B are polynomials of degree r, then (A_{r+1}) holds. The following Theorem applies to this case.

Theorem. (Helffer-Nourrigat-Wang, [2]). *In three dimensions, if (A_r) holds for some $r \geq 0$, then $\sigma_{ess}(H(A)) = (-\infty, -mc^2] \cup [mc^2, \infty)$.*

This result is remarkable, because magnetic fields increasing like polynomials are known to yield Schrödinger operators with compact resolvent [15]. It is clear from Section 6 this result is very specific to three dimensions. Since the essential spectrum may contain embedded eigenvalues it is interesting to know criteria for the absence of eigenvalues as in the next Theorem.

Theorem. (Kalf-Berthier-Georgescu, [1,16]). *If $B(x) \to 0$ and $x \wedge B(x) \to 0$, as $|x| \to \infty$, then the Dirac operator has no eigenvalues λ with $|\lambda| > mc^2$.*

10. Scattering Theory

One of the basic problems in scattering theory is proving asymptotic completeness of the wave operators

$$\Omega_\pm(H, H_0) \equiv \operatorname*{s-lim}_{t \to \pm\infty} e^{iHt} e^{-iH_0 t} P_{cont}(H_0), \tag{72}$$

Supersymmetry implies a relation between the wave operators $\Omega_\pm(H_P(A), H_P(0))$ of the nonrelativistic theory and the relativistic wave operators $\Omega_\pm(H(A), H(0))$. Unfortunately, the Dirac operator is not an "admissible" function of the Pauli operator, therefore we cannot apply directly the invariance principle in order to conclude existence of wave operators in the relativistic case from the existence of the nonrelativistic wave operators. Nevertheless we have the following theorem, where F denotes the projection operator to the subspace belonging to the indicated region of the spectrum of the self-adjoint operator.

Theorem. (Thaller, [9, 17]). *Let $H = Q + m\tau$, $H_0 = Q_0 + m\tau$ be two Dirac operators with supersymmetry. Assume that for all $0 < a < b < \infty$ and for Ψ in some dense subset of $F(a < Q_0^2 < b)\mathcal{H}_{a.c.}(Q_0^2)$ the following condition is satisfied with $k = 1,2$*

$$\|(Q^k - Q_0^k)\, e^{-iQ_0^2 t}\, \Psi\| \le \text{const.}(1 + |t|)^{1-k-\delta}. \qquad (73)_k$$

Then the wave operators $\Omega_\pm(Q^2, Q_0^2)$ and $\Omega_\pm(H, H_0)$ exist, and

$$\Omega_\pm(H, H_0) = \Omega_\pm(Q^2, Q_0^2)\, F(H_0 > 0) + \Omega_\mp(Q^2, Q_0^2)\, F(H_0 < 0) \qquad (74)$$

A proof of this theorem is given in Ref. [17]. Now we apply it to the case of magnetic fields. Note that $Q^2 = H_P(A)$ is just the nonrelativistic Pauli operator. Assume that the magnetic field strength B decays at infinity, such that we have, for some $\delta > 0$,

$$B(x) \le \text{const.}(1 + |x|)^{-3/2-\delta} \qquad (75)$$

Choose the transversal (or Poincaré) gauge

$$A(x) = \int_0^1 s\, B(xs) \wedge x\, ds. \qquad (76)$$

A is uniquely characterized by $A(x) \cdot x = 0$, and $A(x)$ decays like $|x|^{-1/2-\delta}$, as $|x| \to \infty$. Hence the expressions div$A$, A^2, $\vec{\Sigma} \cdot B$ occurring in $Q^2 - Q_0^2$ are all of short-range. The remaining long-range term is $A(x) \cdot p$. It can be written as $A(x) \cdot p = G(x) \cdot (x \wedge p)$, where

$$G(x) = \int_0^1 s\, B(xs)\, ds \qquad (77)$$

satisfies,

$$|G(x)| \le \text{const.}(1 + |x|)^{-3/2-\delta)} \qquad (78)$$

and since the angular momentum $L = x \wedge p$ remains constant under the nonrelativistic free time evolution $\exp(-iQ_0^2 t)$, we easily obtain by a stationary phase argument for Ψ in a suitable dense set (see also [22])

$$\|A(x) \cdot p\, e^{-ip^2 t}\, \Psi\| \le \text{const.}(1 + |t|)^{-3/2-\delta}. \qquad (79)$$

The condition $(73)_1$ simply becomes

$$\|A(x)\, e^{-ip^2 t}\, \Psi\| \leq \text{const.}(1+|\mathrm{t}|)^{-\delta} \tag{80}$$

and is trivially satisfied. Hence we have proven existence of the nonrelativistic and relativistic wave operators. But also asymptotic completeness is true.

Theorem. (Loss-Thaller, [18]). *Let $H(A)$ and $H(0)$ be given as in (2) and assume that the magnetic field strength B satisfies*

$$D^\gamma B(x) \leq \text{const.}(1+|\mathrm{x}|)^{-3/2-\delta-\gamma}, \tag{81}$$

for some $\delta > 0$ and multiindices γ with $|\gamma| = 0, 1, 2$. Then the nonrelativistic and relativistic wave operators in the transversal gauge are asymptotically complete.

Existence of wave operators is usually expected to hold for "short-range potentials", where (each component of) the potential matrix V satisfies

$$|V(x)| \leq \text{const}(1+|\mathrm{t}|)^{-1-\delta}. \tag{82}$$

A famous counter example is the electrostatic Coulomb potential, where $|V(x)|$ decays like $|x|^{-1}$. In this case the wave operators do not exist [19] and one has to introduce modifications of the asymptotic time evolution. For the magnetic fields in the theorem above the potential matrix $-\vec{\alpha} \cdot A$ has a much slower decay. Indeed, previous results in the literature have been obtained only by introducing modifications of the wave operators (see, e.g., Ref. [20] and the references therein). Asymptotic completeness is due to the transversality of A. In another gauge A is not transversal and if ∇g is long-range, then the unmodified wave operators (72) would not exist. Instead, asymptotic completeness holds for $\Omega_\pm(H(A), H(\nabla g))$. These remarks might be of importance, because physicists use almost exclusively the Coulomb gauge instead of the transversal gauge, which is best adapted to scattering theory. Note that although the wave and scattering operators depend on the choosen gauge, the physically observable quantities like scattering cross sections are gauge independent.

In situations like the Aharonov Bohm effect one has used the free asymptotics (e.g., plane waves for the asymptotic description of stationary scattering states), together with the Coulomb gauge, although the vector potential is long-range. But in this case the calculations are justified, because in two dimensions and for rotationally symmetric fields the Coulomb gauge coincides with the transversal gauge (see also Ref. [21], for a discussion).

Under weaker decay conditions on the magnetic field strength the wave operators would not exist in that form, because then the term A^2 occurring in $Q^2 - Q_0^2$ would become long-range. In this case one really needs modified wave operators, similar to the Coulomb case.

The scattering problem is nontrivial even in classical mechanics. From special examples we know that classical paths of particles in magnetic fields satisfying our requirements do not have asymptotes. It is easy to see that the velocity of the particles is asymptotically constant. But if we compare the asymptotic motion of a particle in a magnetic field with a free motion, one would have to add a correction which is transversal to the asymptotic velocity and which increases for $\delta < 1/2$ like $|t|^{1/2-\delta}$. Thus the situation seems to be worse than in the Coulomb problem. There the interacting particles also cannot be asymptotically approximated by free particles, but at least the classical paths do have asymptotes. (The correction in the Coulomb problem increases like $\ln|t|$ and is parallel to the asymptotic velocity). A discussion of these effects in the classical scattering theory with magnetic fields is given in Ref. [22].

References

[1] Berthier, A. and Georgescu, V. On the point spectrum for Dirac operators. J. Func. Anal. 71: 309-338 (1987).

[2] Helffer, B., Nourrigat, J. and Wang, X. P. Sur le spectre de l'equation de Dirac avec champ magnetique. Preprint. (1989).

[3] Hunziker, W. On the nonrelativistic limit of the Dirac theory. Commun. Math. Phys. 40: 215-222 (1975).

[4] Gesztesy, F., Grosse, H. and Thaller, B. A rigorous approach to relativistic corrections of bound state energies for spin-1/2 particles. Ann. Inst. H. Poincaré. 40: 159-174 (1984).

[5] Grigore, D. R., Nenciu, G. and Purice, R. On the nonrelativistic limit of the Dirac Hamiltonian. Preprint, to appear in Ann. Inst. H. Poincaré. (1989).

[6] Cycon, H. L., Froese, R. G., Kirsch, W. and Simon, B. Schrödinger operators with applications to quantum mechanics and global geometry. Springer Verlag. Berlin, Heidelberg, New York, London, Paris, Tokyo 1987.

[7] Leinfelder, H. Gauge invariance of Schrödinger operators and related spectral properties. J. Op. Theory. 9: 163-179 (1983).

[8] Grümm, H. R. Quantum mechanics in a magnetic field. Acta Phys. Austriaca. **53**: 113-131, (1981).

[9] Thaller, B. Normal forms of an abstract Dirac operator and applications to scattering theory. J. Math. Phys. 29: 249-257 (1988).

[10] Aharonov, Y. and Casher, A. Ground state of a spin-1/2 charged particle in a two dimensional magnetic field. Phys. Rev. A19: 2461-2462 (1979).

[11] Avron, J. E. and Seiler, R. Paramagnetism for nonrelativistic electrons and euclidean massless Dirac particles. Phys. Rev. Lett. 42: 931-934 (1979).

[12] Loss, M. and Yau, H. T. Stability of Coulomb systems with magnetic fields III. Zero energy bound states of the Pauli operator. Commun. Math. Phys. 104: 283-290 (1986).

[13] Miller, K. C. Bound states of quantum mechanical particles in magnetic fields. Dissertation, Princeton University. (1982).

[14] Miller, K. and Simon, B. Quantum magnetic Hamiltonians with remarkable spectral properties. Phys. Rev. Lett. 44: 1706-1707 (1980).

[15] Helffer, B. and Mohamed, A. Caractérisation du spectre essentiel de l'opérateur de Schrödinger avec champ magnetique. Ann. Inst. Fourier 38: 95-112 (1988).

[16] Kalf, H. Non-existence of eigenvalues of Dirac operators. Proc. Roy. Soc. Edinburgh. A89: 307-317 (1981).

[17] Thaller, B. Scattering theory of a supersymmetric Dirac operator. Preprint. (1989).

[18] Loss, M. and Thaller, B. Short-range scattering in long-range magnetic fields: The relativistic case. J. Diff. Eq. 73: 225-236 (1988).

[19] Dollard, J. and Velo, G. Asymptotic behaviour of a Dirac particle in a Coulomb field. Il Nuovo Cimento. 45: 801-812 (1966).

[20] Thaller, B. Relativistic scattering theory for long-range potentials of the nonelectrostatic type. Lett. Math. Phys. 12: 15-19 (1986).

[21] Perry, P. Scattering Theory by the Enss Method. Math. Rep. 1, Harwood academic publishers, New York 1983

[22] Loss, M. and Thaller, B. Scattering of particles by long-range magnetic fields. Ann. Phys. 176: 159-180 (1987).

ON THE QUASI-STATIONARY APPROACH TO SCATTERING FOR PERTURBATIONS PERIODIC IN TIME

Dimitri R. Yafaev

Leningrad Department of Math. Inst. (LOMI)
Fontanka 27 Leningrad 191011 USSR

ABSTRACT

Scattering of a plane wave by a time-periodic potential is described by a system of coupled stationary Schrödinger equations. Each equation corresponds to a channel when energy is changed by some integer number. The interaction of a plane wave with a quasi-bound state of a time-periodic potential well is investigated. It is shown that for resonant energies this interaction does not vanish as a coupling constant of a perturbation tends to zero.

INTRODUCTION

The basic concept of scattering theory is that under natural assumptions on a (time-independent) interaction a quantum system can be either in a stationary state or its asymptotic behavior for large time is the same as that for a system without interaction. To a certain extent the same result appears to be true for time-periodic interactions. To see this let $U(t)$, $t \in \mathbb{R}$, $U(0) = I$, be the evolution operator of a system described by a family of self-adjoint operators $H(t)$. Suppose that $H(t) = H_0 + V(t)$ where the operator H_0 corresponds to a free system with the evolution operator $U_0(t) = \exp(-iH_0 t)$ and $V(t + 2\pi) = V(t)$. The unitary operator $\mathcal{U} = U(2\pi)$ plays the role of the self-adjoint Hamiltonian H in the case of time-independent perturbations and is called the monodromy (Floquet) operator. It turns out that under fairly general assumptions on the perturbation $V(t)$ for initial states f, orthogonal to all eigenvectors of $\mathcal{U}$, functions $U(t)f$ have free asymptotics as $t \to \pm\infty$, i.e. there exist $f_\pm$ such that $U(t)f \sim U_0(t)f_\pm$. This is certainly a straightforward generalization of the corresponding result for time-independent perturbations $V(t) = V$ when $U(t) = \exp(-iHt)$ and the eigenvectors of U and H coincide.

A. Boutet de Monvel et al. (eds.), Recent Developments in Quantum Mechanics, 367–380.

There is, however, an important difference with the time-independent case. The Hamiltonian H has absolutely continuous positive spectrum and perhaps some negative eigenvalues. These eigenvalues correspond to bound states of the quantum system with the Hamiltonian H and are stable with respect to small (time-independent) perturbations of H. On the contrary, the spectrum of the monodromy operator U covers the whole unit circle so that its eigenvalues are embedded in the continuous spectrum. Thus it is to be expected that generically U does not have point spectrum at all.

Suppose now that we add to some operator H_1 with negative eigenvalues a time periodic perturbation $\varepsilon V(t)$. The operator $\mathcal{U}_1 = \exp(-2\pi i H_1)$ has eigenvalues embedded in the continuous spectrum but, typically, the monodromy operator $\mathcal{U} = U(2\pi)$ of the family $H(t) = H_1 + \varepsilon V(t)$ has only continuous spectrum. However, it is natural to conjecture that eigenvalues of U_1 do not completely disappear and will be transformed into some kind of resonances of the system with a time-dependent Hamiltonian. In fact, in this situation resonances were correctly defined in [1] as eigenvalues of a certain auxiliary operator.

In this report we shall study this phenomena in the framework of the so-called quasi-stationary approach which is natural from the physical viewpoint [2, 3, 4]. To give an idea of this approach consider scattering of a plane wave of energy λ. In the time-independent case $V(t) = V$ this problem is governed by the stationary Schrödinger equation $H\psi = \lambda\psi$. This is a consequence of the law of energy conservation. Such a law is lacking for time-dependent Hamiltonians.

Nevertheless, for potentials periodic in time the energy λ can only be changed by some integer number. In other words, the quasi-energy, i.e. the energy defined up to an integer, is conserved. This allows us to describe scattering of a plane wave by a time-periodic potential by an infinite system of coupled stationary equations parametrized by $n \in \mathbb{Z}$. Coefficients in the asymptotics of solutions of this system determine the scattering matrix of our problem.

In the case $H(t) = H$ these equations are independent and describe either scattering states (if $\lambda + n \geq 0$) or bound states (if $\lambda + n \leq 0$) of the Hamiltonian H. When a periodic in time perturbation is switched on these equations become coupled so that a non-trivial "interaction" between bound and scattering states arises. This results in the disappearence of proper bound states and in their transformation into resonant states of the time-dependent problem.

This report is organized as follows. In section 1 we give a correct formulation of the scattering problem for perturbations periodic in time. In section 2 the quasi-stationary approach to this problem is described. Here we restrict ourselves to the Schrödinger operator on the half-axis which corresponds to scattering with the orbital quantum number $\ell = 0$ (s-wave scattering). Finally, in section 3 periodic in time zero-range potentials are considered. In this case we develop the perturbation theory in small coupling constant ε for the scattering matrix. This theory has a comparatively standard form for energies λ separated from points $\lambda_1+\kappa$, where λ_1 is an eigenvalue of the unperturbed Hamiltonian H_1 and $\kappa\in\mathbb{Z}$. On the contrary at resonant energies $\lambda = \lambda_1+\kappa$ the asymptotics of the scattering matrix is more complicated and describes the interaction of an incident wave with the bound state of H_1.

1. SCATTERING THEORY FOR TIME-DEPENDENT HAMILTONIANS

Let $H(t)$, $t\in\mathbb{R}$, be a family of self-adjoint operators with common domains $\mathcal{D}(H(t)) = \mathcal{D}$ in a Hilbert space $\mathcal{H}$. Under fairly general assumptions (see, e.g. [5]) on the dependence of $H(t)$ on time t there exists a unique unitary operator $U(t,s)$ (the evolution operator or the propagator) such that $U(t,s)\colon \mathcal{D}\to\mathcal{D}$, the functions $U(t,\tau)f$ for $f\in\mathcal{D}$ are strongly differentiable with respect to t and

$$i\frac{\partial U(t,\tau)f}{\partial t} = H(t)U(t,\tau)f, \qquad\qquad U(\tau,\tau) = I.$$

Uniqueness of the solution to this problem ensures that

$$U(t,\sigma)U(\sigma,\tau) = U(t,\tau)\,.$$

Moreover, if $H(t)$ is periodic, i.e. $H(t+2\pi) = H(t)$, then

$$U(t+2\pi,\tau+2\pi) = U(t,\tau).$$

Clearly, $U(t,\tau) = \exp(-i(t-\tau)H)$ if $H(t) = H$ does not depend on time.

Suppose now that $H(t) = H_0+V(t)$ is a perturbation of a time-independent Hamiltonian H_0 with the absolutely continuous spectrum. More precisely, we assume that the following strong limits (called wave operators) exist

$$W_\pm(\tau) = \operatorname*{s-lim}_{t\to\pm\infty} U^*(\tau+t,\tau)U_0(t)\,, \qquad U_0(t) = \exp(-iH_0t)\,. \tag{1.2}$$

By (1.1) the existence of $W_\pm(\tau)$ for different τ is equivalent. Moreover, the wave operators are connected by the relation

$$W_\pm(\sigma) = U(\sigma,\tau)W_\pm(\tau)U_0(\tau-\sigma)\,. \tag{1.3}$$

In the time-independent case $H(t) = H$ the wave operators do not depend on τ so that (1.3) ensures the intertwining property $HW_\pm = W_\pm H_0$. In the periodic case $H(t) = H(t+2\pi)$, the operators (1.2) also depend periodically on τ, i.e. $W_\pm(\tau+2\pi) = W_\pm(\tau)$. According to (1.3) it follows that

$$U(\tau+2\pi,\tau)W_\pm(\tau) = W_\pm(\tau)U_0(2\pi)\,. \tag{1.4}$$

Thus the ranges $R(W_\pm(\tau))$ of the wave operators are contained in the absolutely continuous subspace $\mathcal{H}^{(a)}(\mathcal{U}(\tau))$ of the so called monodromy operator $\mathcal{U}(\tau) = U(\tau+2\pi,\tau)$. The restriction of $\mathcal{U}(\tau)$ on $R(W_\pm(\tau))$ is unitarily equivalent to $U_0(2\pi)$. According to (1.3) it suffices to study wave operators only for one fixed τ. Below we set $\tau = 0$ and drop it from the notation, i.e. $U(t) = U(t,0)$, $\mathcal{U} = \mathcal{U}(0)$, $W_\pm = W_\pm(0)$.

By definition (1.2) for $f \in R(W_\pm)$ the function $U(t)f$ evolves according to the free asymptotics as $t \to \pm\infty$ that is, the relation

$$\lim_{t\to\pm\infty} \| U(t)f - U_0(t)f_\pm \| = 0\,, \qquad f_\pm = W_\pm^* f \tag{1.5}$$

holds. Therefore the main problem of scattering theory is to give an effective description of the ranges of the wave operators. In the time-independent case this can be done in spectral terms. Namely, under general assumptions, $R(W_\pm)$ coincide with the absolutely continuous subspace $\mathcal{H}^{(a)}(H)$ of the Hamiltonian H. The relation $R(W_\pm) = \mathcal{H}^{(a)}(H)$ is called asymptotic completeness. For arbitrary time-dependent perturbations spectral quantities are lacking. However, for the periodic case the natural conjecture [6] is that

$$R(W_\pm) = \mathcal{H}^{(a)}(\mathcal{U})\,. \tag{1.6}$$

If this is true, the wave operators $W_\pm$ are called complete.

The scattering operator $\mathcal{S}$ is defined by the relation $\mathcal{S} = W_+^* W_-$. Clearly, $\|\mathcal{S}\| \le 1$. Moreover, $\mathcal{S}$ is a unitary operator, if $R(W_+) = R(W_-)$. In the time-independent case the scattering operator commutes with H_0 which follows from the intertwining property $HW_\pm = W_\pm H_0$. This is of course a consequence of energy conservation. Similarly, by (1.4), in the periodic case

$$\mathcal{S} U_0(2\pi) = U_0(2\pi)\mathcal{S}. \tag{1.7}$$

The relation gives the precise formulation of the law of the quasi-energy conservation. We emphasize that (1.7) does not imply that $\mathcal{S}$ commutes with H_0.

The scattering matrix for the family $H(t) = H_0 + V(t)$, $V(t+2\pi) = V(t)$, is defined in the representation for which $U_0(2\pi)$ is diagonal. Suppose for definiteness that H_0 is absolutely continuous, its spectrum $\sigma(H_0)$ has a constant multiplicity ν, $1 \le \nu \le \infty$, and $\sigma(H_0) = \mathbb{R}_+$. Denote by $\tilde{\mathcal{H}} = L_2(\mathbb{R}_+; G)$ the Hilbert space of vector functions $\tilde{f}(\lambda)$ with values in an auxiliary Hilbert space G of dimension ν. By the spectral theorem, there exists a unitary mapping $\Phi : \mathcal{H} \to \tilde{\mathcal{H}}$ such that $\Phi H_0 \Phi^*$ acts as multiplication by an independent variable $\lambda \in \mathbb{R}_+$. Therefore $\Phi U_0(2\pi)\Phi^*$ is multiplication by $\exp(-2\pi i\lambda)$. To construct a diagonal for U_0 representation it suffices now to identify all points $\lambda \in \mathbb{R}_+$ which differ by some integer. Actually, denote by $\mathfrak{g} = \ell_2(G)$ the ℓ_2-space of vectors $\hat{f} = \{f_n\}$, $n = 0, 1, 2, \ldots$, where $f_n \in G$ and set $\hat{\mathcal{H}} = L_2(I; \mathfrak{g})$, $I = [0, 1)$. Clearly, the operator $T : \tilde{\mathcal{H}} \to \hat{\mathcal{H}}$ defined by the formula

$$(T\tilde{f})_n(\theta) = \tilde{f}(n + \theta), \qquad \theta \in [0, 1), \qquad n = 0, 1, 2, \ldots$$

is unitary and $\hat{\Phi} U_0(2\pi)\hat{\Phi}^*$, $\hat{\Phi} = T\Phi$, acts as multiplication by $\exp(-2\pi i \theta)$. Since $\mathcal{S}$ commutes with $U_0(2\pi)$, the spectral theorem ensures that $\hat{\Phi}\mathcal{S}\hat{\Phi}^*$ acts as multiplication by an operator-function $S(\theta) : \mathfrak{g} \to \mathfrak{g}$, $\theta \in [0, 1)$, called the scattering matrix. It means that for $\hat{f} = \{f_n\}$

$$(S(\theta)\hat{f})_n = \sum_{m=0}^{\infty} s_{nm}(\theta) f_m$$

so that the operator $\tilde{\mathcal{S}} = \Phi \mathcal{S} \Phi^*$ acts in the space $\tilde{\mathcal{H}}$ by the formula

$$(\tilde{\mathcal{S}} \tilde{f})(n + \theta) = \sum_{m=0}^{\infty} s_{nm}(\theta) \tilde{f}(m + \theta). \tag{1.8}$$

The unitarity of the scattering operator $\mathcal{S}$ in $\mathcal{H}$ induces the unitarity of the scattering matrix $S(\theta)$ in $\mathfrak{g}$ for all (more precisely, for almost all) $\theta \in [0, 1)$. In particular, it means that

$$\sum_{n=0}^{\infty} s^*_{nm}(\theta) s_{nm}(\theta) = I, \qquad m = 0, 1, 2, \ldots \tag{1.9}$$

Let us now consider the Schrödinger operator with a time-periodic potential. Set $\mathcal{H} = L_2(\mathbb{R}^d)$, $d \geq 1$, $H_0 = -\Delta$ and $H(t) = -\Delta + q(x,t)$ where $q = \bar{q}$, $q(x,t+2\pi) = q(x,t)$ and

$$|q(x,t)| \leq C(1+|x|)^{-1-\varepsilon}, \qquad \varepsilon > 0. \tag{1.10}$$

By C we denote positive constants whose precise values are of no importance for us. Under assumption (1.10) the wave operators exist and, as shown in [7], they are complete, i.e. the relation (1.6) holds. Moreover, the monodromy operator u does not have a singular continuous part so that $R(W_\pm) = \mathcal{H} \ominus \mathcal{H}^{(p)}(\mathcal{U})$ where $\mathcal{H}^{(p)}(\mathcal{U})$ is spanned by eigenvectors of $\mathcal{U}$. Note, however, that we do not know any non-trivial example when $\mathcal{U}$ does have eigenvectors. On the other hand, sufficient conditions for the unitarity of $W_\pm$ may be given. For example, $R(W_\pm) = \mathcal{H}$ if q is repulsive, i.e. $\frac{\partial q}{\partial |x|} \leq 0$ (see [8] where the periodicity of q is not assumed).

Results for Schrödinger operators hold also for the case $\mathcal{H} = L_2(\mathbb{R}_+)$, $H_0 = -\frac{d^2}{dx^2}$, $H(t) = -\frac{d^2}{dx^2} + q(x,t)$ with the boundary condition $u(0) = 0$. In this case Φ is (up to the change of variables) the sine-Fourier transformation

$$\tilde{f}(\lambda) = (\Phi f)(\lambda) = \pi^{-1/2}\lambda^{-1/4} \int_0^\infty f(x) \sin(\lambda^{1/2}x)\, dx, \tag{1.11}$$

$G = \mathbb{C}$ and the relation (1.9) means that

$$\sum_{n=0}^{\infty} |s_{nm}(\theta)|^2 = 1, \qquad m = 0, 1, 2 \ldots \tag{1.12}$$

2. THE QUASI-STATIONARY APPROACH

Our aim here is to give a framework for a description of scattering of a plane wave by a periodic in time potential. We restrict ourselves to scattering with zero orbital quantum number so that the problem is governed by the equation

$$i\frac{\partial u}{\partial t} = -\frac{\partial^2 u}{\partial x^2} + q(x,t)u \tag{2.1}$$

with the boundary condition $u(0,t) = 0$. Suppose as before that q is real, $q(x,t) = q(x,t+2\pi)$ and q vanishes sufficiently rapidly as $x \to \infty$. We shall concentrate on the formal construction of solutions of (2.1) and omit some technical details.

Let us expand the periodic function $q(x, t)$ in a Fourier series

$$q(x,t) = \sum_{n=-\infty}^{\infty} q_n(x) \exp(int) . \tag{2.2}$$

We will look for solutions of equation (2.1) which have a representation of the form

$$u(x, t) = \sum_{n=-\infty}^{\infty} u_n(x) \exp(-i(n+\theta)t) \tag{2.3}$$

where the parameter $\theta \in [0,1)$. Substituting (2.2) and (2.3) into (2.1) and comparing the coefficients of e^{-int} we arrive at a system of ordinary differential equations

$$-u_n''(x) + \sum_{\ell=-\infty}^{\infty} q_{\ell-n}(x)\, u_\ell(x) = (n+\theta)\, u_n(x) \tag{2.4}$$

for functions u_n, $n \in \mathbb{Z}$. The condition $u(0, t) = 0$ requires that $u_n(0) = 0$ for all n. Solutions of the form (2.3) describe a stationary process. In fact, for any $\tau \in \mathbb{R}$

$$(2\pi)^{-1} \int_{\tau}^{\tau+2\pi} |u(x,t)|^2 dt = \sum_{n=-\infty}^{\infty} |u_n(x)|^2 . \tag{2.5}$$

The solution of the system (2.4) corresponding to the plane wave $\exp(-i\,\lambda^{1/2} x)$, $\lambda > 0$, coming from infinity is distinguished by its asymptotics

$$u_n(x, \lambda) \sim \delta_{nm} \exp(-i\,\lambda^{1/2} x) - S_n(\lambda) \exp\left(i(\theta+n)^{1/2} x\right) \tag{2.6}$$

as $x \to \infty$. We suppose always that

$$\lambda = m + \theta$$

where $m = [\lambda]$ is an entire part of λ and $\theta \in [0, 1)$. Moreover, $\delta_{mm} = 1$, $\delta_{nm} = 0$ if $n \neq m$, and

$$i(\theta+n)^{1/2} = -|\,\theta+n\,|^{1/2}, \qquad n \leq -1 .$$

The coefficients (the amplitudes) $S_n(\lambda)$ are determined by the system (2.4). A solution of the system (2.4) with the asymptotics (2.6) exists and the amplitudes $S_n(\lambda)$ are uniquely defined. We omit here the precise formulation and the proof of this assertion. In particular, since the point $\lambda = 0$ is exceptional for the equation $-u'' = \lambda u$ one should avoid all integer energies $\lambda = 0, 1, 2, \ldots$

Note that for the time-independent case when $q(x, t) = q_0(x)$, equations (2.4) become decoupled and are reduced to

$$-u_n'' + q_0 u_n = (\theta + n)\, u_n . \tag{2.7}$$

Thus under assumption (2.6) $u_n(x, \lambda) = 0$ if $n \neq m$, $n \geq 0$ and $S_m(\lambda)$ is the scattering matrix for energy λ. If $n < 0$, then equation (2.7) has a non-trivial solution with the asymptotics (2.6) if $\theta+n$ is an eigenvalue of the operator $H = -\frac{d^2}{dx^2} + q_0(x)$ with the boundary condition $u_n(0) = 0$. In this case $S_n(\lambda)$ is arbitrary.

In the general situation the system (2.4) with the boundary conditions $u_n(0) = 0$ and (2.6) describes the scattering of a plane wave $\exp(-i\,\lambda^{1/2} x)$ by the potential (2.2). Due to the law of the quasi-energy conservation in such a process the energy λ can be changed only by some integer number. Thus, the allowed values of the energy are $\lambda+\kappa$ where $\kappa \in \mathbb{Z}$. We emphasize that for $n \geq 0$ the terms $S_n(\lambda) \exp\left(i(\theta + n)^{1/2} x\right)$ in (2.6) correspond to waves going to infinity with energies $\lambda+\kappa$, $\kappa = n-m$. For $n < 0$ the functions $u_n(x)$ are exponentially decaying and, as we shall see, do not contribute to the scattering matrix.

Let us now establish the connection between the time-dependent approach described in section 1 and the quasi-stationary one. Let $H(t) = -\frac{d^2}{dx^2} + q(x, t)$, $H_0 = -\frac{d^2}{dx^2}$ with the boundary condition $u(0) = 0$ be self-adjoint operators in the space $\mathcal{H} = L_2(\mathbb{R}_+)$. Denote by $\mathcal{S}$ the corresponding scattering operator and by $s_{nm}(\theta)$ elements of the corresponding scattering matrix $S(\theta)$. We shall show that the amplitudes $S_n(\lambda)$ are related to $s_{nm}(\theta)$ by the formula

$$S_n(m + \theta) = s_{nm}(\theta)\, (m + \theta)^{1/4} (n + \theta)^{-1/4} . \tag{2.8}$$

According to (1.5) the action of $\mathcal{S}$ can be described in the following way. Let $f \in \mathcal{H}$ be arbitrary and the solution $u(x,t)$ of (2.1), $u(0,t) = 0$, be chosen in such a way that

$$U(t) \sim U_0(t)\, f \ \text{as} \ t \to -\infty \tag{2.9}$$

(this asymptotic relation should be understood in $\mathcal{H}$). Then

$$U(t) \sim U_0(t) \mathcal{S} f \ \text{as} \ t \to +\infty . \tag{2.10}$$

Clearly, $\mathcal{S}$ is defined by its values on some dense set in $\mathcal{H}$.

First, we find the asymptotics of $U_0(t)f$ as $t \to \pm\infty$. By making use of the sine-Fourier transformation (1.11) we find that

$$\big(U_0(t)f\big)(x) = -i2^{-1}\pi^{-1/2}\int_0^\infty \big(\exp(i\lambda^{1/2}x)-\exp(-i\lambda^{1/2}x)\big)\exp(-i\lambda t)\tilde{f}(\lambda)\,\lambda^{-1/4}\,d\lambda,$$

$$\tilde{f} = \Phi f. \tag{2.11}$$

If, for example, $\tilde{f} \in C_0^\infty(\mathbb{R}_+)$, then integration by parts shows that the term with $\exp(i\lambda^{1/2}x - i\lambda t)$ does not contribute to the asymptotics of (2.11) as $t \to -\infty$. It follows that

$$\big(U_0(t)f\big)(x) \sim i\,2^{-1}\pi^{-1/2}\int_0^\infty \exp(-i\lambda^{1/2}x - i\lambda t)\,\tilde{f}(\lambda)\,\lambda^{-1/4}\,d\lambda, \quad t \to -\infty, \tag{2.12}$$

and similarly,

$$\big(U_0(t)g\big)(x) \sim -i\,2^{-1}\pi^{-1/2}\int_0^\infty \exp(i\lambda^{1/2}x - i\lambda t)\,\tilde{g}(\lambda)\,\lambda^{-1/4}\,d\lambda, \quad t \to +\infty \tag{2.13}$$

Now we construct the solution of the equation (2.1) with the asymptotics (2.12). Let functions $u_n(x,\lambda)$ satisfy the system (2.4) and the boundary conditions (2.6) and $u_n(0,\lambda) = 0$. Note that for all $m = 0, 1, 2, \ldots$ and $\theta \in [0,1)$ the functions

$$u(x,t) = \sum_{n=-\infty}^{\infty} u_n(x,\lambda)\exp(-i(n+\theta)t) \tag{2.14}$$

satisfy equation (2.1) and the boundary condition $u(0,t) = 0$. Integrating the functions (2.14) over θ with some $\psi \in C_0^\infty(I)$ we find that

$$u(x,t) = \sum_{n=-\infty}^{\infty}\int_0^1 u_n(x, m+\theta)\exp(-i(n+\theta)t)\,\psi(\theta)\,d\theta \tag{2.15}$$

also satisfies (2.1) and $u(0,t) = 0$. Let us now look at the asymptotics of the function (2.15) as $t \to \pm\infty$. According to (2.6) terms with $n < 0$ do not contribute to the asymptotics both as $t \to -\infty$ and $t \to +\infty$. The asymptotics of terms with $n \geq 0$ is determined by the asymptotics of functions $u_n(x,\lambda)$ as $x \to \infty$. In the limit $t \to -\infty$ only the incoming plane wave $\exp(-i\lambda^{1/2}x)$ is essential so that only the term with $n = m$ will survive and

$$u(x,t) \sim \int_0^1 \exp(-i(m+\theta)^{1/2}x - i(m+\theta)t)\,\psi(\theta)\,d\theta, \qquad t \to -\infty. \tag{2.16}$$

Similarly, the term with $\exp(-i\lambda^{1/2}x)$ disappears in the limit $t \to +\infty$. Therefore

$$u(x,t) \sim -\sum_{n=0}^{\infty} \int_0^1 \exp\left(i(n+\theta)^{1/2}x - i(n+\theta)t\right) S_n(m+\theta)\psi(\theta)d\theta\,, \qquad t \to +\infty. \quad (2.17)$$

Suppose that in (2.11) $\tilde{f} \in C_0^\infty(m,m+1)$ where m is one of the numbers $0, 1, 2, \ldots$ The sum of such functions over different m is dense in $L_2(\mathbb{R}_+)$. If we set

$$\psi(\theta) = i\, 2^{-1} \pi^{-1/2} \tilde{f}(m+\theta)(m+\theta)^{-1/4},$$

then $\psi \in C_0^\infty(I)$ and, clearly, the asymptotics (2.12) and (2.16) coincide. Similarly, the asymptotics (2.13) and (2.17) coincide if

$$i\, 2^{-1} \pi^{-1/2} \tilde{g}(\lambda) \lambda^{-1/4} = S_n(\lambda + m - n)\, \psi(\lambda - n) \qquad \lambda \in (n, n+1).$$

By (2.9) and (2.10) $\tilde{g} = \tilde{\mathcal{S}} \tilde{f}$ where $\tilde{\mathcal{S}} = \Phi \mathcal{S} \Phi^*$ so that

$$(\tilde{\mathcal{S}}\tilde{f})(\lambda) = \lambda^{1/4} (\lambda + m - n)^{-1/4} S_n(\lambda + m - n) \tilde{f}(\lambda + m - n)\,, \qquad \lambda \in (n, n+1). \quad (2.18)$$

Note now that, if $\tilde{f} \in C_0^\infty(m, m+1)$, then the right-hand side of (1.8) consists of only one term, i.e. $(\tilde{\mathcal{S}}\tilde{f})(n+\theta) = s_{nm}(\theta)\tilde{f}(m+\theta)$. Therefore (2.8) is an immediate consequence of (2.18).

According to (1.12) the equality (2.8) ensures that

$$\lambda^{-1/2} \sum_{n=0}^{\infty} (\lambda+n-m)^{1/2} |S_n(\lambda)|^2 = 1\,, \qquad m = [\lambda]\,.$$

Thus $\lambda^{-1/2} (\lambda+n-m)^{1/2} |S_n(\lambda)|^2$ is the probability of the appearence of the outgoing wave $\exp\left(i(\lambda+n-m)^{1/2}x\right)$ when $\exp(-i\lambda^{1/2}x)$ is scattered by a time-periodic potential. We emphasize that there are no apriori restrictions on values of the amplitudes $S_n(\lambda)$ for negative n. Therefore functions $u_n(x,\lambda)$, exponentially decaying as $x \to \infty$, may be arbitrary large in a bounded domain. We shall see in the next section that this situation actually occurs so that $u_n(x,\lambda)$ with $n < 0$ should be interpreted as a kind of quasi-bound state.

3. RESONANT SCATTERING FOR TIME-PERIODIC PERTURBATIONS

Here we shall consider zero-range perturbations with a "coupling constant" depending periodically on time. Mathematically this problem is governed by the equation

$$i\frac{\partial u}{\partial t} = -\frac{\partial^2 u}{\partial x^2} \qquad x>0 \tag{3.1}$$

with the time-dependent boundary condition

$$u'(0,t) = h(t)u(0,t)\,, \qquad \overline{h(t)} = h(t)\,, \qquad h(t+2\pi) = h(t)\,. \tag{3.2}$$

The detailed proof of the existence of the evolution operator $U(t)$ corresponding to (3.1), (3.2) was given in [9] . As in the preceding section, let $H_0 = -\frac{d^2}{dx^2}$ with the boundary condition $u(0) = 0$. Then the wave operators (1.2) exist. Moreover, the asymptotic completeness (1.6) holds and the monodromy operator $\mathcal{U} = U(2\pi)$ does not have a singular continuous spectrum. These results were obtained in [10] by purely time-dependent tools (by the Enss method).

Here we restrict ourselves to the consideration of the simplest case

$$h(t) = -h_1 + 2\,\varepsilon \cos t. \tag{3.3}$$

Let us show first of all that, as remarked in [11], the operator $\mathcal{U}$ does not have eigenvalues. Indeed, suppose that $\mathcal{U}f = \mu f\,, |\mu| = 1\,, f \in L_2(\mathbb{R}_+)$. Then the function $u(t) = U(t)f$ is represented in the form (2.3) with $\theta \in [0,1]$ defined by $\mu = \exp(-2\pi i\,\theta)$. Functions $u_n(x)$ in (2.3) satisfy equations $-u_n''(x) = (n+\theta)u_n(x)$ and according to (2.5) belong to the space $L_2(\mathbb{R}_+)$. It follows that $u_n(x) = 0$, if $n \geq 0$, and $u_n(x) = R_n \exp\left(-(n+\theta)^{1/2}x\right)$, if $n < 0$. Now the boundary condition (3.2) gives the equation

$$-\sum_{n=-\infty}^{-1} |n+\theta|^{1/2} R_n\, e^{-int} = h(t) \sum_{n=-\infty}^{-1} R_n\, e^{-int}\,.$$

In the case (3.3) it is equivalent to the system

$$(h_0 - |n+\theta|^{1/2}) R_n = \varepsilon\,(R_{n+1} + R_{n-1})\,, \qquad n = 0,-1,-2,\ldots \tag{3.4}$$

with $R_0 = R_1 = 0$. Equation (3.4) with $n = 0$ shows that $R_{-1} = 0$, then equation (3.4) with

$n=-1$ ensures that $R_{-2}=0$ and so on. It follows that $R_n=0$ for all $n=-1,-2,\dots$ and therefore $u=0$.

The quasi-stationary approach to problem (3.1), (3.2) can be developped similarly to section 2. In particular, the amplitudes S_n are introduced by the relation (2.6) which turns now into a precise equality. Our goal here is to describe the asymptotic behavior as $\varepsilon \to 0$ of the amplitudes $S_n(\lambda, \varepsilon)$ corresponding to (3.3). Note that for $\varepsilon = 0$ the function (3.3) does not depend on t so that $U(t) = \exp(-i H_1 t)$ where $H_1 = -\frac{d^2}{dx^2}$ with the boundary condition $u'(0) = -h_1 u(0)$. If $h_1 \le 0$, this operator is positive and, if $h_1 > 0$, it has (exactly one) negative eigenvalue $\lambda_1 = -h_1^2$ with the corresponding eigenfunction $\exp(-h_1 x)$. The scattering matrix for the pair H_0, H_1 equals

$$S^{(1)}(\lambda) = (h_1 - i\lambda^{1/2})(h_1 + i\lambda^{1/2})^{-1}. \tag{3.5}$$

Below we will give asymptotic formulas for $S_n(\lambda, \varepsilon)$ as $\varepsilon \to 0$. We omit all calculations which can be found in [12].

First we consider the non-resonant case when either $h_1 \le 0$ or $h_1 > 0$ but $\lambda - \lambda_1 \notin \mathbb{Z}$. Under both of these assumptions the asymptotic expansions of amplitudes are described by regular perturbation theory. In particular, the amplitude $S_m(\lambda,\varepsilon)$ converges to the scattering matrix $S^{(1)}(\lambda)$, i.e.

$$S_m(\lambda,\varepsilon) = (h_1 - i\lambda^{1/2})(h_1 + i\lambda^{1/2})^{-1} + O(\varepsilon^2). \tag{3.6}$$

The leading term of the corrections to the case $\varepsilon = 0$ is determined by the amplitudes

$$S_{m\pm 1}(\lambda,\varepsilon) = -2i\varepsilon\lambda^{1/2}\left(h_1 + i(\lambda \pm 1)^{1/2}\right)^{-1}(h_1 + i\lambda^{1/2})^{-1} + O(\varepsilon^2) \tag{3.7}$$

corresponding to states with energies $\lambda \pm 1$. Similar relations hold for all $S_n(\lambda, \varepsilon)$. We note only that $S_n(\lambda, \varepsilon) = O(\varepsilon^{|n-m|})$. Thus the probability of excitation of states with energies $\lambda + \kappa$, $\kappa \in \mathbb{Z}$, is proportional to $\varepsilon^{|\kappa|}$.

If $h_1 < 0$ and λ equals one of the resonant energies $\lambda_1 + \kappa$, there arises a non-trivial interaction of the incident wave $\exp(-i\lambda^{1/2}x)$ with the quasi-bound state of the time-dependent well. This interaction does not vanish in the limit $\varepsilon \to 0$. For definiteness we suppose that $0 < h_1 < 1$ and λ approaches the point $\lambda_0 = -h_1^2 + 1$. In this case the resonant interaction is the most significant. Let

$$\Omega(\lambda, \varepsilon) = \left[-h_1 + (1-\lambda)^{1/2} + \varepsilon^2\left(h_1 - (2-\lambda)^{1/2}\right)^{-1}\right](h_1 + i\lambda^{1/2}) + \varepsilon^2.$$

Clearly, $\Omega(\lambda, \varepsilon) \to 0$ as $\lambda \to \lambda_0$ and $\varepsilon \to 0$. It is important, however, that for $\lambda \in [\delta, 1-\delta]$ with arbitrary $\delta > 0$

$$\varepsilon^2 |\, \Omega^{-1}(\lambda, \varepsilon) \,| \le C < \infty .$$

The asymptotics of the amplitudes S_0 and S_{-1} are given by the expressions

$$S_0(\lambda,\varepsilon) = (h_1 - i\,\lambda^{1/2})(h_1 + i\,\lambda^{1/2})^{-1} + 2i\varepsilon^2 \lambda^{1/2}(h_1 + i\,\lambda^{1/2})^{-1}\Omega^{-1}(\lambda, \varepsilon) + O(\varepsilon), \qquad (3.8)$$

$$S_{-1}(\lambda,\varepsilon) = 2i\,\varepsilon\,\lambda^{1/2}\,\Omega^{-1}(\lambda, \varepsilon)\big(1 + O(\varepsilon)\big), \qquad (3.9)$$

which are uniform in $\lambda \in [\delta, 1-\delta]$. Other amplitudes satisfy the bounds $S_n(\lambda,\varepsilon) = O(\varepsilon^n)$, if $n \ge 1$, and $S_n(\lambda, \varepsilon) = O(\varepsilon^{-n-2})$, if $n \le -2$. We emphasize that in a neighbourhood of λ_0 both terms in (3.8) have the same order and $|S_0(\lambda,\varepsilon)| = 1$ up to an error of order ε. If λ is separated from the point λ_0, we can replace $\Omega(\lambda,\varepsilon)$ by $\Omega(\lambda,0)$ so that we recover the relations (3.6), (3.7) (for $m = 0$). In the particular case $\lambda = \lambda_0$ we have that

$$\Omega(\lambda_0, \varepsilon) = b_1\big(h_1 - (1 + h_1^2)^{1/2}\big)^{-1} \varepsilon^2$$

where

$$b_1 = 2h_1 - (1 + h_1^2)^{1/2} + i\,(1 - h_1^2)^{1/2}.$$

Therefore, according to (3.8), (3.9),

$$S_0(\lambda_0, \varepsilon) = \bar{b}_1 b_1^{-1} + O(\varepsilon),$$

$$S_{-1}(\lambda_0, \varepsilon) = 2i\,(1 - h_1^2)^{1/2}\big(h_1 - (1 + h_1^2)^{1/2}\big) b_1^{-1} \varepsilon^{-1} + O(1).$$

As could be expected the amplitude $S_{-1}(\lambda_0, \varepsilon)$ grows infinitely as $\varepsilon \to 0$ which is consistent with the decoupling of bound states and scattering states in this limit. The amplitude $S_0(\lambda_0, \varepsilon)$ has a finite limit $\bar{b}_1 b_1^{-1}$ which is, however, different from the scattering matrix (3.5) at energy λ_0 for the time-independent boundary condition $u'(0) = -h_1 u(0)$. Therefore at energy λ_0 we find an additional resonant phase shift which does not vanish in the limit $\varepsilon \to 0$.

RÉFÉRENCES

[1] Yajima K. (1982), Resonances for the AC-Stark effect, Commun. Math. Phys. 87, 331-352.

[2] Kazanskii A.N., Ostrovskii V.N., Solov'ev E.A. (1976), Passage of low-energy particles through a nonstationary potential barrier and the quasi-energy spectrum, Soviet Physics -JETP 48, N° 2, 254-259.

[3] Manakov N.L., Fainshtein A.G. (1981), Quasistationary quasi-energy states and convergence of perturbation series in a monochromatic field , Theor. Math. Phys. 48, 815-822.

[4] Büttiker M., Landauer R. (1985), Traversal time for tunneling, Physica Scripta 32, 429-434.

[5] Reed M., Simon B. (1975), Methods of modern mathematical physics II, Academic Press, New York.

[6] Schmidt G. (1975), On scattering by time dependent perturbations, Indiana Univ. Math. 24, 925-935.

[7] Yajima K. (1977), Scattering theory for Schrödinger equations with potentials periodic in time, J. Math. Soc. Japan 29, n° 4, 729-743.

[8] Yafaev D.R. (1984), The virial theorem and conditions for unitarity of wave operators in scattering by a time-dependent potential, Proc. Steklov Inst. of Math. 2 (V. 159), 219-226.

[9] Sayapova M.R., Yafaev D.R. (1984), The evolution operator for time-dependent potentials of zero-radius, Proc. Steklov Inst. of Math. 2 (V. 159), 173-180.

[10] Sayapova M.R., Yafaev D.R. (1985), Scattering theory for zero-range potentials periodic in time, Selecta Math. Sov. 4, N° 3, 277-287.

[11] Koratjaev E.L. (1980), On the spectrum of the monodromy operator for the Schrödinger operator with a periodic in time potential, Notes of LOMI Sci. Seminars 96, 101-104, Russian.

[12] Yafaev D.R. (1990), On resonant scattering for time-periodic perturbations, Colloque "Equations aux Dérivées Partielles", Saint-Jean-de-Monts, Publications de l'Université de Nantes.

Existence, uniqueness and some properties of Schrödinger propagators

KENJI YAJIMA

Department of Pure and Applied Sciences
University of Tokyo
3-8-1 Komaba, Meguroku, Tokyo 153, Japan

1. Introduction.

We consider a time dependent Schrödinger equation for quantum particles in an electro-magnetic field:

$$ih\frac{\partial u}{\partial t} = H(t)u = (H_0(t) + V(t,x))u, \quad t \in \mathbf{R}^1, \quad x \in \mathbf{R}^n, \tag{1.1}$$

$$H_0(t) = \sum_{j=1}^{n} \frac{1}{2}(-i\frac{\partial}{\partial x_j} - A_j(t,x))^2, \tag{1.2}$$

where $V(t,x)$ and $A(t,x) = (A_1(t,x), A_2(t,x), ..., A_n(t,x))$ are the electric scalar and magnetic vector potentials of the field. The purpose of this paper is to report, without proofs, some of the author's recent results on
(1) the existence and uniqueness of a unitary propagator, or fundamental solution, $\{U(t,s), t,s \in \mathbf{R}^1\}$ in $\mathsf{H} = L^2(\mathbf{R}^n)$ for (1.1) under most general conditions on $A(t,x)$ and $V(t,x)$; and
(2) the reguralIty and the smoothing properties of the propagator, under various conditions on the potentials.

The equation (1.1) is of fundamental importance in quantum mechanics and has been extensively studied by many mathematicians and physicists almost since its advent (cf. Dirac [**5**]). A basic mathematical question about (1.1), among others, is the existence and the uniqueness of the propagator, that is, the existence of unique dynamics associated with it. When $H(t) = H$ is t-independent, the problem is virtually equivalent to that of selfadjointness of H, via celebrated Stone's theorem, and has been solved to a rather satisfactory stage now (see e.g. Cycon-Froese-Kirsch-Simon[**4**], Reed-Simon[**26**] and references therein). When $H(t)$ is genuinely time dependent, on the other hand, the situation is by far less satisfactory, in spite of much effort devoted to the problem.

A. Boutet de Monvel et al. (eds.), Recent Developments in Quantum Mechanics, 381–394.

One of the methods to study the problem is to apply to (1.1) the abstract theory of evolution equations(cf. Kato[**18**], Tanabe[**29**], Masuda[**22**], Pazy[**25**], Goldstein[**10**] and references therein). This method is very much powerful in various cases and has been widely used. The theory, however, has been originally designed for solving parabolic or hyperbolic equations and, when applied to Schrödinger equations, it often imposes rather severe restrictions on the potentials, which exclude many concrete and important physical examples out of its scope. For example, (1.1) with $A(t,x) = 0$ and $V(t,x) = \sum_{j=1}^{N} |x - g_j(t)|^{-1}$, Coulomb potential with moving centres $g_1(t), ..., g_N(t)$ $(n = 3)$, can not be treated by directly applying the abstract theory to (1.1) (see, however, Hunziker[**13**]). Another method, very much different the first one, is to construct the propagator $U(t,s)$ directly in the form of integral operators, starting from its semi-classical approximation. This method which we call the method of bicharacteristics was first introduced by Fujiwara([**7**] and [**8**]) for studying Schrödinger equations and $\{U(t,s)\}$ was constructed in the form of oscillatory integral operators (OIOs) when $A(t,x) = 0$ and $V(t,s)$ is quadratically bounded at infinity: $|V(t,x)| \le C(1 + |x|)^2$, and Kitada[**20**] and Kitada-Kumanogo[**21**] subsequently genelarized this result and constructed it in the form of Fourier integral operators when $A(t,x)$ is linear in the spatial variables and $V(t,x)$ is quadratically bounded at infinity. This method, however, requires the potentials be C^∞ in the spatial variables.

Thus a theorem on the existence and uniqueness of the propagator of (1.1) for a larger class of potentials including smooth magnetic and singular unbounded scalar ones has been in order, and one of the purposes of this paper is to report on a recent progress on this problem.

Once the existence and the uniqueness of the propagator is established, it is natural to ask its regularity, that is, if $U(t,s)$ preserves certain nice subspaces of H on which $U(t,s)$ is strongly differentialble. We shall report a few results in this direction under various conditions on the potentials.

Another subject to be discussed in this paper is the so called *smoothing property* of the propagator: $U(t,s)f$ becomes smoother than original f after $|t-s| > 0$, which is of interest of its own and has many striking applications. The property may be manifested by various estimates using L^p-norms in space time variables or weighted Sobolev space norms. We shall examine these properties for our propagators and make some comments on estimates for the free propagator.

Before stating our result precisely, we recall the following well known fact: Let $G(t,x)$ be a real smooth function and let T be the gauge transformation:

$$Tu(t,x) = exp(-iG(t,x))u(t,x). \tag{1.3}$$

Then T transforms Eqn. (1.1) into the same one with A and V being replaced by $A + \partial_x G$ and $V - \partial_t G$, respectively. In particular, by taking $G(t,x) = \int_0^t V(s,x)ds$, smooth scalar potentials may always be eliminated from (1.1) by the gauge transformation T by changing A by $A + \partial_x G$.

2. Existence and Uniqueness and L^p-smoothing Property of Propagators.

Our assumptions for the existence and uniqueness of the propagator may be stated as follows. We write by $B(t,x)$ the strength tensor of the magnetic field, i.e. $B(t,x)$ is a matrix with (j,k)-component $B_{jk}(t,x) = (\partial A_k/\partial x_j - \partial A_j/\partial x_k)(t,x)$. For an interval I, $L^{\ell,\theta}(I) = L^\theta(I, L^\ell(\mathbf{R}^n))$, and $L^{\ell,\theta}_{loc}(I) = L^\theta_{loc}(I, L^\ell(\mathbf{R}^n))$. $\partial_x = (\partial/\partial x_1, \cdots, \partial/\partial x_n)$ and for multi-index $\alpha = (\alpha_1, \cdots, \alpha_n)$, $\partial_x^\alpha = (\partial/\partial x_1)^{\alpha_1} \cdots (\partial/\partial x_n)^{\alpha_n}$.

ASSUMPTION (A): (1) For $j = 1, ..., n$, $A_j(t,x)$ is real function of $(t,x) \in \mathbf{R}^{n+1}$ and, for any multi-index α, $\partial_x^\alpha A_j(t,x)$ is C^1 in $(t,x) \in \mathbf{R}^{n+1}$. Moreover, there exists $\epsilon > 0$ such that

$$|\partial_x^\alpha B(t,x)| \le C_\alpha (1+|x|)^{-1-\epsilon}, \qquad |\alpha| \ge 1,$$

$$|\partial_x^\alpha A(t,x)| + |\partial_x^\alpha \partial_t A(t,x)| \le C_\alpha, \quad |\alpha| \ge 1, \quad (t,x) \in \mathbf{R}^{n+1}.$$

(2) For some $p > n/2$ and $\alpha = 2p/(2p-n)$, $V \in L^{p,\alpha}(\mathbf{R}^1) + L^{\infty,1}(\mathbf{R}^1)$.

We set for $k = 0, 1, ...$

$$\Sigma(k) = \{u \in L^2(\mathbf{R}^n) : \sum_{|\alpha|+|\beta| \le k} \|x^\beta \partial_x^\alpha u\|^2 = \|u\|^2_{\Sigma(k)} < \infty\}$$

and $\Sigma(-k)$ is the dual space of $\Sigma(k)$. For Banach spaces X and Y, $B(X,Y)$ stands for the set of bounded operators from X to Y, $B(X) = B(X,X)$. We use C^k_* to indicate the k-times strong continuous differentiability of operator valued functions, e.g. $C^1_*(I^2, B(X,Y))$ is the set of strongly continuously differentiable $B(X,Y)$-valued functions on I^2.

THEOREM 1 ([**32**]). *Let Assumption (A) be satisfied, $q = 2p/p-1$ and $\theta = 4p/n$. Then, $H(t) \in B(\mathbf{H} \cap L^q(\mathbf{R}^n), \Sigma(-2))$ and its part in $\mathbf{H}$ is sefadjoint for a. e. $t \in \mathbf{R}^1$, that is, $H(t)$ with domain $D(H(t)) = \{u \in \mathbf{H} \cap L^q(\mathbf{R}^n) : H(t)u \in \mathbf{H}\}$ is selfadjoint in $\mathbf{H}$. Moreover, there uniquely exists a family of operators $\{U(t,s) : t,s \in \mathbf{R}^2\}$ which satisfies the following properties.*

(1) $U(t,s)$ is unitary in $L^2(\mathbf{R}^n)$ with $U(t,s)U(s,r) = U(t,r)$.

(2) $U(\cdot,\cdot) \in C_*(\mathbf{R}^2, B(\mathbf{H}))$.

(3) For any $s \in \mathbf{R}^1$ and $f \in \mathsf{H}$, $U(\cdot, s)f \in L^{q,\theta}_{loc}(\mathbf{R}^1)$ and with constant $C > 0$ independent of s, f and compact intervals I,

$$\|U(\cdot, s)f\|_{L^{q,\theta}(I)} \le C(1 + |I|)^{1/\theta}\|f\|. \tag{2.1}$$

(4) If $f \in \mathsf{H}$, $U(t, s)f$ is $\Sigma(-2)$-valued absolutely continuous in t and satisfies the equation

$$i(\partial/\partial t)U(t, s)f = H(t)U(t, s)f,$$

in $\Sigma(-2)$ at almost every $t \in \mathbf{R}^1$.

In what follows, the exponents $p > n/2$, $q = 2p/p - 1$ and $\alpha = 2p/(2p - n)$ are reserved to denote those appeared in Assumption (A) and Theorem 1. (However, the p in the terminology L^p-smoothing property and etc. never means this p and α will be often used for denoting multi-indeces. We are sorry for this and hope this will not cause any confusion.) Several remarks are in order.

REMARK(A): Let real function $V_0(t, x)$ be such $\partial_x^\alpha V_0(t, x)$ is continuous in (t, x) for all $|\alpha| \ge 0$ and

$$|\partial_x^\alpha V_0(t, x)| \le C_\alpha, \quad |\alpha| \ge 2.$$

Then, Theorem 1 remains to hold for (1.1) with $V(t, x) + V_0(t, x)$ in place of $V(t, x)$, the latter being assumed to satisfy Assumption (A). This can be easily seen by using the gauge transformation (1.3) with $G(t, x) = \int_0^t V_0(s, x)ds$. Similar remark will apply to theorems to follow, though we shall not mention it explicitly any more.

REMARK(B): Estimate (2.1) of Theorem 1 implies that $U(t, s)f \in L^q(\mathbf{R}^n)$ for $a.e.t \in \mathbf{R}^1$ for any $f \in \mathsf{H}$ and $s \in \mathbf{R}^1$ and is a manifestation of the L^p smoothing property of the propagator $\{U(t, s)\}$. This is a generalization of the following estimate for the free propagator: For any ℓ and θ with $0 \le n(1/2 - 1/\ell) < 1$ and $\theta = 4\ell/n(\ell - 2)$,

$$\|exp(it\Delta/2)\phi\|_{\ell,\theta} \le C\|\phi\|, \tag{2.2}$$

where $\|\cdot\|$ and $\|\cdot\|_{\ell,\theta}$ are the norms of $L^2(\mathbf{R}^n)$ and $L^{\ell,\theta} = L^\theta(\mathbf{R}^1, L^\ell(\mathbf{R}^n))$, respectively, and $\Delta = \partial^2/\partial x_1^2 + \cdots + \partial^2/\partial x_n^2$ is the n-dimensional Laplacian. (2.2) was discovered and proved by Strichartz[**28**] for a restrict range of ℓ and was extended by Ginibre-Velo[**11**] to general $2 \le \ell < 2n/(n-2)$. But it was Kato[**16**] who first observed the smoothing effect of Schrödinger propagator. He showed

$$\int_{-\infty}^{\infty} \|Ae^{it\Delta/2}\phi\|^2 \le C(\|A\|^2_{L^{n-\epsilon}} + \|A\|^2_{L^{n+\epsilon}})\|\phi\|^2, \quad \phi \in \mathsf{H} \tag{2.3}$$

for $A \in L^{n-\epsilon}(\mathbf{R}^n) \cap L^{n+\epsilon}(\mathbf{R}^n)$ $(\epsilon > 0)$. This implied $e^{it\Delta/2}\phi \in D(A)$ for almost every $t \in \mathbf{R}^1$ and he observed that this was a manifestation of the smoothing property of the free propagator $e^{it\Delta/2}$.

In fact we have the following generalization of (2.3).

THEOREM 2 ([**19**]). *Let $A \in L^n(\mathbf{R}^n)$. Then, with constant $C > 0$ independent of $A \in L^n(\mathbf{R}^n)$ and $\phi \in \mathsf{H}$,*

$$\int_{-\infty}^{\infty} \|Ae^{it\Delta/2}\phi\|^2 \leq C\|A\|^2_{L^n(\mathbf{R}^n)}\|\phi\|^2, \quad \phi \in \mathsf{H}. \tag{2.4}$$

Hence we have

$$\|e^{it\Delta/2}\phi\|^2_{L^{2n/n-2}(\mathbf{R}^n, L^2(\mathbf{R}^1))} \leq C\|\phi\|, \quad \phi \in \mathsf{H}. \tag{2.5}$$

Note, however, estimate (2.5) is of type somewhat different from (2.2) and neither of them is stronger than the other. We also remark that (2.2) and (2.5) are estimates global in time and they manifest the local decay property of the free propagator simultaneously. This is in contrast to (2.1) where the right hand side increases with $|I|$. This is, of course, natural in general if we take the possibility of the existence of bound states into account.

REMARK(C): A striking implication of the local decay estimate (2.2) or (2.5) is the existence of the restriction of Fourier transform of $L^{\ell,\theta}$-functions onto the quadratic surface $\{\xi_0 = (1/2)(\xi_1^2 + \cdots + \xi_n^2)\}$. Indeed, (2.2) and (2.5) are respectively equivalent to the estimates

$$\int_{\mathbf{R}^n} |\hat{f}(\xi^2/2, \xi)|^2 d\xi \leq C\|f\|^2_{L^{\ell',\theta'}}, \tag{2.6}$$

$$\int_{\mathbf{R}^n} |\hat{f}(\xi^2/2, \xi)|^2 d\xi \leq C\|f\|^2_{L^{2n/n+2}(\mathbf{R}^n, L^2(\mathbf{R}^1))}, \tag{2.7}$$

where ℓ' and θ' are indeces conjugate to ℓ and θ: $1/\ell + 1/\ell' = 1/\theta + 1/\theta' = 1$. In fact, Strichartz [**28**] deduced (2.2) from (2.6), the latter being proved by a real variable method.

REMARK(D): Let $V(t, x) = V(x)$ be time independent and satisfy (2) of Assumption (A) with some $p > n/2$. Then it is satisfied with any $n/2 < p' \leq p$ and

$U(\cdot, s, h)f \in L^{\ell,\theta}_{loc}(\mathbf{R}^1)$ for any $2 \le \ell < 2n/(n-2)$ and $\theta = \theta(\ell) = 4\ell/n(\ell-2)$. In what follows, $\theta(\ell)$ will stand for this last function.

REMARK(E): The L^p-smoothing property of the propagator is expected to hold, at least in a weaker form, for Schrödinger equations with faster increasing potentials. In fact, if

$$H = -d^2/dx^2 \text{ on } [0,1] \text{ with Dirichlet conditions at } x = 0,\ 1,$$

which is an extreme case that $V_0(x) = \infty$ for $x \in (-\infty, 0) \cup (1, \infty)$ in the terminology of Remark (a), it can be easily checked by explicit computation using Fourier series expansions that

$$exp(-itH)u \in L^4_{loc}(\mathbf{R}^1, L^4([0,1])), \qquad u \in L^2([0,1]),$$

and certain smoothing effect remains even in this extreme case.

REMARK(F): The L^p-smoothing property of the propagator implies an interesting result on the *change of the phase* of (generalized) Fourier coefficients in the expansions by the eigenfunctions of Schrödinger operators. The following is an obvious consequence of Theorem 1 and the fact that

$$\{u \in \mathsf{H} : (-i\partial/\partial x_j - A_j(t,x))u \in \mathsf{H},\ j = 1, 2, \cdots, n\} \subset L^q(\mathbf{R}^n).$$

We refer to Edward[**6**] for references to the related results on standard Fourier series.

COROLLARY 3 ([**32**]). *Let A and V be independent of t and satisfy Assumption (A) and let $H = H(t)$ be the selfadjoint operator in $\mathsf{H} = L^2(\mathbf{R}^n)$ defined by the Friedrichs extension:*

$$D(H) = \{u \in \mathsf{H} : (-i\partial/\partial x_j - A_j(x))u \in \mathsf{H},\ j = 1, \cdots, n,\ Hu \in \mathsf{H}\}.$$

Suppose that H has only pure point spectrum $\{\lambda_j\}_{j=1}^{\infty}$ and let $\{\phi_j\}_{j=1}^{\infty}$ be the associated complete set of orthonormal eigenfunctions of H. Then for any $2 \le \ell < 2n/(n-2)$ and $\{c_j\} \in \ell^2$,

$$\sum_{j=1}^{\infty} exp(-it\lambda_j)c_j\phi_j(x) \in L^{\ell}, \qquad a.e.\, t \in \mathbf{R}^1.$$

3. Regularity of Propagators.

The solution $u(t) = U(t,s)f$, $f \in \mathsf{H}$, obtained in Theorem 1 is *weak solution* in the sense that it satisfies (1.1) only in $\Sigma(-2)$ for $a.e. t \in \mathbf{R}^1$, and it is natural to ask the following question: When data f is nice, say $f \in \Sigma(2)$, is $u(t) = U(t,s)f$ a *strong solution*, that is, does $u \in C_*^1(\mathbf{R}^1, \mathsf{H})$, $u(t) \in D(H(t))$ for every $t \in \mathbf{R}^1$ and satisfy (1.1) in H? This amounts to finding an invariant subspace $\mathcal{D} \subset D(H(t))$ such that $U(t,s)\mathcal{D} \subset \mathcal{D}$, $t, s \in \mathbf{R}^1$, and is called the regularity problem for $\{U(t,s)\}$ and it turns out to be a rather subtle question. We study this problem only under certain stronger conditions on $A(t,x)$ and $V(t,x)$.

We first consider the case that $H_0(t) = H_0$ is independent of t and give a sufficient condition on $V(t,x)$ such that $D(H_0)$ is an invariant subspace.

ASSUMPTION (B): Let p be as in Assumption (A), $\tilde{p} = max(p,2)$, $\alpha_1 = 4p/(4p-n)$, and $p_1 = 2np/(n+4p)$ $(n \geq 5)$, $p_1 > 2p/(p+1)$ $(n = 4)$, $p_1 = 2p/(p+1)$ $(n \leq 3)$. $V(t,x)$ satisfies

$$V(\cdot\,,x) \in C_*(\mathbf{R}^1, L^{\tilde{p}}(\mathbf{R}^n)) + C_*(\mathbf{R}^1, L^\infty(\mathbf{R}^n)),$$

$$\partial_t V(\cdot\,,x) \in L^{p_1,\alpha_1}(\mathbf{R}^1) + L^{\infty,1}(\mathbf{R}^1).$$

We have the following regularity

THEOREM 4 ([**32**]). *Let Assumptions (A) and (B) be satisfied and $H_0(t) = H_0$ be independent of t. Then, $H(t)$ is essentially selfadjoint on $C_0^\infty(\mathbf{R}^n)$ and its closure, which will be denoted by the same symbol, has constant domain*

$$D(H(t)) = D(H_0) = \{u \in \mathsf{H} : H_0 u \in \mathsf{H}\}.$$

Moreover the propagator $U(t,s)$ of Theorem 1 satisfies the following properties:

(1) $U(\cdot,\cdot) \in C_*(\mathbf{R}^2, B(D(H_0))) \cap C_*^1(\mathbf{R}^2, B(D(H_0), \mathsf{H}))$.
(2) If $f \in D(H_0)$, then

$$i\partial_t U(\cdot\,,s,)f = H(\cdot)U(\cdot\,,s)f \in C(\mathbf{R}^1, \mathsf{H}) \cap L^{q,\theta}_{loc}(\mathbf{R}^1),$$

$$i\partial_s U(\cdot\,,s)f = -U(\cdot\,,s)H(s)f \in C(\mathbf{R}^1, \mathsf{H}) \cap L^{q,\theta}_{loc}(\mathbf{R}^1),$$

where we have equipped $D(H_0)$ with graph norm and $\theta = \theta(q) = 4p/n$.

When $H_0(t)$ is genuinely t-dependent, $D(H_0(t))$ is in general wildly changing with t, and it is too ambitious to expect $U(t,s)D(H(s)) \subset D(H(t))$ even when

$H(t) = H_0(t)$. However, if $V(t,x)$ suitably decays at infinity, $\Sigma(2) \subset D(H_0(t))$ can be an invariant subspace for $U(t,s)$. The following theorem tells us how much decay is necessary for this to hold. We write $< x >= (1+|x|^2)^{1/2}$.

ASSUMPTION (C): Let p be as in Assumption (A), $\tilde{p} = max(p,2)$, $\alpha_1 = 4p/(4p-n)$, $2 < \beta < 3$, and $p_1 = 2np/(n+4p)$ $(n \geq 5)$, $p_1 > 2p/(p+1)$ $(n=4)$, $p_1 = 2p/(p+1)$ $(n \leq 3)$. $V(t,x)$ satisfies

$$V(\cdot\,,x) \in \langle x\rangle^{-3} C_*(\mathbf{R}^1, L^{\tilde{p}}(\mathbf{R}^n)) + \langle x\rangle^{-\beta} C_*(\mathbf{R}^1, L^{\infty}(\mathbf{R}^n)),$$

$$\partial_t V(\cdot\,,x) \in \langle x\rangle^{-2}(L^{p_1,,\alpha_1}(\mathbf{R}^1) + L^{\infty,1}(\mathbf{R}^1)).$$

THEOREM 5 ([**32**]). *Let Assumptions (A) and (C) be satisfied. Then $U(t,s)$ of Theorem 1 satisfies $U(\cdot\,,\cdot) \in C_*(\mathbf{R}^2, B(\Sigma(2))) \cap C^1_*(\mathbf{R}^2, B(\Sigma(2),\mathbf{H}))$. If $f \in \Sigma(2)$, then*

$$i\partial_t U(\cdot\,,s)f = H(\cdot)U(\cdot\,,s)f \in C(\mathbf{R}^1,\mathbf{H}) \cap L^{q,\theta}_{loc}(\mathbf{R}^1),$$

$$i\partial_s U(\cdot\,,s)f = -U(\cdot\,,s)H(s)f \in C(\mathbf{R}^1,\mathbf{H}) \cap L^{q,\theta}_{loc}(\mathbf{R}^1),$$

where $\theta = \theta(q) = 4p/n$.

When $H(t) = H_0(t)$, much more is known, and the propagator can be obtained in the form of OIOs. (Recall remark (a) that (1.1) can be reduced to this case if $V(t,x)$ is smooth in x-variables.) We note (cf. Ikebe-Kato [**14**]) that $H_0(t)$ is essentially sefadjoint on $C_0^\infty(\mathbf{R}^n)$ and its closure, which will be denoted by the same symbol, is identical its maximal extension:

$$D(H_0(t)) = \{u \in \mathbf{H} : H_0(t)u \in \mathbf{H}\},$$

in particular, $H_0(t) \supset \Sigma(2)$ for every $t \in \mathbf{R}^1$. The following Theorems 6 and 7 are extensions of Fujiwara[**7**],[**8**], Kitada[**20**] and Kitada-Kumanogo[**21**] mentioned in the introduction.

THEOREM 6 ([**32**]). *Let Assumption (A) be satisfied and $H(t) = H_0(t)$. Then the propagator $U(t,s) = U_0(t,s)$ of Theorem 1 satisfies the following additional properties.*

(1) $U_0(\cdot\,,\cdot) \in C_*(\mathbf{R}^2, B(\Sigma(k))) \cap C^1(\mathbf{R}^2, B(\Sigma(k),\Sigma(k-2)))$ for $k = 0, \pm1, \cdots$.
(2) If $f \in \Sigma(2)$, then $U_0(t,s)f$ satisfies

$$i(\partial/\partial t)U_0(t,s)f = H_0(t)U_0(t,s)f,$$

$$i(\partial/\partial s)U_0(t,s)f = -U_0(t,s)H_0(s)f,$$

in $\mathbf{H}$.
(3) $U_0(t,s)$ maps $\mathcal{S}(\mathbf{R}^n)$ onto $\mathcal{S}(\mathbf{R}^n)$ continuously.

THEOREM 7 ([**32**]). *Let Assumption (A) be satisfied and $H(t) = H_0(t)$. Then there exists a positive number T such that for $0 < |t-s| \le T$, $U_0(t,s)$ can be written in the form of an oscillatory integral operator:*

$$U_0(t,s)f(x) = (2\pi i(t-s))^{-n/2}\int e^{iS(t,s,x,y)}b(t,s,x,y)f(y)dy,$$

where $S(t,s,x,y)$ and $b(t,s,x,y)$ satisfy the following properties.
(1) $S(t,s,x,y)$ is a real solution of the Hamilton-Jacobi equation

$$(\partial S/\partial t)(t,s,x,y) + (1/2)((\partial S/\partial x)(t,s,x,y) - A(t,x))^2 + V(t,x) = 0$$

which is C^1 in (t,s,x,y), C^∞ in (x,y) and satisfies the estimate

$$|\partial_x^\alpha \partial_y^\beta \{S(t,s,x,y) - \frac{(x-y)^2}{2(t-s)}\}| \le C_{\alpha\beta}, \qquad |\alpha+\beta| \ge 2.$$

(2) For any α and β, $\partial_x^\alpha \partial_y^\beta b(t,s,x,y)$ is C^1 in (t,s,x,y) and satisfies

$$|\partial_x^\alpha \partial_y^\beta b(t,s,x,y)| \le C_{\alpha\beta}, \qquad |\alpha+\beta| \ge 0.$$

Theorems 1 to 7, except Theorem 2, are proved in [**32**]. Let us briefly sketch the strategy of the proofs here. We first study Eqn. (1.1) when $H(t) = H_0(t)$. In this case, it is possible to construct the WKB-approximation $E(t,s)$ of the propagator by analysing the Hamiltonian flow corresponding to (1.1). We find that $E(t,s)$ is an OIO, $E(s,s) = 1$ and

$$i\partial_t E(t,s) - H_0(t)E(t,s) = -iG(t,s) \tag{3.1}$$

is also an OIO. If $U_0(t,s)$ does exist, (3.1) can be solved for $E(t,s)$ in the form

$$E(t,s) = U_0(t,s) - \int_s^t U_0(t,r)G(r,s)dr. \tag{3.2}$$

We regard (3.2) as an operator equation for $U_0(t,s)$ and solve it by iteration:
(3.3)

$$U_0(t,s) = E(t,s) + \int_s^t E(t,r)G(r,s)dr + \int_s^t \{\int_r^t E(t,\tau)G(\tau,r)d\tau\}G(r,s)dr + \cdots.$$

We show that the right hand side of (3.3) indeed converges and gives the desired propagator $U_0(t,s)$ in the form of Theorems 6 and 7. We then examine the smoothing property of $\{U_0(t,s) : t,s \in \mathbf{R}^1\}$ and consider the integral equation:

$$u(t) = U_0(t,s)f - i\int_s^t U_0(t,r)V(r)u(r)dr, \tag{3.4}$$

which is equivalent to (1.1). We shall show that the integral equation (3.4) is uniquely solvable in a certain function space of space time variables via contraction mapping theorem and that this solution is the solution of (1.1) desired in Theorem 1. The regularity theorems 4 and 5 are proved by first establishing an a priori estimate for the integral operator in (3.4) and then using the approximation arguments. We refer to [**32**] for the details.

4. Smoothing property in terms of weighted Sobolev norms.

We already stated the smoothing property of the propagator in term of L^p-norms. In this section we present estimates which manifest the smoothing effect of the propagator in terms of local Sobolev norms and state some of their applications. We let $\mathcal{F}$ be the Fourier transform, and define

$$|D|^\alpha = \mathcal{F}^*|x|^\alpha\mathcal{F}.$$

Our first result in this direction is

THEOREM 8 ([**18**]). *Let $0 \le \alpha < 1/2$. Then*

$$\int_{-\infty}^{\infty} \||x|^{\alpha-1}|D|^\alpha e^{it\Delta}u\|^2 dt \le C\|u\|^2, \qquad u \in \mathsf{H}; \tag{4.1}$$

$$\int_{-\infty}^{\infty} \|e^{it\Delta}u\|^2_{H^{1/2}_{-1}} dt \le C\|u\|^2, \qquad u \in \mathsf{H}; \tag{4.2}$$

where

$$H^s_r = \{u \in \mathcal{S}' : \|u\|_{H^s_r} = \|(1+|x|^2)^{r/2}(1-\Delta)^{s/2}u\| < \infty\}$$

is the weighted Sobolev space.

Estimates (4.1) and (4.2) imply rather surprizing facts that $|x|^{\alpha-1}|D|^{\alpha}e^{it\Delta}u \in L^2(\mathbf{R}^n)$, $0 \le \alpha < 1/2$ and that $(1+|x|^2)^{-1/2}(1-\Delta)^{1/4}e^{it\Delta}u \in L^2(\mathbf{R}^n)$ for $a.e.\, t \in \mathbf{R}^1$ only if $u \in L^2(\mathbf{R}^n)$, and are generalizations of resulta in Constantin-Saut[**3**] and Sjölin[**27**] in the sense these estimates simultaneously manifest the local decaying property of the free propagator. Theorem 8 can be deduced from the fact that the operators $|x|^{\alpha-1}|D|^{\alpha}$ and $< x >^{-1} (1-\Delta)^{1/4}$ are $(-\Delta)$-smooth in the sense of Kato[**16**], that is proved in [**18**] by an rather elementary method. Estimates of type (4.2) but local in time were proved by Constantin-Saut[3] for (1.1) with $V(t,x)$ which has some singularity.

For Schrödinger equations with magnetic potentials we have the following generalization of (4.2), which can be deduced from Theorems 6 $\sim$ 7 by using some elementary results on OIOs.

THEOREM 9 ([**31**]). *Suppose that condition (A) be satisfied and let $T_1 > 0$ be sufficiently small. Then, for $0 \le \sigma < 1/2$ and $\rho \in \mathbf{R}^1$, there exists a constant $C_{\rho\sigma} > 0$ such that*

$$(4.3) \qquad \int_{-T_1}^{T_1} \|\langle x\rangle^{-2\sigma-|\rho|}U_0(t+s,s)f\|^2_{H^{\rho+\sigma}}dt \le C_{\rho\sigma}\|f\|^2_{H^{\rho}}, \quad f \in \mathcal{S}(\mathbf{R}^n).$$

Estimate (4.3), together with the fact that $U_0(t,s)u$ satisfies the Schrödinger equation, implies the following *maximal inequality* of Schrödinger type.

THEOREM 10 ([**31**]). *Suppose that condition (A) be satisfied and let $T_1 > 0$ be sufficiently small. Then, for $\gamma > 1/2$, there exists a constant $C > 0$ such that for some $\delta < 3$,*

$$\int sup_{|t|<T_1}|\langle x\rangle^{-\delta}U_0(t,0)f(x)|dx \le C\|f\|^2_{H^{\gamma}}, \quad f \in \mathcal{S}(\mathbf{R}^n).$$

Theorem 10 in turn implies the following summability theorem.

THEOREM 11 ([**31**]). *Suppose that condition (A) be satisfied and let $T_1 > 0$ be sufficiently small. Then for any $f \in H^{\gamma}(\mathbf{R}^n) \cap L^1(\mathbf{R}^n)$ with $\gamma > 1/2$,*

$$lim_{t\to 0}U_0(t,0)f(x) = f(x), \qquad a.e. x \in \mathbf{R}^n.$$

In virtue of (4.2), Theorems 10 and 11 hold with $\gamma = 1/2$ for the free Schrödinger propagator (see also Sjölin [**27**]). The meaning of Theorems 10 and 11 may be understood by considering the particular case discussed in Corollary 3. In this case, $u(t,x) = U_0(t,0)f(x)$ is given by a (generalized) Fourier series

$$u(t,x) = \sum_{j=1}^{\infty} exp(-it\lambda_j)c_j\phi_j(x), \qquad c_j = (f,\phi_j),$$

where $(\cdot,\cdot)$ is the inner product in $\mathbf{H}$. Theorem 10 claims that the maximal inequality holds for $u(t,x)$ in an averaged sense and Theorem 11 says that the series converges as $t \to 0$ to $f(x)$ for a, e, x, if $f \in H^{\gamma}(\mathbf{R}^n) \cap L^1(\mathbf{R}^n)$.

The summability Theorem 11 was first proved by Carleson[**2**] for the free propagator in one dimension with $\gamma = 1/4$, which is shown to be sharp by Dahlberg and Kenig[**5**]. The maximal inequality of Schödinger type, Theorem 10, was first shown by Kenig and Ruiz[**19**] for $n = 1$ with $\gamma = 1/4$ and both of them were generalized to the general n-dimensional free Schrödinger propagators by Sjölin [**27**]. Thus Theorems 10 and 11 generalized Sölin's results to Schrödinger equations with interaction.

References

1. K. Asada and D. Fujiwara, *On some oscillatory integral transformation in* $L^2(\mathbf{R}^n)$, Japanese J. Math. **4** (1978), 299–361.
2. L. Carleson, *Some analytical problems related to statstical mechanics*, "Euclidean Harmonic Analysis" in Lecture Notes in Math. Springer **799** (1979), 5–45.
3. P. Constantin and J. C. Saut, *Local smoothing properties of dispersive equations*, J. of Amer. Math. Soc. **1** (1988), 413–439.
4. H. L. Cycon, R. G. Greose, W. Kirsch and B. Simon, "Schrödinger operators with application to quantum mechanics and global geometry," Springer, Berlin-Heidelberg, 1987.
5. B. E. J. Dahlberg and C. E. Kenig, *A note on the almost everywhere behaviour of solution to the Schrödinger equation*, Lecture Notes in Math. Springer **908** (1982), 205–208.
6. P. A. M. Dirac, "The principle of quantum mechanics," Oxford Univ. Press, Oxford, 1947.
7. R. E. Edward, "Fourier series, a modern introduction," Springer, Berlin-New York, 1982.
8. D. Fujiwara, *A construction of the fundamental solution for the Schrödinger equation*, J. d'Analyse Math. **35** (1979), 41–96.

9. D. Fujiwara, *Remarks on convergence of the Feynmann path integrals*, Duke Math. J. **47**, 559–600.
10. J. Ginibre and G. Velo, *The global Cauchy problem for the non-linear Schrödinger equation revisited*, ann. Inst. H. Poincaré Anal. Non Linéaire **2**, 309–327.
11. J. Goldstein, "Semigroups of linear operators and applications," Oxford Univ. Press, Oxford, 1985.
12. J. S. Howland, *Stationary scattering theory for time dependent Hamiltonians*, Math. Ann. **207** (1974), 315–335.
13. W. Hunziker, *Distortion analyticity and molecular resonance curves*, Ann. l'Inst. H. Poincaré, Phys. Theor. **45** (1986), 339–358.
14. T. Ikebe and T. Kato, *Uniqueness of self-adjoint extension of singular elliptic differential operators*, Arch. Rat. Mech. Ana. **9** (1962), 77–92.
15. T. Kato, *Fundamental properties of Hamiltonian operators of Schrödinger type*, Trans. Amer. Math. Soc. **70** (1951), 195–211.
16. T. Kato, *Wave operators and similarity for some non-selfadjoint operators*, Math. Ann. **162** (1966), 258–279.
17. T. Kato, *Linear evolution equations of "hyperbolic" type*, J. Fac. Sci. Univ. Tokyo Sec. I **17** (1972), 241–258.
18. T. Kato and K. Yajima, *Some examples of smooth operators and the associated smoothing effect*, to appear, Reviews in Math. Physics, **2** (1990).
19. C. E. Kenig and A. Ruiz, *A strong type* (2, 2) *estimatate for a maximal operator associated to the Schrödinger equation*, Trans. Amer. Math. Soc. **280** (1983), 239–246.
20. H. Kitada, *On a construction of the fundamental solution for Schrödinger equation*, J. Fac. Sci. Univ. Tokyo Sec. IA **27** (1980), 193–226.
21. H. Kitada and H. Kumanogo, *A family of Fourier Integral operators and the fundamental solution for a Schrödinger equation*, Osaka J. Math. **18** (1981), 291–360.
22. K. Masuda, "Evolution equations (in Japanese)," Kinokuniya-Shoten, Tokyo, 1979.
23. T. Matsumura and M. Nagase, *On sufficeint conditions for the boundedness of pseuo-differential operators*, Proc. Japan Acad. Ser. A **55** (1979), 293–296.
24. M. Nagase, *The L^p-boundedness of pseudo-differential operators with non-regular symbols*, Comm. P. D. E. **2** (1977), 1045–1061.
25. A. Pazy, "A semigroup of linear operators and applications to partial differential equations," Springer, Berlin-Heidelberg-New York, 1983.
26. M. Reed and B. Simon, "Methods of modern mathematical physics, Vol.II. Fourier ananlysis and selfadjointness," Academic Press, New York, 1977.
27. P. Sjölin, *Regularity of solutions to the Schrödinger equations*, Duke Math. J. **55** (1987), 699–715.
28. R. Strichartz, *Restriction of Fourier transforms to quadratic surfaces and decay of solutions of wave equations*, Duke Math. J. **44** (1977), 705–714.
29. H. Tanabe, "Evolution equations (in Japanese)," Iwanami-Shoten, Tokyo, 1975.

30. K. Yajima, *Existence of solutions for Schrödinger evolution equations*, Commun. Math. Phys. **110** (1987), 415–426.
31. K. Yajima, *On smoothing property of Schrödinger propagators*, to appear in Springer Lect. Notes in Math. Proc. of Kato conference.
32. K. Yajima, *Schrödinger evolution equations with magnetic field*, to appear, J. d'Analyse Math..

Mathematical Physics Studies

Publications:

1. F.A.E. Pirani, D.C. Robinson and W.F. Shadwick: *Local Jet Bundle Formulation of Bäcklund Transformations.* 1979 ISBN 90-277-1036-8
2. W.O. Amrein: *Non-Relativistic Quantum Dynamics.* 1981 ISBN 90-277-1324-3
3. M. Cahen, M. de Wilde, L. Lemaire and L. Vanhecke (eds.): *Differential Geometry and Mathematical Physics.* Lectures given at the Meetings of the Belgian Contact Group on Differential Geometry held at Liège, May 2–3, 1980 and at Leuven, February 6–8, 1981. 1983 ISBN 90-277-1508-4 (pb)
4. A.O. Barut (ed.): *Quantum Theory, Groups, Fields and Particles.* 1983 ISBN 90-277-1552-1
5. G. Lindblad: *Non-Equilibrium Entropy and Irreversibility.* 1983 ISBN 90-277-1640-4
6. S. Sternberg (ed.): *Differential Geometric Methods in Mathematical Physics.* 1984 ISBN 90-277-1781-8
7. J.P. Jurzak: *Unbounded Non-Commutative Integration.* 1985 ISBN 90-277-1815-6
8. C. Fronsdal (ed.): *Essays on Supersymmetry.* 1986 ISBN 90-277-2207-2
9. V.N. Popov and V.S. Yarunin: *Collective Effects in Quantum Statistics of Radiation and Matter.* 1988 ISBN 90-277-2735-X
10. M. Cahen and M. Flato (eds.): *Quantum Theories and Geometry.* 1988 ISBN 90-277-2803-8
11. Bernard Prum and Jean Claude Fort: *Processes on a Lattice and Gibbs Measures.* 1991 ISBN 0-7923-1069-1
12. A. Boutet de Monvel, Petre Dita, Gheorghe Nenciu and Radu Purice (eds.): *Recent Developments in Quantum Mechanics.* 1991 ISBN 0-7923-1148-5

Kluwer Academic Publishers – Dordrecht / Boston / London

Mathematical Physics Studies

[illegible]

Kluwer Academic Publishers – Dordrecht / Boston / London